New Frontiers in Regional Science: Asian Perspectives

Volume 19

Editor in Chief
Yoshiro Higano, University of Tsukuba

Managing Editors
Makoto Tawada (General Managing Editor), Aichi Gakuin University
Kiyoko Hagihara, Bukkyo University
Lily Kiminami, Niigata University

Editorial Board
Yasuhiro Sakai (Advisor Chief Japan), Shiga University
Yasuhide Okuyama, University of Kitakyushu
Zheng Wang, Chinese Academy of Sciences
Hiroyuki Shibusawa, Toyohashi University of Technology
Saburo Saito, Fukuoka University
Makoto Okamura, Hiroshima University
Moriki Hosoe, Kumamoto Gakuen University
Budy Prasetyo Resosudarmo, Crawford School of Public Policy, ANU
Shin-Kun Peng, Academia Sinica
Geoffrey John Dennis Hewings, University of Illinois
Euijune Kim, Seoul National University
Srijit Mishra, Indira Gandhi Institute of Development Research
Amitrajeet A. Batabyal, Rochester Institute of Technology
Yizhi Wang, Shanghai Academy of Social Sciences
Daniel Shefer, Technion - Israel Institute of Technology
Akira Kiminami, The University of Tokyo
Jorge Serrano, National University of Mexico
Binh Tran-Nam, UNSW Sydney
Ngoc Anh Nguyen, Development and Policies Research Center
Thai-Ha Le, RMIT University Vietnam

Advisory Board
Peter Nijkamp (Chair, Ex Officio Member of Editorial Board), Tinbergen Institute
Rachel S. Franklin, Brown University
Mark D. Partridge, Ohio State University
Jacques Poot, University of Waikato
Aura Reggiani, University of Bologna

New Frontiers in Regional Science: Asian Perspectives

This series is a constellation of works by scholars in the field of regional science and in related disciplines specifically focusing on dynamism in Asia.

Asia is the most dynamic part of the world. Japan, Korea, Taiwan, and Singapore experienced rapid and miracle economic growth in the 1970s. Malaysia, Indonesia, and Thailand followed in the 1980s. China, India, and Vietnam are now rising countries in Asia and are even leading the world economy. Due to their rapid economic development and growth, Asian countries continue to face a variety of urgent issues including regional and institutional unbalanced growth, environmental problems, poverty amidst prosperity, an ageing society, the collapse of the bubble economy, and deflation, among others.

Asian countries are diversified as they have their own cultural, historical, and geographical as well as political conditions. Due to this fact, scholars specializing in regional science as an inter- and multi-discipline have taken leading roles in providing mitigating policy proposals based on robust interdisciplinary analysis of multifaceted regional issues and subjects in Asia. This series not only will present unique research results from Asia that are unfamiliar in other parts of the world because of language barriers, but also will publish advanced research results from those regions that have focused on regional and urban issues in Asia from different perspectives.

The series aims to expand the frontiers of regional science through diffusion of intrinsically developed and advanced modern regional science methodologies in Asia and other areas of the world. Readers will be inspired to realize that regional and urban issues in the world are so vast that their established methodologies still have space for development and refinement, and to understand the importance of the interdisciplinary and multidisciplinary approach that is inherent in regional science for analyzing and resolving urgent regional and urban issues in Asia.

Topics under consideration in this series include the theory of social cost and benefit analysis and criteria of public investments, socio-economic vulnerability against disasters, food security and policy, agro-food systems in China, industrial clustering in Asia, comprehensive management of water environment and resources in a river basin, the international trade bloc and food security, migration and labor market in Asia, land policy and local property tax, Information and Communication Technology planning, consumer "shop-around" movements, and regeneration of downtowns, among others.

Researchers who are interested in publishing their books in this Series should obtain a proposal form from Yoshiro Higano (Editor in Chief, higano@jsrsai.jp) and return the completed form to him.

More information about this series at http://www.springer.com/series/13039

Saburo Saito • Kosuke Yamashiro
Editors

Advances in Kaiyu Studies

From Shop-Around Movements Through Behavioral Marketing to Town Equity Research

Editors
Saburo Saito
Fukuoka University
Fukuoka, Fukuoka, Japan

Kosuke Yamashiro
Nippon Bunri University
Oita, Oita, Japan

ISSN 2199-5974   ISSN 2199-5982   (electronic)
New Frontiers in Regional Science: Asian Perspectives
ISBN 978-981-13-1738-5   ISBN 978-981-13-1739-2   (eBook)
https://doi.org/10.1007/978-981-13-1739-2

Library of Congress Control Number: 2018961394

© Springer Nature Singapore Pte Ltd. 2018
This work is subject to copyright. All rights are reserved by the Publisher, whether the whole or part of the material is concerned, specifically the rights of translation, reprinting, reuse of illustrations, recitation, broadcasting, reproduction on microfilms or in any other physical way, and transmission or information storage and retrieval, electronic adaptation, computer software, or by similar or dissimilar methodology now known or hereafter developed.
The use of general descriptive names, registered names, trademarks, service marks, etc. in this publication does not imply, even in the absence of a specific statement, that such names are exempt from the relevant protective laws and regulations and therefore free for general use.
The publisher, the authors and the editors are safe to assume that the advice and information in this book are believed to be true and accurate at the date of publication. Neither the publisher nor the authors or the editors give a warranty, express or implied, with respect to the material contained herein or for any errors or omissions that may have been made. The publisher remains neutral with regard to jurisdictional claims in published maps and institutional affiliations.

This Springer imprint is published by the registered company Springer Nature Singapore Pte Ltd.
The registered company address is: 152 Beach Road, #21-01/04 Gateway East, Singapore 189721, Singapore

*To all students who participated
in the on-site interview surveys
of consumer Kaiyu behaviors*

Preface

We have been carrying out many empirical and theoretical studies on consumer shop-around behaviors for three decades in Japan and other parts of Asia. Consumer shop-around behaviors also are referred to as *Kaiyu* in Japanese, a term widely used in several fields such as city planning, marketing, real estate, tourism, and regional policy. This book demonstrates how our *Kaiyu* research has evolved from the original idea to the present state and envisages the prospective *Kaiyu* studies in the age of big data and the Internet of Things (IoT).

The distinguishing feature of our research is that we regard *Kaiyu* as consumers' simultaneous decisions sequentially made while undertaking their shop-arounds as to which shops they visit, for what purpose, and how much they spend there. This is a sharp contrast to much research on trip chains, which only deals with spatial movements. As a result, our studies first enabled one to empirically explore the relationships between consumer shop-around movements and money flows among shopping sites within a city center retail environment. Furthermore, we have uncovered the roles of various urban policies and facilities inexplicit so far by revealing how they contribute to the turnover of the whole town through stimulating *Kaiyu*. This gives us a universal means of evaluation for urban revitalization policy. Actually, most research instances presented in this book more or less are concerned with the ex post evaluation of various kinds of city center revitalization policies implemented in actual cities.

Thus, we have renovated the scope of consumer shop-around studies from shop-around movements in the context of city planning, through shopping behavioral marketing, to economic evaluation of urban revitalization policy. In fact, almost all of our research are directed to extending the frontier of shop-around studies based on the on-site empirical interview surveys of consumer shop-around behaviors at real retail environments conducted by ourselves in the last 20 years. These research efforts brought us to rethinking the goal of urban development and the value of city from consumer's micro-behavioral viewpoints and led to a new concept of town equity.

This book demonstrates step by step these conceptual developments by showing our concrete research examples. It begins with discussing some meta-theoretic concept to form our shop-around research, reports our various studies done at cities in Japan, and leads to the demonstration of the econometric method to evaluate urban transport policy from the perspective of consumer shop-arounds. The book also introduces a new concept of town equity and an emerging view of goal of urban development, and shows the future direction of research on consumer shop-around behaviors in the age of big data and the Internet of Things (IoT) by exemplifying the experimental use of smartphones for *Kaiyu* studies.

Most of our concrete research examples are concerned with how scientifically to carry out the ex post evaluation of various urban revitalization policies. These concrete research examples, as will soon be apparent, deeply rely on the statistical methodologies we developed to correct the choice-based sampling biases inherent in consumer's micro-behavioral *Kaiyu* data obtained from on-site sampling surveys of consumer shop-around behaviors. This is another distinctive feature of our *Kaiyu* studies.

As is well known, when conducting the on-site random sampling survey, we inherently face with the problem of choice-based sampling biases since the more frequent visitors are more likely to be chosen as a sample. Moreover, our on-site surveys choose respondents at random from the visitors at several sampling points in the middle of their shop-arounds. Since the frequency of visits to these sampling points differs among respondents, the pooled use of shop-around data from different sampling points causes the choice-based sampling bias. In short, our on-site surveys are characterized as the sampling of shopping trip chains of respondents chosen randomly at sampling points by intercepting their shopping chains in the middle of their shop-arounds. This property requires us to develop a theory for on-site sampling survey.

This book focuses on the applications of these statistical methods to ex post evaluation of urban revitalization policies. The methodological development of these statistical methods themselves is discussed in detail in a separate volume in this series. Also our *Kaiyu* studies are concerned with the ex ante evaluation of urban policy, which deeply is related to the modeling and forecasting of *Kaiyu*. The topics of modeling and forecasting *Kaiyu* are taken up in detail in a separate volume in this series. The readers are encouraged to be referred to these volumes for the advances in statistical methods for on-site sampling surveys and modeling and forecasting *Kaiyu*.

Fukuoka, Japan   Saburo Saito

Contents

1 **Introduction: A Meta-theoretic Evaluation Framework for *Kaiyu* Studies** 1
Saburo Saito

Part I Policy Evaluation from *Kaiyu* Movements

2 **How Did the Large-Scale City Center Retail Redevelopment Change Consumer Shop-Around Behaviors?: A Case of the City Center District at Fukuoka City, Japan** 13
Saburo Saito and Masakuni Kakoi

3 **Evaluating Municipal Tourism Policy from How Visitors Walk Around Historical Heritage Area: An Evaluation of the "Walking Path of History" of Dazaifu City, Japan, Based on Visitors' *Kaiyu* Behavior** 47
Saburo Saito, Masakuni Iwami, Mamoru Imanishi, and Kosuke Yamashiro

4 **How Did the Extension of Underground Shopping Mall Vitalize *Kaiyu* Within City Center?** 69
Saburo Saito, Masakuni Kakoi, and Masakuni Iwami

Part II Some Characteristics of *Kaiyu*

5 **Occurrence Order of Shop-Around Purposes** 91
Saburo Saito

6 ***Kaiyu* Distance Distribution Function at Downtown Space** 111
Saburo Saito and Hiroyuki Motomura

7 **The Factors Determining Staying Time of *Kaiyu* in City Center** ... 131
Saburo Saito, Kosuke Yamashiro, and Mamoru Imanishi

| 8 | Little's Formula and Parking Behaviors | 145 |

Saburo Saito, Kosuke Yamashiro, Masakuni Iwami, and Mamoru Imanishi

Part III Economic Effects by Accelerating *Kaiyu*

| 9 | The Economic Effects of City Center 100-Yen Circuit Bus | 165 |

Saburo Saito and Kosuke Yamashiro

| 10 | Time Value of Shopping | 189 |

Saburo Saito and Kosuke Yamashiro

| 11 | Roles of City Center Cafés and Their Economic Effects on City Center: A Consumer Behavior Approach Focusing on *Kaiyu* | 217 |

Saburo Saito, Masakuni Iwami, and Kosuke Yamashiro

Part IV Economic Effects by Increasing Visitors

| 12 | The Economic Effects of Opening a New Subway Line on City Center Commercial District | 241 |

Saburo Saito and Kosuke Yamashiro

| 13 | Did an Introduction of a New Subway Line Increase the Frequency of Visits to City Center? | 261 |

Saburo Saito and Kosuke Yamashiro

| 14 | To What Extent Did the Woodworks Festival Attract People? | 273 |

Saburo Saito, Kosuke Yamashiro, Masakuni Iwami, and Mamoru Imanishi

| 15 | How Did the Effects of the Festival Held on Main Street Spread Over Other Districts Within a City Center? | 297 |

Saburo Saito, Kosuke Yamashiro, and Masakuni Iwami

Part V *Kaiyu* Marketing and Value of Visit to City Center

| 16 | Did the Grand Renewal Opening of Department Store Enhance the Visit Value of Customers? | 317 |

Saburo Saito, Kosuke Yamashiro, and Masakuni Iwami

| 17 | A New Entry of Large Variety Shop Increases the Value of City Center? | 341 |

Saburo Saito, Kosuke Yamashiro, and Masakuni Iwami

Part VI Emerging View of the Goal of Urban Development

| 18 | The Concept of Town Equity and the Goal of Urban Development | 361 |

Saburo Saito

19 City Center Parking Policy: A Business Model Approach 369
Saburo Saito, Kosuke Yamashiro, and Masakuni Iwami

Part VII Information and Consumer *Kaiyu* Behaviors

20 Exploring Information Processing Behaviors of Consumers in the Middle of Their *Kaiyu* with Smartphone 397
Mamoru Imanishi, Kosuke Yamashiro, Masakuni Iwami, and Saburo Saito

Part VIII Urban Policy and Consumer Welfare

21 Travel Demand Function of Korean Tourists to Kyushu Region, Japan 419
Saburo Saito, Hiroyuki Motomura, and Masakuni Iwami

22 Direct Approach to Estimating Welfare Changes Brought by a New Subway Line 437
Kosuke Yamashiro and Saburo Saito

Index 461

Contributors

Mamoru Imanishi Department of Business and Economics, Nippon Bunri University, Oita City, Japan

Masakuni Iwami Fukuoka University Institute of Quantitative Behavioral Informatics for City and Space Economy (FQBIC), Fukuoka, Japan

Masakuni Kakoi Faculty of Economics, Fukuoka University, Fukuoka, Japan

Fukuoka University Institute of Quantitative Behavioral Informatics for City and Space Economy (FQBIC), Fukuoka, Japan

Hiroyuki Motomura Department of Business and Economics, Nippon Bunri University, Oita City, Japan

Saburo Saito Faculty of Economics, Fukuoka University, Fukuoka, Japan

Fukuoka University Institute of Quantitative Behavioral Informatics for City and Space Economy (FQBIC), Fukuoka, Japan

Kosuke Yamashiro Department of Business and Economics, Nippon Bunri University, Oita City, Japan

Chapter 1
Introduction: A Meta-theoretic Evaluation Framework for *Kaiyu* Studies

Saburo Saito

Abstract A meta-theoretic viewpoint is presented for the evaluation of the urban development policy. Three kinds of typical evaluation schemes are extracted that drive various urban studies to evaluate urban development policies. They are the ideal city type, the activity effect type, and the city formation system type evaluation schemes. Advances in *Kaiyu* studies are driven by the evaluative framework based on the basic and the extended forms of the activity effect evaluation scheme.

Keywords Evaluation scheme · Ideal city type · Activity effect type · City formation system type · *Kaiyu* · Urban development policy evaluation · City system · Physical system · Activity system · Social decision-making system

1 An Evaluative Viewpoint for Urban Development and Management Policy Research

In this chapter we present some meta-theoretic consideration of how we incorporate the evaluative viewpoints into our *Kaiyu* studies. As *Kaiyu* studies inherently are concerned with urban development and management policy, implicitly or explicitly, they necessarily convey some evaluation. Here we compare some implicit or explicit frameworks of the evaluation contained in various urban studies. In fact, each research of our *Kaiyu* studies, while not explicitly stated, stands on either one of evaluation schemes presented below. Our discussion becomes meta-theoretic in the sense that we evaluatively compare various evaluation schemes contained in the area of city planning, urban development, and city management.

S. Saito (✉)
Faculty of Economics, Fukuoka University, Fukuoka, Japan

Fukuoka University Institute of Quantitative Behavioral Informatics for City and Space Economy (FQBIC), Fukuoka, Japan
e-mail: saito@fukuoka-u.ac.jp

As a start, consider the following. Urban development intends to change a town into something new. If it intends to change a town from its current state S_0 to a new state S_1, it is implied that the state S_1 is preferable to the state S_0. Of course, such an evaluation will be different depending on who evaluates this. Therefore, if the preferences of the evaluators are different, the conclusion that the state S_1 is preferable to the state S_0 will not necessarily be the same among them but will naturally differ.

However, even though the preferences of individuals differ, it is also true that there is a certain way of thinking – a framework and viewpoint for evaluating a town. Here, we discuss and compare the ways of thinking which typically appear in city planning research when evaluating a town. We call those ways of evaluative thinking as evaluation schemes.

1.1 City As a City Formation System

Before discussing the evaluation schemes, we must share the viewpoint of seeing how the city is working and functioning as a system. Below, we provide one perspective to see a city in that way.

There are various angles to view a city.

Physical System

First, the place and the space where the city is located are necessary. WEB space on the Internet is also taken up as a virtual space, but here we consider the city (town) that exists in a real space. The first things to recall on the real space are hard facilities such as roads and railroads, houses and offices, and commercial facilities. Let's call this "physical system."

Next, the real space and the hard facility by themselves do not have any functions. There are various entities that use them. They may be a resident, a company engaged in production activities, consumers who go shopping, visitors who enjoy sightseeing and leisure, and so on. Activities of these various entities utilizing the space and the hard facilities bring functions into the real space and the hard facilities.

Activity System

Activities of various entities on the real physical system will be called "activity system." The activity system can be considered as representing various transactions among entities such as exchanging goods, services, money, information, and even energy on the physical system as a platform. For example, suppose a consumer bought clothes at a shop in a town. From the viewpoint of the consumer, that event proceeds as follows: the consumer moved to the shop by spending the travel expense and time of the movement to the shop where the shopping is carries out, chose the clothes from the merchandise being displayed at the shop, and bought them by paying the price to the shop owner. From the shop owner side, the owner purchased the merchandise, used the place of a commercial facility called a shop, displayed the merchandise at the shop, and sold the clothes to the consumer by receiving the price from the consumer. In other words, the shop owners and consumers, who are the subjects in the activity system, have exchanged goods and money they hold. Thus the event of the shopping is seen as an example of transactions between the two entities in the activity system.

On the other hand, from a different viewpoint, the activities of both shop owners and consumers can be considered in a unified way as the production activities of what outputs they produce from what they inputs. According to this framework, the consumer is regarded as producing shopping behavior as the output by using the inputs such as the travel expense, the time required for the movement to the store, the time spent shopping, the information acquisition cost, and the price paid to the purchased good. Of course, the shop owner can be thought of as a producer who produces the sale of goods as outputs by using stocks of goods and shops as inputs in the same way as ordinary enterprise production activities.

Until now, we have looked at the physical system and the activity system as the systems that constitute the city (town), but from the viewpoint of city planning and management, another point of view is necessary.

Social Decision-Making System

The viewpoint is social decision-making system. Who builds a new physical system and who modifies and updates this by what procedure? It is also a viewpoint of who induces various actors in the activity system to change their actions in a new direction by what kind of procedure. For example, consider an entity trying to construct a compound shopping building. Certainly, when completed, the entity becomes one of the various actors in the activity system since the entity will raise revenue from the completed shopping building by providing products and services to the visitors there. Also the activity of construction itself might well be regarded as one of the activities in the activity system.

However, as usual in city planning and management, there are rules to determine what kind of building can be built in a specified place under what building coverage rate and what floor area ratio. These rules are generally called "institutions." While we do not go into details of the contents of these rules, the point we should note here is that these institutions and rules inherently are generated from some social decision-making system. Thus the activity of constructing the compound shopping building is considered as the outcome of some social decision-making system. More specifically, we call the procedure or mechanism where plural entities in the activity system, while mutually related, collectively decide by themselves the construction and modification of the physical system as "social decision-making system."

To summarize, we conceptualize the city (town) as a complex system consisting of three mutually interrelated subsystems: the physical system composed of hard facilities in the real space, the activity system composed of activities conducted by various entities on the physical system, and the social decision-making system formed by various actors in the activity system for generating, modifying, maintaining, and developing the physical system.

City As a City Formation System

The one reason why we take a view like this is that we would like to see the city (town) not as a static system, but as a self-evolving dynamic system such that it internalizes a mechanism to change itself. In addition, we wish to see a city is the result of the accumulated decisions made by numerous people involved therein. Hence the essential point of capturing the city as a complex system composed of three subsystems is in the viewpoint that many entities in the city are involved and interrelated with each other, collectively decide, and cooperate to form a city. We call this complex system "city formation system."

1.2 Three Evaluation Schemes

We provide a framework to see a city as a self-evolving complex system which internalizes a mechanism to change itself and call it the city formation system.[1] The city as a complex system is conceptualized as consisting of three interrelated systems: the physical system of hard facilities in real space, the activity system of various actors over the physical system, and the social decision-making system for modifying and maintaining these two systems.

We defined the evaluation schemes in city planning studies as the modes of thinking for evaluating urban development which reside explicitly or implicitly in each individual study. From among those modes of thinking, three evaluation schemes such as the ideal city type, activity effect type, and city formation system type can be extracted as types by focusing on the differences in how they put their emphases on which part in the above three systems composing a complex city system when viewing and evaluating a town.

Below, I explain them in order.

---

[1]Three evaluation schemes for the first time were introduced and discussed by Kumata and Saito [1]. Recently Saito [5] revisited the three evaluation schemes to review the past, ongoing, and future of *Kaiyu* studies in the new age of ICT (Information and Communication Technology) and IoT (Internet of Things).

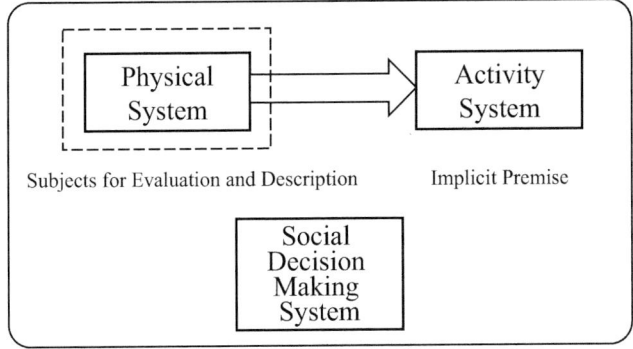

Fig. 1.1 Ideal city evaluation scheme

1.3 Ideal City Type Evaluation Scheme

The ideal city type evaluation scheme is still commonly used for evaluating towns. In particular, it is implicitly used when adopting a method such as selecting a public facility through a design competition.

As shown in Fig. 1.1, in the ideal city type evaluation scheme, the physical system is the objective for evaluation. This evaluation scheme is characterized as its feature that criteria for evaluation and the object for observation and description are limited to this physical system (such as the shape, color, and design of facilities).

In this scheme, the social decision-making component of the city formation system is simply referred to as a means of creating a physical system or is totally ignored. With regard to the activity system, there is an implicit premise that "The optimization of the physical system brings the optimization of the activity system at the same time." However, the factual bases and procedures for verifying whether or not the premise is true are not explicitly indicated.

The evaluation based on the ideal city type evaluation scheme proceeds as which state of the physical system is better than which state. Thus the subjects to be evaluated in the ideal city type evaluation scheme are the different states of the hard facilities of the physical system, and the criteria for evaluating them are also described by the physical system itself, which is the feature of this evaluation scheme.

Regarding the mode of thinking classified into the ideal city type evaluation scheme, there is a long history of this mode of thinking as in the ideal city theory, and even today, not explicitly stated, land use plan such as zoning planned to be set without a clear evaluation criterion can be said as an example of the ideal city type evaluation scheme.

Furthermore, there exist various indicators of cities' status such as "rankings of cities"[2] that are often reported in the mass media and the recent "evaluation of city

---

[2] For example, the journal by the City Planning Institute of Japan recently organized special topics on evaluation and rankings of cities [2].

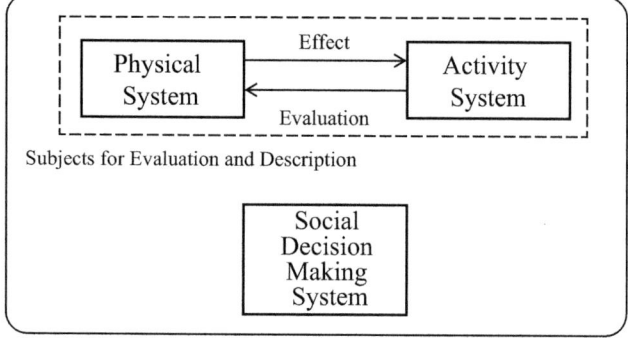

Fig. 1.2 Activity effect evaluation scheme

structure,"[3] in which the interrelationships between hard facilities within a city are statistically analyzed. The mode of evaluative thinking in all of these efforts can be classified as the ideal city type evaluation scheme since they all based their evaluation on the indicators created from the physical system of city.

1.4 Activity Effect Type Evaluation Scheme

The next is the activity effect type evaluation scheme. This evaluation scheme is shown in Fig. 1.2. This scheme focuses on the activity system, which is an implicit assumption in the ideal city type. This evaluation scheme explicitly describes the interactions between the physical system and the activity system and incorporates them into the evaluation. There are several variants of this activity effect type evaluation scheme.

Basic Form of Activity Effect Evaluation Scheme

The premise of the basic form is that "the purpose of the physical system is to optimize the activities carried out on the physical system." This evaluation scheme is a mode of evaluative thinking typically seen in architectural design. In the mid-1950s, this evaluation scheme once was employed for evaluating the arrangements of the rooms in public housing by employing the living style of the separation

---

[3]The Ministry of Land, Infrastructure, and Transport recently published the handbook for evaluating urban structures on their website [3].

of eating and sleeping spaces as a criterion for the modern living style.[4] Note that here the arrangements of rooms or floor plans correspond to some states of the physical system and the living style corresponds to some state of activities in the activity system. Actually the evaluative study was carried out by conducting surveys of living behaviors of residents living in various kinds of floor plans to observe whether or not they use separate rooms for eating and sleeping. From the observed results, the floor plans are evaluated by how the floor plans contributed to inducing the residents' living behaviors to comply with the criterial living style of the separation of eating and sleeping spaces.

In the basic form of the activity effect type evaluation scheme, changing the state of the physical system becomes a policy, and the mode of evaluative thinking is concerned with a policy evaluation in which the policy is evaluated by its effects on the activity system such as how the change of the physical system improved the state of the activity system. While its evaluation criteria are placed in the state of the activity system, the original aim is the evaluation of the state of the physical system and the ways that it changes. In this sense, the simplest basic form can be said to be an evaluation scheme incorporated into an architectural design or such one employed as an emergency evacuation plan for a large facility.

Physical=Activity Interdependence Extension Form

In the activity effect type evaluation scheme, the object of observation and description is the interaction between the physical and activity systems. In this evaluation scheme, many entities or actors in the activity system can be regarded as consumers and visitors who decide their activities autonomously by and for themselves. If we think like this, many entities supplying hard facilities in the physical system also can be regarded as maintaining and renovating their facilities in the physical system autonomously in response to the requirements of various entities in the activity system. Thus, we are led to the understanding that the changes of the physical system cause the changes in the activity system and also the changes of the activity system will cause the changes in the physical system.

Therefore, while the basic form of the activity effect evaluation scheme considered the one-way causal effect from the physical system to the activity system, we can expand it into the form that includes both way causal relation of interdependence between the physical and activity systems. We refer to this expanded form of the activity effect evaluation scheme as the physical=activity interdependence extension form of the activity effect evaluation scheme. A typical instance of this physical=activity interdependence extension form can be seen as the integrated model of land use and transportation that attempts to model the process where

---

[4]Refer to the writings of Nishiyama on his methodology of the survey of living behaviors of residents in houses, in particular, House Planning in Vol.1 by Nishiyama [4].

transportation, land use, housing locations, and so on change simultaneously while being interdependent with each other.

Policy Extension Form

Furthermore, in this activity effect evaluation scheme if we include policy measures that intend to affect directly the behaviors of actors in the activity system such as information provision for sales promotion, this activity effect evaluation scheme can be extended into a framework for evaluating the effects of a wide range of policies on the activity system. We call this framework as the policy extension form of the activity effect evaluation scheme. This policy extension form can be applied to the evaluation of the policies not limited to changes in physical systems but expanded to such policies as community programs, welfare plans, events, regional promotion policies, marketing, and so on.[5]

1.5 City Formation System Type Evaluation Scheme

The final evaluation scheme is the city formation system type evaluation scheme. This evaluation scheme stands on the concept that grasps a town as the city formation system, as shown in Fig. 1.3. It is an evaluative framework that explicitly describes and evaluates the city formation system consisting of three systems: the physical system, activity system, and social decision-making system.

The purpose to conceptualize a city as a city formation system lies in the view that the city formation system includes a mechanism by which the city itself modifies the physical and activity systems as a social decision-making system. That is to say, as shown in Fig. 1.3, it is the point of view that the policy to modify the physical and activity systems is itself the output of the social decision-making system.

In other words, in this framework the social decision-making system introduces policies into the physical and activity systems, feedback from their effects is provided to the social decision-making system, and policies continue to be created and developed.

---

[5] One of the earliest attempts to evaluate the city center space from consumers' *Kaiyu* behaviors is Saito [6], which is based on the basic form of activity effect evaluation scheme. Other *Kaiyu* studies based on this basic form of activity effect evaluation scheme will be presented in Part II of this book. The policy extension form of activity effect evaluation scheme started with a study on estimating the economic effect of the 100-yen city center circuit bus. Yamashiro [7] characterized the study as a consumer behavior approach to urban development. *Kaiyu* studies based on the consumer behavior approach will be discussed in Part III to Part IV of this book. Further developments of the policy extension form of activity effect evaluation scheme to various city center revitalization policies such as *Kaiyu* marketing are the themes of Part V to Part VIII of this book.

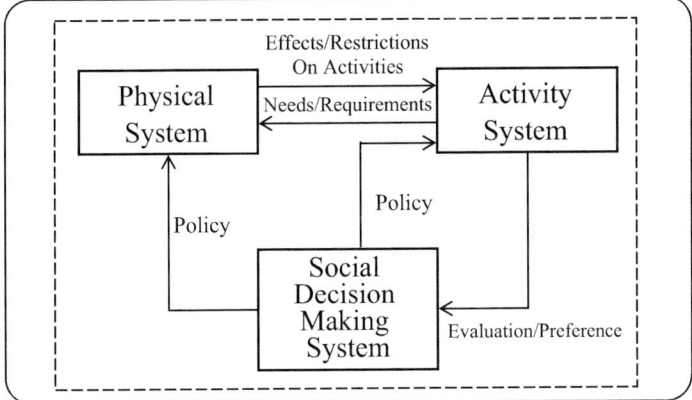

Fig. 1.3 City formation system type evaluation scheme

Therefore, in the city formation system type evaluation scheme, from the long-term perspective, the physical system is recognized as a product of the social decision-making system. In the long run, the town is considered to be formed as a stack of decisions made by the people involved in it.

The concept of urban development policy and planning assumed by this evaluation scheme is a self-adapting social process that sets goals, creates policies, evaluates the results, and, based on this, changes and achieves the goals even by forming and modifying new measures, organizations, and decision-making procedures.

Entities involved in the social decision-making system also make up the activity system, but the difference can be said to be their involvement in maintaining and developing the mechanism of urban development itself.

This concludes the discussion on the evaluation schemes.

References

1. Kumata Y, Saito S (1975) An approach to theory of design for planning organization. Urban Probl Res 27(2):44–62. (in Japanese)
2. City Planning Institute of Japan (2015) Special topics: evaluation and rankings of cities. City Plan Rev 64(1):3–70. (in Japanese)
3. Ministry Of Land, Infrastructure, and Transport (2014) Handbook for evaluating urban structure (in Japanese). http://www.mlit.go.jp/common/001104012.pdf (in Japanese)
4. Nishiyama U (1967–1968) The collected works of Uzo Nishiyama, vol 1–4. Keiso Shobo. (in Japanese)
5. Saito S (2017) Evaluation of a city. In: Kawahara Y, Saito S (eds) Chapter 2 in social city. Foundation of the promotion of the open university of Japan. (in Japanese)

6. Saito S (1988) Assessing the space structure of central shopping district viewed from consumers' shop-around behaviors: a case study of the midtown district of Saga City. Fukuoka Univ Econ Rev 33(1):47–108. (in Japanese)
7. Yamashiro K (2012) A study on the evaluation of transport policy based on consumer behavior approach. Doctoral Dissertation (Economics-1402), Fukuoka University. (in Japanese)

Part I
Policy Evaluation from *Kaiyu* Movements

Chapter 2
How Did the Large-Scale City Center Retail Redevelopment Change Consumer Shop-Around Behaviors?: A Case of the City Center District at Fukuoka City, Japan

Saburo Saito and Masakuni Kakoi

Abstract At the city center of Fukuoka City, a core city of Northern part of Kyushu island in Japan, large-scale retail developments have completed one after another within 2 years of 1996–1998. They include openings of two new department stores and one new large commercial complex and one enlargement of shop-floor area of an existing department. As a result, the total increase of shop-floor area becomes 13.4 ha. If restricted to the central commercial district of "Tenjin," the total shop-floor space of the department stores has drastically been expanded to 2.7 times the previous one. Such a rush of large-scale retail developments is quite a rare event. Thus, the case of Fukuoka City gives a valuable occasion to investigate the effect of redevelopment on the structure of city center since, to our knowledge, few empirical research have been carried out to identify the effect of redevelopment based on a behavioristic standpoint.

This study has conducted three on-site surveys of consumer shop-around behaviors before and after the actual developments at the city center of Fukuoka City and identified their effect on the structure of city center as the changes in consumer shop-around behaviors among retail establishments using Markov shop-around model.

Keywords Shop-around behavior · *Kaiyu* · Center of gravity · City center structure · Retail redevelopment · Development effect on consumer behaviors

---

This chapter is based on the paper by Saburo Saito, Takaaki Nakashima, and Masakuni Kakoi [1], "Identifying the Effect of City Center Retail Redevelopment on Consumer's Shop-around Behavior: An Empirical Study on Structural Changes of City Center at Fukuoka City," *Studies in Regional Science*, vol. 29, pp. 107–130, 1999 (in Japanese).

S. Saito (✉) · M. Kakoi
Faculty of Economics, Fukuoka University, Fukuoka, Japan

Fukuoka University Institute of Quantitative Behavioral Informatics for City and Space Economy (FQBIC), Fukuoka, Japan
e-mail: saito@fukuoka-u.ac.jp; kakoi@econ.fukuoka-u.ac.jp

1 Purpose of This Study

In the city center district of Fukuoka City in Japan, following the opening of Canal City Hakata, a commercial complex, in April 1996, large-scale redevelopments of commercial facilities were completed one after another. These included the opening of the new department store, *Iwataya* Z-side, in September 1996; the opening of an expanded sales floor of the department store, *Daimaru* Elgala, in March 1997; and the opening of the new department store, Fukuoka *Mitsukoshi*, in October 1997. Furthermore, in March 1999, the opening of a large-scale commercial establishment at Hakata Riverain was planned, as a part of the *Shimokawabata* district redevelopment. Between 1996 and 1998, the large-scale commercial redevelopment increased the sales floor space of the entire city center district by 134,000 m^2. As far as the Tenjin district, which is the central commercial district at Fukuoka City, is concerned, before the large-scale commercial redevelopment, the sales floor space of department stores was just 55,000 m^2, but after the redevelopment, it increased by 2.7 times larger than before to 149,000 m^2.

Recently, in many local cities, due to the advancement of motorization and locations of large stores in suburban areas, the decline of central commercial areas is becoming a problem. This is not only a problem from the perspective of industrial policy or securing employment in the central shopping streets but also there is the sense of crisis that losing the central commercial district in a local city is equivalent to losing most of its urban appeal. Fukuoka City is not an exception in this widespread trend toward motorization and suburbanization. However, what characterizes the city center retail environment of Fukuoka City is that, in anticipation of such the trend, the merchants and developers themselves proactively took risks and made improvements to large-scale city center commercial facilities, and surprisingly enough for Fukuoka City, the result of their efforts is becoming a strategy that drastically increases the attractiveness of the city center with looking at the East Asian market, more specifically anticipating the competition between mega cities in East Asia.

In any case, it is very rare for such large-scale commercial redevelopments to be carried out in the downtown area of a regional core city within such a short time period. Thus, Fukuoka City gives a very valuable case for understanding the effects that the large-scale commercial redevelopment can have on urban structures. In particular, as one of the effects that these large-scale developments can have on the city center structure of Fukuoka City, the present city center structure concentrated to Tenjin in a monopolar way is expected to change to the one with planar spread. In addition, due to the opening of large commercial facilities clustered in the south, a big change such as the phenomenon of the southward movement of the "center of gravity" within the Tenjin district has been pointed out.

In this study we focus on shop-around behavior, which is the behavior of consumers who walk around in the city center district. While taking advantage of this valuable opportunity for case study, the aim of this study is to analyze quantitatively and empirically the changes in the city center structure due to large-scale developments in terms of changes in consumer shop-around behavior by conducting surveys of shop-around behavior three times, before, in-between, and after these large-scale developments. To date, there have been no attempts to verify the effects of large-scale developments empirically from the actual behavioral changes of consumers standing on the behavioristic viewpoint. This study likely represents the first such attempt not only in Japan, but in the world.

There are three streams in the literature on consumer shop-around behavior: (1) studies on pedestrian flow [2–4], (2) studies on trip chains [5–19], and (3) studies on the modification and extension of the Huff model [20, 21]. Borger and Timmermans [22] present a detailed review of previous studies published in English on pedestrian flows and trip chains in commercial districts. A detailed review of existing research published in Japanese is found in Saito and Ishibashi [23]. Here, we outline the studies that relate directly to the present study.

Fukami [2, 3] focuses on the fact that pedestrian flow in commercial districts such as traffic volume in front of the stores greatly influences the total sales of individual stores. He developed a Monte Carlo simulation model to forecast the pedestrian flow from the standpoint that such a forecast is indispensable for evaluating the development plans of a commercial district. Hanson's [6, 7] viewpoint is quite similar. It is interesting that there was a similar viewpoint of this problem in the West during the same period. Inspired by Fukami, Saito [12] considered consumer shop-around behavior as a concrete manifestation that the agglomeration effect of commercial facilities reflects on consumer behavior and developed a shop-around Markov model with an infinite number of "shop-arounds" as its measurement model. This model is reported by Sakamoto [16]. On the other hand, Takeuchi [21] introduced simultaneous behaviors with main and subordinate purposes into the Huff model and increased its predictive power for the amount of sales of eating and drinking sectors. Motivated by Takeuchi's research, Saito [20] reconsidered trips of purchasing within the wider framework of shop-around behavior and developed a disaggregated hierarchical choice Huff model that considers one-step consumer shop-around behavior from large retail stores to small ones.

Saito and Ishibashi [24] introduced explanatory variables into the shop-around Markov model which had until that time been limited to describing the current state and developed the model that can estimate changes of shop-around patterns under the condition of a given number of incoming visitors when the configuration pattern of the commercial establishments had changed due to redevelopment. After that, by developing a Poisson model based on on-site surveys for forecasting the number of visitors [25], a shop-around Markov model that allows for the increase in the number of attracted visitors due to redevelopment was constructed [26]. Additionally, a shop-around Markov model with a monetary basis was developed [27–29].

Various extensions have been made of the shop-around Markov model, but these are models that forecast changes in shop-around behavior due to redevelopment. To date, there has been no study to take up an actual redevelopment, observe consumer shop-around behaviors before and after the redevelopment, and then measure and verify the changes of the consumer shop-around behaviors. The significance of the present study is of course to measure the changes of shop-around behavior due to the redevelopment standing on this behavioristic viewpoint. But this study also has the purpose to demonstrate that the shop-around Markov model is an effective tool for measuring and verifying the development effects that are reflected on the consumer shop-around behaviors in the sense that the model first made possible such measurement and verification. In fact, as will be seen below, the three on-site surveys conducted in this study are different in their sampling points, questionnaire items, and the numbers of samples collected. Thus, a simple comparison among these surveys cannot clearly determine the effect by the redevelopment. However, by the use of the tool of our shop-around Markov model which made possible to analyze data with different survey points and samples in the same framework, this study became possible for the first time to measure the effect of the redevelopment in terms of the changes of consumer behaviors.

From the above discussion, the purpose of this study is to strictly define the shop-around behavior of consumers; to actually observe their shop-around behavior three times, before, in-between, and after the large-scale redevelopment; and to identify and verify the effect of the large-scale redevelopment as reflected in consumer shop-around behaviors. In particular, the phenomenon of the southward movement of the "center of gravity" of the Tenjin viewed from consumers' movements within the Tenjin district is specified and verified by using the shop-around Markov model.

2 Large-Scale Redevelopment at City Center Commercial District of Fukuoka City and Surveys of Consumer Shop-Around Behaviors

2.1 Outline of Large-Scale Redevelopment at Downtown Fukuoka

Figure 2.1 is a map of the city center district of Fukuoka City and the location of the large-scale redevelopment sites. This figure also shows the zone division of the city center district, which is used in the analysis. In this study, the city center district of Fukuoka City is defined as a block surrounded by 50-m-width roads (i.e., *Watanabe* Street, *Sumiyoshi* Street, *Taihaku* Street, and *Showa* Street) that contains the areas of Tenjin, Hakata Station, and Canal City Hakata. Fukuoka City is a twin city. It includes two core business districts in its city center district. They are the Tenjin district, which is the connecting point between the Nishitetsu Fukuoka station terminal and the municipal subway, and the Japan Railways (JR) Hakata Station district, where the Japan Railways (JR) and the municipal subway connect.

A rush of redevelopments started with the appearance in April 1996 of Canal City Hakata, a giant commercial complex with a sales floor space of 41,959 m^2 and a total floor space of 235,000 m^2, at the *Sumiyoshi* district located in between those two core business districts where the former National Railway Hakata Station was located. Subsequently, the department store Iwataya Z-Side opened in September

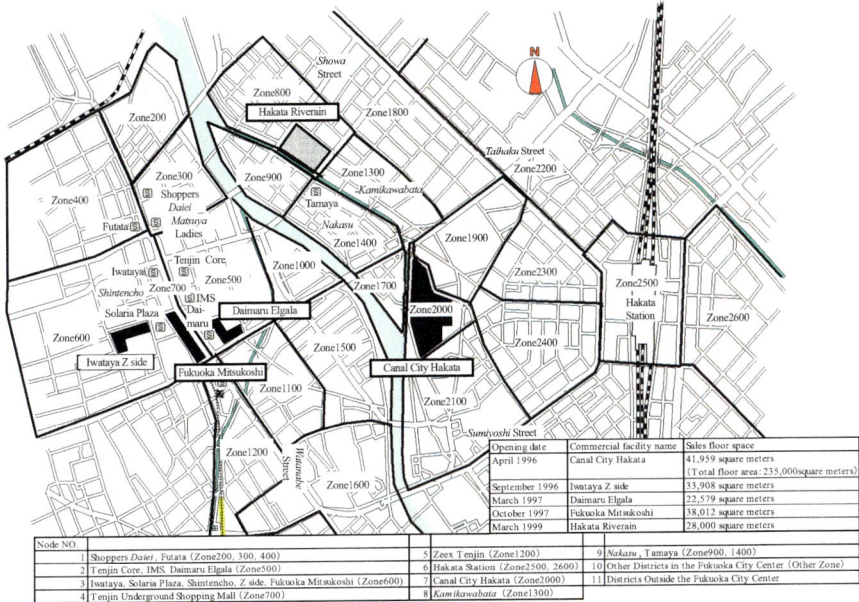

Fig. 2.1 Map of the city center district of Fukuoka City, including large-scale redevelopment

1996, with a sales floor space of 33,908 m². Next, Daimaru Elgala, a municipal redevelopment project by Fukuoka City, opened in March 1997 with an expanded sales floor space of 18,800 m². In addition, Fukuoka Mitsukoshi opened in October 1997, with a sales floor space of 38,012 m². The opening of Hakata Riverain was planned in March 1999, with a sales floor space of 28,000 m².

2.2 Surveys of Consumer Shop-Around Behaviors at the City Center of Fukuoka City

The surveys of consumer shop-around behaviors at city center of Fukuoka City were interview surveys in which the respondents were randomly sampled among visitors to the city center and asked about their shop-around behaviors, that is, the behavior of walking around the city center district. The three surveys were conducted at the following points of time: The first was in September 1996 which was after the opening of Canal City Hakata and immediately before the opening of *Iwataya* Z-Side; the second was in July 1997, after the opening of *Daimaru* Elgala; and the third was in February 1998, after the opening of Fukuoka *Mitsukoshi*.

The first survey was conducted on September 15, 16, and 17 [Sunday, Monday (transferred holiday), and Tuesday], 1996; the second, on July 18, 19, and 20 (Friday, Saturday, and Sunday), 1997; and the third, on January 30 and 31, and February

1 (Friday, Saturday, and Sunday), 1998. Each survey was conducted over a 3-day period. The survey items include places respondents visited on the survey day, purposes done there, and expenditure spent there if any in the order of their occurrence. For the first survey, four survey sampling points were set up at Shoppers *Daiei*, Solaria Plaza, Canal City Hakata, and the Hakata Station Concourse. As for the second survey, the two sampling points, *Iwataya* Z-Side and *Daimaru* Elgala, were added to the previous four sampling points. As for the third survey, Fukuoka *Mitsukoshi* was added to the survey sampling sites of the first and second surveys, which resulted in setting up a total of seven sampling points. In all surveys, the survey time was from 11 a.m. to 8 p.m. We were able to collect 475, 1482, and 1817 valid samples, respectively.

The details of the survey method are as follows: We set up several survey points in major commercial facilities at the city center district of Fukuoka City, prepared desks and chairs at each survey point, and conducted interview surveys with questionnaire sheets of about 15 min to samples randomly sampled from the visitors who visited survey points. In Fig. 2.1, the names of the commercial facilities are shown where the survey points were set up.

In order to ensure the random sampling, we extracted samples from each survey point at even intervals as much as possible. Due to the survey time period and the commercial facilities that were used as survey points, the point of time when the samples were surveyed was when they were in the middle of undertaking shop-around behavior on that survey day. Therefore, in the survey, we asked the respondents about their future shop-around behaviors as well, in addition to their shop-around behavior history on that day until the moment of the interview.

Figure 2.1 shows that the whole city center district was divided into 25 zones. In the analysis, collapsing these zones further into 10 nodes and adding a hypothetical node outside the city center district, we set up 11 nodes in total. These zone divisions and the definition of each node are also shown in Fig. 2.1. As is clear from Fig. 2.1, the nodes and the survey points do not have any one-to-one correspondence.

3 Method of Measuring the Effect of Large-Scale Redevelopment Using the Shop-Around Markov Model

3.1 The Shop-Around Markov Model

Consumer behaviors of walking around the city center commercial district are called shop-around (*Kaiyu*) behaviors. Figure 2.2 is a conceptual diagram of shop-around behavior. It shows that a consumer who departs from home and walks around shopping sites in the order of A, B, and C within a city center commercial district. The shopping site A is the entry point into the city center commercial district, and the walk-arounds at the second and subsequent steps are defined as "shop-around." The shopping sites visited by shop-around are B and C. The "walk-around" is defined to

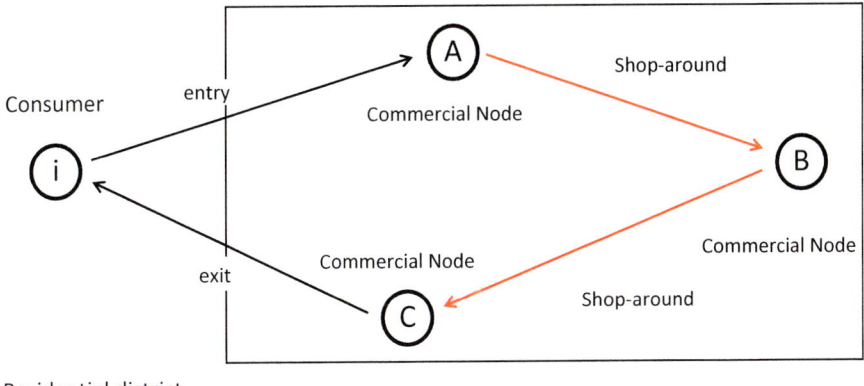

Fig. 2.2 Conceptual diagram of shop-around behavior

occur when either the visited place or the behavioral purpose changes. The shop-around Markov model is a model that expresses consumer shop-around behaviors as a stationary absorbing Markov chain, assuming that consumer shop-around behaviors depart from home, repeat an infinite number of shop-arounds, and finally return home. Let H_0 denote the home node (absorption state) and H the departure node. Also denote the set of commercial facility nodes as I. In the case of Fig. 2.2, I denotes the set of {A, B, C}. We express consumer shop-around behaviors by the state transition probability P

$$P = \begin{matrix} H_0 \\ H \\ I \end{matrix} \begin{bmatrix} 1 & 0 & 0 \\ 0 & 0 & P_{HI} \\ P_{IH_0} & 0 & P_{II} \end{bmatrix} \quad (2.1)$$

If the initial state of "being at home at start" is expressed as the initial distribution, it can be expressed as $\pi^0 = [0\ 1\ \mathbf{0}]$ ($\mathbf{0}$ is a zero vector). If shop-around or P is repeated infinitely after the initial state, the state distribution becomes $\pi^\infty = [1\ 0\ \mathbf{0}] = \pi^0 P^\infty$, which is representing that consumers who departed from home finally are all "coming back home," the state of which is expressed as $\pi^\infty = [1\ 0\ \mathbf{0}]$.

Let us define the "shop-around effect" as a scale for measuring how much shop-around takes place. To simplify, the shop-around effect is a scale that shows how many visits one trip entered into the city center district brings to each commercial facility by shop-around. Specifically, it is defined as the expected total frequency of visits to each commercial facility from other commercial facilities after the second step by shop-arounds.

By using the partitioned matrix of P to obtain the n-th higher-order transition probability $P^{(n)}$, the element in $P^{(n)}$ which is the same position as P_{HI} in P is known to be $P_{HI}P_{II}^{n-1}$. Since this represents the probability of visiting each commercial facility at the n-th step, by adding all the subsequent probabilities after the second step, we obtain the shop-around effect RE as the following:

$$\begin{aligned} \text{RE} &= P_{\text{HI}}P_{\text{II}} + P_{\text{HI}}P_{\text{II}}^2 + \cdots \\ &= P_{\text{HI}}P_{\text{II}}(I - P_{\text{II}})^{-1} \end{aligned} \quad (2.2)$$

The formula, $P_{\text{II}}(I-P_{\text{II}})^{-1}$, which is obtained by replacing P_{HI} with identity matrix, can be interpreted as a scale showing how much shop-around visit frequency is brought to other commercial facilities when one trip enters each commercial facility.

The above expression for shop-around effect RE is formulated in terms of probabilities such as their sum. If we would like to express the shop-around effect in terms of the number of visitors to each commercial facility, we have only to replace the entry choice probability P_{HI} with the entry visit frequency F_{HI}, which is the vector composed of the number of visitors to each commercial facility as entrance. Consequently, the shop-around effect will be obtained as the number of visitors who visited each commercial facility by shop-around, which is called the shop-around visit frequency. Therefore, the total visit frequency TVST to each commercial facility can be decomposed into the entry visit frequency and the shop-around visit frequency and expressed as follows:

$$\begin{aligned} \text{TVST} &= F_{\text{HI}}(I - P_{\text{II}})^{-1} \\ &= F_{\text{HI}} + F_{\text{HI}}P_{\text{II}}(I - P_{\text{II}})^{-1} \end{aligned} \quad (2.3)$$

To apply the shop-around Markov model to real data, the transition probability P must be calculated from the observed results of actual consumers' shop-around behaviors obtained from surveys. To do this, we must aggregate the observed results. The aggregation method we used here is a usual method in which we first treat each consumer shop-around behavior as a "linked trip" expressed as a sequence of nodes the consumer visited, then decompose it into the "Unlinked Trips" between two consecutive nodes, and finally aggregate them. If the aggregated result is represented by the observed frequency matrix T, it can be expressed as follows:

$$T = \begin{matrix} H_0 \\ H \\ I \end{matrix} \begin{bmatrix} N & 0 & 0 \\ 0 & 0 & F_{\text{HI}}^{\text{OBS}} \\ F_{\text{IH}_0}^{\text{OBS}} & 0 & F_{\text{II}}^{\text{OBS}} \end{bmatrix} \quad P^{\text{OBS}} = \begin{matrix} H_0 \\ H \\ I \end{matrix} \begin{bmatrix} 1 & 0 & 0 \\ 0 & 0 & P_{\text{HI}}^{\text{OBS}} \\ P_{\text{IH}_0}^{\text{OBS}} & 0 & P_{\text{II}}^{\text{OBS}} \end{bmatrix} \quad (2.4)$$

Here, N is the total number of samples, and P^{OBS} is the observed transition probability obtained by converting T into a probability matrix. It should be noted that $F_{\text{HI}}^{\text{OBS}}$ is the observed entry visit frequency and column sum $F_{\text{II}}^{\text{OBS}}$ is the observed shop-around visit frequency RE^{OBS}.

As will be described in Sect. 3.2, the validity of the shop-around Markov model is in the reproducibility theorem, which states that the observed shop-around visit frequency RE^{OBS} is identical to the shop-around effect RE^{PRED} forecasted by

Eq. (2.2), using the observed entry visit frequency F_{HI}^{OBS} and the observed shop-around choice probability P_{II}^{OBS} [23, 24, 30, 31].

3.2 Aggregation Validity of the Shop-Around Markov Model

Stationarity and infinitely repeated shop-around is the premise of the shop-around Markov model, but this is only an approximation of actual shop-around behaviors among consumers. Actually, it has been reported that the observed consumer shop-around behavior does not follow a stationary process since the probability of returning home increases as the shop-around steps extend [13, 15]. (Also refer to Chaps.5 and 6 of this book) Then, there would arise a question of whether or not the approximation by the shop-around Markov model is valid. For this, we established the following theorem which shows in some way the validity of the shop-around Markov model as an aggregated model (for proof, see Saito and Ishibashi [23]):

Reproducibility Theorem: The observed shop-around visit frequency RE^{OBS} coincides with the shop-around effect RE^{PRED} forecasted by equation (2.2) using the observed entry visit frequency F_{HI}^{OBS} and the observed shop-around choice probability P_{II}^{OBS}.

This theorem shows that even if the individual-level shop-around consumer behavior was following any non-stationary processes or a series of any random stochastic processes, the observed shop-around visit frequency, which is the observed aggregated result, equals the forecasted shop-around effect calculated by the infinite number of stationary Markov chains based on the observed shop-around choice probability and the observed entry visit frequency. In this sense, Saito and Ishibashi [23] refer to this theorem as the observed aggregate stationarity theorem.

This reproducibility theorem states that the observed shop-around visit frequency, which is the aggregated result of the survey, can be completely recovered by the forecasted value of the shop-around visit frequency via the shop-around Markov model. This is done by using the observed shop-around choice probability and the number of observed entry visitors, both of which were obtained through the survey.

However, the observed entry visit frequency, which is a result of the survey, is at most the number of samples collected by the survey, whereas the actual entry visit frequency is in hundreds of thousands of units. Therefore, the basic way of thinking in this study is that instead of the observed entry visit frequency obtained from the survey, we use the estimated value of the actual entry visit frequency estimated separately and the observed shop-around choice probability to obtain the forecasted value of the shop-around visit frequency via the shop-around Markov model and employ this forecasted value as the estimated value of the actual shop-around visits. If the observed shop-around choice probability obtained from the survey sufficiently and accurately reflects the actual shop-around pattern, the estimated value of the shop-around visit frequency must recover the actual shop-around visit frequency via the reproducibility theorem.

Therefore, in this study, the actual entry visit frequency in each survey period is estimated separately. On the other hand, the forecasted value of shop-around visit frequency via the shop-around Markov model using the observed shop-around choice probability at the time of each survey is assumed to be the estimated value of the actual shop-around visit frequency at the time of each survey. We then compare these to measure the effect of the large-scale redevelopment.

However, to simply compare the estimated values for each year does not lead to the measurement of the true effect of the large-scale redevelopment. In the next section, we address this point and describe how to solve this problem.

3.3 Measuring the Effect of the Large-Scale Redevelopment

There are several methods to measure the effect of the large-scale redevelopment on shop-around behaviors by using the shop-around Markov model. The simplest method is to directly compare the shop-around choice probability. This is done in Sect. 5.1. The next method is a way of investigating the change in shop-around effect that each commercial facility brings to the other commercial facilities. This is the task of Sect. 5.2. The former presents an attempt to measure directly how shop-around behavior changes by a change in the shop-around choice probability, and the latter presents an attempt to compare the changes in the city center structure that are reflected in the change in the shop-around effect.

The question is how to measure the effect of the large-scale redevelopment. This is the purpose of this study. It is not possible to measure the effect of the large-scale redevelopment by simply comparing the aggregated results of the surveys at three points in time. This is because the survey sampling points and the number of samples collected differ among the three surveys, according to the survey design. This problem is one that always occurs in similar inter-temporal surveys. Therefore, the task in Sect. 5.3 is to make an interannual comparison based on the actual total visit frequency and shop-around visit frequency, via a method that uses the actual entry visit frequency (as mentioned in Sect. 3.2). This is meaningful, if it is possible to observe the total visit frequency and describe how it changes over time. However, this method does not measure the true effect of the large-scale redevelopment. This is because it is necessary to remove the effect of the increase in the entry visit frequency due to the large-scale redevelopment in order to measure the true effect of large-scale redevelopment.

All of this is due to the following reason. In this study, we capture the effect of the large-scale redevelopment on the total visit frequency as a combined effect, that is to say, the effect of increasing the number of people entering through redevelopment and the effect of changes in consumer shop-around behavior through redevelopment. The purpose of this study is to measure changes in shop-around behaviors caused by the large-scale redevelopment. Because of this, we call the true effect on the total visit frequency due to the large-scale redevelopment as the effect resulting only from the changes in the shop-around behaviors caused by the large-scale redevelopment.

Hence, in order to measure this, it is necessary to equalize the entry visit frequency for the interannual comparison, eliminate the effect of the changes in the number of people entering, and extract only the development effect owing to the changes in shop-around behaviors.

Therefore, to measure the true effect of the large-scale redevelopment on a frequency basis, we made the following ingenuity. If there is no redevelopment, it is thought that consumer shop-around behaviors will be the same as before. This means that the observed shop-around choice probability does not change. Therefore, if the entry visit frequency after redevelopment were given under the observed shop-around choice probability before redevelopment, the predicted value $\text{TVST}^{\text{PRED}}$ of the total visit frequency to each commercial facility would become as follows.

$$\text{TVST}^{\text{PRED}} = F_{\text{HI}}^{a} + F_{\text{HI}}^{a} P_{\text{II}}^{b} \left(I - P_{\text{II}}^{b}\right)^{-1} \quad (2.5)$$

Here, the superscript letters a and b refer to after and before development, respectively. The second term in this formula is the predicted value of shop-around visit frequency, which is denoted as RE^{PRED}. On the other hand, the measured TVST^{OBS} of the total visit frequency after development is as follows.

$$\text{TVST}^{\text{OBS}} = F_{\text{HI}}^{a} + F_{\text{HI}}^{a} P_{\text{II}}^{a} \left(I - P_{\text{II}}^{a}\right)^{-1} \quad (2.6)$$

As before, the second term is the actual measured value of shop-around visit frequency, which is denoted as RE^{OBS}.

These differences, namely, ($\text{TVST}^{\text{OBS}} - \text{TVST}^{\text{PRED}}$) and ($\text{RE}^{\text{OBS}} - \text{RE}^{\text{PRED}}$), are the true effects of large-scale redevelopment on total visit frequency and shop-around visit frequency, respectively. In other words, by fixing the entry visit frequency at one point and using the shop-around choice probability before and after development, it is possible to measure the true effect on the shop-around visit frequency or the total visit frequency to each commercial facility due to the large-scale redevelopment.

3.4 Center of Gravity by Visit Frequency

Most of the measurements of the center of gravity of a commercial district are calculated by using weights that correspond to the sales floor area of commercial facilities within the district. We would like to propose the concept of a new center of gravity by visit frequency based on the aim of this behavioristic study, which is to capture the change in the city center retail structure due to the large-scale redevelopment from the changes in shop-around behaviors. The purpose of calculating the center of gravity is that we wish to express the changes of total visit frequency to commercial facilities in the Tenjin district over 3 year as clearly as possible to represent as the movement of one locational point of the center of gravity over these years.

The method of calculating the center of gravity by total visit frequency is as follows: As commercial facilities in the Tenjin district, we counted 17 stores in 1996, 19 stores in 1997, and 20 stores in 1998. The calculation of the center of gravity of 1996 is shown below. It is the same in 1997 and 1998.

$$\text{X coordinate} = \frac{\sum_i \text{TVST}_i^{96} X_i}{\sum_i \text{TVST}_i^{96}} \qquad \text{Y coordinate} = \frac{\sum_i \text{TVST}_i^{96} Y_i}{\sum_i \text{TVST}_i^{96}} \qquad (2.7)$$

TVST_i^{96}, total visit frequency to store i in 1996
X_i, X coordinate of store i
Y_i, Y coordinate of store i
i, store in the Tenjin area

4 Results of the Surveys of Shop-Around Behavior in the Downtown Fukuoka Area

Tables 2.1, 2.2, and 2.3 show the observed shop-around choice probability obtained from the results of the three surveys of shop-around behavior at the city center of Fukuoka City (i.e., September 1996, July 1997, and January 1998). As described in Sect. 3.1, the observed shop-around choice probability is an ordinary maximum likelihood estimate obtained from the aggregated result of "Unlinked Trips" derived by pooling respondents' shop-around data from all the survey sampling points and decomposing them into these "Unlinked Trips" between two consecutive nodes.[1]

---

[1]The surveys conducted here belong to a class of on-site random sampling surveys which randomly extract the respondents at places the respondents visited. These on-site random sampling surveys necessarily generate the choice-based sampling biases unless we set up sampling points at all the commercial facilities and collect the numbers of samples at those facilities in proportion to the total numbers of visitors to those facilities. Later in 2001, we developed the consistent estimation method to estimate consumer shop-around patterns based on data obtained from the on-site random sampling survey. (Cf. Saito et al. [32]; Saito and Nakashima [33]) However, here we employed the usual method.

2 How Did the Large-Scale City Center Retail Redevelopment Change...

Table 2.1 Shop-around choice probability (According to the 1996 survey)

	Node	1	2	3	4	5	6	7	8	9	10	11	Exit	Total (%)	Total (frequency)
Shoppers Daiei, Futata (zone 200, 300, 400)	1	48.07	5.61	6.67	16.14	0.00	0.00	0.00	0.00	0.00	0.00	0.35	23.16	100.00	285
Tenjin Core, IMS, *Daimaru*/Tenjin Fukuoka, *Daimaru* Elgala (zone 500)	2	6.49	40.08	18.32	16.03	0.38	0.38	1.15	0.00	0.00	0.76	0.76	15.65	100.00	262
*Iwataya*A-side, Solaria Plaza, *Shintencho* Shopping Street (zone 600)	3	6.71	9.49	40.51	7.87	0.23	0.23	1.39	0.00	0.23	0.46	1.16	31.71	100.00	432
Tenjin Underground Shopping Mall (zone 700)	4	28.74	23.35	25.15	0.60	0.60	0.00	0.60	0.00	0.60	0.00	0.00	20.36	100.00	167
Zeex Tenjin (zone 1200)	5	0.00	0.00	50.00	0.00	33.33	0.00	0.00	0.00	0.00	0.00	0.00	16.67	100.00	6
Hakata Station district (zone 2500, 2600)	6	0.00	0.33	0.00	0.00	0.00	34.53	7.49	0.00	0.00	0.98	0.33	56.35	100.00	307
Canal City Hakata (zone 2000)	7	1.10	1.32	0.66	0.00	0.00	1.98	65.20	3.30	0.00	3.30	0.44	22.69	100.00	454
Kamikawabata district (zone 1300)	8	0.00	0.00	0.00	0.00	0.00	0.00	23.08	46.15	7.69	1.54	0.00	21.54	100.00	65
Nakasu district, Fukuoka Tamaya (zone 900, 1400)	9	6.67	13.33	6.67	0.00	0.00	0.00	0.00	13.33	6.67	6.67	0.00	46.67	100.00	15
Other districts in the Fukuoka City Center	10	0.00	5.66	5.66	0.00	1.89	13.21	20.75	1.89	0.00	9.43	1.89	39.62	100.00	53
Districts outside the Fukuoka City Center	11	0.00	0.00	1.43	0.00	0.00	2.86	0.00	0.00	0.00	0.00	5.71	90.00	100.00	70
Entry	12	7.27	7.42	20.76	6.67	0.00	27.42	15.00	2.58	1.06	3.64	8.18	0.00	100.00	1429

Table 2.2 Shop-around choice probability (According to the 1997 survey)

	Node	1	2	3	4	5	6	7	8	9	10	11	Exit	Total (%)	Total (frequency)
Shoppers Daiei, Futata (zone 200, 300, 400)	1	47.73	6.10	7.67	13.77	0.00	0.00	0.78	0.00	0.31	0.00	0.16	23.47	100.00	639
Tenjin Core, IMS, Daimaru Tenjin Fukuoka, Daimaru Elgala[a] (zone 500)	2	3.23	48.86	16.73	9.38	0.38	0.19	1.58	0.00	0.32	0.32	0.00	19.01	100.00	1578
Iwataya A-side, Solaria Plaza, Shintencho Shopping Street, Iwataya Z-Side[b] (zone 600)	3	3.47	15.00	48.36	5.80	0.48	0.00	0.72	0.00	0.30	0.18	0.12	25.58	100.00	1673
Tenjin Underground Shopping Mall (zone 700)	4	12.57	20.23	13.37	40.46	0.23	0.00	0.00	0.00	0.00	0.00	0.11	13.03	100.00	875
Zeex Tenjin (zone 1200)	5	0.00	19.51	24.39	7.32	19.51	0.00	0.00	0.00	0.00	0.00	0.00	29.27	100.00	41
Hakata Station district (zone 2500, 2600)	6	0.00	0.00	0.00	0.00	0.00	38.27	2.21	0.00	0.34	2.38	0.17	56.63	100.00	588
Canal City Hakata (zone 2000)	7	0.45	1.79	0.75	0.00	0.00	1.94	43.82	1.49	0.15	3.28	0.15	46.20	100.00	671
Kamikawabata district (zone 1300)	8	0.00	0.00	0.00	0.00	0.00	0.00	25.58	44.19	4.65	0.00	0.00	25.58	100.00	43
Nakasu district, Fukuoka Tamaya (zone 900, 1400)	9	3.03	18.18	0.00	6.06	3.03	3.03	12.12	9.09	12.12	0.00	0.00	33.33	100.00	33
Other districts in the Fukuoka City Center	10	0.00	6.72	0.00	0.75	0.00	10.45	14.93	0.00	0.00	6.72	0.75	59.70	100.00	134
Districts outside the Fukuoka City Center	11	0.00	0.00	1.54	1.54	0.00	1.54	0.00	0.00	0.00	3.08	10.77	81.54	100.00	65
Entry	12	6.16	16.93	23.20	10.04	0.89	18.37	15.93	0.61	0.67	4.38	2.83	0.00	100.00	4186

[a]Expansion in March, 1997
[b]Opened in September, 1996

Table 2.3 Shop-around choice probability (According to the 1998 survey)

	Node	1	2	3	4	5	6	7	8	9	10	11	Exit	Total (%)	Total (frequency)
Shoppers Daiei, Futata (zone 200, 300, 400)	1	36.06	7.85	15.03	16.86	0.33	0.00	0.17	0.00	0.33	0.33	0.50	22.54	100.00	599
Tenjin Core, IMS, Daimaru Tenjin Fukuoka, Daimaru Elgala (zone 500)	2	3.27	39.58	22.47	10.26	0.44	0.19	1.51	0.00	0.44	0.76	0.25	20.83	100.00	1589
Iwataya A-side, Solaria Plaza, Shintencho Shopping Street, Iwataya Z-Side, Fukuoka Mitsukoshi[a] (zone 600)	3	2.38	11.98	54.20	5.31	0.38	0.10	0.41	0.00	0.10	0.45	0.03	24.67	100.00	3146
Tenjin Underground Shopping Mall (zone 700)	4	9.34	16.67	16.92	45.96	0.08	0.00	0.00	0.00	0.00	0.00	0.08	10.94	100.00	1188
Zeex Tenjin (zone 1200)	5	4.92	19.67	37.70	1.64	11.48	0.00	3.28	0.00	0.00	1.64	0.00	19.67	100.00	61
Hakata Station district (zone 2500, 2600)	6	0.00	0.31	0.42	0.10	0.00	37.33	6.26	0.00	0.10	1.67	0.42	53.39	100.00	959
Canal City Hakata (zone 2000)	7	0.35	2.44	2.09	0.00	0.12	4.65	55.35	1.16	0.00	1.16	0.00	32.67	100.00	860
Kamikawabata district (zone 1300)	8	0.00	1.72	0.00	0.00	0.00	0.00	22.41	44.83	5.17	3.45	0.00	22.41	100.00	58
Nakasu district, Fukuoka Tamaya (zone 900, 1400)	9	2.63	10.53	5.26	0.00	0.00	0.00	2.63	7.89	10.53	7.89	0.00	52.63	100.00	38
Other districts in the Fukuoka City Center	10	1.20	9.04	9.04	1.20	1.20	10.84	6.02	1.81	1.20	22.29	0.60	35.54	100.00	166
Districts outside the Fukuoka City Center	11	3.36	0.84	0.00	0.00	0.00	4.20	0.00	0.00	0.00	0.00	5.04	86.55	100.00	119
Entry	12	5.56	11.85	30.82	8.73	1.22	22.43	10.96	0.67	0.67	2.91	4.17	0.00	100.00	5467

[a]Opened in October 1997

5 Measurement of Structural Changes in the City Center District due to Large-Scale Redevelopment: Verification of the Southward Movement Phenomenon in Tenjin District

5.1 Interannual Comparison of Shop-Around Choice Probability

In Tables 2.1, 2.2, and 2.3, we directly compare the observed shop-around choice probabilities for the years 1996, 1997, 1998, respectively.

Let us look at Tables 2.1 and 2.2. Between 1996 and 1997, *Daimaru* Elgala opened at node 2, and *Iwataya* Z-Side at node 3. Nodes 2 and 3 correspond to Tenjin South district, and node 1 corresponds to Tenjin North district. Looking at the diagonal elements, we can see that the staying probability (sojourn probability) in the area of nodes 2 and 3, where redevelopment occurred, increased. In addition, we can see that the shop-around probability from the redeveloped nodes 2 and 3 to node 1 in Tenjin North, which was not redeveloped, decreased.

Let us compare Tables 2.2 and 2.3 and examine the effect of Fukuoka *Mitsukoshi's* opening. Here too, the sojourn probability at the redeveloped node 3 is increasing. The shop-around probability from node 1 and node 2 to node 3 increased, and the shop-around probability from node 3 to node 2 decreased.

Based on the above, a general tendency can be observed: the staying probability increases within the development site. Additionally, the shop-around probability from Tenjin North district to the development site of Tenjin South district increases, and the decrease in the shop-around probability from the south to the north suggests the existence of a southward movement phenomenon in the flow of people.

5.2 Interannual Comparison of the Shop-Around Effect

In this section, we make an interannual comparison from the viewpoint of how much one trip to each commercial district brings the shop-around effect to other commercial district. Tables 2.4, 2.5, and 2.6 show the results of the shop-around effect on each district listed in the column of the table, as caused by one trip to each commercial district in the row of the table, as determined by using the observed shop-around probability of 1996, 1997, and 1998.

Looking at the total shop-around effect, it consistently increased from 1996 to 1998 in nodes 1, 2, 3, and 4 in the Tenjin district where development was conducted except one movement of node 2 from 1997 to 1998. To see the circumstances during this period in more detail, we focus on nodes 2 and 3 and make an interannual comparison of the mutual shop-around effect as shown in Figs. 2.3 and 2.4.

2 How Did the Large-Scale City Center Retail Redevelopment Change...

Table 2.4 Shop-around effect to each column district, caused by one trip to each row district (According to the 1996 survey)

	Node	1	2	3	4	5	6	7	8	9	10	11	Total	Shop-around effect on other sites
Shoppers Daiei, Futata (zone 200, 300, 400)	1	1.370	0.535	0.663	0.524	0.010	0.011	0.066	0.006	0.005	0.011	0.022	3.223	1.853 (57%)
Tenjin Core, IMS, *Daimaru* Tenjin Fukuoka, *Daimaru* Elgala (zone 500)	2	0.683	1.099	0.969	0.526	0.021	0.027	0.148	0.012	0.007	0.029	0.033	3.553	2.454 (69%)
Iwataya A-side, Solaria Plaza, Shintencho Shopping Street (zone 600)	3	0.513	0.507	1.049	0.327	0.014	0.020	0.127	0.011	0.008	0.020	0.032	2.628	1.579 (60%)
Tenjin Underground Shopping Mall (zone 700)	4	0.981	0.782	0.950	0.367	0.021	0.015	0.107	0.010	0.012	0.017	0.023	3.283	2.917 (89%)
Zeex Tenjin (zone1200)	5	0.385	0.380	1.537	0.245	0.510	0.015	0.096	0.008	0.006	0.015	0.024	3.221	2.711 (84%)
Hakata Station district (zone 2500, 2600)	6	0.020	0.030	0.024	0.010	0.001	0.545	0.369	0.024	0.002	0.031	0.008	1.066	0.521 (49%)
Canal City Hakata (zone 2000)	7	0.133	0.140	0.134	0.055	0.005	0.121	2.115	0.200	0.017	0.121	0.021	3.062	0.947 (31%)
Kamikawabata district (zone 1300)	8	0.107	0.124	0.117	0.047	0.004	0.064	1.400	0.987	0.164	0.099	0.012	3.127	2.139 (68%)
Nakasu district, Fukuoka Tamaya (zone 900, 1400)	9	0.327	0.407	0.367	0.148	0.008	0.034	0.293	0.293	0.097	0.102	0.013	2.090	1.992 (95%)
Other districts in the Fukuoka City Center	10	0.119	0.210	0.258	0.074	0.035	0.259	0.816	0.092	0.009	0.142	0.033	2.046	1.904 (93%)
Districts outside the Fukuoka City Center	11	0.008	0.009	0.032	0.005	0.000	0.047	0.013	0.001	0.000	0.001	0.061	0.178	0.117 (66%)

Table 2.5 Shop-around effect to each column district, caused by one trip to each row district (According to the 1997 survey)

	Node	1	2	3	4	5	6	7	8	9	10	11	Total	Shop-around effect on other sites
Shoppers Daiei, Futata (zone 200, 300, 400)	1	1.183	0.775	0.764	0.704	0.011	0.007	0.069	0.004	0.014	0.007	0.006	3.542	2.360 (67%)
Tenjin Core, IMS, *Daimaru* Tenjin Fukuoka, *Daimaru* Elgala*[a]* (zone 500)	2	0.373	1.563	1.052	0.597	0.021	0.014	0.101	0.005	0.015	0.015	0.003	3.758	2.195 (58%)
Iwataya A-side, Solaria Plaza, Shintencho Shopping Street, *Iwataya* Z-Side*[b]* (zone 600)	3	0.334	0.953	1.427	0.468	0.021	0.008	0.070	0.004	0.013	0.011	0.005	3.313	1.886 (57%)
Tenjin Underground Shopping Mall (zone 700)	4	0.664	1.253	1.068	1.138	0.019	0.008	0.065	0.003	0.011	0.009	0.006	4.242	3.104 (73%)
Zeex Tenjin (zone 1200)	5	0.252	1.024	1.087	0.481	0.255	0.007	0.052	0.003	0.009	0.008	0.003	3.180	2.924 (92%)
Hakata Station district (zone 2500, 2600)	6	0.004	0.017	0.009	0.006	0.000	0.630	0.080	0.003	0.007	0.045	0.004	0.804	0.174 (22%)
Canal City Hakata (zone 2000)	7	0.038	0.119	0.081	0.037	0.001	0.070	0.829	0.050	0.007	0.067	0.004	1.302	0.472 (36%)
Kamikawabata district (zone 1300)	8	0.035	0.113	0.068	0.044	0.005	0.038	0.870	0.831	0.099	0.032	0.002	2.137	1.306 (61%)
Nakasu district, Fukuoka Tamaya (zone 900, 1400)	9	0.216	0.707	0.374	0.322	0.050	0.074	0.375	0.198	0.154	0.018	0.002	2.489	2.335 (94%)
Other districts in the Fukuoka City Center	10	0.039	0.216	0.099	0.067	0.002	0.195	0.310	0.009	0.003	0.089	0.010	1.038	0.949 (91%)
Districts outside the Fukuoka City Center	11	0.019	0.046	0.064	0.047	0.001	0.035	0.014	0.000	0.001	0.039	0.121	0.387	0.266 (69%)

[a]Expansion in March, 1997
[b]Opened in September, 1996

2 How Did the Large-Scale City Center Retail Redevelopment Change... 31

Table 2.6 Shop-around effect to each column district, caused by one trip to each row district (According to the 1998 survey)

	Node	1	2	3	4	5	6	7	8	9	10	11	Total	Shop-around effect on other sites
Shoppers Daiei, Futata (zone 200, 300, 400)	1	0.769	0.716	1.254	0.812	0.017	0.013	0.051	0.004	0.012	0.025	0.013	3.686	2.917 (79%)
Tenjin Core, IMS, *Daimaru* Tenjin Fukuoka, *Daimaru* Elgala (zone 500)	2	0.259	1.160	1.406	0.630	0.019	0.023	0.100	0.005	0.014	0.034	0.008	3.657	2.497 (68%)
Iwataya A-side, Solaria Plaza, Shintencho Shopping Street, *Iwataya* Z-Side, Fukuoka *Mitsukoshi*[a] (zone 600)	3	0.216	0.734	1.817	0.485	0.018	0.016	0.061	0.003	0.008	0.027	0.005	3.389	1.573 (46%)
Tenjin Underground Shopping Mall (zone 700)	4	0.454	1.021	1.535	1.338	0.016	0.014	0.059	0.003	0.009	0.023	0.008	4.480	3.143 (70%)
Zeex Tenjin (zone 1200)	5	0.260	0.866	1.631	0.441	0.143	0.024	0.142	0.006	0.008	0.047	0.005	3.572	3.429 (96%)
Hakata Station district (zone 2500, 2600)	6	0.011	0.047	0.071	0.023	0.001	0.621	0.239	0.007	0.003	0.040	0.008	1.071	0.450 (42%)
Canal City Hakata (zone 2000)	7	0.044	0.181	0.252	0.074	0.006	0.180	1.306	0.051	0.005	0.045	0.002	2.145	0.839 (39%)
Kamikawabata district (zone 1300)	8	0.042	0.202	0.226	0.077	0.005	0.093	0.974	0.853	0.110	0.114	0.002	2.697	1.845 (68%)
Nakasu district, Fukuoka Tamaya (zone 900, 1400)	9	0.109	0.378	0.448	0.153	0.006	0.039	0.194	0.170	0.132	0.133	0.003	1.766	1.634 (93%)
Other districts in the Fukuoka City Center	10	0.102	0.408	0.602	0.199	0.023	0.249	0.260	0.052	0.024	0.309	0.011	2.239	1.930 (86%)
Districts outside the Fukuoka City Center	11	0.065	0.047	0.060	0.035	0.001	0.072	0.013	0.000	0.001	0.003	0.054	0.352	0.298 (85%)

[a]Opened in October 1997

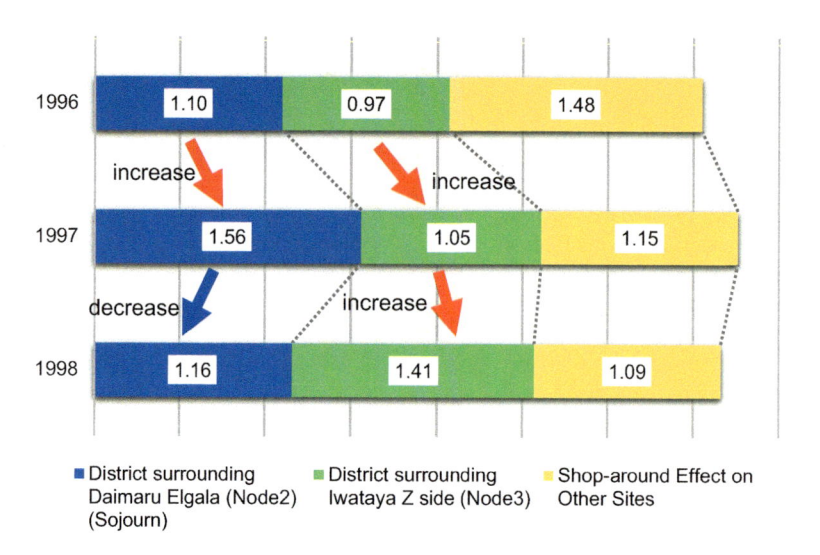

Fig. 2.3 Shop-around effect caused by one trip to the district surrounding *Daimaru* Elgala (node 2)

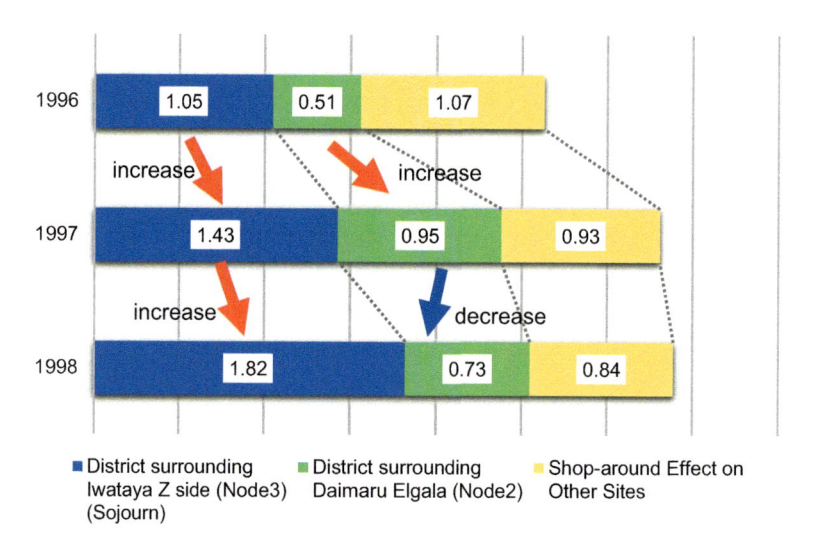

Fig. 2.4 Shop-around effect caused by one trip to the district surrounding Z-Side (node 3)

It should be noted that development was carried out in both nodes 2 and 3 between 1996 and 1997. In Figs. 2.3 and 2.4, we can see that the shop-around effect on the other district is increasing for each district. In contrast, the development of Fukuoka *Mitsukoshi* occurred only at node 3 from 1997 to 1998. As a result, although the staying (sojourn) effect, that is, the shop-around effect on its own district for node 3 and the shop-around from node 2 increased, one can see that the shop-around effect from node 3 to node 2, as well as the staying (sojourn) effect for node 2, decreased.

When constructing a forecast model, it is essential to use a model that can correctly explain these movements.

5.3 Interannual Comparison of Shop-Around Visit Frequency and Total Visit Frequency

In this section, we estimate the actual shop-around visit frequency and total visit frequency by using the number of visitors entering the city center commercial district which is estimated separately, and we compare these interannually.

Regarding the number of visitors entering the city center commercial district of Fukuoka City, we used the forecasting model of Saito et al. [25], which is constructed based on data obtained from the first on-site sampling survey of consumer shop-around behaviors (September 1996). (Cf. [34]) The forecast is based on the Poisson regression model which removes the choice-based sampling bias accrued to the on-site random sampling surveys, in which the respondents are sampled at random at the destination places they visited. It is a forecast of the number of people incoming from the metropolitan area, which has a total population of about 1.7 million and is a commuting zone where the commuting rate among the permanent employed residents who commute for work and school to Fukuoka City exceeds 20%. According to this forecast, the number of people who visit the Tenjin district for shopping, eating, and leisure was 105,402 person per day before the opening of *Iwataya* Z-Side, 117,895 after opening *Iwataya* Z-Side, 124,691 after expanding the sales floor space at *Daimaru* Elgala, and 137,939 after the opening of Fukuoka *Mitsukoshi*. The estimated numbers of people who visit the Hakata Station and Canal City Hakata areas for shopping, eating, and leisure were 33,488 and 26,627 person per day, respectively.

In the shop-around Markov model, needed is not the total number of visitors who enter the city center district but each number of visitors who visit every commercial district node as entrance. Thus, we need to determine the entry visit frequency by allocating above total entrance numbers to each commercial district node. As for the allocation to each commercial district node in the Tenjin district, we used the entrance choice model of Saito and Ishibashi [23, 24] and decided to use the above estimated values for the Hakata Station and Canal City Hakata areas as they are composed of one district.

We suppose that the increase in the number of entrants due to development simply will appear as an increase in the entry visit frequency to the site where development took place. Therefore, we determine the entry visit frequency in 1997, in addition to that in 1996, for the increase in the number of people entering due to the opening of *Iwataya* Z-Side (i.e., 12,493 person/day) as an increase in the entry visit frequency to the commercial district of node 3 and the increase in the number of visitors entering due to *Daimaru* Elgala (i.e., 6796 person/day) as an increase in the commercial district of node 2. Likewise, an increase of 13,284 person/day on account of Fukuoka *Mitsukoshi* is recorded as an increase in the entry visit frequency to the commercial district of node 3.

Tables 2.7, 2.8, and 2.9 show the estimated results of the shop-around visit frequency and total visit frequency in each year by using the entry visit frequency for each year. Additionally, in Tables 2.7, 2.8, and 2.9, to show the measurement error of the shop-around choice probability, we also show a 90% confidence interval for the shop-around visit frequency obtained by bootstrap. More specifically, according to the observed shop-around choice probabilities of 1996, 1997, and 1998 in Tables 2.1, 2.2, and 2.3, respectively, a resampling equal to the size of the frequency in the rightmost column of the table is conducted, the maximum likelihood estimate of the shop-around choice probability is obtained from this resample, and the shop-around effect is measured. This is repeated 100 times to determine the 90% confidence interval of the shop-around effect that corresponds to Tables 2.4, 2.5, and 2.6. The 90% confidence interval of the shop-around visit frequency is then obtained by multiplying this by the entry visit frequency. One can see that the precision of the measurements for 1997 and 1998, as well as that of nodes 2 and 3, which have a large number of visitors entering, is high compared to those of 1996, which has a small number of samples.

Now, let us see how the visit frequency has changed due to the development of *Iwataya* Z-Side and Daimaru Elgala, according to Tables 2.7 and 2.8. In Elgala (node 2) and Z-Side (node 3), the shop-around visit frequency, the staying, and total visit frequency have increased. Let us also look at Table 2.9, which contains the estimation results for 1998. At node 3, where Fukuoka *Mitsukoshi* was opened, there has been a significant increase since 1997 (i.e., 17,275 person/day of shop-around visit frequency, 93,415 of total visit frequency, and 62,893 of staying). On the other hand, looking at node 2 which was not developed between 1997 and 1998, as compared to the estimation results in 1997, the shop-around visit frequency increased by 4486 person/day, but the staying decreased significantly. As a result, the total visit frequency decreased by 21,255 person/day.

In node 1 of the Tenjin North district, where there was no development between 1996 and 1998, the shop-around visit frequency and the total visit frequency declined. The total visit frequency there decreased by 20,720 person/day between 1996 and 1997 and by 17,100 between 1997 and 1998, for a total decrease of 37,820 between 1996 and 1998.

Based on the results of the above analysis, we found that the total visit frequency to node 1 of the Tenjin North district, which had not been developed, decreased. From these results, it might be possible to confirm the southward movement phenomenon of the flow of people in the Tenjin district. However, such a simple comparison would likely to lead to underestimate the effect of the decrease in the number of visitors to the Tenjin North district caused by the redevelopment. This is because, as of 1998, the effect of the increased number of entrants between 1996 and 1998 was added to the estimate of total visit frequency. To measure the true effect on the visit frequency due to the redevelopment, it is necessary to remove the effect of the increased number of entrants. This is the task to take up in Sect 5.4.

Table 2.7 Estimated results of entry visit frequency, shop-around visit frequency, and total visit frequency (According to the 1996 survey)

	Node	Entry visit frequency	Shop-around visit frequency (excluding sojourn)	Sojourn	Shop-around visit frequency	Shop-around visit frequency 90% confidence interval		Total visit frequency
Shoppers Daiei, Futata (zone 200, 300, 400)	1	17,252	34,198	47,626	81,824	66,547	99,059	99,075
Tenjin Core, IMS, *Daimaru* Tenjin Fukuoka, *Daimaru* Elgala (zone 500)	2	31,459	35,776	44,966	80,742	64,709	100,736	112,200
Iwataya A-side, Solaria Plaza, Shintencho Shopping Street (zone 600)	3	46,681	44,425	62,037	106,462	88,292	123,680	153,143
Tenjin Underground Shopping Mall (zone 700)	4	4059	46,030	302	46,332	38,670	54,361	50,391
Zeex Tenjin (zone 1200)	5	2030	1233	1632	2865	832	8065	4895
Hakata Station district (zone 2500, 2600)	6	33,488	4259	19,906	24,165	19,125	30,891	57,654
Canal City Hakata (zone 2000)	7	26,627	14,061	76,226	90,287	70,608	109,154	116,914
Kamikawabata district (zone 1300)	8	4500	5743	8780	14,523	7915	21,877	19,023
Nakasu district, Fukuoka Tamaya (zone 900, 1400)	9	10,000	2120	866	2986	831	5508	12,985
Other districts in the Fukuoka City Center	10	0	7150	745	7895	4025	11,842	7895
Districts outside the Fukuoka City Center	11	0	3828	232	4060	1758	6123	4060
Total		176,096	198,823	263,317	462,140	363,312	571,296	638,236

Table 2.8 Estimated results of entry visit frequency, shop-around visit frequency, and total visit frequency (According to the 1997 survey)

	Node	Entry visit frequency	Shop-around visit frequency (excluding sojourn)	Sojourn	Shop-around visit frequency	Shop-around visit frequency 90% confidence interval		Total visit frequency
Shoppers Daiei, Futata (zone 200, 300, 400)	1	17,252	23,703	37,399	61,103	53,986	69,491	78,355
Tenjin Core, IMS, *Daimaru* Tenjin Fukuoka, *Daimaru* Elgala[a] (zone 500)	2	38,255	57,018	91,023	148,040	135,970	163,472	186,295
Iwataya A-side, Solaria Plaza, *Shintencho* Shopping Street, *IwatayaZ*-Side[b] (zone 600)	3	59,174	49,299	101,568	150,868	138,471	162,163	210,042
Tenjin Underground Shopping Mall (zone 700)	4	4059	41,736	31,116	72,852	65,092	81,988	76,911
Zeex Tenjin (zone 1200)	5	2030	2314	1053	3367	1921	5008	5397
Hakata Station District (zone 2500, 2600)	6	33,488	2640	22,393	25,033	21,239	29,115	58,521
Canal city Hakata (zone 2000)	7	26,627	11,923	30,063	41,986	34,513	49,291	68,613
Kamikawabata district (zone 1300)	8	4500	2298	5382	7680	4805	12,482	12,180
Nakasu district, Fukuoka Tamaya (zone 900, 1400)	9	10,000	2331	1701	4032	1672	6704	14,032
Other districts in the Fukuoka City Center	10	0	4634	334	4968	3312	6417	4968
Districts outside the Fukuoka City Center	11	0	700	85	785	211	1366	785
Total		195,385	198,597	322,116	520,714	461,192	587,497	716,098

[a]Expansion in March, 1997
[b]Opened in September, 1996

Table 2.9 Estimated results of entry visit frequency, shop-around visit frequency, and total visit frequency (According to the 1998 survey)

	Node	Entry visit frequency	Shop-around visit frequency (excluding sojourn)	Sojourn	Shop-around visit frequency	Shop-around visit frequency 90% confidence interval		Total visit frequency
Shoppers Daiei, Futata (zone 200, 300, 400)	1	17,252	21,914	22,088	44,003	38,081	49,955	61,255
Tenjin Core, IMS, *Daimaru* Tenjin Fukuoka, *Daimaru* Elgala (zone 500)	2	38,255	61,504	65,330	126,835	114,983	138,004	165,040
Iwataya A-side, Solaria Plaza, *Shintencho* shopping Street, *Iwataya* Z-Side, Fukuoka *Mitsukoshi*[a] (zone 600)	3	72,422	66,574	164,461	231,035	215,475	246,700	303,457
Tenjin Underground Shopping Mall (zone 700)	4	4059	43,612	40,543	84,155	74,986	91,725	88,214
Zeex Tenjin (zone 1200)	5	2030	2362	569	2931	1942	3835	4961
Hakata Station district (zone 2500, 2600)	6	33,488	5515	23,233	28,748	25,053	32,516	62,236
Canal city Hakata (zone 2000)	7	26,627	11,477	47,232	58,709	50,814	67,624	85,336
Kamikawabata district (zone 1300)	8	4500	2201	5444	7645	4834	11,520	12,145
Nakasu district, Fukuoka Tamaya (zone 900, 1400)	9	10,000	2013	1413	3426	1815	5070	13,426
Other districts in the Fukuoka City Center	10	0	6392	1833	8225	5240	11,267	8225
Districts outside the Fukuoka City Center	11	0	1202	64	1266	614	2031	1266
Total		208,633	224,766	372,212	596,978	533,837	660,247	805,561

[a]Opened in October 1997

5.4 Measurement of the Effect of Large-Scale Redevelopment on Visit Frequency

In this section, by using the method proposed in Sect. 3.3, we measure the true effect of the large-scale redevelopment.

Tables 2.10 and 2.11 show the analytical results. The columns in Table 2.10 named as "prediction" show the predicted values of the shop-around and the total visit frequency for each row commercial district which are calculated by using the entry visit frequency of 1997 and the observed shop-around choice probability of 1996. By their construction, these figures imply the amounts of shop-around and total visit frequency each commercial district would have had at the time of 1997 if the developments of *Iwataya* Z-Side and *Daimaru* Elgala had not been carried out. Alternatively, supposing that each commercial district is given the same entry visit frequency as 1997, they correspond to how much shop-around and total visit frequency each commercial district would have attracted if the consumer shop-around had continued to be the same as the year of 1996.[a] Expansion in March 1997 [b] Opened in September 1996 [a] Opened in October 1997

On the other hand, those columns in Table 2.10 named as "realization" show the estimated actual values of the shop-around and the total visit frequency for each commercial district calculated by using both of the entry visit frequency and the observed shop-around choice probability at the time of 1997. They are the estimates of actual realizations.

As shown in Table 2.10, in nodes 2 and 3 (where Daimaru Elgala and Iwataya Z-Side are located), the estimated actual values become larger than the predicted values without the developments for either the shop-around visit frequency or the total visit frequency. The true effects on total visit frequency for nodes 2 and 3 due to the development of *Iwataya* Z-Side and *Daimaru* Elgala increased by 53,499 and 24,713 person/day, respectively. In contrast, the effect of the development on the total visit frequency for node 1 of the Tenjin North district where there were no development decreased by 31,770 person/day between 1996 and 1997.

Table 2.11 shows the results of the same analysis before and after the opening of Fukuoka Mitsukoshi. The true effect (1997–1998) corresponding to the total visit frequency to each commercial district after Fukuoka *Mitsukoshi's* opening was a decrease of 21,503 person/day at node 1 of Tenjin North district. Additionally, node 2 where there were no development also saw a decrease of 33,757 person/day. In contrast, at node 3 where Fukuoka Mitsukoshi opened saw an increase of 61,317.

Furthermore, in the last column of Table 2.11, the total true effect from 1996 to 1998 obtained in the same way is shown for each commercial district.

In conclusion, the true effect of these developments during 1996–1998 on the total visit frequency was a decrease of 55,634 person/day for node 1 of Tenjin North district (where no development had been carried out), an increase of 25,634 at node 2 (where Daimaru Elgala had been developed), and an increase of 91,028 at node 3 (where *Iwataya* Z-Side and Fukuoka *Mitsukoshi* had been developed).

Table 2.10 Comparison of shop-around visit frequency and total visit frequency: effect of Z-side's and Elgala's development (1996–1997)

	Node	Shop-around visit frequency (1997 realization)	Shop-around visit frequency (1997 prediction)	Effect of redevelopment 1996-1997 (shop-around visit freq.)	Total visit frequency (1997 realization)	Total visit frequency (1997 prediction)	Effect of redevelopment 1996-1997 (total visit freq.)
Shoppers Daiei, Futata (zone 200, 300, 400)	1	61,103	92,873	−31,770	78,355	110,125	−31,770
Tenjin Core, IMS, *Daimaru* Tenjin Fukuoka, *Daimaru* Elgala [a] (zone 500)	2	148,040	94,541	53,499	186,295	132,796	53,499
Iwataya A-side, Solaria Plaza, Shintencho Shopping Street, Iwataya Z-Side[b] (zone 600)	3	150,868	126,154	24,714	210,042	185,328	24,714
Tenjin Underground Shopping Mall (zone 700)	4	72,852	53,996	18,856	76,911	58,055	18,856
Zeex Tenjin (zone 1200)	5	3367	3176	191	5397	5206	191
Hakata station district (zone 2500, 2600)	6	25,033	24,597	436	58,521	58,085	436
Canal city Hakata (zone 2000)	7	41,986	92,884	−50,898	68,613	119,511	−50,898
Kamikawabata district (zone 1300)	8	7680	14,734	−7054	12,180	19,234	−7054
Nakasu district, Fukuoka Tamaya (zone 900, 1400)	9	4032	3132	900	14,032	13,132	900
Other districts in the Fukuoka City Center	10	4968	8347	−3379	4968	8347	−3379
Districts outside the Fukuoka City Center	11	785	4686	−3901	785	4686	−3901

[a]Expansion in March 1997
[b]Opened in September 1996

Table 2.11 Comparison of shop-around visit frequency and total visit frequency: effect of Fukuoka Mitsukoshi's opening (1997–1998)

	Node	Shop-around visit frequency (1998 realization)	Shop-around visit frequency (1998 prediction)	Effect of Redevelopment 1997–1998 (shop-around visit freq.)	Total visit frequency (1998 realization)	Total visit frequency (1998 prediction)	Effect of redevelopment 1997–1998 (total visit freq.)	Effect of redevelopment 1996–1998 (total visit freq.)
Shoppers Daiei, Futata (zone 200, 300, 400)	1	44,003	65,506	−21,503	61,255	82,758	−21,503	−55,634
Tenjin Core, IMS, *Daimaru* Tenjin Fukuoka, *Daimaru* Elgala (zone 500)	2	126,835	160,592	−33,757	165,040	198,797	−33,757	25,634
Iwataya A-side, Solaria plaza, *Shintencho* Shopping Street, *Iwataya* Z-Side, Fukuoka *Mitsukoshi*[a] (zone 600)	3	231,035	169,718	61,317	303,457	242,140	61,317	91,028
Tenjin Underground Shopping Mall (zone 700)	4	84,155	79,020	5135	88,214	83,079	5135	25,848
Zeex Tenjin (zone 1200)	5	2931	3641	−710	4961	5671	−710	−424
Hakata Station district (zone 2500, 2600)	6	28,748	25,135	3614	62,236	58,623	3613	3894
Canal city Hakata (zone 2000)	7	58,709	42,913	15,796	85,336	69,540	15,796	−35,856
Kamikawabata district (zone 1300)	8	7645	7733	−88	12,145	12,233	−88	−7228
Nakasu district, Fukuoka Tamaya (zone 900, 1400)	9	3426	4207	−780	13,426	14,207	−781	189
Other districts in the Fukuoka City Center	10	8225	5109	3116	8225	5109	3116	−390
Districts outside the Fukuoka City Center	11	1266	847	419	1266	847	419	−3846

[a]Opened in October 1997

Tenjin North District — Sojourn: 47,626

Canal City Hakata — Sojourn: 76,266

Tenjin South District — Sojourn: 142,093

Hakata Station District

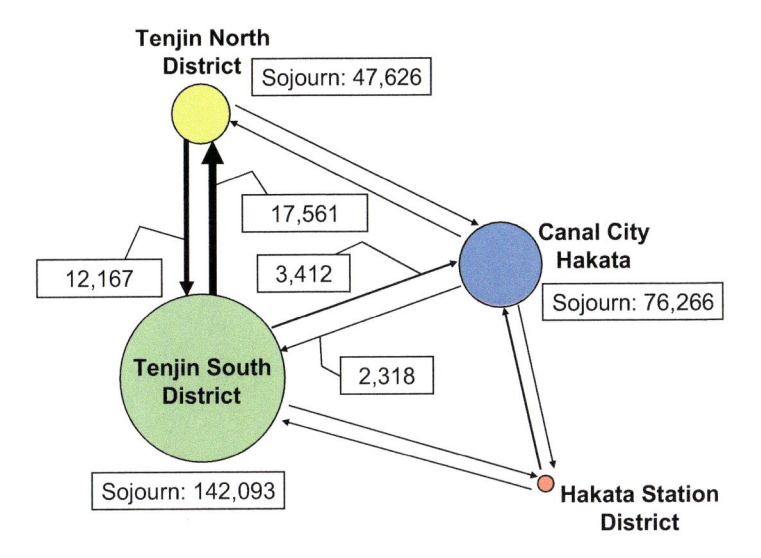

17,561
12,167
3,412
2,318

Fig. 2.5 Shop-around pattern (According to the 1996 survey) unit: people per day

5.5 Measuring the Effect of the Large-Scale Redevelopment on Shop-Around Pattern

In the shop-around Markov model, it is possible to estimate not only the total visit frequency to each commercial district but also how people move around between commercial districts, that is, the shop-around pattern, by using the following formula:[2]

$$\text{diag}(TVST)P_{II} \tag{2.8}$$

Here, diag(TVST) denotes a diagonal matrix with the vector TVST as its diagonal element.

Figures 2.5, 2.6, and 2.7 depict the estimated results of shop-around patterns for each survey year. Here, to simplify the figures, among all nodes, we picked up the three nodes (nodes 1, 2, and 3) of the Tenjin district, which is a major commercial cluster in the city center commercial district of Fukuoka City, Hakata Station district (node 6), and Canal City Hakata (node 7), and then divide the Tenjin district into the Tenjin North district of node 1 and the Tenjin South district composed of nodes 2 and 3 combined. Arrows indicate shop-around movements; circles indicate staying (sojourns), which imply in this case consumers' shop-around movements among commercial facilities located in the same district; and their thickness and size

---

[2] Alternatively, we can use another formula, $\text{diag}(P_{HI})P_{II}(I - P_{II})^{-1}$.

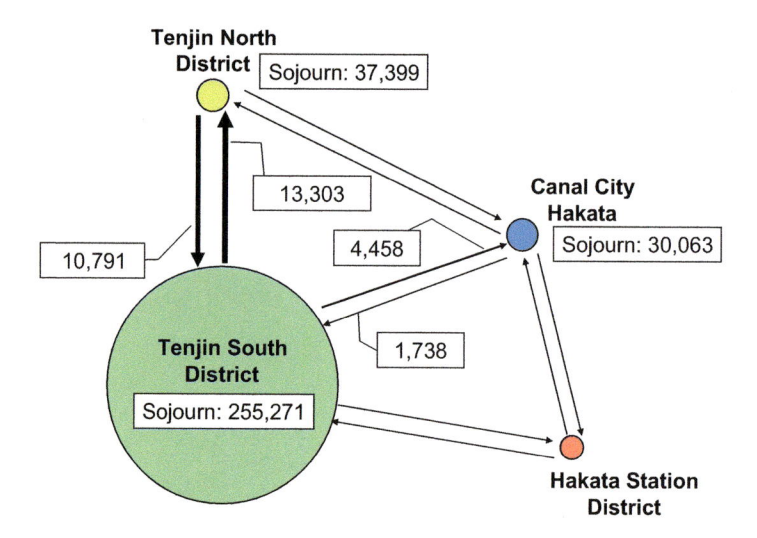

Fig. 2.6 Shop-around pattern (According to the 1997 survey) unit: people per day

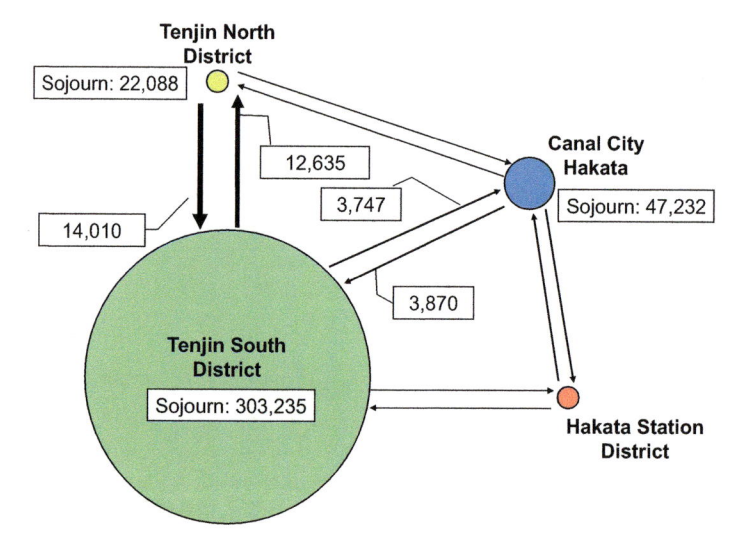

Fig. 2.7 Shop-around pattern (According to the 1998 survey) unit: people per day

indicate the magnitude of frequency. It should be noted that all of the redevelopments discussed in this paper were carried out in the Tenjin South district.

From these figures, it becomes obvious that while staying in the Tenjin South district increased every year the development took place (i.e., the size of 142,093 person/day in 1996 increased to 303,235 in 1998), staying in the Tenjin North district continues to decrease (i.e., the size of 47,626 person/day in 1996 decreased to 22,088 in 1998, or by more than 50%). It is also clear that the number of shop-around visitors going from the Tenjin South district to the Tenjin North district is also steadily decreasing. Additionally, from the above results, we can confirm the southward movement phenomenon of the flow of people within the Tenjin district.

5.6 Southward Movement of the Center of Gravity in the Tenjin District As Derived from the Visit Frequency

In this section, we calculate the center of gravity based on the visit frequency proposed in Sect. 3.4 to investigate how the center of gravity of the Tenjin district has moved due to the redevelopment. The center of gravity calculated here is a center of gravity based on visit frequency, which is a new simple activity-based concept. Therefore, by illustrating the changes of the center of gravity, we would like to show clearly how the center of activity in the Tenjin district has shifted. In particular, we would like to show that the center of activity in Tenjin district was moving southward year by year.

Figure 2.8 shows the change of the center of gravity in the Tenjin district based on the total visit frequency. The center of gravity located at the northern end of Nishitetsu Fukuoka Station in 1996 had moved southward by approximately 105 m after the opening of Fukuoka *Mitsukoshi* in 1998. From this, it can be said that the center of gravity of the Tenjin district calculated by the visit frequency was moving southward as each redevelopment was carried out. Thus, the southward movement phenomenon of the Tenjin area's center of gravity is verified.

To tell the truth, this result has a sequel. Later in 2002, one news was reported that the point of the highest appraisal land price[3] in Fukuoka City had continued to be the place located by the *Iwataya* Department store for last 38 years, but it was replaced for the first time with the place located by the Solaria Plaza to move southward. This fact gives some evidence that the land price followed up consumers' shop-around behaviors within the Tenjin district with a delay of 4 years. Figure 2.8 also shows this movement of the highest appraisal land price.

6 Conclusion and Future Research

To date, there have been no studies that verified the effect of the large-scale development on human behavior from a behavioristic perspective. The most significant contribution of this research is that we actually have carried out this from the viewpoint of shop-around behavior. The purpose of this research was that based on

---

[3]National Tax Agency, Japan, annually publicizes the appraisal land price called *Rosenka*, which is based on appraising land values according to standard values of adjoining streets and used for the appraisal of inheritance and fixed property taxes.

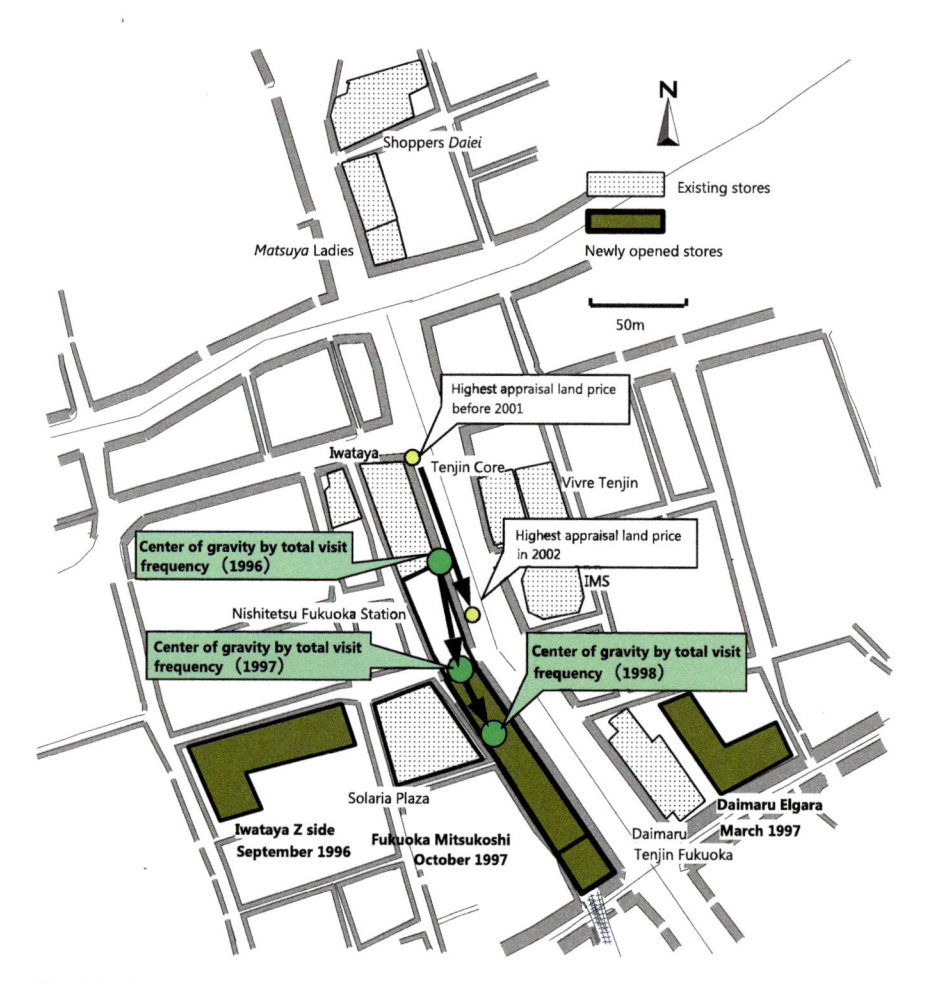

Fig. 2.8 Changes in the center of gravity within the Tenjin district derived by the total visit frequency

giving a precise definition to "shop-around behavior," by taking up a concrete example of large-scale commercial redevelopments in the city center commercial district of Fukuoka City, we empirically observed shop-around behaviors before and after large-scale redevelopments and verified the effect of the large-scale redevelopment on the shop-around behaviors by clarifying the change in the structure of the city center commercial district such as the southward movement phenomenon of the center of gravity in the Tenjin district.

From the above results of our analysis, we believe that we have been able to verify structural changes in the city center due to redevelopment by using the simple but effective tool of the shop-around Markov model.

While our analysis of this research was limited to a descriptive one that used only observed shop-around choice probability, various themes such as modeling with explanatory variables, extending from the number of people to a monetary basis, and theoretical analysis of what kinds of development patterns would trigger the largest shop-around effect are left for future research.

References

1. Saito S, Nakashima T, Kakoi M (1999) Identifying the effect of city center retail redevelopment on consumer's shop-around behavior: an empirical study on structural changes of city center at Fukuoka city. Stud Reg Sci 29:107–130. (in Japanese)
2. Fukami T (1974) A study on pedestrian flows in a commercial district Part 1. Pap City Plan, pp 43–48. (in Japanese)
3. Fukami T (1977) A study on pedestrian flows in a commercial district Part 2. Pap City Plan, pp 61–66. (in Japanese)
4. Hagishima S, Mitsuyoshi K, Kurose S (1987) Estimation of pedestrian shopping trips in a neighborhood by using a spatial interaction model. Environ Plan A 19:1139–1152
5. Hanson S (1979) Urban travel linkages: a review. In: Hensher DA, Stopher PR (eds) Behavioral travel modeling. Croom-Helm, London, pp 81–100
6. Hanson S (1980a) The importance of the multi-purpose journey to work in urban travel behavior. Transportation 9:229–248
7. Hanson S (1980b) Spatial diversification and multipurpose travel: implications for choice theory. Geogr Anal 12:245–257
8. Kondo K, Kitamura R (1987) Time-space constraints and the formation of trip chains. Reg Sci Urban Econ 17:49–65
9. Kondo K (1987) Transport behavior analysis. Koyo Shobo, Kyoto. (in Japanese)
10. Lerman SR (1979) The use of disaggregate choice models in semi-Markov process models of trip chaining behavior. Transp Sci 13:273–291
11. O'Kelly ME (1981) A model of the demand for retail facilities, incorporating multistop, multipurpose trips. Geogr Anal 13:134–148
12. Saito S (1983) Present situation and challenges for the commercial districts in Nobeoka area, In: Committee for Modernizing Commerce Nobeoka Region Section (ed) The report of regional plan for modernizing commerce Nobeoka area, pp 37–96. (in Japanese)
13. Saito S (1988a) Duration and order of purpose transition occurred in the shop-around trip chain at a Midtown District. Pap City Plan, pp 55–60. (in Japanese)
14. Saito S (1988b) Assessing the space structure of central shopping district viewed from consumers' shop-around behaviors: a case study of the midtown district of Saga city. Fukuoka University Economic Review 33:47–108. (in Japanese)
15. Saito S, Sakamoto T, Motomura H, Yamaguchi S (1989) Parametric and non-parametric estimation of distribution of consumer's shop-around distance at a midtown district. Pap City Plan, pp 571–576. (in Japanese)
16. Sakamoto T (1984) An absorbing Markov chain model for estimating consumers' shop-around effect on shopping districts. Pap City Plan, pp 289–294. (in Japanese)
17. Wrigley N, Dunn R (1984a) Stochastic panel-data models of urban shopping behaviour: 1. Purchasing at individual stores in a single city. Environ Plan A 16:629–650
18. Wrigley N, Dunn R (1984b) Stochastic panel-data models of urban shopping behaviour: 2. Multistore purchasing patterns and the Dirichlet model. Environ Plan A 16:759–778
19. Sasaki T (1972) Estimation of person trip patterns through Markov chains. In: Newell GF (ed) Traffic flow and transportation. Elsevier, New York
20. Saito S (1984) A disaggregate hierarchical huff model with considering consumer's shop-around choice among commercial districts: developing SCOPES (Saga Commercial Policy Evaluation System). Planning and Public Management 13:73–82. (in Japanese)

21. Takeuchi S (1981) A model for location planning of commercial functions. Plan Public Manag, pp 25–33. (in Japanese)
22. Borgers A, Timmermans H (1986) City centre entry points, store location patterns and pedestrian route choice behaviour: a microlevel simulation model. Socio Econ Plan Sci 20:25–31
23. Saito S, Ishibashi K (1992b) A Markov chain model with covatiates to forecast consumer's shopping trip chains within a central commercial district. Paper presented at the The Fourth World Congress of the Regional Science Association International, Palma de Mallorca, Spain
24. Saito S, Ishibashi K (1992a) Forecasting consumer's shop-around behaviors within a city center retail environment after its redevelopments using Markov chain model with covariates. Pap City Plan 27:439–444. (in Japanese)
25. Saito S, Kakoi M, Nakashima T (1999) On-site Poisson regression modeling for forecasting the number of visitors to city center retail environment and its evaluation. Stud Reg Sci 29:55–74. (in Japanese)
26. Saito S, Kumata Y, Ishibashi K (1995) A choice-based Poisson regression model to forecast the number of shoppers: its application to evaluating changes of the number and shop-around pattern of shoppers after city center redevelopment at Kitakyushu city. Pap City Plan 30:523–528. (in Japanese)
27. Ishibashi K (1998) A study of an evaluation model for development program in central commercial district using consumer shop-around behavior. Doctoral dissertation. The Graduate School of Science and Engineering, Tokyo Institute of Technology. (in Japanese)
28. Ishibashi K, Saito S, Kumata Y (1998a) A disaggregate Markov shop-around model to forecast sales of retail establishments based on the frequency of shoppers' visits: its application to city center retail environment at Kitakyusyu city. Pap City Plan 33:349–354. (in Japanese)
29. Ishibashi K, Saito S, Kumata Y (1998b) Forcasting sales of shopping sites by use of the frequency-based disaggregate Markov shop-around model of consumers: its application to central commercial district at Kitakyushu City. Presented at the 5th Summer Institute of the Pacific Regional Science Conference Organization (PRSCO), Nagoya, Japan
30. Saito S (1997a) Analysis of the structure of city center space viewed from consumer shop-around behavior. Abstracts of The 1997 Spring National Conference of ORSJ, pp 20–21. (in Japanese)
31. Saito S (1997b) Urban development viewed from consumer shop-around behavior. Panelist at open symposium at 25th annual meeting. In: Abstracts of The 25th annual meeting of the Behaviormetric Society, Sendai, pp 22–25. (in Japanese)
32. Saito S, Nakashima T, Kakoi M (2001) The consistent OD estimation for on-site person trip survey. Stud Reg Sci 31:191–208. (in Japanese)
33. Saito S, Nakashima T (2003) An application of the consistent OD estimation for on-site person trip survey: estimating the shop-around pattern of consumers at Daimyo district of Fukuoka city, Japan. Stud Reg Sci 33:173–203. (in Japanese)
34. Saito S (1998) Report of Surveys of consumer shop-around behavior at midtown of Fukuoka city 1996–1998. (in Japanese)

Chapter 3
Evaluating Municipal Tourism Policy from How Visitors Walk Around Historical Heritage Area: An Evaluation of the "Walking Path of History" of Dazaifu City, Japan, Based on Visitors' *Kaiyu* Behavior

Saburo Saito, Masakuni Iwami, Mamoru Imanishi, and Kosuke Yamashiro

Abstract Dazaifu City, located in Fukuoka, Japan, is famous for its unique position in Japanese history from the seventh century. Thus, there remain many historical and cultural heritages in its administrative division. Among many tourist attractions, the places that attract largest numbers of tourists are the Dazaifu *Tenman-gu* Shrine and Kyushu National Museum. Dazaifu City government tries to utilize these cultural and historical assets and sets out the comprehensive policy to regard a whole Dazaifu City itself as "museum everywhere as it is." Under this general strategic policy, city officials devised a policy of "walking path of history," which connects many places of interest in the town they would like to introduce to tourists. They intended to have tourists use the walking path of history and visit many attractive places on the walking path. But up to now, most of tourists do not know the walking path so

---

This chapter is based on the paper, Booyoung Park, Saburo Saito, Masakuni Iwami, Mamoru Imanishi, and Kosuke Yamashiro [1], "An evaluation of walking path of history of Dazaifu City from visitors' *Kaiyu* behaviors," presented at the 48th Annual Meeting of Japan Section of Regional Science Association International (JSRSAI), 2011, which is modified for this chapter.

S. Saito (✉)
Faculty of Economics, Fukuoka University, Fukuoka, Japan

Fukuoka University Institute of Quantitative Behavioral Informatics for City and Space Economy (FQBIC), Fukuoka, Japan
e-mail: saito@fukuoka-u.ac.jp

M. Iwami
Fukuoka University Institute of Quantitative Behavioral Informatics for City and Space Economy (FQBIC), Fukuoka, Japan
e-mail: miwami@econ.fukuoka-u.ac.jp

M. Imanishi · K. Yamashiro
Department of Business and Economics, Nippon Bunri University, Oita City, Japan
e-mail: imanishimm@nbu.ac.jp; yamashiroks@nbu.ac.jp

that there are few citizens or tourists who use the walking path of history. In this study, we address the problem of how we can induce tourists to utilize the walking path of history and propose means to solve the problem while clarifying the obstacles to attain the goal based on tourists' *Kaiyu* behaviors within Dazaifu City.

Keywords Tourism · Municipal policy · Evaluation · *Kaiyu* · Historical heritage · Shop-around Markov model · Dazaifu City · Fukuoka

1 Purpose

Dazaifu City in Japan is a historic city with more than a thousand year history and has many historical heritages all over its administrative area. It is located in the northernmost part of Kyushu Island, the most southwest of the four islands that composed Japan. Dazaifu City has major tourist attractions such as the Dazaifu *Tenman-gu* Shrine, Kyushu National Museum, and *Sando* shopping street (the street leading to the shrine). They are attracting many visitors every year. Dazaifu City administration puts emphasis on the establishment of tourist information centers and tourism promotion projects. Also, the city administration is undertaking various activities for developing its tourism industry such as activities to provide tourist guide services by volunteers with hospitality and to conduct surveys on tourist trends. Among the important tourism policies the city administration puts a force on is the "walking path of history" policy. We focus on this policy. The walking path of history is a 7.2-km-long sightseeing course that connects Dazaifu City's major historical sites and is set in a way that people can move around the path on foot or by rental bicycles.

The motivation for the municipal to set up the policy is as follows. Since the city has a long history, all over the city area, there are many historic remains, literal historical sites, and famous places related to historical events. So the city administration would like to preserve and maintain them by turning its whole area into a "museum everywhere as it is."

While the policy of the "walking path of history" was produced and implemented from the above background, the municipal at present has yet to precisely grasp how many visitors or tourists go around the walking path of history in order to see how well the policy attains its goal. The city administration is afraid that only few tourists use the walking path of history. Therefore, grasping the current situation has become an issue of concern for furthering this policy that plans to have many tourists visit many historic sites.

Thus, our aim of this research is to directly address questions such as how many people are stopping by and how they are visiting the walking path of history and to clarify the current situation and problem of the walking path of history from tourists' behavior. Furthermore, based on these, we aim to evaluate the walking path of history as a tourism policy and to propose its improvement measures.

The authors already have been trying to evaluate tourism policies based on micro-behavioral data of tourists at the destination area of attractions, which we call the

Kaiyu approach (Cf. [4]). The characteristic of the *Kaiyu* approach is that the evaluation of policies is performed by focusing on the tourists' behaviors of how many and how well they move around or sightseeing around like consumers' shop-around behavior within the destination area of tourist attractions.

In the research area of tourism policy, there have been few previous studies to evaluate tourism policy based on the analysis of tourists' behaviors, for instance, studies on how tourists' behaviors change depending on the differences in tourism policies.

Thus, we set up our purpose of this research as follows: (1) we first carried out two surveys, the 1st On-site Survey of Sightseeing and *Kaiyu* Behavior of Visitors to Dazaifu City and the Survey of Pedestrian Traffic Volume in Dazaifu City. (2) Based on data from these surveys, we estimate the actual number of visitors who enter Dazaifu City and (3) identify the actual numbers of their movements among tourist attractions within the city. (4) In particular, we specify how many visitors actually use the walking path of history. (5) With these, we evaluate the present outcome of the policy of the walking path of history and (6) make a policy proposal for increasing the number of users of the walking path of history.

2 Analysis Framework

2.1 Outline of the "Walking Path of History"

The walking path of history is one of the various efforts the municipal is tackling toward the development of tourism in Dazaifu City.

The walking path of history is a 7.2-km-long walking road connecting historical sites between the Dazaifu *Tenman-gu* Shrine and the *Mizuki* Ruins (ancient castle walls). Those historical sites on the path include major sites representing Dazaifu such as the ruins of *Chikuzen Kokubun-ji* Temple, ruins of the old Dazaifu Government Office, *Gakko-in* Ruins (ruins of ancient school), *Kaidan-in* Temple, *Kanzeon-ji* Temple, and so on. This feature makes the walking path of history a fascinating great walking path.

Figure 3.1 is the guide map for the walking path of history created by the Dazaifu *Fureai* Museum (a museum that promotes cultural contact in Dazaifu City), which helps people explore the above historical sites along the walking path of history.

2.2 Data Used

The data used in this research was obtained from the 1st On-site Survey of Sightseeing and *Kaiyu* Behavior of Visitors to Dazaifu City conducted on May 29 (Saturday) and May 30 (Sunday), 2010, and the data of the Survey of Pedestrian Traffic Volume in Dazaifu City conducted on May 29, 2010 (Saturday).

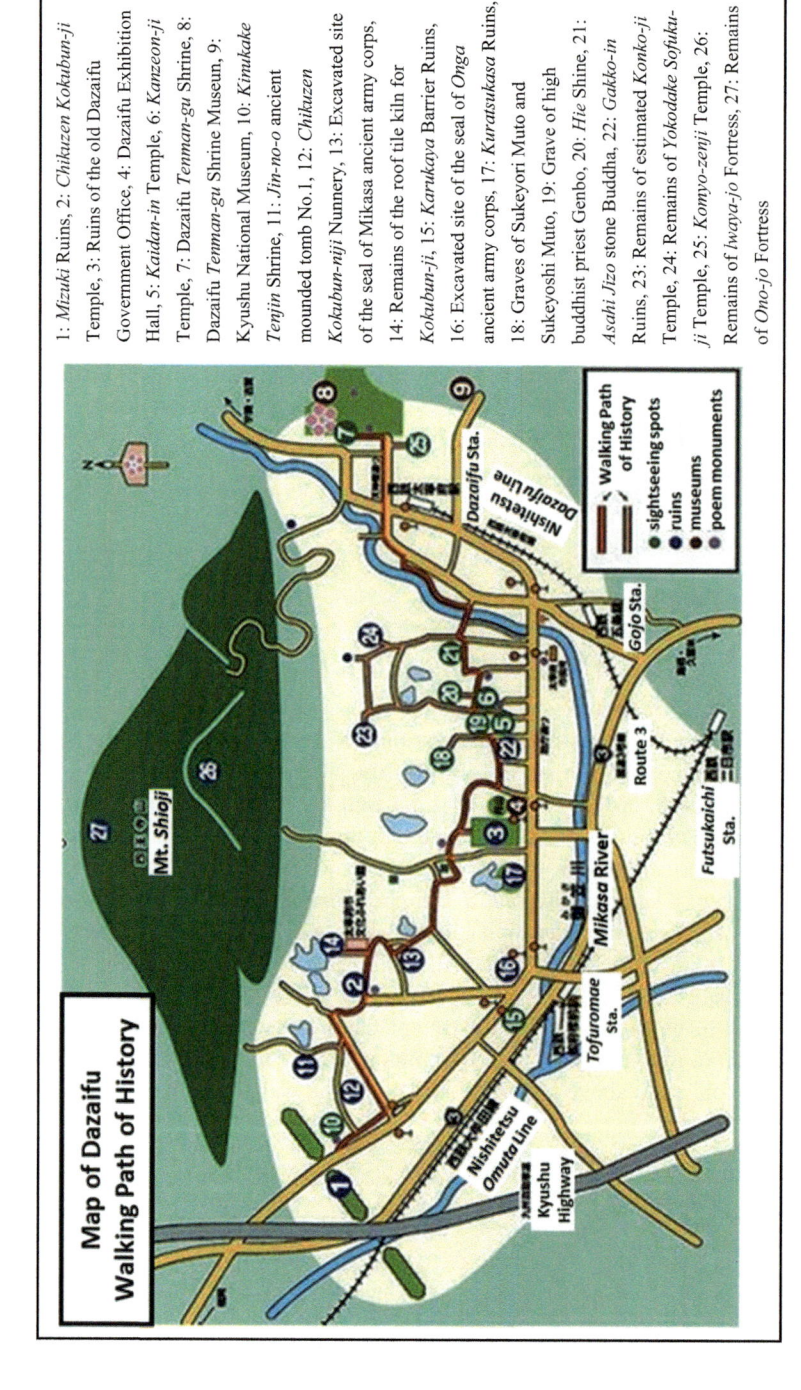

Fig. 3.1 Map of the walking path of history in Dazaifu City [5]

1: *Mizuki* Ruins, 2: *Chikuzen Kokubun-ji* Temple, 3: Ruins of the old Dazaifu Government Office, 4: Dazaifu Exhibition Hall, 5: *Kaidan-in* Temple, 6: *Kanzeon-ji* Temple, 7: Dazaifu *Tenman-gu* Shrine, 8: Dazaifu *Tenman-gu* Shrine Museum, 9: Kyushu National Museum, 10: *Kimukake Tenjin* Shrine, 11: *Jin-no-o* ancient mounded tomb No.1, 12: *Chikuzen Kokubun-niji* Nunnery, 13: Excavated site of the seal of Mikasa ancient army corps, 14: Remains of the roof tile kiln for *Kokubun-ji*, 15: *Karukaya* Barrier Ruins, 16: Excavated site of the seal of *Onga* ancient army corps, 17: *Kuratsukasa* Ruins, 18: Graves of Sukeyori Muto and Sukeyoshi Muto, 19: Grave of high buddhist priest Genbo, 20: *Hie* Shine, 21: *Asahi Jizo* stone Buddha, 22: *Gakko-in* Ruins, 23: Remains of estimated *Konko-ji* Temple, 24: Remains of *Yokodake Sofuku-ji* Temple, 25: *Komyo-zenji* Temple, 26: Remains of *Iwaya-jo* Fortress, 27: Remains of *Ono-jo* Fortress

3 Evaluating Municipal Tourism Policy from How Visitors Walk Around... 51

The 1st On-site Survey of Sightseeing and *Kaiyu* Behavior of Visitors to Dazaifu City is an approximately 15-min questionnaire interview survey, in which several sampling points are set; the respondents are sampled at random from the visitors to those sampling points; and the respondents are asked about their sightseeing and *Kaiyu* behaviors on the day of the survey.

Regarding sightseeing and *Kaiyu* behaviors, we asked the respondents which places they visited for what purpose and how much they spent there in the order of those occurrences. The survey sampling sites were the Dazaifu *Tenman-gu* Shrine, the Kyushu National Museum, the *Sando* shopping street, and the *Kanzeon-ji* Temple. In total, a number of 377 valid completed samples were collected.

In the Survey of Pedestrian Traffic Volume in Dazaifu City, several survey points were set up in the streets of Dazaifu City in order to measure how many people passed through, and toward which direction, at intervals of 15 min. In particular, as for Dazaifu *Tenman-gu* Shrine, all of its entrances are set as the survey points in order to measure the total of the net incoming visitors to the shrine by counting all the visitors coming into and out of the shrine. The setup of these survey points is depicted in Fig. 3.2 shown later.

Table 3.1 shows the outline of the 1st On-site Survey of Sightseeing and *Kaiyu* Behavior of Visitors to Dazaifu City. Table 3.2 summarizes the Survey of Pedestrian Traffic Volume in Dazaifu City.

Table 3.1 The 1st On-site Survey of Sightseeing and *Kaiyu* Behavior of Visitors to Dazaifu City: overview

Date	2010.05.29 (Sat), 2010.05.30 (Sun) 11:00–17:00
Sampling points	The Dazaifu *Tenman-gu* Shrine, Kyushu National Museum
	Sando shopping street (2 points), the *Kanzeon-ji* Temple
Collected samples	377 samples
Survey method	(1) On-site survey for visitors at sampling points
	(2) Random sampling from the visitors
	(3) 15–20-min interview questionnaire survey
Questionnaire items	(1) Demographic items (residence area, age, gender, professions, etc.)
	(2) Record of *Kaiyu* on the day (places visited, purposes done, expenditure there)
	(3) Frequency of visits to Dazaifu City, other attractions in the city
	(4) Evaluation of tourism policies by the municipal

Table 3.2 The Survey of Pedestrian Traffic Volume in Dazaifu City: overview

Date	2010.05.29 (Sat) 11:00–17:00
Survey points	(1) Dazaifu *Tenman-gu* Shrine: 13 points (every entrance/exit)
	(2) *Sando* shopping street: 1 point
	(3) *Kanzeon-ji* Temple: 2 points
Survey method	Subject for counting: pedestrian, people riding bicycle
	Counting method: count the number of pedestrians passing survey point by direction for every 15 min with manual numerator

2.3 Method

The analysis of this research proceeds as the following steps.

Step 1. From the *Kaiyu* behavior data obtained from the 1st On-site Survey of Sightseeing and *Kaiyu* Behavior of Visitors to Dazaifu City, we estimate the observed origin destination frequency matrix among districts within the city.

Step 2. We estimate the observed origin destination density matrix by dividing each cell of the observed origin destination frequency matrix by the number of total samples.[1]

Step 3. We obtain the total number of visitors entering Dazaifu *Tenman-gu* Shrine from the Survey of Pedestrian Traffic Volume in Dazaifu City.

Step 4. We expand the density of total visits to Dazaifu *Tenman-gu* Shrine obtained at step 2 by the total number of visitors entering Dazaifu *Tenman-gu* Shrine obtained at step 3 to obtain the estimated total number of net incoming visitors to Dazaifu City.

Step 5. By transforming the observed origin destination frequency matrix into the *Kaiyu* (shop-around) choice probability matrix and applying *Kaiyu* (shop-around) Markov model, we estimate the number of visitors who visit and utilize the walking path of history to evaluate the policy and make a policy proposal.

2.4 Kaiyu *Markov Model*

The *Kaiyu* Markov model is a model that formulates consumer's shop-around or *Kaiyu* behavior as a stationary absorbing Markov chain [3]. Consumer's shop-around behavior (hereafter, *Kaiyu*) can be thought of as a loop-shaped trip in which the consumer departs from home, visits shops, and finally returns home. More precisely, *Kaiyu* is defined as the movement among shopping sites after the first stop. The first visit to some facility directly from home is called the entrance (entry). The *Kaiyu*, in other words, is visits to the shopping sites after the entrance. The *Kaiyu* Markov model expresses this loop-shaped *Kaiyu* behavior as an infinitely repeated Markov chain. Let H_0 be the state of returning home (absorption state) and H the state of departure. Let I denote the set of sites of commercial facilities, that is, tourist attractions in this research. Let P denote the state transition probability matrix for the *Kaiyu* Markov model. The transition probability matrix P can be expressed as a partitioned matrix as follows:

---

[1]While the consistent estimation method developed by Saito and Nakashima [2] should have been used, we employed here the usual OD density calculation method.

$$P = \begin{array}{c} H_0 \\ H \\ I \end{array} \begin{bmatrix} 1 & 0 & 0 \\ 0 & 0 & P_{HI} \\ P_{IH_0} & 0 & P_{II} \end{bmatrix}$$

Here, P_{HI} represents the entrance choice probability, P_{II} represents the *Kaiyu* (shop-around) choice probability, and P_{IH_0} represents the returning home probability.

We define the *Kaiyu* effect as a measure for how much *Kaiyu* occurs. Speaking in a word, the *Kaiyu* effect is the effect of how many visits one visit to a certain facility would induce to each of other facilities. More specifically, it is the total expected visit frequency to each facility through *Kaiyu* visits after the entrance.

To obtain the *n*-th-order transition probability matrix of the partitioned matrix P, the element corresponding to P_{HI} becomes $P_{HI}P_{II}^{n-1}$. Since this is the probability of visit to each facility after the *n*-th step from home, the *Kaiyu* effect RE is formulated as the sum of all these probabilities after the second step.

$$\begin{aligned} RE &= P_{HI}P_{II} + P_{HI}P_{II}^2 + \cdots \\ &= P_{HI}P_{II}(I - P_{II})^{-1} \end{aligned}$$

The formula $P_{II}(I - P_{II})^{-1}$ obtained by replacing P_{HI} with identity matrix is a measure showing how much visit frequency would be brought to all other facilities by *Kaiyu* behaviors when one trip enters each facility.

Up to now, we have discussed the *Kaiyu* effect in terms of probability, but to represent this in terms of frequency, we have only to replace the entrance choice probability P_{HI} with the entrance visit frequency F_{HI}. The *Kaiyu* effect by frequency basis becomes the number of people who visited each facility by *Kaiyu*. Therefore, the total frequency of visits to each facility TVST can be decomposed into the entrance visit frequency and the *Kaiyu* visit frequency and can be expressed as follows:

$$\begin{aligned} \text{TVST} &= F_{HI}(I - P_{II})^{-1} \\ &= F_{HI} + F_{HI}P_{II}(I - P_{II})^{-1} \end{aligned}$$

A frequency matrix F_{II}, which indicates how many people moved from facility i to facility j, is called a *Kaiyu* pattern and is obtained as follows:

$$F_{II} = \text{diag}(F_{HI})P_{II}(I - P_{II})^{-1}$$

Here, $\text{diag}(x)$ represents a diagonal matrix with a vector x as its diagonal element.

3 What Places Do Visitors to Dazaifu City Visit and How Do They *Kaiyu*?

3.1 Total Number of Incoming Visitors to Dazaifu *Tenman-gu Shrine*

As shown in Fig. 3.2, all the entrances and exits of Dazaifu *Tenman-gu* Shrine are set as the survey points for the Survey of Pedestrian Traffic Volume in Dazaifu City. As a result, from the survey we obtained the total number of incoming visitors entering Dazaifu *Tenman-gu* Shrine (total inflow population). Figure 3.3 shows the result.

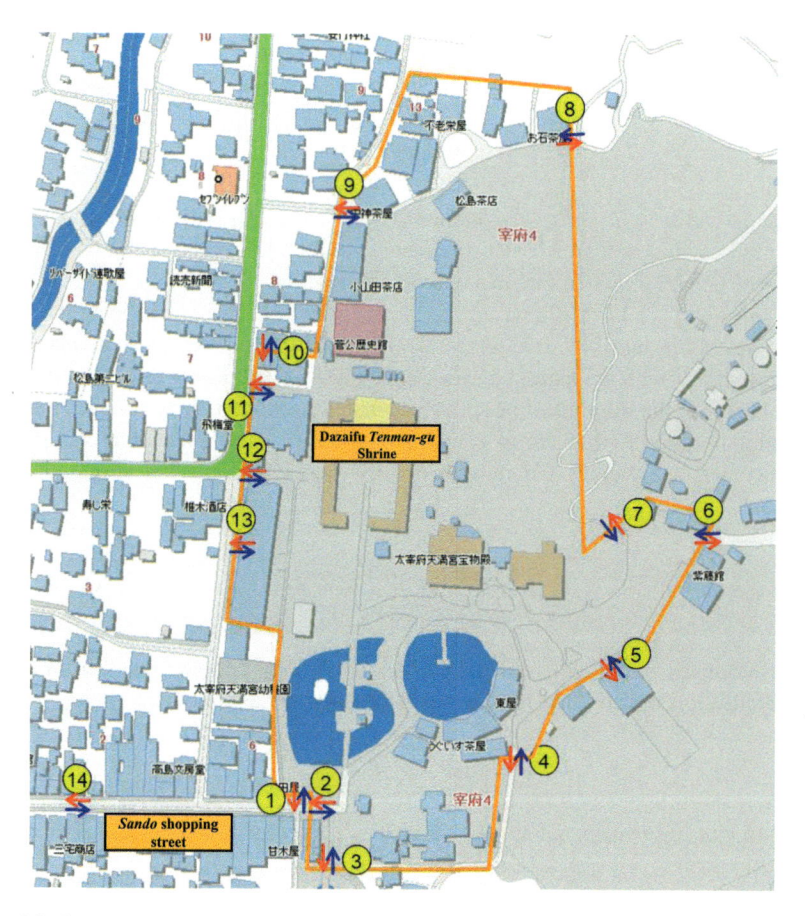

Fig. 3.2 Survey points of the Survey of Pedestrian Traffic Volume at the Dazaifu *Tenman-gu* Shrine and the *Sando* shopping street area (Cf. [6])

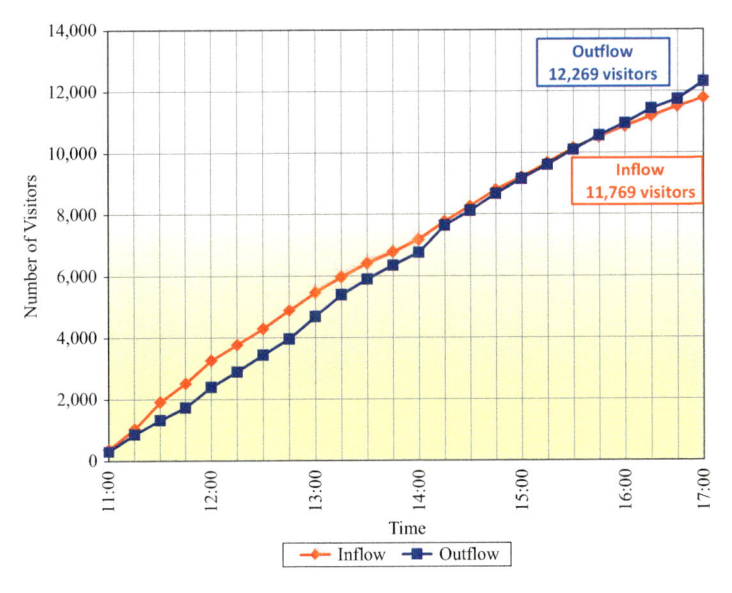

Fig. 3.3 Total numbers of inflow and outflow visitors to and from Dazaifu *Tenman-gu* Shrine

The inflow population is high in the morning, but outflow population increases from around 2:00 p.m. The total inflow population to Dazaifu *Tenman-gu* Shrine was 11,769, and the total outflow was 12,296.

3.2 Estimation of the Number of Visitors Entering Dazaifu City

We now estimate the actual number of incoming visitors who enter Dazaifu City, using the observed origin destination density matrix of *Kaiyu* movements among tourist attractions. The procedure is as follows.

First, as shown in Table 3.3, we made the observed origin destination frequency matrix of *Kaiyu* movements among tourist attractions from the 1st On-site Survey of Sightseeing and *Kaiyu* Behavior of Visitors to Dazaifu City.

Then, the observed origin destination density matrix was created by dividing each cell of Table 3.3 by the number of total 324 samples. Here, the numbers of diagonal elements, which mean "staying," were removed. This is shown in Table 3.4.

Table 3.3 Observed origin destination frequency matrix

Area		Node	1	2	3	4	5	6	7	8	9	10	11	12	13	14	15	16	17	18	19	20	21	22	23	Total
1000	Dazaifu *Tenman-gu* Shrine	1	40	21	65	10	0	4	1	1	0	0	0	0	1	0	0	0	0	0	0	0	0	0	159	302
2000	Kyushu National Museum	2	27	4	23	3	0	1	1	0	0	0	0	0	0	0	0	0	0	0	0	0	0	0	52	111
3000	*Sando* shopping street (*Tenman-gu* Shrine side)	3	46	13	53	22	0	1	2	0	0	0	0	0	0	0	0	0	0	1	0	0	0	0	68	206
4000	*Sando* shopping street (Dazaifu Station side)	4	10	2	13	8	0	0	1	0	0	0	0	0	0	0	0	0	0	0	0	0	0	0	31	65
5000	District around North of *Sando* Shopping Street	5	2	1	0	0	0	0	0	0	0	0	0	0	0	0	0	0	0	0	0	0	0	0	0	3
6000	District around South of *Sando* shopping street	6	0	0	0	0	0	0	0	0	0	0	0	0	0	0	0	0	0	0	0	0	0	0	6	6
7000	District around West of *Sando* shopping street	7	3	0	1	1	0	0	0	1	0	0	0	0	0	0	0	0	0	0	0	0	0	0	3	9
8000	District around Dazaifu Souvenir shop	8	0	0	0	0	0	0	0	0	1	0	0	0	0	0	0	0	0	0	0	0	0	0	1	2
9000	District around West of Gojo 1 (Fifth Ave. 1)	9	0	0	0	0	0	0	0	0	0	0	0	1	0	0	0	0	0	0	0	0	0	0	0	1
10000	District around East of Gojo 1 (Fifth Ave. 1)	10	0	0	0	0	0	0	0	0	0	0	0	0	0	0	0	0	0	0	0	0	0	0	0	0
11000	District around *Chikaishi-dai* High School	11	0	0	0	0	0	0	0	0	0	0	0	0	0	0	0	0	0	0	0	0	0	0	1	1
12000	District around welfare facilities	12	0	0	0	0	0	0	0	0	0	0	0	0	1	0	0	0	0	0	0	0	0	0	0	1
13000	District around *Kanzeon-ji* Temple	13	0	0	1	0	0	0	0	0	0	0	0	0	0	2	0	0	0	0	0	0	0	0	1	4
14000	District around Ruins of the old Dazaifu Government Office	14	1	0	0	0	0	0	0	0	0	0	0	0	0	0	1	0	0	0	0	0	0	0	0	2
15000	District around Sakamoto	15	0	0	0	0	0	0	0	0	0	0	0	0	0	0	0	1	0	0	0	0	0	0	0	1
16000	District around *Kokubun-ji* (Ancient National Temple Remains)	16	0	0	0	0	0	0	0	0	0	0	0	0	0	0	0	0	0	0	0	0	0	0	1	1
17000	District around Gojo Station (Fifth Ave.)	17	0	0	0	0	0	0	0	0	0	0	0	0	0	0	0	0	0	0	0	0	0	0	0	0
18000	District around Dazaifu City Hall	18	0	0	1	0	0	0	0	0	0	0	0	0	0	0	0	0	0	0	0	0	0	0	1	2
19000	District around South of *Suzaku-oji* (Ancient Main Road)	19	0	0	0	0	0	0	0	0	0	0	0	0	0	0	0	0	0	0	0	0	0	0	0	0
20000	District around North of *Suzaku-oji* (Ancient Main Road)	20	0	0	0	0	0	0	0	0	0	0	0	0	0	0	0	0	0	0	0	0	0	0	0	0
21000	District around *Tofuro-mae* Station	21	0	0	0	0	0	0	0	0	0	0	0	0	0	0	0	0	0	0	0	0	0	0	0	0
99999	Outside survey target area	22	1	0	0	0	0	0	0	0	0	0	0	0	0	0	0	0	0	0	0	0	0	0	0	1
199	Home	23	172	70	49	21	3	0	0	0	0	0	1	0	0	0	1	0	0	1	0	0	0	1	0	324
	Total		302	111	206	65	3	6	9	2	1	0	1	1	4	2	1	1	0	2	0	0	0	1	324	1042

Table 3.4 Observed origin destination density matrix

Area		Node	1	2	3	4	5	6	7	8	9	10	11	12	13	14	15	16	17	18	19	20	21	22	23	Total
1000	Dazaifu *Tenman-gu* Shrine	1	–	0.065	0.201	0.031	0.000	0.012	0.003	0.003	0.000	0.000	0.000	0.000	0.003	0.000	0.000	0.000	0.000	0.000	0.000	0.000	0.000	0.000	0.491	0.809
2000	Kyushu National Museum	2	0.083	–	0.071	0.009	0.000	0.003	0.003	0.000	0.000	0.000	0.000	0.000	0.000	0.000	0.000	0.000	0.000	0.000	0.000	0.000	0.000	0.000	0.160	0.330
3000	*Sando* shopping street (*Tenman-gu* Shrine side)	3	0.142	0.040	–	0.068	0.000	0.003	0.006	0.000	0.000	0.000	0.000	0.000	0.000	0.000	0.000	0.000	0.000	0.003	0.000	0.000	0.000	0.000	0.210	0.472
4000	*Sando* shopping street (Dazaifu Station side)	4	0.031	0.006	0.040	–	0.000	0.000	0.003	0.000	0.000	0.000	0.000	0.000	0.000	0.000	0.000	0.000	0.000	0.000	0.000	0.000	0.000	0.000	0.096	0.176
5000	District around North of *Sando* shopping street	5	0.006	0.003	0.000	0.000	–	0.000	0.000	0.000	0.000	0.000	0.000	0.000	0.000	0.000	0.000	0.000	0.000	0.000	0.000	0.000	0.000	0.000	0.000	0.009
6000	District around South of *Sando* shopping street	6	0.000	0.000	0.000	0.000	0.000	–	0.000	0.000	0.000	0.000	0.000	0.000	0.000	0.000	0.000	0.000	0.000	0.000	0.000	0.000	0.000	0.000	0.019	0.019
7000	District around West of *Sando* shopping street	7	0.009	0.000	0.003	0.003	0.000	0.000	–	0.003	0.000	0.000	0.000	0.000	0.000	0.000	0.000	0.000	0.000	0.000	0.000	0.000	0.000	0.000	0.009	0.028
8000	District around Dazaifu Souvenir shop	8	0.000	0.000	0.000	0.000	0.000	0.000	0.000	–	0.003	0.000	0.000	0.000	0.000	0.000	0.000	0.000	0.000	0.000	0.000	0.000	0.000	0.000	0.003	0.006
9000	District around West of Gojo 1 (Fifth Ave. 1)	9	0.000	0.000	0.000	0.000	0.000	0.000	0.000	0.000	–	0.000	0.000	0.003	0.000	0.000	0.000	0.000	0.000	0.000	0.000	0.000	0.000	0.000	0.000	0.003
10000	District around East of Gojo 1 (Fifth Ave. 1)	10	0.000	0.000	0.000	0.000	0.000	0.000	0.000	0.000	0.000	–	0.000	0.000	0.000	0.000	0.000	0.000	0.000	0.000	0.000	0.000	0.000	0.000	0.000	0.000
11000	District around *Chikushi-dai* High School	11	0.000	0.000	0.000	0.000	0.000	0.000	0.000	0.000	0.000	0.000	–	0.000	0.000	0.000	0.000	0.000	0.000	0.000	0.000	0.000	0.000	0.000	0.003	0.003
12000	District around welfare facilities	12	0.000	0.000	0.000	0.000	0.000	0.000	0.000	0.000	0.000	0.000	0.000	–	0.003	0.000	0.000	0.000	0.000	0.000	0.000	0.000	0.000	0.000	0.000	0.003
13000	District around *Kanzeon-ji* Temple	13	0.000	0.000	0.003	0.000	0.000	0.000	0.000	0.000	0.000	0.000	0.000	0.000	–	0.006	0.000	0.000	0.000	0.000	0.000	0.000	0.000	0.000	0.003	0.012

(continued)

Table 3.4 (continued)

Area		Node	1	2	3	4	5	6	7	8	9	10	11	12	13	14	15	16	17	18	19	20	21	22	23	Total
14000	District around Ruins of the old Dazaifu Government Office	14	0.003	0.000	0.000	0.000	0.000	0.000	0.000	0.000	0.000	0.000	0.000	0.000	0	–	0.003	0.000	0.000	0.000	0.000	0.000	0.000	0.000	0.000	0.006
15000	District around Sakamoto	15	0.000	0.000	0.000	0.000	0.000	0.000	0.000	0.000	0.000	0.000	0.000	0.000	0.000	0.000	–	0.003	0.000	0.000	0.000	0.000	0.000	0.000	0.000	0.003
16000	District around *Kokubun-ji* (Ancient National Temple Remains)	16	0.000	0.000	0.000	0.000	0.000	0.000	0.000	0.000	0.000	0.000	0.000	0.000	0.000	0.000	0.000	–	0.000	0.000	0.000	0.000	0.000	0.000	0.003	0.003
17000	District around Gojo Station (Fifth Ave.)	17	0.000	0.000	0.000	0.000	0.000	0.000	0.000	0.000	0.000	0.000	0.000	0.000	0.000	0.000	0.000	0.000	–	0.000	0.000	0.000	0.000	0.000	0.000	0.000
18000	District around Dazaifu City Hall	18	0.000	0.000	0.003	0.000	0.000	0.000	0.000	0.000	0.000	0.000	0.000	0.000	0.000	0.000	0.000	0.000	0.000	–	0.000	0.000	0.000	0.000	0.003	0.006
19000	District around South of *Suzaka-oji* (Ancient Main Road)	19	0.000	0.000	0.000	0.000	0.000	0.000	0.000	0.000	0.000	0.000	0.000	0.000	0.000	0.000	0.000	0.000	0.000	0.000	–	0.000	0.000	0.000	0.000	0.000
20000	District around North of *Suzaka-oji* (Ancient Main Road)	20	0.000	0.000	0.000	0.000	0.000	0.000	0.000	0.000	0.000	0.000	0.000	0.000	0.000	0.000	0.000	0.000	0.000	0.000	0.000	–	0.000	0.000	0.000	0.000
21000	District around *Tofuro-mae* Station	21	0.000	0.000	0.000	0.000	0.000	0.000	0.000	0.000	0.000	0.000	0.000	0.000	0.000	0.000	0.000	0.000	0.000	0.000	0.000	0.000	–	0.000	0.000	0.000
99999	Outside Survey Target Area	22	0.003	0.000	0.000	0.000	0.000	0.000	0.000	0.000	0.000	0.000	0.000	0.000	0.000	0.000	0.000	0.000	0.000	0.003	0.000	0.000	0.000	–	0.000	0.003
199	Home	23	0.531	0.216	0.151	0.065	0.009	0.000	0.012	0.000	0.000	0.000	0.003	0.000	0.006	0.000	0.000	0.000	0.000	0.003	0.000	0.000	0.000	0.003	–	1.000
Total			**0.809**	0.330	0.472	0.176	0.009	0.019	0.028	0.006	0.003	0.000	0.003	0.003	0.012	0.006	0.003	0.003	0.000	0.006	0.000	0.000	0.000	0.003	1.000	2.892

3 Evaluating Municipal Tourism Policy from How Visitors Walk Around...

Next, by using the actual number of visitors at one place, we estimate the total number of net incoming visitors to the entire Dazaifu City with using the above origin destination density matrix.

According to the Survey of Pedestrian Traffic Volume, the total actual number of inflow population into the *Tenman-gu* Shrine was 11,769 persons per day, so we use this number.

From Table 3.4, it is seen that when the density of incoming visitors to Dazaifu City is set to 1, the density of total visit frequency, the sum of entrance and *Kaiyu* visit frequency, to the *Tenman-gu* Shrine becomes 0.8086. With this, we can estimate the number of incoming visitors to the entire Dazaifu City from the following relation:

$$\text{Total in flow population into } Tenman\text{-}gu \text{ (person/day)} :$$
$$\text{Visitors incoming to Dazaifu City (person/day)} = 11,769 : x$$
$$= 0.8086 : 1.000$$

After all, the value of the number of incoming visitors to the entire Dazaifu City 14,554 (person/day) is obtained by dividing 11,769 by 0.8086.

As for this estimated result, we think that it is almost accurate. If we presume that this number is the annual average number of incoming visitors into Dazaifu City person per day, the annual incoming visitors into the city become about 5.3 million persons per year. It is argued that the number of annual incoming visitors to Dazaifu City is about 7 million persons per year. While there is a difference by about 2 million between two numbers, the Dazaifu *Tenman-gu* Shrine is famous in Japan for its attraction of New Year's worshipers. The number of New Year's worshipers to the Dazaifu *Tenman-gu* Shrine is publicized as about 2 million people, which fills the above difference.

3.3 Comparison of Estimated Number of Visitors with Real Counted Numbers

Our estimation method to estimate the total number of incoming visitors to the entire Dazaifu City requires the actual number of visitors to at least one facility. For this, we used the actual number of inflow population into the Dazaifu *Tenman-gu* Shrine obtained by conducting the pedestrian counting survey. Fortunately, another facility, the Kyushu National Museum, has real observed numbers of visitors to their facility counted by people counting system. They kindly provide us with those numbers of visitors counted on 2 days of our surveys.

Let us now estimate the number of incoming visitors to Dazaifu City using the real counted number of visitors to the Kyushu National Museum.

When the 1st On-site Survey of Sightseeing and *Kaiyu* Behavior of Visitors to Dazaifu City was conducted, the number of visitors to the Kyushu National Museum was 3698 on Saturday, May 29, 2010, and 4698 on May 30, 2010. From now on, we are going to use the average of them as the number of daily visitors to the Museum in the period of 2-day survey.

As a result, the total number of daily visitors to the Kyushu National Museum is 4198 (person/day). From Table 3.4 of the observed origin destination density matrix of *Kaiyu* movements, we see that when setting the density of the total number of net incoming visitors to the entire Dazaifu City equal to 1, the density of total visit frequency to the Kyushu National Museum is 0.330. Therefore, by using the above method, the number of net incoming visitors to the entire Dazaifu City can be estimated as 12,721 persons per day.

On the other hand, given the counted number of inflow visitors to the *Tenman-gu* Shrine used, the number of incoming visitors to the Kyushu National Museum is estimated as 4806 person/day, which is a little bit overestimated but can be said relatively a good estimate.

3.4 Estimated Result of Kaiyu Movements Among Attractions Within Dazaifu City

Table 3.5 gives the observed entrance choice probability P_{HI} and *Kaiyu* choice probability matrix P_{II}. By using the total number of net incoming visitors to Dazaifu City, 14,554 persons, and the observed entrance choice probability, we obtain the estimate of the entrance visit frequency F_{HI}. Then by using *Kaiyu* probability P_{II} and the entrance visit frequency F_{HI}, we can estimate a matrix of the actual numbers of *Kaiyu* movements among tour attractions within the city. The result is shown in Table 3.6. Its main features are as follows.

Looking at the number of entrance visitors from home to each attraction or area, the area with the largest entrance visitors is the Dazaifu *Tenman-gu* Shrine with 7726 person/day. The next is the Kyushu National Museum (3144 person/day), which shows that the number of entrance visitors directly from home to the area becomes larger where major attractions are located.

Figure 3.4 depicts how visitors move around or *Kaiyu* among tour attractions within Dazaifu City in a diagram, which visualizes the result of *Kaiyu* Markov model.

From Fig. 3.4, we see that the largest movement by *Kaiyu* is the one from Dazaifu *Tenman-gu* Shrine to the *Sando* shopping street (Dazaifu *Tenman-gu* Shrine side) with 3532 persons. The second largest one is the movement from the Kyushu National Museum to the *Sando* shopping street (Dazaifu *Tenman-gu* Shrine side) with 1496 persons.

It is apparent that most of *Kaiyu* movements within the city occur among the areas of the Dazaifu *Tenman-gu* Shrine, the Kyushu National Museum, and the *Sando* shopping street and less movements occur beyond these areas. The walking path of history is seen not to attract many visitors.

Table 3.5 Observed entrance and *Kaiyu* probability P_H, P_{II}

Area		Node	1	2	3	4	5	6	7	8	9	10	11	12	13	14	15	16	17	18	19	20	21	22	23	Total
1000	Dazaifu Tenman-gu Shrine	1	0.132	0.070	0.215	0.033	0.000	0.013	0.003	0.003	0.000	0.000	0.000	0.000	0.003	0.000	0.000	0.000	0.000	0.000	0.000	0.000	0.000	0.000	0.526	1.000
2000	Kyushu National Museum	2	0.243	0.036	0.207	0.027	0.000	0.009	0.009	0.000	0.000	0.000	0.000	0.000	0.000	0.000	0.000	0.000	0.000	0.000	0.000	0.000	0.000	0.000	0.468	1.000
3000	Sando shopping street (Tenman-gu Shrine side)	3	0.223	0.063	0.257	0.107	0.000	0.005	0.010	0.000	0.000	0.000	0.000	0.000	0.000	0.000	0.000	0.000	0.000	0.005	0.000	0.000	0.000	0.000	0.330	1.000
4000	Sando shopping street (Dazaifu Station side)	4	0.154	0.031	0.200	0.123	0.000	0.000	0.015	0.000	0.000	0.000	0.000	0.000	0.000	0.000	0.000	0.000	0.000	0.000	0.000	0.000	0.000	0.000	0.477	1.000
5000	District around North of Sando shopping street	5	0.667	0.333	0.000	0.000	0.000	0.000	0.000	0.000	0.000	0.000	0.000	0.000	0.000	0.000	0.000	0.000	0.000	0.000	0.000	0.000	0.000	0.000	0.000	1.000
6000	District around South of Sando shopping street	6	0.000	0.000	0.000	0.000	0.000	0.000	0.000	0.000	0.000	0.000	0.000	0.000	0.000	0.000	0.000	0.000	0.000	0.000	0.000	0.000	0.000	0.000	1.000	1.000
7000	District around West of Sando shopping street	7	0.333	0.000	0.111	0.111	0.000	0.000	0.000	0.111	0.000	0.000	0.000	0.000	0.000	0.000	0.000	0.000	0.000	0.000	0.000	0.000	0.000	0.000	0.333	1.000
8000	District around Dazaifu Souvenir Shop	8	0.000	0.000	0.000	0.000	0.000	0.000	0.000	0.000	0.500	0.000	0.000	0.000	0.000	0.000	0.000	0.000	0.000	0.000	0.000	0.000	0.000	0.000	0.500	1.000
9000	District around West of Gojo 1 (Fifth Ave. 1)	9	0.000	0.000	0.000	0.000	0.000	0.000	0.000	0.000	0.000	0.000	0.000	1.000	0.000	0.000	0.000	0.000	0.000	0.000	0.000	0.000	0.000	0.000	0.000	1.000

(continued)

Table 3.5 (continued)

Area	Node	1	2	3	4	5	6	7	8	9	10	11	12	13	14	15	16	17	18	19	20	21	22	23	Total
10000	District around East of Gojo 1 (Fifth Ave. 1)	**10**	0.000	0.000	0.000	0.000	0.000	0.000	0.000	0.000	0.000	0.000	0.000	0.000	0.000	0.000	0.000	0.000	0.000	0.000	0.000	0.000	0.000	0.000	0.000
11000	District around Chikushi-dai High School	**11**	0.000	0.000	0.000	0.000	0.000	0.000	0.000	0.000	0.000	0.000	0.000	0.000	0.000	0.000	0.000	0.000	0.000	0.000	0.000	0.000	0.000	1.000	1.000
12000	District around welfare facilities	**12**	0.000	0.000	0.000	0.000	0.000	0.000	0.000	0.000	0.000	0.000	0.000	0.000	1.000	0.000	0.000	0.000	0.000	0.000	0.000	0.000	0.000	0.000	0.000	1.000
13000	District around Kanzeon-ji Temple	**13**	0.000	0.000	0.250	0.000	0.000	0.000	0.000	0.000	0.000	0.000	0.000	0.000	0.000	0.500	0.000	0.000	0.000	0.000	0.000	0.000	0.000	0.000	0.250	1.000
14000	District around Ruins of the old Dazaifu Government Office	**14**	0.500	0.000	0.000	0.000	0.000	0.000	0.000	0.000	0.000	0.000	0.000	0.000	0.000	0.000	0.500	0.000	0.000	0.000	0.000	0.000	0.000	0.000	0.000	1.000
15000	District around Sakamoto	**15**	0.000	0.000	0.000	0.000	0.000	0.000	0.000	0.000	0.000	0.000	0.000	0.000	0.000	0.000	0.000	1.000	0.000	0.000	0.000	0.000	0.000	0.000	0.000	1.000
16000	District around Kokubun-ji (Ancient National Temple Remains)	**16**	0.000	0.000	0.000	0.000	0.000	0.000	0.000	0.000	0.000	0.000	0.000	0.000	0.000	0.000	0.000	0.000	0.000	0.000	0.000	0.000	0.000	0.000	1.000	1.000
17000	District around Gojo Station (Fifth Ave.)	**17**	0.000	0.000	0.000	0.000	0.000	0.000	0.000	0.000	0.000	0.000	0.000	0.000	0.000	0.000	0.000	0.000	0.000	0.000	0.000	0.000	0.000	0.000	0.000	0.000

18000	District around Dazaifu City Hall	18	0.000	0.000	0.500	0.000	0.000	0.000	0.000	0.000	0.000	0.000	0.000	0.000	0.000	0.000	0.000	0.000	0.000	0.000	0.000	0.500	1.000
19000	District around South of *Suzaku-oji* (Ancient Main Road)	19	0.000	0.000	0.000	0.000	0.000	0.000	0.000	0.000	0.000	0.000	0.000	0.000	0.000	0.000	0.000	0.000	0.000	0.000	0.000	0.000	0.000
20000	District around North of *Suzaku-oji* (Ancient Main Road)	20	0.000	0.000	0.000	0.000	0.000	0.000	0.000	0.000	0.000	0.000	0.000	0.000	0.000	0.000	0.000	0.000	0.000	0.000	0.000	0.000	0.000
21000	District around *Tofuro-mae* Station	21	0.000	0.000	0.000	0.000	0.000	0.000	0.000	0.000	0.000	0.000	0.000	0.000	0.000	0.000	0.000	0.000	0.000	0.000	0.000	0.000	0.000
99999	Outside survey target area	22	1.000	0.000	0.000	0.000	0.000	0.000	0.000	0.000	0.000	0.000	0.000	0.000	0.000	0.000	0.000	0.000	0.000	0.000	0.000	0.000	1.000
199	Home	23	0.531	0.216	0.151	0.065	0.009	0.000	0.012	0.000	0.000	0.000	0.003	0.000	0.006	0.000	0.000	0.003	0.000	0.000	0.000	0.003	0.000	1.000

Table 3.6 Estimated frequency matrix of tourists' *Kaiyu* movements

Area		Node	1	2	3	4	5	6	7	8	9	10	11	12	13	14	15	16	17	18	19	20	21	22
1000	Dazaifu *Tenman-gu* Shrine	1	–	1,006	3,532	860	0	160	88	41	20	0	0	20	51	26	13	13	0	18	0	0	0	0
2000	Kyushu National Museum	2	1,443	–	1,496	351	0	58	56	11	5	0	0	5	10	5	2	2	0	7	0	0	0	0
3000	*Sando* shopping street (*Tenman-gu* Shrine side)	3	1,103	326	–	487	0	35	49	9	4	0	0	4	8	4	2	2	0	18	0	0	0	0
4000	*Sando* Shopping Street (Dazaifu Station side)	4	354	92	444	–	0	8	24	4	2	0	0	2	3	1	1	1	0	2	0	0	0	0
5000	District around North of *Sando* shopping street	5	136	60	61	15	–	3	2	1	0	0	0	0	1	0	0	0	0	0	0	0	0	0
6000	District around South of *Sando* shopping street	6	0	0	0	0	0	–	0	0	0	0	0	0	0	0	0	0	0	0	0	0	0	0
7000	District around West of *Sando* shopping street	7	99	13	72	35	0	2	–	20	10	0	0	10	10	5	3	3	0	0	0	0	0	0
8000	District around Dazaifu Souvenir Shop	8	0	0	0	0	0	0	0	–	0	0	0	0	0	0	0	0	0	0	0	0	0	0
9000	District around West of Gojo 1 (Fifth Ave. 1)	9	0	0	0	0	0	0	0	0	–	0	0	0	0	0	0	0	0	0	0	0	0	0
10000	District around East of Gojo 1 (Fifth Ave. 1)	10	0	0	0	0	0	0	0	0	0	–	0	0	0	0	0	0	0	0	0	0	0	0
11000	District around *Chikushi-dai* High School	11	0	0	0	0	0	0	0	0	0	0	–	0	0	0	0	0	0	0	0	0	0	0
12000	District around Welfare Facilities	12	0	0	0	0	0	0	0	0	0	0	0	–	0	0	0	0	0	0	0	0	0	0
13000	District around *Kanzeon-ji* Temple	13	40	6	45	7	0	1	1	0	0	0	0	0	–	44	22	22	0	0	0	0	0	0
14000	District around Ruins of the old Dazaifu Government Office	14	0	0	0	0	0	0	0	0	0	0	0	0	0	–	0	0	0	0	0	0	0	0
15000	District around Sakamoto	15	0	0	0	0	0	0	0	0	0	0	0	0	0	0	–	0	0	0	0	0	0	0
16000	District around *Kokubun-ji* (Ancient National Temple Remains)	16	0	0	0	0	0	0	0	0	0	0	0	0	0	0	0	–	0	0	0	0	0	0
17000	District around Gojo Station (Fifth Ave.)	17	0	0	0	0	0	0	0	0	0	0	0	0	0	0	0	0	–	0	0	0	0	0
18000	District around Dazaifu City Hall	18	11	3	35	5	0	0	0	0	0	0	0	0	0	0	0	0	0	–	0	0	0	0
19000	District around South of *Suzaku-oji* (Ancient Main Road)	19	0	0	0	0	0	0	0	0	0	0	0	0	0	0	0	0	0	0	–	0	0	0
20000	District around North of *Suzaku-oji* (Ancient Main Road)	20	0	0	0	0	0	0	0	0	0	0	0	0	0	0	0	0	0	0	0	–	0	0
21000	District around *Tofuro-mae* Station	21	0	0	0	0	0	0	0	0	0	0	0	0	0	0	0	0	0	0	0	0	–	0
99999	Outside survey target area	22	58	6	20	5	0	1	0	0	0	0	0	0	0	0	0	0	0	0	0	0	0	–
199	Home	23	7,726	3,144	2,201	943	135	0	180	0	0	45	45	0	90	0	0	0	0	45	0	0	0	45

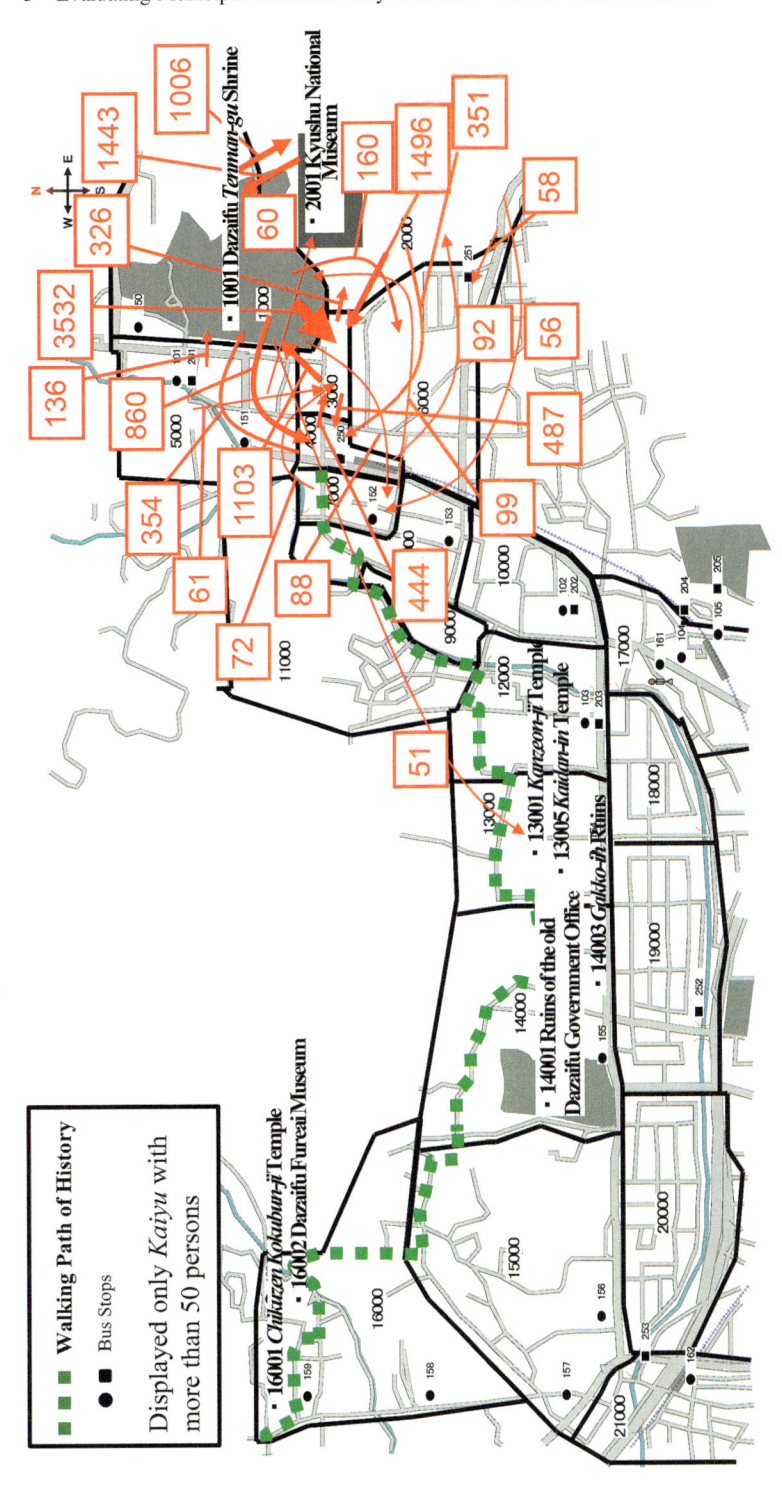

3 Evaluating Municipal Tourism Policy from How Visitors Walk Around... 65

Fig. 3.4 Estimated result of *Kaiyu* movements in Dazaifu City using the *Kaiyu* Markov model (unit: person/day)

4 Evaluating Tourism Policy of Walking Path of History

The broken line in Fig. 3.4 is the walking path of history. As can be seen from the figure, *Kaiyu* is concentrated around the Dazaifu *Tenman-gu* Shrine, the Kyushu National Museum, and the *Sando* shopping street. Despite the desire of the municipal to have many people visit other locations such as *Kanzeon-ji* Temple, which has a national treasure, and the ruins of the ancient Dazaifu Government Office, this goal of the policy of the walking path of history has not been achieved.

The problem is that the walking path of history does not attract many visitors in contrast to the areas near Dazaifu *Tenman-gu* Shrine.

To examine the reason why people are not using the walking path of history might lead to possible causes, questions, and ideas for improvements. Some of them are listed as follows. First, the walking path of history seems not clearly linked to the areas near Dazaifu *Tenman-gu* Shrine where many people are attracted. Does the walking path start in the vicinity of the Dazaifu *Tenman-gu* Shrine, the *Sando* shopping street, and the Kyushu National Museum, where most people gather, and is it appealing to many visitors? The length of the walking path is 7.2 km; is there a way for visitors coming by car to return to the starting point? Since it is too long, should it be divided into several short sections, with a theme for each one, to try to increase the value of the walking experience? Also, is there a way to provide information using mobile ICT devices, so that visitors will know what they will be able to see in each place?

5 Conclusion and Further Research

We have estimated the total number of net incoming visitors into the entire Dazaifu City for the first time. With this total number of net incoming visitors to the city, we also for the first time were able to estimate the actual number of people's movements by their *Kaiyu* behavior among sightseeing sites within the city on the real counted number basis.

With these estimated results, we clarified that visitors' *Kaiyu* movements are concentrated around the areas of the Dazaifu *Tenman-gu* Shrine, the Kyushu National Museum, and the *Sando* shopping street. Less *Kaiyu* movements occur beyond these areas. In particular, not so many visitors are using the walking path of history. This implies that the goal of the municipal tourism policy of the walking path of history that intends to have many people visit many historical sites on the path has not been attained. By examining the reason why people are not using the path, we have proposed several ideas to improve the present situation.

It should be noted that the motivation behind this research is to present a concrete framework to evaluate a tourism policy by its effects on tourists' behaviors, more specifically, its effects on the tourists' *Kaiyu* behaviors among sightseeing places. We believe that we have shown that our *Kaiyu* method is effective to evaluate a specific policy like the walking path of history. Based on the *Kaiyu* approach, we also have proposed some ideas to improve the policy of the walking path of history.

However, also should be noted is that these are not just ideas. They can be experimentally implemented, and it is possible to check what measures are effective by evaluating the change in the number of people actually walking and the number of visitors moving around Dazaifu City by their *Kaiyu* behaviors.

As for further research, this time we have yet to apply the consistent estimation method [2] for the on-site survey to the estimation of the number of net incoming visitors to the entire Dazaifu City and the number of visitors' *Kaiyu* movements within the city. Thus, the next step to further this research is to perform the consistent estimation method to refine those estimates.

Acknowledgments We would like to express our deep gratitude to Mr. Ryoma Yoshitake, Mr. Nobuhide Beppu, Ms. Yurika Nakamura, Ms. Kanako Kondō, and Mr. Kyohei Kaida for their cooperation in carrying out this research.

References

1. Park B, Saito S, Iwami M, Imanishi M, Yamashiro K (2011) An evaluation of walking path of history of Dazaifu city from visitors' *Kaiyu* behaviors. Paper presented at the 48th Annual Meeting of Japan Section of Regional Science Association International (JSRSAI). (in Japanese)
2. Saito S, Nakashima T (2003) An application of the consistent OD estimation for on-site person trip survey: estimating the shop-around pattern of consumers at Daimyo district of Fukuoka City, Japan. Stud Reg Sci 33(3):173–203. (in Japanese)
3. Saito S, Ishibashi K (1992) Forecasting consumer's shop-around behaviors within a city center retail environment after its redevelopments using Markov chain model with covariates. Pap City Plan 27:439–444. (in Japanese)
4. Saito S, Park B, Sato T, Imanishi M, Cai J, Yu H, Yamashiro K (2010) Sightseeing *Kaiyu* behaviors of Dazaifu city visitors. Paper presented at the 47th Annual Meeting of Japan Section of Regional Science Association International (JSRSAI). (in Japanese)
5. Dazaifu Fureai Museum (Museum that promotes cultural contact in Dazaifu City). http://dazaifu.mma.co.jp
6. Zenrin Co. Ltd (2003) Zenrin digital map Z6

Chapter 4
How Did the Extension of Underground Shopping Mall Vitalize *Kaiyu* Within City Center?

Saburo Saito, Masakuni Kakoi, and Masakuni Iwami

Abstract In February 2005, the Tenjin underground shopping mall, located in the center of Tenjin commercial district of Fukuoka City, was extended, and its shop floor space was increased. This development project was designated as a private urban revitalization project by the Urban Renaissance Headquarters of Japanese Government. It is expected to attract more customers to the Tenjin district, make it the main axis of the shop-arounds or *Kaiyu* flows within the district, and link the underground shops of the department stores together. The purpose of this study is to clarify how much shop-arounds or *Kaiyu* flows among the department stores adjoining to the underground shopping mall were vitalized by the extension of the underground shopping mall. This study also aims at verifying how much the redevelopment project of the underground shopping mall has contributed to the revitalization of the city center commercial district through enhancing *Kaiyu* within the district.

Keywords Underground shopping mall · *Kaiyu* · City center revitalization · Department store · Markov model · Urban development · Evaluation · Tenjin · Fukuoka

---

This chapter is based on the paper, Saburo Saito, Masakuni Kakoi, Ryo Nakamura, and Masakuni Iwami [5], "Evaluation of Underground Shopping Mall Development from the Viewpoint of Vitalizing *Kaiyu* within a City Center: A case of Tenjin district of Fukuoka City," *Papers of the 23rd Annual Meeting of The Japan Association for Real Estate Sciences,* vol. 23, pp. 91–96, 2007, which is modified for this chapter.

S. Saito (✉) · M. Kakoi
Faculty of Economics, Fukuoka University, Fukuoka, Japan

Fukuoka University Institute of Quantitative Behavioral Informatics for City and Space Economy (FQBIC), Fukuoka, Japan
e-mail: saito@fukuoka-u.ac.jp; kakoi@econ.fukuoka-u.ac.jp

M. Iwami
Fukuoka University Institute of Quantitative Behavioral Informatics for City and Space Economy (FQBIC), Fukuoka, Japan
e-mail: miwami@econ.fukuoka-u.ac.jp

1 Purpose

In February 2005, the Tenjin underground shopping mall, located in the center of Tenjin commercial district of Fukuoka City, was extended, and its shop floor space was increased. Tenjin district is one of the largest retail accumulations in western part of Japan. This development to extend the underground shopping mall was carried out at the same time as the opening of Fukuoka City Subway Nanakuma Line, whose terminal station, Tenjin Minami Station, is connected to the south end of the extended area of the underground shopping mall. This development project has also been designated by the Urban Renaissance Headquarters of Japanese Government as one of private urban revitalization project plans stipulated by the new law, Act on Special Measures concerning Urban Reconstruction enacted in 2002.

Changing the configuration of the commercial facilities by this kind of redevelopment affects consumers' decision-making of their shopping behaviors in a downtown commercial district. This expansion project is expected to attract more customers to the Tenjin district and give a main axis of *Kaiyu* flow (consumers' shop-around flow) within the Tenjin district. In particular, the characteristics of the extended area of the underground shopping mall are that its east side is adjoining to the underground shops of the one department store and its west side is connecting to the underground shops of the other two department stores. Thus, the opening of the extended underground shopping mall is highly expected to stimulate the *Kaiyu* among these three department stores.

This study addresses to evaluate these developments by measuring the change of consumer shop-around behaviors. We developed the evaluation framework of urban development based on consumer shop-around behaviors and have been conducting many on-site surveys of consumer shop-around behaviors in the downtown at various cities. Especially, since 1996, we have been conducting these surveys every year in the downtown of Fukuoka City. (Cf. [1, 4], also see Chap. 2 of this book)

In this study, we apply the shop-around Markov model by Saito and Ishibashi [2] to the data obtained from the surveys conducted before and after the development of the underground shopping mall (in 2004 and 2005) and evaluate the change of *Kaiyu* flow. The shop-around Markov model is a model to formulate consumers' shop-around behaviors as a stationary absorbing Markov chain, which can estimate and predict the number of visitors to each commercial facility in the downtown commercial district. The number of visitors to each commercial facility is called the "total visits" to that facility. The salient feature of this model is that it can decompose the "total visits" into the "entry visits" and "shop-around visits." The entry visits to the commercial facility is the number of visitors who first shopped at that facility in the downtown. The "shop-around visits" to the commercial facility is the number of visitors who shopped at that facility from other commercial facilities by their shop-around behavior. Additionally, the shop-around Markov model can define and estimate the "shop-around effect" that serves as a scale for measuring how much one entry visit to the one facility would bring the shop-around visits to the other facilities.

The purpose of this study is to verify to what extent the extension of the Tenjin underground shopping mall enhanced the *Kaiyu* flows among commercial facilities, especially among three department stores at the city center of Fukuoka City.

While we focus on the effect of the extension of the Tenjin underground shopping mall, we had an opportunity for examining another large-scale development in the Tenjin commercial district in the previous year of the Tenjin underground shopping mall project. That is the relocation and opening of *Iwataya's* new store in March 2004. Iwataya is a long-established local department store in the Tenjin district.

We compare the difference in the changes of *Kaiyu* flows caused by these two projects so that we evaluate the redevelopment project of the underground shopping mall. For measuring the changes in *Kaiyu* flows by *Iwataya's* project, we use the data obtained from the consumer shop-around surveys in 2003 and 2004.

By doing this, this study also aims at considering how much the redevelopment project of the underground shopping mall has contributed to the improvement of the city attractiveness of Fukuoka City.

2 Framework of the Analysis

2.1 Tenjin Area in Fukuoka City

In this study, we investigate the impacts of the two redevelopment projects in the Tenjin area from the viewpoint of consumers' shop-around behaviors.

Figure 4.1 shows the map of Tenjin area in Fukuoka City, Japan. This figure shows that *Iwataya* had relocated its former store to the new department store located at the southeast part of Tenjin district in March 2004 and the Tenjin underground shopping mall had expanded to the south and increased its shop floor space in February 2005.

Before its relocation, *Iwataya* had two stores, *Iwataya* A-side as its former main building and *Iwataya* Z-side. *Iwataya* fell into bankruptcy in 1999 and made its management reconstruction plan. According to this plan, *Iwataya* closed A-side in February 2004 and constructed a new department store located opposite north to Z-side across the street[1]. *Iwataya* made the former Z-side as its new main store and a new building as a new annex store. By this grand renewal relocated opening, Iwataya restarted its department store business running as two stores as one unit. Also, by the project of underground shopping mall, the west side of the extended area of the underground shopping mall is connecting to the underground shops of the Fukuoka *Mitsukoshi* department store. Also, the east side of the underground shopping mall is adjoining to the underground shops of the Fukuoka Tenjin *Daimaru* department store and the *Daimaru* Elgala.

---

[1] As for the circumstances and histories during this period, please refer to the introduction of Chap. 16 of this book and references therein.

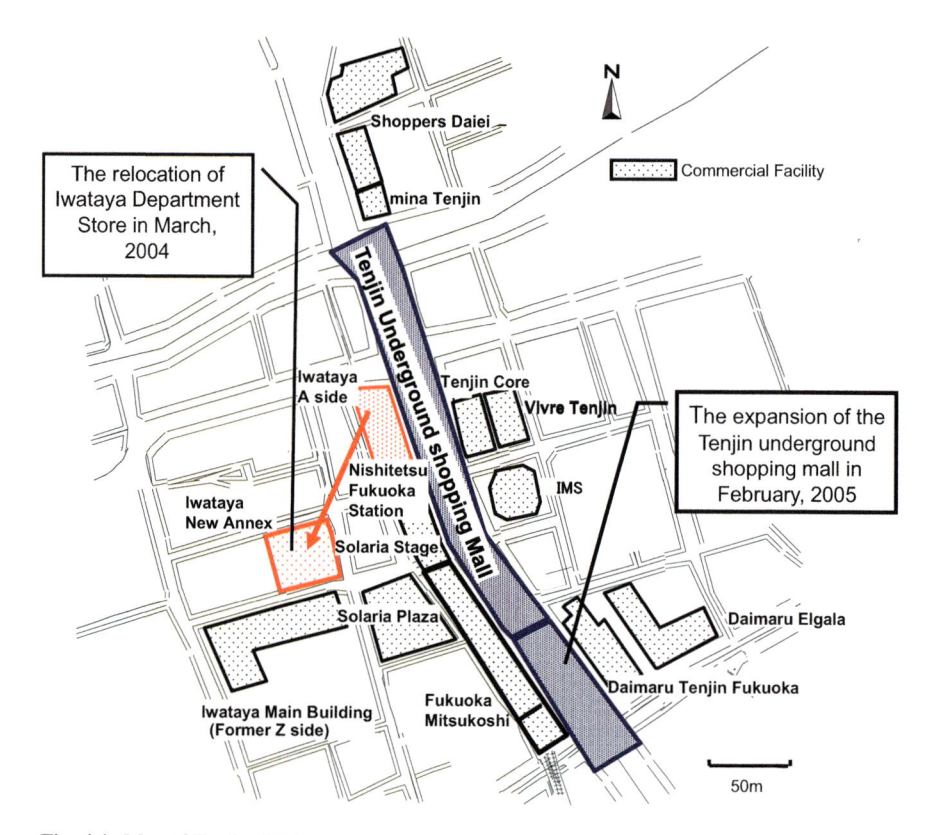

Fig. 4.1 Map of Tenjin district

2.2 Data Used

In this study, we use data collected from three surveys of consumer shop-around behaviors conducted in 2003, 2004, and 2005. These surveys were planned and implemented by the Fukuoka University Institute of Quantitative Behavioral Informatics for City and Space Economy (FQBIC). The outline of the surveys is shown in Table 4.1.

In the tenth survey of consumer shop-around behaviors at the city center of Fukuoka City, we added a survey sampling site at the Tenjin underground shopping mall. However, to consistently handling the data of the three surveys, we removed samples collected in the Tenjin underground shopping mall. Consequently, we used 808 samples obtained from the tenth survey. Gender and age ratios of samples of these surveys are shown in Figs. 4.2 and 4.3.

Table 4.1 Outline of the surveys of consumer shop-around behaviors

	The eighth survey of consumer shop-around behavior at the city center of Fukuoka City	The ninth survey of consumer shop-around behavior at the city center of Fukuoka City	The tenth survey of consumer shop-around behavior at the city center of Fukuoka City
Date of survey	2003.06.27, 2003.06.28, 2003.06.29	2004.7.10, 2004.7.11	2005.7.2, 2005.7.3
Survey time	12:00–19:00	12:00–19:00	12:00–19:00
Survey points	Shoppers Daiei, Solaria Plaza, *Iwataya* Z-side, *Daimaru* Tenjin Fukuoka, Fukuoka *Mitsukoshi*, Canal City Hakata, Hakata Riverain, Hakata Station Concourse	Shoppers Daiei, Solaria Plaza, *Iwataya* Z-side, *Daimaru* Tenjin Fukuoka, Fukuoka *Mitsukoshi*, Canal City Hakata, Hakata Riverain, Hakata Station Concourse	Shoppers Daiei, Solaria Plaza, *Iwataya* Z-side, *Daimaru* Tenjin Fukuoka, Fukuoka *Mitsukoshi*, Canal City Hakata, Hakata Riverain, Hakata Station Concourse, Tenjin underground shopping mall
Number of samples	1007 samples	682 samples	1038 samples
Main questionnaire items	1. Sample profiles (residence, age, gender, occupation, etc.)		
	2. Shop-around history (places visited, purposes done there, and expenditure there if any)		
	3. Travel time to the city center of Fukuoka City and mode of transportation		
	4. Visit frequency to Fukuoka city center, main commercial districts in Fukuoka city center, and main commercial facilities in Fukuoka city center		
Survey method	1. On-site sampling survey		
	2. Samples drawn at random from visitors at sampling places in the Fukuoka city center		
	3. Interview with questionnaire for 15–20 min		

2.3 *Shop-Around Markov Model*

The consumer shop-around behavior is defined as the movement by consumers who walk around in the downtown area for shopping, leisure, and eating purposes. Suppose that a consumer left home and walked around the shopping sites in a downtown area in the order of A, B, and C. Site A is the entrance to the commercial area, and the subsequent movements to B and C are defined as "shop-around."

The shop-around Markov model is a model expressed as a stationary absorbing Markov chain. This model assumes consumer shopping behavior that consumer leaves home and returns home after repeating infinite number of shop-arounds in downtown shopping sites.

H_0 is the returning home node (absorbing state), H is the departure node, the set of nodes corresponding to n commercial facilities is expressed as I, and the consumer shop-around behavior is represented by a state transition probability matrix P.

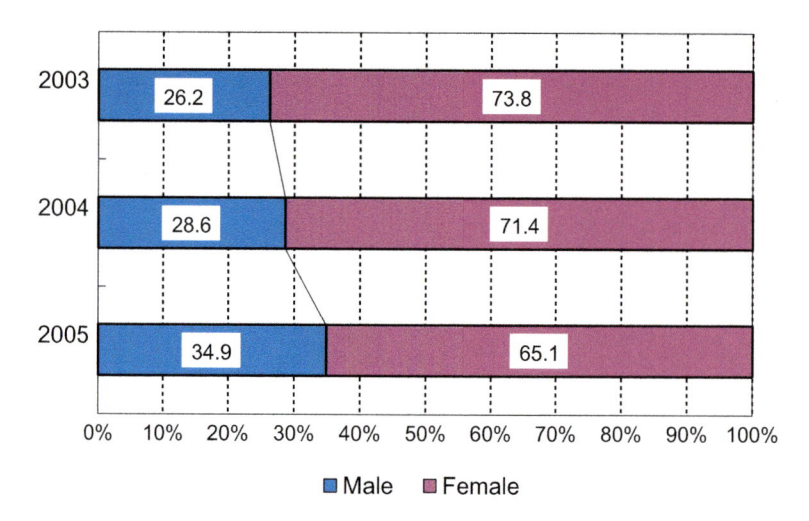

Fig. 4.2 Gender

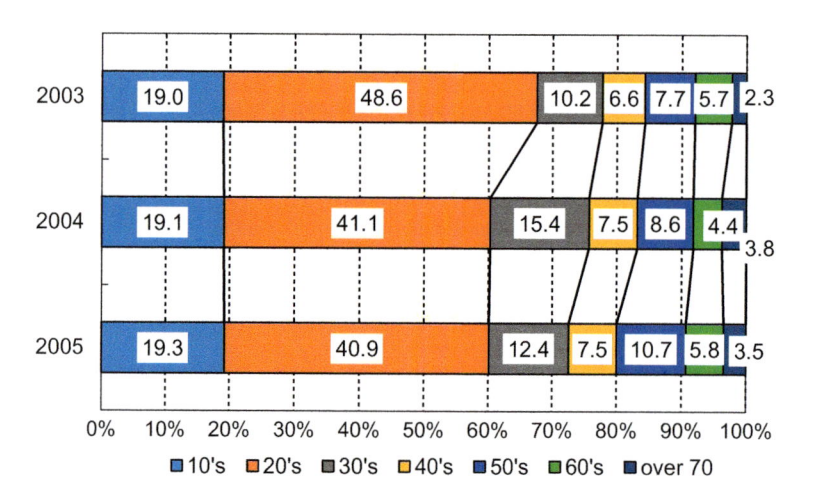

Fig. 4.3 Age

$$P = \begin{matrix} H_0 \\ H \\ I \end{matrix} \begin{bmatrix} 1 & 0 & 0 \\ 0 & 0 & P_{\text{HI}} \\ P_{\text{IHo}} & 0 & P_{\text{II}} \end{bmatrix}$$

2.4 The Definition of Shop-Around Effect

The shop-around effect is a concept that tries to measure how much shop-around is induced. Here we give its precise definition. The shop-around effect is a scale that measures how many shop-around visits to each commercial facility would be induced if one entry trip was given to the downtown area.

By using the partitioned transition probability matrix of P, we can obtain the n-th order transition probability, whose component that corresponds to P_{HI} becomes $P_{HI}P_{II}^{n-1}$. Since this is the probability of visiting each commercial facility node at the n-th step, by adding all the subsequent probabilities after the second step we obtain the shop-around effect RE:

$$RE = P_{HI}P_{II} + P_{HI}P_{II}^2 + \cdots$$
$$= P_{HI}P_{II}(I - P_{II})^{-1}$$

Furthermore, $P_{II}(I - P_{II})^{-1}$, obtained by replacing P_{HI} with an identity matrix, serves as a scale that measures how many shop-around visits one entry visit to each commercial facility would bring to other commercial facilities.

Thus far, we have considered "shop-around effect" on a probability basis. Let us discuss it on the frequency basis. In order to express "shop-around effect" on the visit frequency basis to each commercial facility node, we just replace the entry choice probability P_{HI} with the entry visit frequency F_{HI}. The number of people who visited each commercial facility by shop-around is the shop-around visit frequency. Therefore, the total visit frequency (TVST) to each commercial facility can be decomposed into the entry visits and shop-around visits. TVST can be expressed as follows.

$$\text{TVST} = F_{HI}(I - P_{II})^{-1}$$
$$= F_{HI} + F_{HI}P_{II}(I - P_{II})^{-1}$$

In order to actually apply the shop-around Markov model, the transition probability P must be measured from the observed data of consumers' shop-around behaviors. To do this, we employed the aggregation method which decomposes each consumer shop-around behavior represented as a linked trip, which is a chain of nodes the consumer shop-arounds, into unlinked trips between two successive nodes and aggregates them. By this aggregation method, we obtain the transition probability.

If we denote the aggregation result as an observed frequency matrix T, it can be expressed as follows:

$$T = \begin{matrix} H_0 \\ H \\ I \end{matrix} \begin{bmatrix} N & 0 & 0 \\ 0 & 0 & F_{HI}^{OBS} \\ F_{IHo}^{OBS} & 0 & F_{II}^{OBS} \end{bmatrix}$$

$$P^{OBS} = \begin{matrix} H_0 \\ H \\ I \end{matrix} \begin{bmatrix} 1 & 0 & 0 \\ 0 & 0 & P_{HI}^{OBS} \\ P_{IHo}^{OBS} & 0 & P_{II}^{OBS} \end{bmatrix}$$

Here, N is the total number of samples, and P^{OBS} is the observed transition probability obtained by converting T into a probability matrix. The F_{HI}^{OBS} is observed entry visits, and the column sum of F_{II}^{OBS} is the observed shop-around visits RE^{OBS}.

3 Measurement of Shop-Around Effects and Their Interannual Comparison

In this section, we conduct an interannual comparison of shop-around effects. Tables 4.2, 4.3, and 4.4 show the shop-around effects in 2003, 2004, and 2005. The table shows how much shop-around effects one entry visit to each row site has brought to each column site. In addition, in each row at the rightmost column in the tables (the next to the rightmost column in Tables 4.3 and 4.4), we give the estimate of the shop-around effect on other column sites except its own site, that is, the total of the shop-around effects by each row site on all the other column sites excluding its own site. Based on these results, we conduct the interannual comparisons as follows. The main findings revealed by these comparisons are as follows:

1. *Interannual comparison between 2003 and 2004*

The shopping sites that increased their shop-around effect on other sites from 2003 to 2004 were the underground shopping mall, *Daimaru*, *Iwataya* New Annex, Solaria Plaza, Hakata Riverain, the northern part of Tenjin district (Tenjin North), and *Imaizumi* district. This fact can be seen from the rightmost column of Table 4.3, which shows the ratio of the shop-around effect on other sites in 2004 to that in 2003.

2. *Interannual comparison between 2004 and 2005*

In all shopping sites except the underground shopping mall, their shop-around effects on other sites were increasing, especially in Fukuoka *Mitsukoshi*, *Daiei*, Canal City, and *Imaizumi*, where the shop-around effects on other sites have increased by 1.4 times or more. This is also seen from the rightmost column of Table 4.4, which shows the ratio of the shop-around effect on other sites in 2005 to that in 2004.

It should be noted that the interannual comparisons between consumer shop-around effects imply that the comparisons were made in terms of probability. In

Table 4.2 Shop-around effects by one visit to the one site of the rows on the other sites of the columns (2003)

Node		1	2	3	4	5	6	7	8	9	10	11	12	13	14	15	16	17	18	19	Total	Shop-around effect on other sites
1	Tenjin underground shopping mall	0.213	0.216	0.269	0.068	0.209	0.322	0.171	0.295	0.109	0.616	0.118	0.062	0.219	0.078	0.229	0.021	0.131	0.098	0.136	3.581	3.369
2	Fukuoka *Mitsukoshi*	0.155	0.248	0.255	0.042	0.184	0.268	0.076	0.125	0.059	0.155	0.135	0.029	0.182	0.049	0.218	0.027	0.128	0.087	0.322	2.744	2.495
3	*Daimaru* Tenjin Fukuoka	0.103	0.246	0.534	0.027	0.150	0.184	0.100	0.120	0.048	0.111	0.113	0.033	0.171	0.040	0.169	0.016	0.093	0.115	0.126	2.499	1.965
4	Iwataya A-side	0.308	0.317	0.350	0.254	0.265	0.313	0.088	0.143	0.135	0.214	0.104	0.065	0.237	0.043	0.197	0.021	0.098	0.105	0.127	3.388	3.134
5	Iwataya Z-side	0.165	0.208	0.185	0.025	0.307	0.297	0.080	0.130	0.068	0.155	0.132	0.025	0.198	0.044	0.413	0.035	0.110	0.055	0.121	2.752	2.445
6	Solaria Plaza	0.149	0.185	0.167	0.028	0.245	0.462	0.096	0.153	0.064	0.155	0.105	0.024	0.163	0.052	0.284	0.039	0.101	0.059	0.113	2.644	2.182
7	IMS	0.334	0.235	0.330	0.066	0.226	0.344	0.150	0.403	0.109	0.343	0.152	0.051	0.267	0.066	0.289	0.025	0.105	0.093	0.136	3.726	3.576
8	Tenjin Core	0.349	0.206	0.225	0.060	0.224	0.298	0.181	0.676	0.137	0.445	0.116	0.038	0.298	0.067	0.287	0.024	0.095	0.081	0.105	3.913	3.237
9	Solaria Stage	0.284	0.196	0.338	0.052	0.297	0.431	0.138	0.225	0.134	0.317	0.165	0.039	0.316	0.054	0.305	0.041	0.108	0.109	0.123	3.672	3.539
10	Shoppers Daiei	0.292	0.140	0.164	0.043	0.121	0.179	0.080	0.168	0.060	0.955	0.085	0.039	0.170	0.136	0.203	0.016	0.089	0.062	0.101	3.103	2.148
11	Canal City Hakata	0.049	0.088	0.083	0.011	0.051	0.083	0.023	0.062	0.028	0.079	1.259	0.030	0.067	0.027	0.117	0.011	0.249	0.175	0.097	2.590	1.331
12	Hakata Riverain	0.053	0.087	0.072	0.027	0.051	0.113	0.024	0.048	0.024	0.087	0.120	0.299	0.097	0.036	0.070	0.009	0.201	0.229	0.102	1.748	1.449
13	Tenjin South	0.182	0.193	0.269	0.051	0.183	0.290	0.126	0.216	0.072	0.241	0.092	0.049	0.274	0.052	0.232	0.020	0.120	0.096	0.106	2.865	2.591
14	Tenjin North	0.116	0.127	0.183	0.021	0.136	0.156	0.047	0.127	0.035	0.496	0.111	0.064	0.114	0.183	0.166	0.014	0.166	0.088	0.086	2.436	2.253
15	*Daimyo*	0.122	0.187	0.204	0.027	0.272	0.259	0.067	0.146	0.052	0.147	0.112	0.035	0.159	0.058	0.611	0.086	0.107	0.060	0.096	2.805	2.195
16	*Imaizumi*	0.101	0.270	0.190	0.023	0.210	0.211	0.054	0.099	0.039	0.183	0.085	0.029	0.226	0.076	0.396	0.096	0.086	0.093	0.136	2.605	2.509
17	Hakata	0.081	0.092	0.123	0.023	0.072	0.113	0.035	0.075	0.038	0.082	0.269	0.050	0.092	0.040	0.121	0.011	0.695	0.107	0.101	2.221	1.526
18	Other districts in the Fukuoka city center	0.119	0.137	0.156	0.020	0.100	0.207	0.044	0.111	0.065	0.144	0.320	0.160	0.112	0.057	0.160	0.031	0.244	0.296	0.094	2.577	2.280
19	Districts outside the Fukuoka city center	0.047	0.084	0.100	0.010	0.073	0.087	0.023	0.040	0.017	0.072	0.092	0.022	0.102	0.023	0.071	0.006	0.183	0.039	0.099	1.189	1.090

Table 4.3 Shop-around effects by one visit to the one site of the rows on the other sites of the columns (2004)

	Node	1	2	3	4	5	6	7	8	9	10	11	12	13	14	15	16	17	18	19	Total	Shop-around effect on other sites	Shop-around effect on other sites (2004/2003)
Tenjin underground shopping mall	1	0.232	0.350	0.408	0.091	0.286	0.482	0.145	0.264	0.116	0.299	0.123	0.069	0.276	0.044	0.222	0.018	0.062	0.058	0.110	3.655	3.423	1.02
Fukuoka Mitsukoshi	2	0.091	0.302	0.313	0.054	0.213	0.258	0.055	0.123	0.063	0.112	0.112	0.040	0.228	0.035	0.237	0.014	0.043	0.048	0.143	2.486	2.184	0.88
Daimaru Tenjin Fukuoka	3	0.105	0.289	0.590	0.046	0.170	0.270	0.087	0.122	0.076	0.107	0.114	0.036	0.262	0.022	0.191	0.020	0.036	0.071	0.116	2.730	2.140	1.09
Iwataya New Annex	4	0.118	0.245	0.331	0.286	0.554	0.440	0.102	0.112	0.097	0.127	0.156	0.037	0.304	0.020	0.296	0.019	0.037	0.046	0.117	3.444	3.158	1.01
Iwataya Main Building	5	0.082	0.181	0.261	0.120	0.375	0.402	0.075	0.099	0.072	0.085	0.093	0.032	0.277	0.023	0.326	0.019	0.043	0.039	0.093	2.697	2.322	0.95
Solaria Plaza	6	0.138	0.301	0.248	0.057	0.237	0.708	0.112	0.183	0.103	0.107	0.130	0.048	0.282	0.036	0.306	0.037	0.040	0.063	0.119	3.255	2.547	1.17
IMS	7	0.206	0.280	0.420	0.095	0.247	0.377	0.233	0.271	0.092	0.134	0.155	0.065	0.302	0.020	0.264	0.018	0.057	0.052	0.091	3.381	3.148	0.88
Tenjin Core	8	0.264	0.269	0.304	0.058	0.235	0.453	0.219	0.551	0.121	0.148	0.204	0.044	0.279	0.022	0.295	0.019	0.054	0.051	0.123	3.713	3.162	0.98
Solaria Stage	9	0.151	0.337	0.318	0.084	0.227	0.458	0.097	0.238	0.189	0.136	0.118	0.045	0.276	0.021	0.233	0.017	0.037	0.096	0.139	3.218	3.030	0.86
Shoppers Daiei	10	0.190	0.130	0.233	0.050	0.143	0.198	0.110	0.154	0.072	0.372	0.104	0.048	0.181	0.026	0.177	0.011	0.053	0.049	0.054	2.354	1.982	0.92
Canal City Hakata	11	0.032	0.091	0.102	0.023	0.052	0.082	0.022	0.044	0.022	0.061	1.205	0.035	0.136	0.006	0.086	0.006	0.051	0.083	0.048	2.187	0.982	0.74
Hakata Riverain	12	0.079	0.097	0.161	0.024	0.111	0.111	0.051	0.052	0.037	0.071	0.347	0.505	0.159	0.010	0.128	0.007	0.030	0.265	0.116	2.360	1.856	1.28
Tenjin South	13	0.118	0.210	0.278	0.062	0.203	0.313	0.093	0.140	0.090	0.155	0.155	0.066	0.296	0.016	0.239	0.021	0.037	0.064	0.101	2.659	2.362	0.91
Tenjin North	14	0.189	0.197	0.246	0.096	0.139	0.193	0.063	0.159	0.054	0.156	0.111	0.115	0.306	0.165	0.135	0.010	0.031	0.105	0.124	2.593	2.428	1.08
Daimyo	15	0.071	0.144	0.164	0.066	0.245	0.336	0.074	0.087	0.078	0.089	0.172	0.049	0.206	0.013	0.494	0.050	0.033	0.045	0.076	2.492	1.998	0.91
Imaizumi	16	0.070	0.321	0.285	0.046	0.244	0.329	0.057	0.089	0.057	0.082	0.099	0.036	0.335	0.016	0.373	0.250	0.032	0.042	0.166	2.929	2.679	1.07
Hakata	17	0.095	0.139	0.150	0.020	0.082	0.103	0.033	0.099	0.043	0.064	0.174	0.048	0.101	0.009	0.078	0.020	0.239	0.112	0.124	1.732	1.493	0.98
Other districts in the Fukuoka city center	18	0.083	0.180	0.235	0.029	0.109	0.138	0.053	0.067	0.077	0.066	0.391	0.210	0.205	0.025	0.138	0.009	0.056	0.205	0.092	2.367	2.162	0.95
Districts outside the Fukuoka city center	19	0.030	0.090	0.094	0.014	0.072	0.082	0.020	0.047	0.032	0.055	0.073	0.035	0.069	0.006	0.082	0.005	0.067	0.060	0.190	1.120	0.930	0.85

Table 4.4 Shop-around effects by one visit to the one site of the rows on the other sites of the columns (2005)

	Node	1	2	3	4	5	6	7	8	9	10	11	12	13	14	15	16	17	18	19	Total	Shop-around effect on other sites	Shop-around effect on other sites (2005/2004)
Tenjin underground shopping mall	1	0.510	0.457	0.434	0.110	0.188	0.346	0.131	0.213	0.131	0.285	0.156	0.048	0.260	0.047	0.190	0.008	0.146	0.096	0.099	3.856	3.345	0.98
Fukuoka *Mitsukoshi*	2	0.466	0.614	0.342	0.121	0.180	0.318	0.088	0.180	0.091	0.109	0.134	0.042	0.296	0.028	0.245	0.017	0.130	0.108	0.162	3.673	3.059	1.40
Daimaru Tenjin Fukuoka	3	0.423	0.336	0.821	0.082	0.189	0.243	0.162	0.170	0.074	0.098	0.195	0.044	0.282	0.022	0.170	0.006	0.133	0.105	0.114	3.668	2.847	1.33
Iwataya New Annex	4	0.406	0.350	0.353	0.311	0.386	0.271	0.076	0.173	0.088	0.097	0.116	0.077	0.280	0.023	0.279	0.025	0.098	0.079	0.083	3.572	3.261	1.03
Iwataya Main Building	5	0.325	0.319	0.286	0.164	0.439	0.328	0.086	0.169	0.081	0.088	0.132	0.040	0.332	0.028	0.303	0.008	0.114	0.077	0.097	3.418	2.979	1.28
Solaria Plaza	6	0.310	0.290	0.242	0.082	0.240	0.598	0.077	0.177	0.085	0.095	0.133	0.056	0.277	0.031	0.308	0.012	0.121	0.094	0.139	3.367	2.769	1.09
IMS	7	0.583	0.434	0.456	0.085	0.224	0.367	0.456	0.299	0.162	0.133	0.139	0.041	0.313	0.028	0.249	0.008	0.114	0.082	0.096	4.268	3.812	1.21
Tenjin Core	8	0.501	0.361	0.299	0.084	0.158	0.382	0.139	0.572	0.173	0.159	0.250	0.057	0.359	0.028	0.187	0.007	0.149	0.099	0.119	4.083	3.511	1.11
Solaria Stage	9	0.421	0.409	0.286	0.119	0.199	0.496	0.145	0.234	0.146	0.105	0.134	0.044	0.303	0.025	0.287	0.009	0.126	0.111	0.115	3.715	3.569	1.18
Shoppers Daiei	10	0.775	0.289	0.287	0.085	0.141	0.223	0.100	0.153	0.078	1.252	0.139	0.043	0.212	0.148	0.128	0.005	0.122	0.125	0.073	4.377	3.125	1.58
Canal City Hakata	11	0.136	0.092	0.111	0.028	0.044	0.073	0.043	0.068	0.038	0.041	1.411	0.037	0.081	0.029	0.052	0.002	0.306	0.103	0.097	2.792	1.381	1.41
Hakata Riverain	12	0.257	0.164	0.249	0.037	0.082	0.134	0.050	0.082	0.070	0.062	0.195	0.362	0.180	0.014	0.164	0.003	0.167	0.355	0.104	2.731	2.370	1.28
Tenjin South	13	0.331	0.274	0.288	0.066	0.156	0.278	0.094	0.166	0.074	0.123	0.172	0.072	0.302	0.031	0.198	0.005	0.147	0.091	0.087	2.956	2.654	1.12
Tenjin North	14	0.336	0.166	0.215	0.043	0.089	0.229	0.055	0.091	0.047	0.283	0.239	0.074	0.336	0.030	0.103	0.004	0.091	0.066	0.056	2.553	2.523	1.04
Daimyo	15	0.256	0.201	0.194	0.069	0.234	0.257	0.060	0.110	0.089	0.095	0.123	0.029	0.210	0.031	0.454	0.012	0.109	0.071	0.079	2.684	2.230	1.12
Imaizumi	16	0.706	0.433	0.270	0.073	0.144	0.233	0.079	0.141	0.082	0.146	0.172	0.049	0.197	0.029	0.312	0.132	0.153	0.255	0.270	3.876	3.744	1.40
Hakata	17	0.227	0.150	0.137	0.036	0.077	0.121	0.037	0.092	0.037	0.079	0.370	0.075	0.130	0.019	0.094	0.003	0.373	0.136	0.142	2.335	1.961	1.31
Other districts in the Fukuoka city center	18	0.230	0.155	0.166	0.034	0.084	0.151	0.054	0.103	0.042	0.058	0.407	0.125	0.145	0.015	0.114	0.003	0.260	0.386	0.202	2.735	2.350	1.09
Districts outside the Fukuoka city center	19	0.168	0.124	0.157	0.030	0.089	0.101	0.036	0.094	0.044	0.047	0.244	0.051	0.144	0.012	0.115	0.003	0.285	0.189	0.417	2.350	1.933	2.08

other words, the comparisons are carried out based on the observed shop-around probabilities from 2003 to 2005. We will make comparisons in terms of frequency in the next section.

4 Estimation of the Numbers of Shop-Around Visits and Total Visits and Their Interannual Comparison

In this section, we estimate the actual numbers of shop-around visits and total visits by using the estimates of the number of visitors entering to the city center of Fukuoka City, which were separately estimated, and make a comparison from 2003 to 2005 in terms of frequency.

We employ the forecast model, which was developed by Saito et al. [3] based on survey data of the first survey of September 1996, for predicting the number of people entering the city center of Fukuoka City. The forecast is based on the weighted Poisson regression model that removes the choice-based sampling bias which inherently occurs since our surveys of consumer shop-around behaviors are conducted as on-site surveys, which are carried out at survey sites set up in the city center. The forecast is based on the estimation of the number of people visiting the city center from their residence in the Fukuoka metropolitan area, which has a total population of approximately 1.7 million people. The metropolitan area is defined as the area in which the percentage of people who commute to Fukuoka City for work and study is more than 20% of employed permanent residents. According to the forecasted numbers for the year 2000, the numbers of visitors were 142,377 people to the Tenjin area, 42,976 to the area around Hakata Station, 26,627 to the area of Canal City Hakata, and 19,285 to the area of Hakata Riverain.

We employed the forecast predicted by Saito et al. [3] for the year 2000 as the estimates of the total numbers of entry visits for the years from 2003 to 2005. The implication of this is that we have ignored the increases in the number of incoming visitors which would be caused by these redevelopment projects of the extension of the underground shopping mall and the renewal opening of the relocated department store and inevitably caused by the opening of the new subway line. The reason why we do this is that we intend to extract the pure net effect of the spatial structural change caused by these two retail-related redevelopment projects. To extract the net effect, we must fix the total entry visit frequency for all years to compare. However, the distributions of the frequency of entry visits over commercial districts for the years of 2003–2005 were estimated by a separate model which considers the changes in allocation of shop floor space among commercial facilities.

In Tables 4.5, 4.6, and 4.7, we give the estimated results of running shop-around Markov model for 2003–2005 by using the above estimate for 2000 as the entry visits for 2003–2005 and by using the observe shop-around probabilities in 2003–2005. In the tables, commercial districts other than the Tenjin district are summarized in others as node 15.

4 How Did the Extension of Underground Shopping Mall Vitalize *Kaiyu*...

Table 4.5 Estimated results of entry visits, shop-around visits, and total visits (2003)

Node		1	2	3	4	5	6	7	8	9	10	11	12	13	14	15	Total
Tenjin underground shopping mall	1	2693	2736	3405	860	2649	4084	2171	3740	1379	7811	2776	989	2902	270	6911	45,376
Fukuoka *Mitsukoshi*	2	3746	6007	6161	1012	4440	6493	1841	3030	1429	3743	4391	1176	5263	658	16,978	66,370
Daimaru Tenjin Fukuoka	3	1645	3909	8484	426	2385	2927	1583	1915	758	1764	2717	629	2688	250	7636	39,718
Iwataya A-side	4	1063	1097	1211	876	916	1083	305	496	466	741	821	150	683	73	1728	11,708
Iwataya Z-side	5	1442	1824	1623	215	2685	2600	704	1137	596	1356	1730	382	3614	304	3881	24,093
Solaria Plaza	6	2535	3159	2845	471	4178	7879	1637	2606	1098	2637	2776	892	4834	669	6856	45,072
IMS	7	1618	1139	1595	319	1095	1664	727	1951	525	1661	1292	321	1399	119	2602	18,027
Tenjin Core	8	2251	1332	1449	388	1445	1923	1170	4363	886	2873	1926	431	1848	154	2804	25,242
Solaria Stage	9	1440	996	1714	264	1505	2185	699	1142	677	1607	1604	274	1545	206	2752	18,611
Shoppers Daiei	10	2622	1257	1469	387	1087	1609	720	1509	539	8577	1530	1226	1828	142	3376	27,878
Tenjin South	11	2854	3024	4208	802	2872	4550	1968	3387	1132	3773	4292	814	3640	316	7252	44,884
Tenjin North	12	615	676	968	110	718	826	247	675	187	2627	606	968	880	72	2730	12,906
Daimyo	13	1516	2328	2539	337	3388	3223	829	1816	643	1829	1974	716	7596	1065	5101	34,901
Imaizumi	14	163	436	307	36	338	340	87	159	63	296	365	123	640	155	693	4202
Other districts in the Fukuoka city center	15	5828	7969	8901	1784	5472	9257	2562	5799	2822	7301	7634	3141	9655	913	119,076	198,114
a Entry visits		12,671	24,190	15,896	3456	8755	17,048	4838	6451	5068	8985	15,666	5299	12,441	1613		142,377
b Shop-around visits (excluding sojourn)		29,339	31,882	38,396	7412	32,488	42,764	16,524	29,362	12,522	40,019	32,143	11,265	41,419	5212	119,076	370,747
c Sojourn visits		2693	6007	8484	876	2685	7879	727	4363	677	8577	4292	968	7596	155		55,979
d Shop-around visits		32,032	37,889	46,880	8288	35,173	50,643	17,251	33,725	13,199	48,597	36,435	12,232	49,014	5367		426,726
e Total visits		44,703	62,080	62,777	11,744	43,928	67,692	22,089	40,176	18,267	57,581	52,101	17,531	61,455	6980		569,103
Shop-around ratio (b/a)		2.315	1.318	2.415	2.145	3.711	2.508	3.415	4.552	2.471	4.454	2.052	2.126	3.329	3.232		2.604
Shop-around ratio (c/a)		0.213	0.248	0.534	0.254	0.307	0.462	0.150	0.676	0.134	0.955	0.274	0.183	0.611	0.096		0.393
Shop-around ratio (d/a)		2.528	1.566	2.949	2.398	4.018	2.971	3.566	5.228	2.604	5.409	2.326	2.308	3.940	3.328		2.997
Shop-around ratio (e/a)		3.528	2.566	3.949	3.398	5.018	3.971	4.566	6.228	3.604	6.409	3.326	3.308	4.940	4.328		3.997

Table 4.6 Estimated results of entry visits, shop-around visits, and total visits (2004)

	Node	1	2	3	4	5	6	7	8	9	10	11	12	13	14	15	Total
Tenjin underground shopping mall	1	3506	5278	6162	1366	4311	7268	2184	3980	1750	4519	4172	670	3357	271	6363	55,157
Fukuoka *Mitsukoshi*	2	1229	4065	4209	731	2865	3472	742	1658	850	1503	3071	469	3191	190	5198	33,441
Daimaru Tenjin Fukuoka	3	1968	5396	11,023	851	3177	5043	1622	2273	1421	1996	4891	415	3564	370	6964	50,975
Iwataya New Annex	4	310	642	869	751	1455	1154	268	294	254	333	797	53	777	50	1032	9037
Iwataya Main Building	5	1097	2439	3508	1611	5045	5414	1013	1330	967	1139	3726	307	4388	252	4045	36,281
Solaria Plaza	6	2622	5728	4715	1076	4517	13,464	2122	3481	1965	2034	5374	692	5819	706	7614	61,928
IMS	7	609	827	1241	282	730	1115	689	800	271	396	891	60	780	53	1243	9984
Tenjin Core	8	2600	2651	2988	575	2314	4458	2152	5424	1191	1458	2749	213	2904	187	4680	36,545
Solaria Stage	9	1242	2762	2610	686	1864	3754	797	1954	1547	1114	2266	175	1912	141	3573	26,393
Shoppers Daiei	10	1555	1063	1907	413	1172	1620	903	1267	595	3050	1483	216	1449	89	2523	19,305
Tenjin South	11	1932	3452	4555	1013	3337	5136	1524	2303	1483	2548	4860	259	3925	351	6930	43,607
Tenjin North	12	496	516	644	253	366	505	166	417	142	410	804	433	355	27	1271	6804
Daimyo	13	766	1554	1775	714	2649	3641	798	939	842	967	2233	137	5348	541	4074	26,977
Imaizumi	14	69	316	280	45	240	324	56	87	56	81	329	16	367	246	369	2882
Other districts in the Fukuoka city center	15	6474	10,259	12,294	1927	7052	8745	2994	6419	3155	5720	10,999	760	8113	1164	92,125	178,200
a Entry visits		15,091	13,450	18,699	2624	13,450	19,027	2953	9842	8201	8201	16,403	2624	10,826	984		142,377
b Shop-around visits (excluding sojourn)		22,968	42,882	47,757	11,544	36,050	51,649	17,340	27,199	14,941	24,214	43,786	4444	40,900	4391		390,065
c Sojourn visits		3506	4065	11,023	751	5045	13,464	689	5424	1547	3050	4860	433	5348	246		59,449
d Shop-around visits		26,474	46,947	58,779	12,295	41,095	65,113	18,029	32,623	16,488	27,264	48,646	4877	46,248	4637		449,514
e Total visits		41,564	60,397	77,479	14,919	54,546	84,140	20,981	42,465	24,690	35,465	65,049	7501	57,074	5621		591,891
Shop-around ratio (b/a)		1.522	3.188	2.554	4.399	2.680	2.714	5.873	2.764	1.822	2.952	2.669	1.693	3.778	4.462		2.740
Shop-around ratio (c/a)		0.232	0.302	0.589	0.286	0.375	0.708	0.233	0.551	0.189	0.372	0.296	0.165	0.494	0.250		0.418
Shop-around ratio (d/a)		1.754	3.490	3.143	4.685	3.055	3.422	6.106	3.315	2.010	3.324	2.966	1.858	4.272	4.712		3.157
Shop-around ratio (e/a)		2.754	4.490	4.143	5.685	4.055	4.422	7.106	4.315	3.010	4.324	3.966	2.858	5.272	5.712		4.157
Total visits (2004 − 2003)		−3139	−1683	14,702	3175	10,618	16,448	−1108	2289	6422	−22,116	12,948	−10,030	−4381	−1358		22,788

Table 4.7 Estimated results of entry visits, shop-around visits, and total visits (2005)

Node	1	2	3	4	5	6	7	8	9	10	11	12	13	14	15	Total
Tenjin underground shopping mall	1	16,179	14,488	13,753	3473	5975	10,967	4164	6767	4152	9028	8255	1496	6012	247	17,274	122,230
Fukuoka *Mitsukoshi*	2	7919	10,424	5800	2047	3063	5405	1498	3061	1544	1856	5034	476	4163	290	9791	62,373
Daimaru Tenjin Fukuoka	3	6582	5230	12,780	1274	2948	3787	2516	2644	1155	1522	4390	335	2648	91	9195	57,098
Iwataya New Annex	4	1380	1189	1199	1056	1310	920	258	586	298	330	952	78	948	86	1545	12,135
Iwataya Main Building	5	4412	4330	3880	2232	5964	4457	1165	2302	1100	1196	4515	387	4123	110	6266	46,438
Solaria Plaza	6	4561	4272	3560	1203	3531	8796	1129	2598	1257	1395	4078	461	4538	180	7999	49,557
IMS	7	1650	1228	1292	240	634	1038	1291	846	458	376	885	78	705	22	1339	12,082
Tenjin Core	8	2126	1532	1271	357	672	1621	588	2427	736	674	1526	121	794	29	2861	17,336
Solaria Stage	9	2384	2313	1617	676	1125	2806	823	1326	826	597	1717	142	1627	50	3002	21,031
Shoppers Daiei	10	4609	1716	1704	504	836	1328	593	911	462	7441	1262	879	761	31	2980	26,018
Tenjin South	11	4501	3728	3907	898	2117	3782	1280	2256	1012	1667	4108	418	2688	74	7726	40,164
Tenjin North	12	571	282	365	72	151	389	94	155	80	481	570	51	175	6	894	4335
Daimyo	13	2966	2338	2249	805	2713	2982	694	1279	1032	1101	2437	361	5268	141	4786	31,152
Imaizumi	14	600	368	229	62	122	198	67	120	70	124	167	24	265	112	763	3291
Other districts in the Fukuoka city center	15	18,326	12,048	13,646	2994	6049	9734	3731	7339	3929	5673	11,225	1864	8586	231	121,962	227,337
a Entry visits		31,702	16,983	15,568	3397	13,587	14,719	2831	4246	5661	5944	13,587	1698	11,605	849		142,377
b Shop-around visits (excluding sojourn)		62,586	55,062	54,474	16,835	31,248	49,415	18,600	32,190	17,284	26,021	47,014	7119	38,032	1589		457,469
c Sojourn visits		16,179	10,424	12,780	1056	5964	8796	1291	2427	826	7441	4108	51	5268	112		76,723
d Shop-around visits		78,765	65,486	67,254	17,891	37,212	58,211	19,891	34,617	18,110	33,461	51,122	7170	43,300	1701		534,192
e Total visits		110,467	82,470	82,822	21,288	50,799	72,930	22,722	38,863	23,771	39,406	64,709	8868	54,905	2550		676,569
Shop-around ratio (b/a)		1.974	3.242	3.499	4.956	2.300	3.357	6.571	7.582	3.053	4.378	3.460	4.192	3.277	1.871		3.213
Shop-around ratio (c/a)		0.510	0.614	0.821	0.311	0.439	0.598	0.456	0.572	0.146	1.252	0.302	0.030	0.454	0.132		0.539
Shop-around ratio (d/a)		2.485	3.856	4.320	5.267	2.739	3.955	7.027	8.153	3.199	5.629	3.763	4.222	3.731	2.003		3.752
Shop-around ratio (e/a)		3.485	4.856	5.320	6.267	3.739	4.955	8.027	9.153	4.199	6.629	4.763	5.222	4.731	3.003		4.752
Total visits (2005 − 2004)		68,903	22,073	5343	6368	−3747	−11,210	1740	−3602	−918	3941	−340	1367	−2169	−3071		84,678

In Tables 4.8, 4.9, and 4.10, we give the results of the analysis when only three department stores are extracted and analyzed in the same way as Tables 4.5, 4.6, and 4.7, but in order to directly evaluate the shop-around among these three different department stores, calculations are made excluding for sojourn, that is, the shop-around movements within the same department store.

The main findings from these analyses will be discussed in Sects. 4.1 and 4.2.

Table 4.8 Estimated results of entry visits, shop-around visits, and total visits among department stores (2003)

	Node	1	2	3	4	Total
Fukuoka *Mitsukoshi*	1		6161	1012	4440	11,614
Daimaru Tenjin Fukuoka	2	3909		426	2385	6720
Iwataya A-side	3	1097	1211		916	3224
Iwataya Z-side	4	1824	1623	215		3662
a Entry visits		24,109	15,896	3456	8755	52,216
b Shop-around visits (excluding sojourn)		6830	8995	1653	7742	25,220
c Sojourn visits		6007	8484	876	2685	18,053
d Shop-around visits		12,837	17,479	2530	10,427	43,273
e Total visits		36,946	33,375	5986	19,182	95,489
Shop-around magnification (b/a)		0.28	0.57	0.48	0.88	0.48
Shop-around magnification (c/a)		0.25	0.53	0.25	0.31	0.35
Shop-around magnification (d/a)		0.53	1.10	0.73	1.19	0.83
Shop-around magnification (e/a)		1.53	2.10	1.73	2.19	1.83

Table 4.9 Estimated results of entry visits, shop-around visits, and total visits among department stores (2004)

	Node	1	2	3	4	Total
Fukuoka *Mitsukoshi*	1		4209	731	2865	7805
Daimaru Tenjin Fukuoka	2	5396		851	3177	9424
Iwataya New Annex	3	642	869		1455	2965
Iwataya Main Building	4	2439	3508	1611		7558
a Entry visits		13,450	18,699	2624	13,450	48,223
b Shop-around visits (excluding sojourn)		8477	8586	3193	7497	27,753
c Sojourn visits		4065	11,023	751	5045	20,883
d Shop-around visits		12,541	19,608	3944	12,542	48,636
e Total visits		25,991	38,307	6568	25,992	96,859
Shop-around magnification (b/a)		0.63	0.46	1.22	0.56	0.58
Shop-around magnification (c/a)		0.30	0.59	0.29	0.38	0.43
Shop-around magnification (d/a)		0.93	1.05	1.50	0.93	1.01
Shop-around magnification (e/a)		1.93	2.05	2.50	1.93	2.01

Table 4.10 Estimated results of entry visits, shop-around visits, and total visits among department stores (2005)

	Node	1	2	3	4	Total
Fukuoka *Mitsukoshi*	1		5800	2047	3063	10,910
Daimaru Tenjin Fukuoka	2	5230		1274	2948	9452
Iwataya New Annex	3	1189	1199		1310	3698
Iwataya Main Building	4	4330	3880	2232		10,442
a Entry visits		16,983	15,568	3397	13,587	49,535
b Shop-around visits (excluding sojourn)		10,749	10,879	5552	7322	34,503
c Sojourn visits		10,424	12,780	1056	5964	30,224
d Shop-around visits		21,174	23,659	6608	13,286	64,727
e Total visits		38,157	39,227	10,005	26,873	114,262
Shop-around magnification (b/a)		0.63	0.70	1.63	0.54	0.70
Shop-around magnification (c/a)		0.61	0.82	0.31	0.44	0.61
Shop-around magnification (d/a)		1.25	1.52	1.95	0.98	1.31
Shop-around magnification (e/a)		2.25	2.52	2.95	1.98	2.31

4.1 Interannual Comparisons of Visits, Shop-Around Visits, and Total Visits

1. *Interannual comparison of total visits and shop-around visits between 2003 and 2004*

    (a) The "total visits" increased at *Daimaru*, *Iwataya* New Annex, *Iwataya* Main Building, Solaria Plaza, Tenjin Core, Solaria Stage, and the southern part of Tenjin district (Tenjin South) from 2003 to 2004 (Tables 4.5 and 4.6).
    (b) The "shop-around visits" increased at Fukuoka *Mitsukoshi*, *Daimaru*, *Iwataya* New Annex, *Iwataya* Main Building, Solaria Plaza, IMS, Solaria Stage, and Tenjin South from 2003 to 2004 (Tables 4.5 and 4.6).

2. *Interannual comparison of total visits and shop-around visits between 2004 and 2005*

    (c) The "total visits" increased at all commercial nodes except for *Iwataya* Main Building, Solaria Plaza, Tenjin Core, Solaria Stage, Tenjin South, *Daimyo*, and *Imaizumi* from 2004 to 2005 (Tables 4.6 and 4.7).
    (d) The "shop-around" visit frequencies increased at all commercial area nodes except for *Iwataya* Main Building, Solaria Plaza, *Daimyo*, and *Imaizumi* from 2004 to 2005 (Tables 4.6 and 4.7).

Look at the total numbers of visitors in the Tenjin district from 2003 to 2004. They increase from 569,103 persons or visits per day to 591,891 visits per day (1.04 times). Whereas the increase of total visits was 22,788 people per day (1.04 times) from 2003 to 2004, the increase from 2004 to 2005 was from 591,891 to 676,569 people per day (1.14 times), which is the much larger increase of 84,678 people than the previous year.

4.2 Interannual Comparisons of Visits, Shop-Around Visits, and Total Visits Among Department Stores

Here we focus on the shop-around among three department stores. Tables 4.8, 4.9, and 4.10 extracted the shop-arounds among department stores from the respective Tables 4.5, 4.6, and 4.7. In this study, we try to evaluate the redevelopment projects from the perspective of how those redevelopment projects stimulate the *Kaiyu* behaviors within the city center. Here we restrict the area surrounded by three department stores and investigate how each redevelopment enhances the *Kaiyu* behaviors within this area.

By comparing Tables 4.8 and 4.9, we are able to evaluate the effect of the *Iwataya* project on the change of *Kaiyu* flow. Also, we can examine the effect of the extension of the underground shopping mall by comparing Tables 4.9 and 4.10. Main findings are as follows:

1. *Interannual comparison of total visits and shop-around visits among department stores between 2003 and 2004*

    (a) The "total visits" increased at all commercial nodes except for Fukuoka *Mitsukoshi* from 2003 to 2004.
    (b) The "shop-around visits" increased at all commercial nodes except for Fukuoka *Mitsukoshi* from 2003 to 2004.
    (c) The number of people moving between two stores of *Iwataya*, i.e., between the former A-side and Z-side in 2003 and between the former Z-side and the new annex store in 2004, significantly increased from 2003 to 2004.

2. *Interannual comparison of total visits and shop-around visits among department stores between 2004 and 2005*

    (d) The "total visits" increased at all commercial nodes from 2004 to 2005.
    (e) The "shop-around visits" increases at all commercial nodes from 2004 to 2005.

First we note the change of the number of shop-arounds within the two stores of *Iwataya*. The relocated renewal opening of *Iwataya* New Annex can be thought of as the movement of the former A-side toward the much closer place to Z-side. Thus, this movement is expected to enhance *Kaiyu* between these two stores. In fact, a significant increase of the number of shop-arounds among these two stores is observed from 2003 to 2004. The number of people moving from the former A-side (new annex in 2004) to the *Iwataya* Main Building (the former Z-side) increased from 916 per day to 1455 per day. Similarly, the number of that from the *Iwataya* Main Building (the former Z-side) to the former A-side (new annex in 2004) increased from 215 per day to 1611 per day. Hence, we see that the effect, which might be called as the proximity effect, had clearly been observed (Tables 4.8 and 4.9).

Changing our attention to the total number of visitors among all department stores from 2003 to 2004, there was an increase from 95,489 people per day to 96,859 people per day, an increase of 1370 (1.01 times), whereas from 2004 to 2005, it was 114,262 people per day, an increase of 17,403 people (1.18 times).

Therefore, while the development due to the relocation and opening of *Iwataya* had an impact on just neighboring commercial facilities, it can be said that the impact of the underground shopping mall development has been spread much wider.

4.3 Evaluating the Development Projects by Kaiyu *Index*

Finally, we evaluate development projects based on a "*Kaiyu* index." Here, the "*Kaiyu* index" is the ratio of dividing the number of shop-around visits excluding sojourn (staying) by the number of entry visits.

This index represents how many shop-around visits one entry visit would bring about by consumer shop-around behaviors. The *Kaiyu* index for each department store is shown at the row expressed as (b/a) in Tables 4.8, 4.9, and 4.10.

Look at the columns of total in these tables. The *Kaiyu* indices among three department stores for 2003, 2004, and 2005 are 0.48, 0.58, and 0.70, respectively.

Thus, we see that the both projects had the effect of vitalizing *Kaiyu* among three department stores.

5 Conclusion

In this study, we examine how the extension of underground shopping mall vitalized *Kaiyu* within city center retail environment as well as the relocated renewal opening of the long-established local department store. We evaluate these projects by relying on the concept of shop-around effect derived by the shop-around Markov model.

We found that the "shop-around effect" caused by one commercial facility to other facilities increased from 2004 to 2005 in almost all commercial facilities, whereas just some commercial facilities made a positive shop-around effect to other districts from 2003 to 2004. Also, both "shop-around visits" and "total visits" from 2004 to 2005 increased significantly compared to 2003–2004. A similar tendency was also observed in the same analysis of only department stores.

From the above, we conclude the development of the underground shopping mall had spread over a wider area and had a greater impact on vitalizing *Kaiyu* within the city center of Fukuoka City than the relocated renewal opening of the long-established local department store.

References

1. Saito S (1988) Assessing the space structure of central shopping district viewed from consumers' shop-around behaviors: a case study of the midtown district of Saga City. Fukuoka Univ Econ Rev 33:47–108. (in Japanese)
2. Saito S, Ishibashi K (1992) Forecasting consumer's shop-around behaviors within a city center retail environment after its redevelopments using Markov chain model with covariates. Pap City Plan 27:439–444. (in Japanese)
3. Saito S, Kakoi M, Nakashima T (1999) On-site Poisson regression modeling for forecasting the number of visitors to City Center retail environment and its evaluation. Stud Reg Sci 29:55–74. (in Japanese)
4. Saito S, Nakashima T, Kakoi M (1999) Identifying the effect of city center retail redevelopment on consumer's shop-around behavior: an empirical study on structural changes of City Center at Fukuoka City. Stud Reg Sci 29:107–130. (in Japanese)
5. Saito S, Kakoi M, Nakamura R, Iwami M (2007) Evaluation of underground shopping mall development from the viewpoint of vitalizing *Kaiyu* within a city center: a case of Tenjin district of Fukuoka City, Papers of the 23rd Annual Meeting of The Japan Association for Real Estate Sciences, vol 23, pp 91–96. (in Japanese)

Part II
Some Characteristics of *Kaiyu*

Chapter 5
Occurrence Order of Shop-Around Purposes

Saburo Saito

Abstract Recent studies have been paying much attention on trip-chaining behavior. However, shop-around trip-chaining behavior at a midtown district has not yet been fully explored. In particular, there is a lack of empirical facts about its purpose transition. Based on the citizen survey conducted by Saga City Government, this research analyzes how many purpose transitions occur before quitting and constructs ratio scale to account for what type of purpose is preferred in what steps of shop-around. Findings are quitting rate of shop-around increases as steps proceed; purpose transition probability is non-stationary; and there are two kinds of purposes which are more likely to occur early or later.

Keywords Shop-around · *Kaiyu* · Purpose transition · Occurrence order · Saga City · Positive reciprocal matrix · Maximum eigenvalue · Ratio scale · Decomposition · Trip chain · Quitting rate

1 The Purpose of the Study

In this study, I use the term shop-around behavior to refer to consumers' behaviors of walking around multiple commercial areas or multiple areas within a downtown space.

This chapter is based on the paper, Saburo Saito [17] "Duration and Order of Purpose Transition Occurred in the Shop-around Trip Chain at a Midtown District," *Papers on City Planning*, No.23, pp. 55–60, 1988, in Japanese with English abstract. Small modifications are made at this time for this chapter.

S. Saito (✉)
Faculty of Economics, Fukuoka University, Fukuoka, Japan

Fukuoka University Institute of Quantitative Behavioral Informatics for City and Space Economy (FQBIC), Fukuoka, Japan
e-mail: saito@fukuoka-u.ac.jp

So far, there have been several studies related to consumers' shop-around behaviors in downtown commercial areas. They can be broadly divided into three streams.

The first is the study on the pedestrian flows at the street network level in downtown commercial areas [3–5]. The second is the analysis of transportation behavior of trip chaining, which is a relatively new research area [6]. The third is the research related to disaggregation and expansion of the Huff model [8–10, 12, 14, 16].

The study on the pedestrian flows can be said to have been focusing on the prediction of pedestrian flows in front of stores since they play a critical role in commercial area planning. Therefore, with respect to the mechanism of pedestrian's decision-making, this study stream has no intention for clarifying the decision-making process of behavioral entity fully but just deals with pedestrian's decision-making only related to the level of street choice.

On the other hand, both the study of a trip chaining and the disaggregate Huff model have an intention to clarify the decision-making mechanism by behavioral entity. If we dare to state a difference between them, studies on trip chains aim to elucidate the structure of daily, weekly, and yearly travel behavior and contribute to the transportation planning with a wider target area such as metropolitan areas. In contrast, studies on the disaggregate Huff model put an emphasis on trip chains in a downtown district to evaluate the downtown space structure and commercial area planning.

This study arises from the research on the disaggregate Huff model and aims to elucidate consumer shop-around behaviors, which are trip-chaining behaviors in a downtown district. There are several previous studies focusing on consumer shop-around behaviors in a downtown district. Saito [9, 10] developed a disaggregated Huff model which incorporates consumers' multistage choices concerning which city they choose to go shopping, which shops, large retail stores or small shops, they choose to shop at, and whether they do shop-around. Traditionally, Huff model is concerned with only consumer's shopping destination choice. Thus the paper by Saito [9, 10] extended the traditional Huff model into a disaggregated Huff model incorporating consumer's multistage choices as well as consumer's shop-around behavior. While Saito [9, 10] considered just one-step shop-around from large retailers to small retailers, Saito [8] also developed an infinite step shop-around model, in which the consumer shop-around behavior is formulated as an absorbing stationary Markov chain and defined the shop-around effect as the visit frequency to each shopping site by consumer shop-around behaviors. He proposed to employ it as the criteria to evaluate the downtown commercial space structure. The model is reported in Sakamoto [14]. However, these studies do not deal with purpose transitions as shop-arounds proceed. While Saito [9, 10] introduces non-homogeneity in the sense that the model is disaggregate and the transition probability between commercial areas varies depending on personal attributes, both are not non-stationary models in which the transition probability changes as a shop-around goes through.

On the other hand, though the traditional Huff model has focused on a shopping trip which is generated under the main purpose of purchasing, Takeuchi [16] dealt with shopping trips which are generated under other main purpose than purchasing, such as working and schooling, and modified the traditional Huff model that can consider a shopping trip where the purpose of purchasing occurs as a secondary purpose. The modified Huff model gave an improved prediction of turnovers for some retail sector by using the probability of simultaneous occurrence of main and secondary behavior purposes. Although the attention was paid on the purposes of shopping trip here, Takeuchi did not deal with how the behavior purpose changes while people shop-around.

In general, shop-around behaviors consist of choosing behavior purposes and selecting commercial areas as destinations. Further, since all people finally go back home in shop-around, how people choose purposes and destinations necessarily changes as they proceed their shop-arounds. Thus studies so far have drawbacks in that they either ignore the purpose transitions or assume the stationary probability of transition between commercial areas and quitting shop-around.

When focusing on decision-making for shop-around, it is natural to assume that people first choose a purpose and then decide where to go. Then the fact that people change their purposes as they go through shop-around should be the main cause to change the choice probability among commercial areas and to make the transition probability non-stationary.

Hence, if we consider a shop-around behavior as the derived demand for carrying out some chosen purpose, it becomes important to know how people change their purposes as they go through shop-around for understanding the mechanism of consumer shop-around behaviors.

From the above, this study aims to detect the regularity in the order of occurrence of shop-around purposes based on data of citizens' shop-around behaviors at the downtown space of Saga City, Japan, obtained from Saga City Citizen Opinion Survey conducted by Saga City Government.

2 Method

2.1 Survey Data

The data used in this study is obtained from Saga City Citizen Opinion Survey conducted by Saga City Government in February 1986, which is a questionnaire survey of citizen's opinions and behaviors for randomly sampled 3100 citizens distributed and collected by mail. The number of collected samples is 1453 samples, which is 47.5% recovery rate (Cf. [11]). The questionnaire items include the question item asking the respondents about their behavior when they visited the downtown district of Saga City. The number of valid samples for this question item collected is 918 samples. The question item uses the recall method to ask the respondents about their most recent visit to the downtown district of Saga City, excluding for the purpose of commuting to work or school. In the question item, we provided the list of 22 behavior purposes as shown in Table 5.1 and the list of

Table 5.1 Shop-around purposes

Purpose number	Behavior purposes
1	Shopping for clothing
2	Shopping for daily necessities
3	Shopping for grocery
4	Shopping for furniture, electric appliances
5	Shopping for hobby items (stationary, sport equipment, books, music instruments, cameras, etc.)
6	Having meals
7	Drinking (alcohol)
8	Coffee/tea
9	Entertainment
10	Movie
11	Attending cultural classes
12	Joining events
13	Going to government offices
14	Going to banks
15	Walking/jogging
16	Sending someone
17	Meeting someone by appointment
18	Passing by (by walking, by bicycle)
19	Working/schooling
20	Parking getting on and off
21	Other purposes
22	Departing/returning home

22 partitioned areas (places) in the downtown and had the respondents record their behavior during that visit by picking up the places they visited and purposes done there from the above lists in the order of their occurrence.

2.2 Definition of Terms Used in the Analysis

We give the definitions of the terms used in the analysis here.

The term "trip" is defined as referring to the entire course of travel in which people start their shop-around trip, visit several places, and end their trip by returning home. It refers to a loop-shaped home-to-home trip visiting other places than home, that is, a multi-stop, multipurpose linked trip.

The term "walk around" is roughly referred to as moving around areas in the question item. To give the more precise definition, the question item is designed to define that "walk around" occurred when either changes in visited places or changes in purposes done there occurred. This is because we would like to define the "shop-around" or "walk around" as happened not only when the places visited changed but also when the purposes changed even in the same place.

The term "step" in the "walk around" is defined as a ranked random variable, which identifies the order of occurrence of the event of "walk around" in such a way as referring to the first occurrence of "walk around" as the first step, the second as the second step of "walk around," and so on. If a certain step in the "walk around" event involves a change of the place given the same purpose, "purpose staying" occurs, and if the purpose changes given the same place, "staying" in that place is considered to occur.

The concept of "purpose transition step" is obtained if we limit the "walk around" event to purpose change. In this case, from the definition of "walk around," the probability of transition to the same purpose becomes zero, which turns out to be a Markov-type transition probability excluding "purpose staying." All of the following analyses are Markov-type analyses limited only to purpose change events.

The term "entrance area" is defined by the first destination or the first visited place in the downtown district. The definition of the first destination is the place where some behavior purpose other than "passing by" and "commuting to work or school, and so on," has occurred for the first time. This is because we would like to give the entrance area the characteristics of the first destination. Thus even if some place is actually the first entry place to the downtown district, the place is not regarded as the entrance area if the purpose to enter the place is just "passing by."

The term "exit area" is defined as the final destination just like the entrance area. The final destination refers to the place visited last with some behavior purpose other than "passing by" and "commuting to work or school, and so on."

The "shop-around route" is defined as a sample path of walking around from the entrance area to the exit area, which is the sequence of visited places by sample's shop-around.

The "shop-around step" is the step in the "walking round" that counts the number of steps from the entrance area which is set to 0th step. That is, it equals the number of steps excluding the visit to the entrance area. If the event is restricted to the purpose change, the step is defined as the "shop-around purpose (transition) step." As we are concerned only with the purpose change below, we also call this as the shop-around step.

Let $X_{ijk}(t)$ denote a random variable, which takes the value 1 when sample i changes its purpose from purpose j to purpose k in the t-th shop-around step and takes the value 0, otherwise. The matrix $N_{jk}(t)$ formed by summing $X_{ijk}(t)$ over all samples is referred to as the "association matrix of shop-around purpose transitions."

Among the behavior purposes listed in Table 5.1, the purposes of "passing by" and "parking" are referred to as derived behavior purposes. These purposes are generated as means to carry out other purposes and may arise as a consequence of modal choice and the spatial structure of the downtown district. The behavior purposes from 1 to 17 in the table are called *non-derived behavior purposes*. We analyze the purpose changes limiting to those among non-derived purposes.[1]

2.3 Procedure for the Analysis

We wish to analyze how the selection of purposes changes as the shop-around steps proceed.

For the purpose, first, we examine the ways of how the probability of returning home (or the probability to continue the shop-around) changes as the shop-around steps proceed and whether they differ depending on the trip characteristics.

Second, when the probability of returning home changes, it is apparent that the transition probability among all purposes becomes non-stationary. But, we examine whether or not the transition probability among shop-around purposes excluding returning home is stationary since it is unknown.

Finally, there might be some shop-around purposes more likely to occur in the early steps of the shop-around. Also there might be some purposes more likely to occur in the later steps of the shop-around. We would like to compare the behavior purposes by exploring which purpose is more likely to be selected than other purposes at each step of shop-arounds. To do this, we construct a ratio scale to measure the order of purpose occurrence from the observed association matrix of shop-around purposes and analyze how it changes as the shop-around steps proceed.

---

[1]The classification of derived and non-derived purposes is derived from focusing on the fact that derived purposes occur as a means to carry out other purposes. In other words, it tries to distinguish between a behavior where the behavior itself is performed as the purpose and a behavior where the behavior is performed as the means to carry out other purposes. It is the same as the distinction between discretionary and mandatory purposes. While purposes such as "picking up" someone or going to meet someone by "appointment" can occur by themselves even if there are no other planned purposes, the purposes of "parking" and "passing by" are assumed to have some other planned purposes. Hence, they are classified as non-derived and derived purposes, respectively.

The difference between the analysis of all purposes and the analysis restricted to the derived purposes is easy to understand through the following example. Assume that shop-around involves the route such as purpose 1 → passing by → purpose 1. In this case, as for the analysis limited to the non-derived purposes excluding "passing by," the number of shop-around steps does not increase and stays equal to one because purpose 1 did not change since the purposes before and after "passing by." are the same as purpose 1 so that no purpose transition is considered to occur.

In contrast, as for all purposes, the purpose transitions occurred two times, from purpose 1 to "passing by" and from "passing by" to purpose 1 so that the number of shop-around steps increases by 2. If one is concerned with how many different discretionary purposes occur in one trip, it is necessary to conduct the analysis limited to non-derived purposes. The reason for including the purpose of "passing by" is to apply this purpose to the evaluation of downtown spatial structure [12].

5   Occurrence Order of Shop-Around Purposes

3   Stationarity of the Transition Probability of Shop-Around Purpose

3.1   Purpose Transition Probability by Shop-Around Steps

In the current data, the observed maximum number of shop-around purpose transition steps is 12. It is not a good idea to directly investigate all the cells in the transition probability matrix at each step of shop-arounds because the number of cells becomes quite large and the number of 0 cells increases. Therefore, focusing on the rows or origins of the purpose transition matrix, we picked up three purposes, "shopping for clothing," "having meals," and "passing by," and checked how these three purposes are transitioned to other purposes at each step of shop-arounds. Figure 5.1 depicts the result. Figure 5.1a–c, respectively, corresponds to the origin purposes, "shopping for clothing," "having meals," and "passing by." Line graphs in each figure represent the transition probabilities from the given origin purpose to the 22 destination purposes on the horizontal axis, respectively, by shop-around step. Look at Fig. 5.1a. As regards "shopping for clothing," the transition probabilities to "shopping for grocery," "having meals," "passing by," and "parking" are high, and the transition probability of "returning home" also becomes higher after the third step vis-à-vis the first and second steps. As regards "having meals," the transition probabilities to "shopping for hobby items," "shopping for grocery," "passing by," and "parking" are high, and the probability of "returning home" increases from the second step onwards. In Fig. 5.1c, when "passing by" is the origin purpose, from the definition of the exit area whose next purpose is fixed "returning home," the probability of "returning home" becomes zero. However, the transition probability to "parking" increases sequentially as the step proceeds, which can be considered the preparation for returning home.

3.2   Analysis of the Probability of "Returning Home"

Let us apply the concept of survival analysis (Cf. [2, 15]). Considering time t as a shop-around step, the survival function $S(t)$ in the survival analysis is the ratio of people who are still shop-around at step t. We call it the shop-around continuing function.

The hazard $h(t)$ corresponds to the "returning home" probability at step t. As $h(t) = -d \log S(t)/dt$, it is possible to know whether the probability of "returning home" is stationary by checking whether h is a constant or $-\log S(t)$ is linear with respect to t.

Figure 5.2 depicts the clustered column chart of shop-around continuing functions for all and for non-derived purposes. Of course, the shop-around continuing function of all purposes is higher than that of non-derived purposes because it includes purposes of "passing by" and "parking." Further, as is evident for

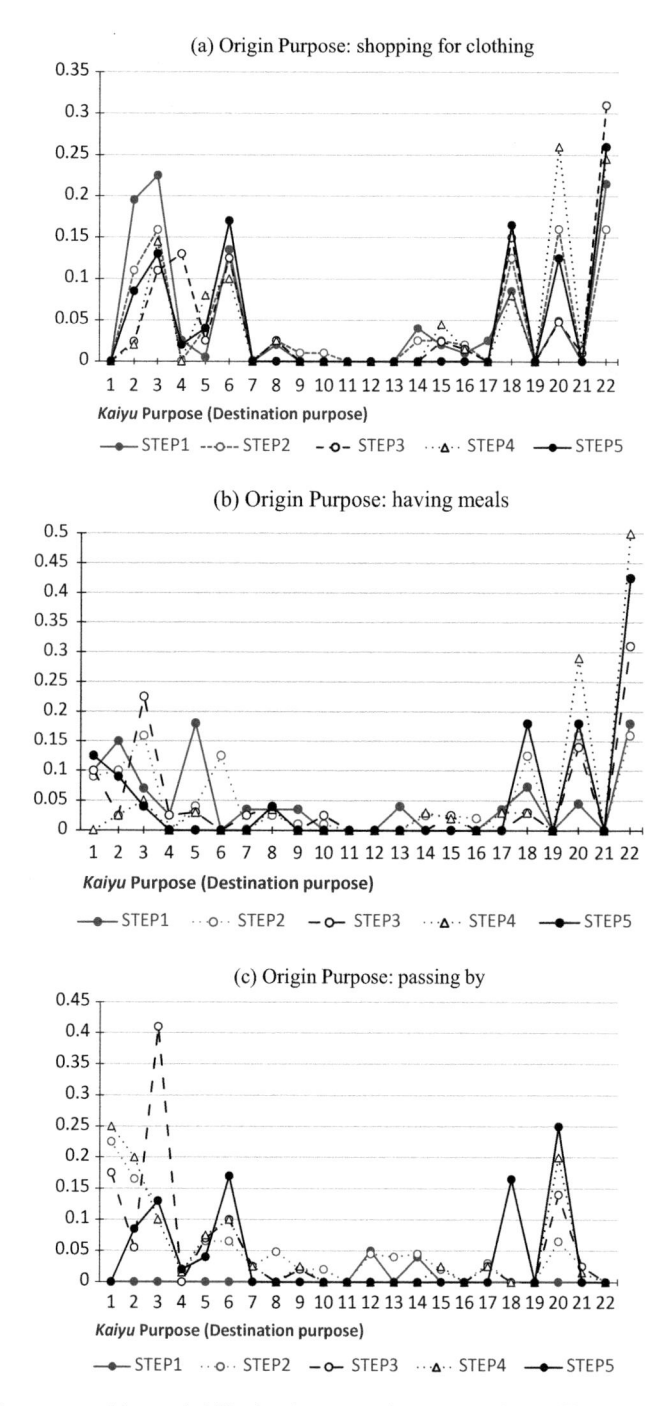

Fig. 5.1 Purpose transition probability by shop-around step under three different origin purposes, (**a**) shopping for clothing, (**b**) having meals, and (**c**) passing by

5 Occurrence Order of Shop-Around Purposes

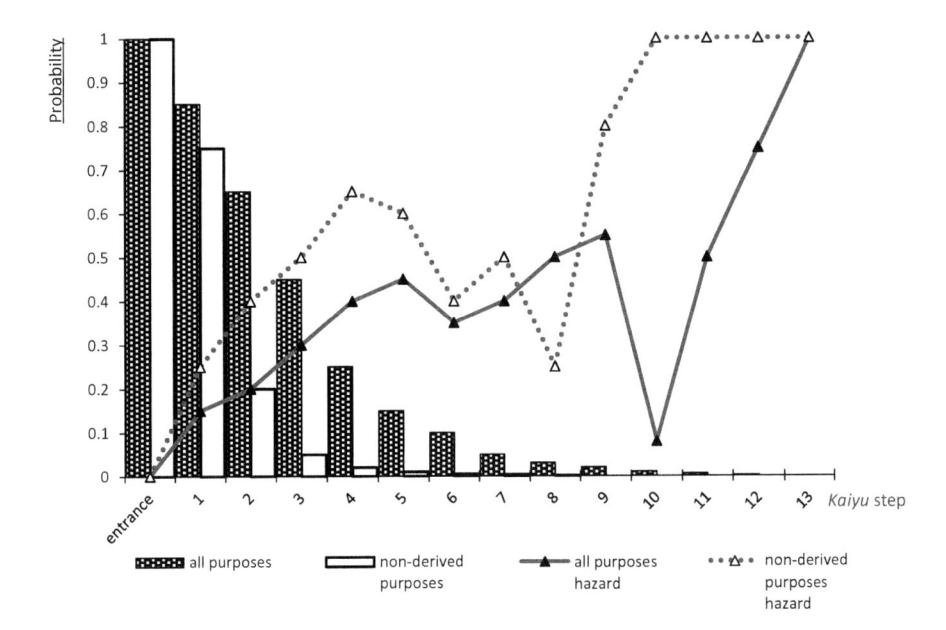

Fig. 5.2 Shop-around continuing functions (all purposes and non-derived purposes) with hazard functions

non-derived purposes, the hazard abruptly increases in the second step. Even for all purposes, hazards increase gradually, which indicates that the probability of "returning home" increases as the shop-around step increases.

Next, we compared the shop-around continuing functions between different trip characteristics. Figure 5.3 compares the shop-around continuing functions between two groups classified by travel modes, using private cars and buses and others. As for the case of all purposes, the shop-around continuing functions for the two groups are significantly different since the purpose of "parking" is included. Therefore, when limited to non-derived purposes, no significant difference was found. The trip characteristics which led to significant differences when limited to non-derived purposes were whether the arrival time of the trip is in the morning or in the afternoon and whether the trip is made on weekday or on weekend[2] (Figs. 5.4 and 5.5).

---

[2]The Wilcoxon test of the homogeneity of stratified survival functions was performed. The results are shown here. As the Wilcoxon statistic compares the two survival functions, it becomes a chi-square distribution with one degree of freedom. To show the results as (χ^2-value:p-value), they become by mode (43.01: 0.0001), by arrival time (4.20: 0.0404), and by weekday/weekend (6.83: 0.0090).

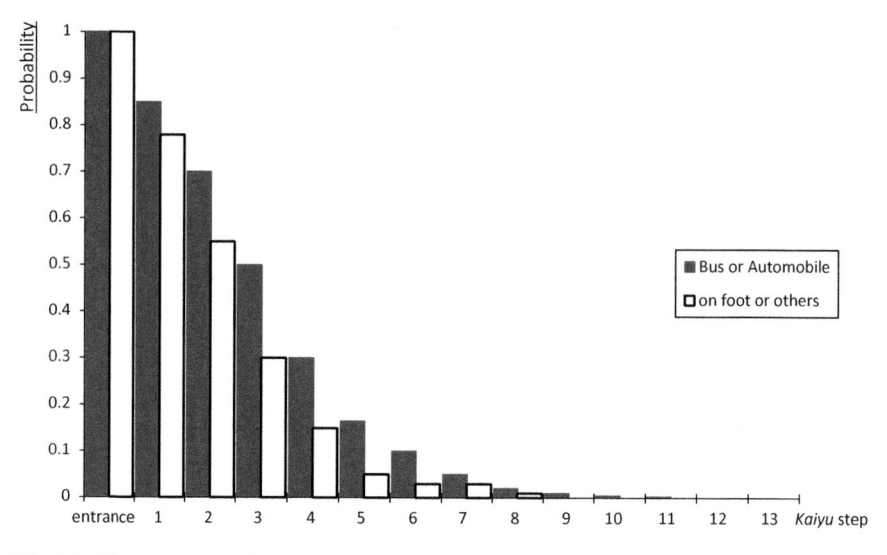

Fig. 5.3 Shop-around continuing function by mode (all purposes)

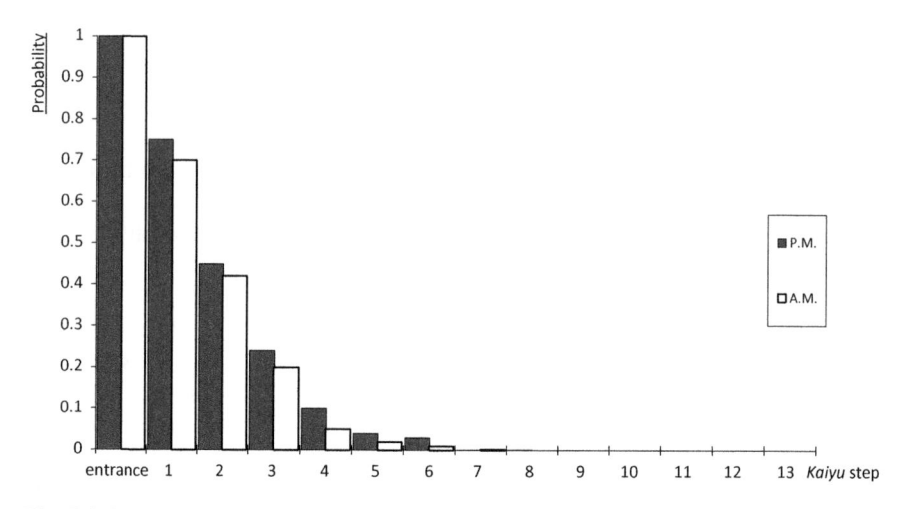

Fig. 5.4 Shop-around continuing function by arrival time (non-derived purposes)

3.3 Stationarity of Purpose Transition Probability Excluding the Purpose of "Returning Home"

As for the two cases of all purposes and non-derived purposes, we carry out the likelihood ratio test for testing the stationarity of purpose transition probability with

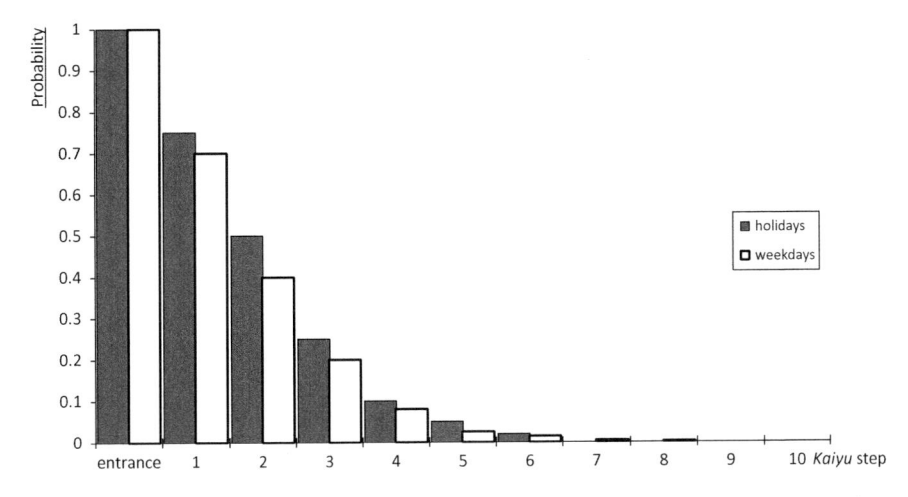

Fig. 5.5 Shop-around continuing function by weekdays and weekends (non-derived purposes)

excluding the purpose of "returning home." In both cases, the stationarity is rejected at a significance level of 0.1%.[3]

4 Analysis of the Occurrence Order of Shop-Around Purpose by Ratio Scale Decomposition Method

4.1 Shop-Around Purpose Distribution by Shop-Around Steps

From the above results, the "returning home" probability and the purpose transition probability excluding "returning home" were both non-stationary. If there are purposes that become more likely to be selected and those that become less likely to be selected as the shop-around steps proceed, those purposes can be considered as the main cause for the purpose transition probability to become non-stationary.

In Table 5.2, we give the frequency table of the occurrence of 17 non-derived shop-around purposes by shop-around step for all samples. The bottom row represents the shop-around continuing function of 17 non-derived shop-around purposes. From this table, we see that "shopping for clothing" has the highest frequency at the entrance step. In contrast, purposes, such as "shopping grocery," "having meals,"

---

[3]Refer to Amemiya ([1], p. 417) for the method of the likelihood ratio test. The shop-around steps from the sixth step and above is collapsed into one step, and the degree of freedom is calculated by deducting the number of random zeroes, which gives DF $= 515$ and $\chi^2 = 92$ for all purposes excluding "returning home," and DF $= 418$ and $\chi^2 = 831$ for non-derived purposes, with $p < 0.001$ for both.

Table 5.2 Frequency of occurrence of purpose by shop-around step (non-derived purposes). For each row of *Kaiyu* purpose, the upper row shows frequency and the lower row percentage

		Kaiyu Step									
		Entrance	1 step	2 step	3 step	4 step	5 step	6 step	7 step	8 step	9 step	Total
Kaiyu purpose	1	290	115	66	18	7	3	0	0	0	0	499
		58.12	23.05	13.23	3.61	1.40	0.60	0.00	0.00	0.00	0.00	100
	2	120	113	63	21	7	5	1	2	0	0	332
		36.14	34.04	18.98	6.33	2.11	1.51	0.30	0.60	0.00	0.00	100
	3	118	146	102	45	18	4	6	1	3	0	443
		26.64	32.96	23.02	10.16	4.06	0.90	1.35	0.23	0.68	0.00	100
	4	22	15	4	5	1	1	1	0	0	0	49
		44.90	30.61	8.16	10.20	2.04	2.04	2.04	0.00	0.00	0.00	100
	5	58	50	26	16	8	2	1	0	0	0	161
		36.02	31.06	16.15	9.94	4.97	1.24	0.62	0.00	0.00	0.00	100
	6	40	88	65	34	5	2	2	2	0	1	239
		16.74	36.82	27.20	14.23	2.09	0.84	0.84	0.84	0.00	0.42	100
	7	22	12	0	5	0	0	0	0	0	0	39
		56.41	30.77	0.00	12.82	0.00	0.00	0.00	0.00	0.00	0.00	100
	8	9	24	13	17	3	2	0	1	0	0	69
		13.04	34.78	18.84	24.64	4.35	2.90	0.00	1.45	0.00	0.00	100
	9	5	7	6	2	1	0	0	0	0	0	21
		23.81	33.33	28.57	9.52	4.76	0.00	0.00	0.00	0.00	0.00	100

5 Occurrence Order of Shop-Around Purposes

											Total
10	10	7	6	1	1	0	0	0	0	0	25
	40.00	28.00	24.00	4.00	4.00	0.00	0.00	0.00	0.00	0.00	100
11	7	3	2	1	1	0	0	0	0	0	14
	50.00	21.43	14.29	7.14	7.14	0.00	0.00	0.00	0.00	0.00	100
12	39	17	10	2	2	1	1	0	0	0	72
	54.17	23.61	13.89	2.78	2.78	1.39	1.39	0.00	0.00	0.00	100
13	35	12	5	4	1	1	1	0	0	0	59
	59.32	20.34	8.47	6.78	1.69	1.69	1.69	0.00	0.00	0.00	100
14	58	31	10	6	1	1	0	0	0	0	107
	54.21	28.97	9.35	5.61	0.93	0.93	0.00	0.00	0.00	0.00	100
15	11	7	9	13	3	2	1	0	1	0	47
	23.40	14.89	19.15	27.66	6.38	4.26	2.13	0.00	2.13	0.00	100
16	16	4	7	4	4	0	0	1	0	0	36
	44.44	11.11	19.44	11.11	11.11	0.00	0.00	2.78	0.00	0.00	100
17	46	16	5	6	4	1	0	0	1	0	79
	58.23	20.25	6.33	7.59	5.06	1.27	0.00	0.00	1.27	0.00	100
Total	906	667	399	200	67	25	14	7	5	1	2291
	39.55	29.11	17.42	8.73	2.92	1.09	0.61	0.31	0.22	0.04	100
Continuing function	1	0.736	0.44	0.221	0.074	0.028	0.015	0.008	0.006	0.001	

"drinking coffee," and "entertainment," have their peak after the entrance. Behavior purposes that consistently drop after the highest occurrence at the entrance step include shopping for "daily necessities," "hobby items," "movies," "cultural classes," "joining events," going to "government offices," "banks," and "appointments."

Even from this analysis, behavior purposes such as "grocery shopping," "meals," and "coffee" are seen to be likely to occur in the second half of shop-around steps. However, the result of this analysis is just the frequency table of occurrence of 17 non-derived shop-around purposes for each shop-around step. Thus it has the disadvantage of not utilizing the information of the occurrence order among shop-around purposes. The goal of this section is to make a scale that enables the comparison of purposes from their easiness to be selected at each shop-around step by utilizing this order information.

4.2 Method

Under the presumption that the result of pairwise comparison between n objects is represented by a positive reciprocal matrix, Saaty [7] derived a ratio scale construction method using the eigenvector corresponding to the Frobenius root as a ratio scale.

The main points are as follows. If an evaluator has the ratio scale $x = (w_1, \ldots, w_n)$ over n objects and consistently performs a pairwise comparison, a positive reciprocal matrix $W = (w_i/w_j)$ is obtained where w_i/w_j represents the i-th row and j-th column element of the matrix W. Here, the reciprocal matrix is referred to as the matrix $A = (a_{ij})$ that satisfies $a_{ij} = 1/a_{ji}$ for all elements. As for the case of W, it holds that rank $(W) = 1$ and its maximum eigenvalue and eigenvector are n and x, respectively, that is to say, $Wx = nx$.

From this fact, if we were given a positive reciprocal matrix W, which is a result of a pairwise comparison carried out by a consistent evaluator, the evaluator's ratio scale x would be recovered by the eigenvector of W corresponding to its maximum eigenvalue. As pairwise comparisons are not normally consistent, the maximum eigenvalue $\lambda(A)$ of the resultant positive reciprocal matrix A of the pairwise comparison is larger than n, and $m = (\lambda(A) - n)/(n - 1)$ is taken as a measure of consistency.

This is the main point of Saaty's [7] method. On the other hand, Saito [13] derives the following ratio scale decomposition theorem, which shows that the ratio scale of the entire group can be decomposed into those ratio scales contributed by group's constituents, from the result of the perturbation of a positive reciprocal matrix by the use of the reciprocal matrix.

Here let the elementwise product (Hadamard product) of the matrix denote by "*" and let the elementwise division denote by "/."

Ratio Scale Decomposition Theorem Let $A, A_1, \ldots, A_s$ be positive reciprocal matrices of pairwise comparisons. $Aw = \lambda(A)w$, $A_k x_k = \lambda(A_k)x_k$, $k = 1, \ldots, s$.

5 Occurrence Order of Shop-Around Purposes

Further, put $W = (w_i/w_j)$, $W_k = (x_{ki}/x_{kj})$, $k = 1, \ldots, s$. $i, j = 1, \ldots, n$. Define $P = A/(W_1 * W_2 * \ldots * W_s)$ and $Py = \lambda(P)y$. Then the following holds.

$$\lambda(A) = \lambda(P)$$
$$A(x_1 * \cdots * x_s * y) = \lambda(P)(x_1 * \cdots * x_s * y)$$

From this proposition, the ratio scale w of A can be decomposed, by putting $w' = (x_1'^* \ldots * x_s') = (x_1 * \ldots * x_s)^{1/s}$, as follows.

$$w = w' + (w - w') = (x_1'^* \ldots * x_s') + (w - w')$$

If all groups $k = 1,\ldots,s$ make the same evaluations, w' equals w.

In this section, we interpret this group as a shop-around step and the entire group as the full shop-around steps, and then apply this theorem.

We make positive reciprocal matrices $A_1, \ldots, A_s$ from the association matrices of shop-around purpose transition $(N_{ij}(t))$ at the respective shop-around steps, $t = 1, \ldots, s$ as below.

$$A_t = \left((N_{ij}(t) + 0.5)/(N_{ji}(t) + 0.5)\right), t = 1, \ldots, s$$

The positive reciprocal matrix of A for the full shop-around steps is made in the same way from the association matrix of shop-around purpose transition (N_{ij}) for all steps, which is the sum of $(N_{ij}(t))$ over t. Here, 0.5 is added because there are random zeros.

As apparent from the method for making A_t, the matrix A_t becomes a positive reciprocal matrix, whose eigenvector corresponding to the largest eigenvalue can be interpreted as a ratio scale for the shop-around purpose occurrence order, which takes the larger value at the shop-around purpose that is more likely to be selected earlier at shop-around step t than other shop-around purposes.

The decomposition result is shown in Table 5.3.

4.3 Results of the Analysis for All Behavior Purposes

Figure 5.6 depicts the three line graphs which correspond to the ratio scales (components of eigenvector) of all shop-around purposes at the three shop-around steps, the first, second, and third steps, while setting the ratio scale for the purpose of "shopping for clothing" to 1 for the normalization. From the figure, the line graph for the first step shows that the purpose of "parking" has the highest priority. At the third step, it is seen that the priority of purposes such as shopping for "daily necessities" and "grocery" and having "meals" and "coffee" becomes increased.

Table 5.3 Decomposition of the ratio scale of the occurrence order of shop-around purposes by shop-around step (non-derived purposes)

Kaiyu step		1	2	3	4	5	6+	Residual	All steps	Multiple product	
Max eigenvalue		24.748	22.862	21.599	19.75	18.475	18.193	23.904	23.904		
Consistency		0.484	0.366	0.287	0.172	0.092	0.075	0.431	0.431		
No. of samples		667	399	200	67	25	27		1385		
Eigenvector		X1	X2	X3	X4	X5	X6+		W	W'	W–W'
Kaiyu purpose	1	4.59	4.26	8.96	5.55	5.60	7.52	3.57	4.29	6.02	−1.73
	2	4.08	5.93	8.84	7.47	5.64	5.52	3.87	5.65	6.21	−0.56
	3	4.81	4.08	7.53	5.86	7.45	4.60	5.76	5.03	5.70	−0.67
	4	6.72	7.11	3.87	5.99	5.21	6.49	5.78	6.37	5.93	0.44
	5	3.70	5.82	5.87	6.31	7.18	7.26	4.17	5.10	6.03	−0.93
	6	3.99	5.88	9.81	6.92	6.38	6.59	2.53	4.97	6.53	−1.56
	7	5.40	7.01	3.63	6.63	5.41	5.50	7.64	6.10	5.62	0.48
	8	4.15	6.47	3.11	8.01	6.35	5.50	6.36	4.36	5.48	−1.12
	9	4.15	5.05	5.55	5.06	5.41	5.50	8.70	4.49	5.22	−0.73
	10	5.09	5.66	5.21	4.79	5.41	5.50	8.59	5.41	5.40	0.01
	11	6.36	4.64	5.05	4.81	5.41	5.50	10.44	6.55	5.40	1.15
	12	8.85	6.35	7.63	5.47	5.92	6.00	4.59	11.31	6.78	4.53
	13	7.51	5.74	6.05	5.98	5.21	5.92	4.35	6.15	6.18	−0.03
	14	5.62	8.80	4.43	5.97	5.14	6.00	4.87	5.78	6.00	−0.22
	15	4.35	6.11	5.74	5.73	7.02	5.49	5.73	5.68	5.83	−0.15
	16	10.72	3.65	4.81	4.18	5.41	5.84	7.93	5.80	5.54	0.26
	17	9.91	7.43	3.87	5.27	5.83	5.28	5.12	6.97	6.14	0.83

5 Occurrence Order of Shop-Around Purposes

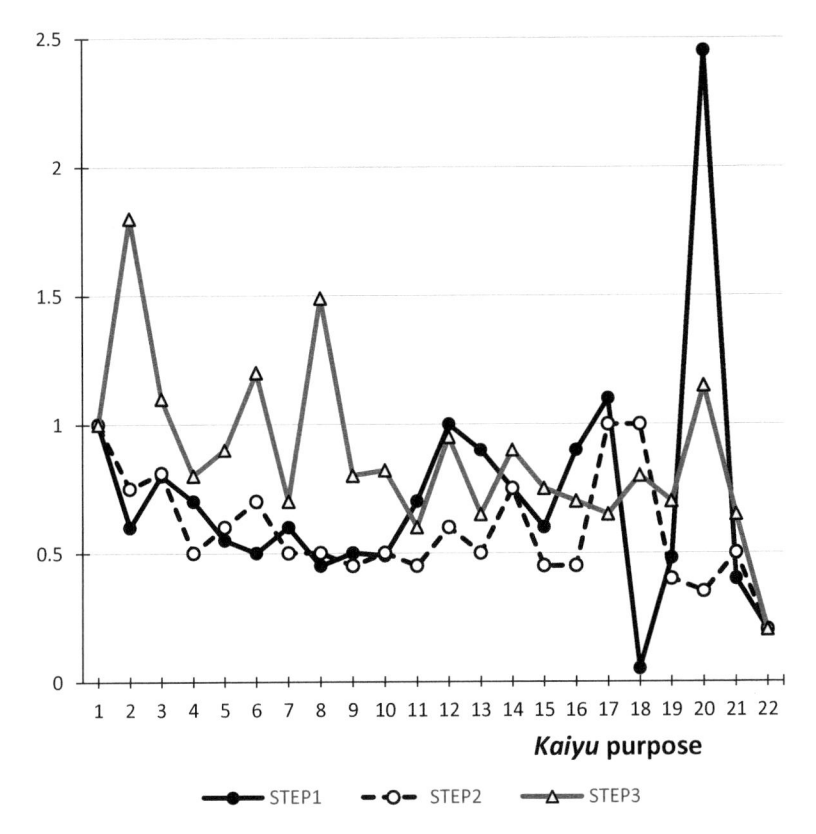

Fig. 5.6 Ratio scale for occurrence order of shop-around purposes by shop-around step (all purposes)

4.4 Results of the Analysis Limited to Non-derived Purposes

Table 5.3 shows the results of decomposition of the ratio scale for all steps into each step. Since the difference between w and w' is large, the table shows that the priority of the purpose occurrence order are changing over the shop-around steps.

In Fig. 5.7, classifying the purposes into two groups, those purposes that are likely to occur at early steps and those at later steps, we illustrate line graphs of ratio scales for those purposes to describe how the ratio scales of those purposes change as the shop-around steps proceed. To do this, for normalizing we set all the ratio scales at the first shop-around step to be equal to 1. From Fig. 5.7a, we see that purposes such as "shopping for furniture," "attending cultural classes," "joining events," and "meeting by appointment" consistently are taking values less than 1 from the second step and beyond, which means these purposes are likely to occur at the early shop-around steps. Conversely, from Fig. 5.7b, behavior purposes, such as "shopping for grocery," "having meals," "coffee," and shopping for "hobby items," are taking values larger than 1 from the second step and beyond, which means these purposes

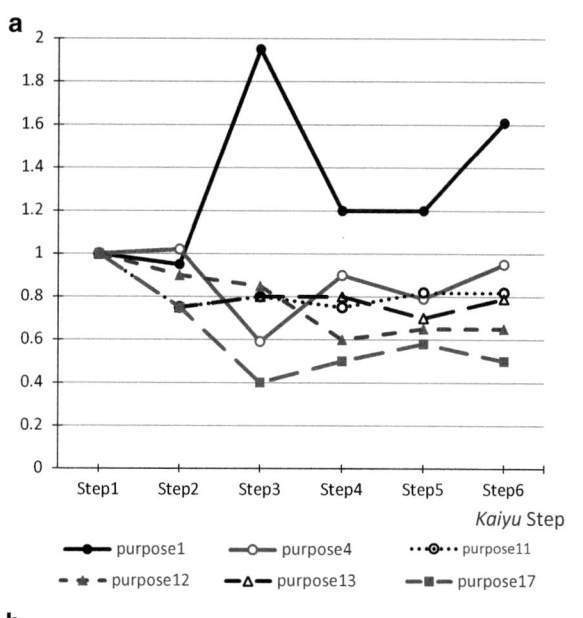

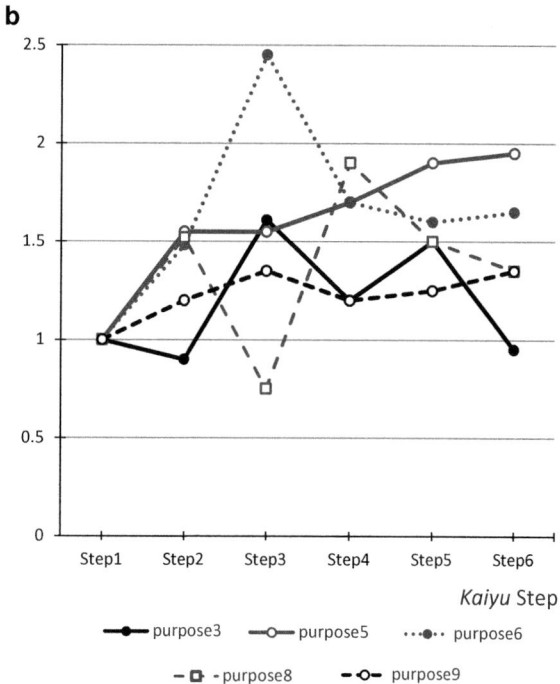

Fig. 5.7 Step profile of ratio scale of occurrence order of shop-around purpose

are likely to occur at the later shop-around steps. A noticeable exception is the purpose of "shopping for clothing" that is likely to occur at every shop-around step.

5 Conclusion and Future Challenges

According to the above analysis, the purpose transition probability is non-stationary, even if excluding the purpose of "returning home." Further, as for the cause to make the purpose transition probability to be non-stationary, we were able to show that there are two kinds of shop-around purposes which are likely to occur either at the early stage of shop-around steps or the later stage of those steps. We also found that the significant differences in the shop-around continuing functions arise from the differences in trip characteristics. Hence the purpose transition probability can be thought of as being heterogeneous. In this study, we only have conducted analyses of the aggregate data. Thus, future research should be carried out by the analyses at the non-aggregate or individual level in which trip characteristics and individual attributes are introduced as explanatory variables.

References

1. Amemiya T (1985) Advanced econometrics. Blackwell, Oxford
2. Cox DR, Oakes D (1984) Analysis of survival data. Chapman and Hall, London
3. Fukami T (1974) A study on pedestrian flows in commercial district part 1. Pap City Plan:43–48. (in Japanese)
4. Fukami T (1977) A study on pedestrian flows in a commercial district part 2. Pap City Plan:61–66. (in Japanese)
5. Hagishima S, Mitsuyoshi K, Kurose S (1987) Estimation of pedestrian shopping trips in a neighborhood by using a spatial interaction model. Environ Plan A 19:1139–1152
6. Kondo K (1987) Transport behavior analysis. KoyoShobo. (in Japanese)
7. Saaty TL (1980) The analytic hierarchy process: planning, priority setting, resource allocation. McGraw-Hill, New York
8. Saito S (1983) Present situation and challenges for the commercial districts in Nobeoka Area. In: Committee for Modernizing Commerce Nobeoka Region Section (ed) The report of regional plan for modernizing commerce Nobeoka area. pp 36–96. (in Japanese)
9. Saito S (1984) A disaggregate hierarchical Huff model with considering consumer's shop-around choice among commercial districts: developing SCOPES (Saga Commercial Policy Evaluation System). Plan Publ Manag 13:73–82. (in Japanese)
10. Saito S (1984) A survey report on the modified Huff model: the development of SCOPES (Saga Commercial Policy Evaluation System) (in Japanese): development of SCOPE (Saga Commercial Policy Evaluation System). Saga City Government, Saga City. (in Japanese)
11. Saito S (1986) Analysis report of Saga citizen opinion survey: City attractiveness and policy demand of Saga City, Saga City Government. (in Japanese)
12. Saito S (1988) Assessing the space structure of central shopping district viewed from consumers' shop-around behaviors: a case study of the Midtown District of Saga City. Fukuoka Univ Econ Rev 33:47–108. (in Japanese)

13. Saito S (1988) Ratio scaling and scale decomposition of rank ordered choice set data: an application of positive reciprocal matrix theory. Fukuoka Univ Rev Econ 33:1–46. (in Japanese)
14. Sakamoto T (1984) An absorbing Markov chain model for estimating consumers' shop-around effect on commercial districts. Pap City Plan:289–294. (in Japanese)
15. SAS Institute (1985) Proc LIFETEST. In: SAS user's guide: statistics
16. Takeuchi S (1981) A model for location planning of commercial functions. Plan Publ Manag:25–33. (in Japanese)
17. Saito S (1988) Duration and order of purpose transition occurred in the shop-around trip chain at a midtown district. Pap City Plan:55–60. (in Japanese)

Chapter 6
Kaiyu Distance Distribution Function at Downtown Space

Saburo Saito and Hiroyuki Motomura

Abstract Previous studies once had raised a question of what caused the difference between Japan and the United States regarding their estimated values for the exponent of distance in the Huff Model. Japan's smaller values were explained by referring to the proximate locations of retail facilities in Japan. On the other hand, there exists empirical research that has examined consumers' walking distances in a shopping district. The research seems to implicitly assume that human physiological constraints determine such distances. If so, why do proximate locations decrease distance resistance? With the intent to link the above two research streams, we estimate a shop-around distance distribution function to clarify how a consumer changes the hazard to quit during the shop-around trip.

Keywords Shop-around · *Kaiyu* · Distance distribution function · Survival analysis · Huff Model · Distance resistance · City center · Shopping district · Hazard · Saga City

---

This chapter is based on the paper, Saburo Saito, Toru Sakamoto, Hiroyuki Motomura, and Seiji Yamaguchi [27], "Parametric and Non-parametric Estimations of Distribution of Consumer's Shop-around Distance at Midtown District," *Papers on City Planning*, Vol. 24, pp.571–576, 1989.

S. Saito (✉)
Faculty of Economics, Fukuoka University, Fukuoka, Japan

Fukuoka University Institute of Quantitative Behavioral Informatics for City and Space Economy (FQBIC), Fukuoka, Japan
e-mail: saito@fukuoka-u.ac.jp

H. Motomura
Department of Business and Economics, Nippon Bunri University, Oita City, Japan
e-mail: motomura@nbu.ac.jp

1 Introduction

This study defines "shop-around behavior" as consumers' behavior of walking around multiple places within a downtown district. Studies of the Huff Model once had discussed the differences in measured values of distance resistance between Japan and the United States [4, 8]. Those studies had evidenced that distance resistance is high in the United States and low in Japan. The reason for this was explained by the argument that commercial sites are located more proximately to each other in Japan than in the United States. On the other hand, other studies had measured the walking distance in commercial sites using a tracking survey method. Therefore, this issue seems to have been argued with the implicit assumption that pedestrians have physiological and physical constraints, such as walking distances of 700 m to 1 km. However, the careful consideration of these two arguments reveals that they contradict each other. If we assume that walking distance depends only on the pedestrian's physiological factors, it is unlikely that distance resistance will decreased by the fact that the commercial sites are proximate to each other.

While consumers are constrained due to their individual attributes, it is also natural to assume that, in terms of shop-around behavior and in the shop-around process, consumers will decide whether to extend their shop-around distance by comparing the shop-around distance up to that moment with the attractiveness of the shop-around destinations ahead. Statistically, it should be considered that consumers in the shop-around process may gradually increase their "returning home" hazard or risk regarding whether or not they decide to go home while considering their past shop-around history and the shop-around destinations ahead.

Thus, the primary issue in such an assumption involves obtaining the distribution function of the shop-around distance, and not the average walking or shop-around distances. This is because hazards and distribution functions correspond one-to-one, as discussed later in Sect. 2.3. Further, though it is impossible to reproduce the distribution with only information on averages, knowing the shop-around distance's distribution function reveals how the "returning home" hazard changes according to the shop-around distance.

Previous research on shop-around behavior includes the following: studies on pedestrian flow [2, 3], studies on trip chains [6, 7, 10, 23–25], studies on the non-aggregation/expansion of the Huff Model [15, 16, 22], and studies on shop-around effects and the occurrence order of shop-around purposes [9, 11–13, 17–19]; however, no studies directly address the shop-around distance distribution. Studies related to such a distribution include analyses of the distribution of the number of shop-around steps as well as the number of stops in a trip chain. The latter case has led to the construction of a theory regarding the numerical distribution of such stops, and empirical analyses have been conducted [6, 24, 25]. However, the theoretical distribution is derived from the stationarity assumption that the returning home probability is constant, and the trip distance distribution is not addressed. On the other hand, a survey of shop-around behaviors in a downtown district reveals that distributions of shop-around steps and purposes are nonstationary [12, 18].

According to a recently developed survival analysis, it is now possible to estimate the returning home hazard function as a function of the shop-around distance and influencing factors, which is the same as estimating the shop-around distance distribution function because there is the one-to-one correspondence between hazard and distribution functions. Therefore, a check of whether the returning home hazard function is constant relative to the distance can determine whether the shop-around process is stationary where returning home probability is constant. Further, if explanatory variables such as the shop-around history and shop-around destination's attractiveness are introduced in the shop-around process, this should enable one to estimate how consumers, while considering these, change returning home hazards in the shop-around process.

Therefore, in this study we aim to estimate the shop-around distance distribution functions based on the survival analysis with the parametric method and the nonparametric method, which does not specify the class of distribution functions, and simultaneously investigate what factors influence the shop-around distance distribution function. Regarding these influence factors, variables such as shop-around history and the shop-around destinations' attractiveness are characterized as time-varying covariates that change depending on the shop-around distance. To introduce these time-varying variables, however, existing statistical methods must be expanded. Thus, in this study, we limit the analysis to the introduction of the time-invariant impact factors such as individual attributes and trip characteristics. We would like to leave the introduction of time-varying covariates as a topic for future research.

2 Framework of the Analysis

2.1 Data Used

The data used are approximately 900 valid samples collected from Saga City Citizen Opinion Survey conducted by Saga City Government in February 1985, which asked the respondents about their shop-around behaviors in the downtown district of Saga City, Japan, as one of its question items.

We asked respondents to recall their most recent visit to the downtown district, excluding work and school commutes, and record their shop-around behaviors as choices from among 22 districts (Fig. 6.1) and 22 behavior purposes (Table 6.1) in the downtown district, which were presented as options in the questionnaire item. The respondents were asked to record shop-around places (destinations they visited), and the shop-around purposes (why they visited there) in the order of walking around [14].

As Fig. 6.1 illustrates, we consider the city center district of Saga City to be a zone composed of 20 districts in the city center district, excluding places of work and residence, that range from 2.5 km north to south and 1 km east to west.

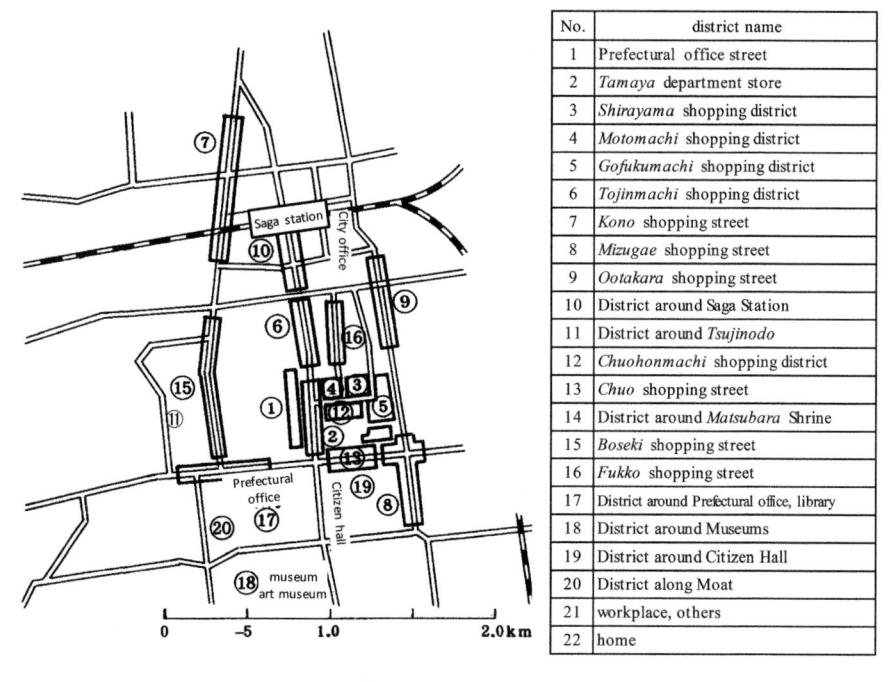

No.	district name
1	Prefectural office street
2	*Tamaya* department store
3	*Shirayama* shopping district
4	*Motomachi* shopping district
5	*Gofukumachi* shopping district
6	*Tojinmachi* shopping district
7	*Kono* shopping street
8	*Mizugae* shopping street
9	*Ootakara* shopping street
10	District around Saga Station
11	District around *Tsujinodo*
12	*Chuohonmachi* shopping district
13	*Chuo* shopping street
14	District around *Matsubara* Shrine
15	*Boseki* shopping street
16	*Fukko* shopping street
17	District around Prefectural office, library
18	District around Museums
19	District around Citizen Hall
20	District along Moat
21	workplace, others
22	home

Fig. 6.1 Map of the city center district at Saga City

2.2 Definition of Shop-Around (Kaiyu) and Its Measurement

First, we give the definitions of several terms. "Trip" is defined as a loop-shaped spatial movement in which a person departs from home, shops around, and returns home again. "Walking around" defined by the question item in the shop-around behavior survey involves moving between areas, but if a change occurs in the purpose of one's shop-around behavior, even when the person is still at the same area, we consider this as a walk-around in that same area. In other words, the event of "walking around" is considered to occur when either the shop-around area or purpose changes. We define the "walk-around step" as a ranked random variable, numbered as "1st step," ..., "s-th step," in the order that the "walk-around" events occur. The random variable M_i represents the total number of "walk-around" steps for sample i, which equals the total number of times that "walk-around" events occur from the moment that sample i enters the downtown district until the moment that sample i exits the downtown district. The behavior purposes of "passing by" and "parking" in Table 6.1 are referred to as "derived behavioral purposes." These purposes have been derivatively generated to conduct other behavior purposes and can be said to arise from the result of modal choice or from the constraints by the

Table 6.1 Shop-around purposes

	Behavior purposes
1	Shopping for clothing
2	Shopping for daily necessities
3	Shopping for grocery
4	Shopping for furniture, electric appliances
5	Shopping for hobby items (stationary, sport appliances, books, music instruments, cameras, etc.)
6	Having meals
7	Drinking (alcohol)
8	Coffee/tea
9	Entertainment
10	Movie
11	Attending cultural classes
12	Joining events
13	Going to government offices
14	Going to banks
15	Walking/jogging
16	Sending someone
17	Meeting someone by appointment
18	Passing by (by walking, bicycle)
19	Commuting to work, go to school
20	Parking/getting on and off/car, bus, train
21	Other purposes
22	Departing/returning home

downtown spatial structure. We refer to the behavior purposes from 1 to 17 in Table 6.1 as non-derived behavior purposes. If appropriate, we carry out the analysis of purpose transitions limited to these non-derived purposes.

The "entry district" into the downtown district is defined as the place where the behavior purposes other than "passing by," "work and school," and "other purpose," first appear in the downtown district. Similarly, the "exit district" from the downtown district is the place where the behavior purposes other than "passing by," "work and school," and "other purpose" appear last in the downtown district, which is the final destination in the shop-around. The "shop-around route" is a sample path of the walking from the entry district to the exit district. However, when doing the analysis limited only to non-derived purposes, the district visited for "parking" is not considered either an entry or exit district. The "shop-around step" t is defined as the walk-around step counting from the entry district. The total number of shop-around steps, N_i for sample i, is defined as the total number of times that walk-around events occur in the shop-around route of sample i.

The variable $X_{ijk}(t)$ is defined as a random dummy variable that assumes 1 when sample i moves from district j to district k at the t-th "shop-around step," and 0 otherwise. The variable $Y_{ijk}(s)$ is defined as a similar random dummy variable at the s-th "walk-around step."

Since no description of the shop-around distance is included in the survey item, we first created a distance matrix $D = (D_{jk})$ of the above 20 districts by 20 districts on the map to obtain each sample's shop-around distance.

From the survey results, as we know the walking around route for each sample, we obtain the estimated values of each sample's shop-around distance by using the distance matrix D. Specifically, denoting the shop-around distance for sample i by SDIST_i, it is given as follows.

$$\text{SDIST}_i = \sum_{t=1}^{N_i} \sum_j \sum_k X_{ijk}(t) \cdot D_{jk} \tag{6.1}$$

Similarly, the total walk-around distance TDIST_i is given as follows.

$$\text{TDIST}_i = \sum_{s=1}^{M_i} \sum_j \sum_k Y_{ijk}(s) \cdot D_{jk} \tag{6.2}$$

2.3 Estimation Method of the Shop-Around (Kaiyu) Distance Distribution Function

First, we give the definitions for the shop-around distance distribution function $F(u)$ and the total distance distribution function $G(u)$.

$$F(u) = \text{Prob}(\text{SDIST} > u) \tag{6.3}$$

$$G(u) = \text{Prob}(\text{TDIST} > u) \tag{6.4}$$

These functions are equal to 1 minus the cumulative distribution function and called as survival functions.

The method to estimate the survival function used in this study is Kaplan-Meier nonparametric estimation method, Cox's proportional hazard model, and the parametric exponential Weibull model, which are briefly described below [1, 5, 26]. Additionally, SAS was used for the actual calculation [20, 21]. While we explain these using the random variable T for "survival time until death" according to a usual survival analysis, it should be noted that we replaced "death" with "returning home" in this study and treat the survival time as a random variable SDIST (TDIST) of the shop-around distance (total distance) until returning home.

The survival analysis involves considering the random variable T of survival time. First, the hazard function $h(t)$ is defined as follows:

$$h(t) = \lim_{d \to 0} \text{Prob}(t \leq T < t+d | T \geq t)/d \tag{6.5}$$

From this definition, the following relationship holds between $h(t)$ and the survival function $F(t)$ of the random variable T and the density function $f(t)$:

$$h(t) = \frac{f(t)}{F(t)} = -\frac{d}{dt}\log F(t), \quad F(t) = \exp\left(-\int_0^t h(u)du\right) \tag{6.6}$$

When T is a discrete random variable which takes discrete values of $a_j, j = 1, \ldots, n$ with the probabilities $f_j, j = 1, \ldots, n$, the hazard h_j at a_j time point becomes as follows:

$$h_j = \frac{f_j}{F(a_j)}, \quad F(t) = \prod_{a_j < t}(1 - h_j) \tag{6.7}$$

Suppose that this distribution follows the parameter θ. Assume that we obtained data with a survival time t_i and a censoring time c_i from n independent samples. Note that the likelihood of a sample that is censored at the time c_i becomes $F(c_i) = \text{Prob}(t \geq c_i)$ because the sample survived at least until the time of c_i. Then the likelihood in this case can be expressed as follows. Here, u and c represent samples not subject to and subject to censoring, respectively.

$$L = \prod_u f(t_i; \theta) \prod_c F(c_i; \theta) \tag{6.8}$$

Set $x_i = \min(t_i, c_i)$. Denote by $r(u)$ the number of samples (risk set) that survived at the time u. If we take the logarithm to summarize the above formula with respect to time, we obtain the following log-likelihood formula.

$$\log L = \sum_u \log(h(x_i; \theta)) - \int_0^\infty r(u)h(u; \theta)du \tag{6.9}$$

In the case of a discrete random variable, letting r_j be the risk set at time point a_j and d_j be the number of "deaths" at that time, we obtain the following.

$$\log L = \sum_j \{d_j \cdot \log(h_j(\theta)) + (r_j - d_j) \cdot \log(1 - h_j(\theta))\} \tag{6.10}$$

The Kaplan-Meier estimator is obtained by replacing the hazard h_j in Eq. (6.7) with its maximum likelihood estimator d_j/r_j derived from Eq. (6.10). According to the value of h_j, the distribution function of Eq. (6.7) can take an arbitrary shape of a step function. Thus, it is called as a nonparametric method. In this study, the survival functions are estimated for different categories of each individual attribute (trip characteristics), and a Wilcoxon test is conducted to determine whether these shapes differ. This also is used to examine the shape of the hazard.

The proportional hazard model can be said a semi-parametric model. We decompose the hazard of Eq. (6.5) into a part dependent on the explanatory variable vector z and a part dependent on time and express this as follows:

$$h(t;z) = h_0(t)b(z;\beta), \quad b(z;\beta) = \exp(z'\beta) \tag{6.11}$$

In this proportional hazard mode, if the partial likelihood concept is used, the term of time has disappeared, and the effect β of the explanatory variable can be estimated by information only on the occurrence order of events without specifying the shape of $h_0(t)$. By utilizing this feature, the purpose of using the proportional hazard model in this study is not to specify the shape of $h_0(t)$ but to extract influence factors that affect hazards.

The exponential Weibull model is a parametric estimation model that specifies the term $h_0(t)$ in the expression (6.11) concretely as at^{a-1}, which sets the hazard function as follows:

$$h(t;z) = at^{a-1}b(z;\beta), \quad b(z;\beta) = \exp(z'\beta) \tag{6.12}$$

Depending on whether a is greater than one, we can determine whether the hazard increases as time elapses.

2.4 Procedure of the Analysis

Interpreting "death" as "returning home" and survival time as shop-around distance, the purpose of this study is to apply the method of survival analysis to the estimation of the shop-around distance distribution function in Eq. (6.3) as well as the total distance distribution function in Eq. (6.4). Simultaneously, we investigate various factors such as demographic attributes of individuals who shop-around and trip characteristics to determine which of these factors influences the shape of the distance distribution function.

The procedure of the analysis proceeds as follows. (1) We perform Kaplan-Meier nonparametric estimations for different categories of each influence factor and examine whether the shapes of the estimated distributions of the number of shop-around steps (hereafter, shop-around step distributions) and shop-around distance distributions differ between these categories. At the same time, we investigate how the hazard changes as the shop-around distance (step) increases. The reason why the shop-around step distributions also are estimated is to confirm whether a similar tendency can be obtained for the observed number of steps because the shop-around distance is not actual measurement, but an estimated value. (2) Next, we estimate the proportional hazard model which introduces as explanatory variables the factors of individual's attributes and trip characteristics into its hazard functions. We then extract factors with a significant effect on the shop-around distance distribution. (3) Finally, we estimate the parametric shop-around distance distribution function using an exponential Weibull model. Depending on whether its shape parameter is greater than one, we can determine whether the returning home hazard increases as the shop-around distance increases. Factors of individual's attributes and trip characteristics introduced as explanatory variables include the personal attributes of age, gender, years of residence, elementary school district, and annual expenditure for leisure and the trip characteristics of arrival time, mode, weekday/weekend, the number of people going out together, and initial purpose. All are categorical variables.

3 Analysis of Shop-Around (*Kaiyu*) Step Distribution

3.1 Method

A Markov-type analysis is employed, which excludes the events of "staying at the same district" from all shop-around movements, and we analyze two cases: all purposes (parking is the entry district) and non-derived purposes (parking is not the entry district). The Kaplan-Meier estimates for shop-around step distributions are obtained for different categories of each influence factor in each of these two cases. We then perform Wilcoxon tests to determine whether the shapes of the shop-around step distributions differ between different categories of influence factors such as individual's attributes and trip characteristics.

3.2 Results of the Analysis

Figure 6.2 displays the estimated results of the shop-around step distributions and the returning home hazards for all purposes and non-derived purposes in the entire group. Naturally, for all purposes the number of steps increases by the step of parking as the entry district so that the proportion of those who continue shop-around at each step for all purposes is higher than that for non-derived purposes. An observation of the shape of the returning home hazard reveals that it becomes high as the shop-around step advances. Thus it is seen that the two shop-around step distributions for all and non-derived purposes are nonstationary distributions, in which the returning home probability increases as the shop-around step extends.

Table 6.2 shows the Wilcoxon test results. As for the all purposes, the automobile mode, as anticipated, had the higher average number of shop-around steps with a statistically significant difference in distributions. Also the statistically significant differences in distributions with the higher average number of steps are observed in the cases in which more than one person went out together, the day of the shop-

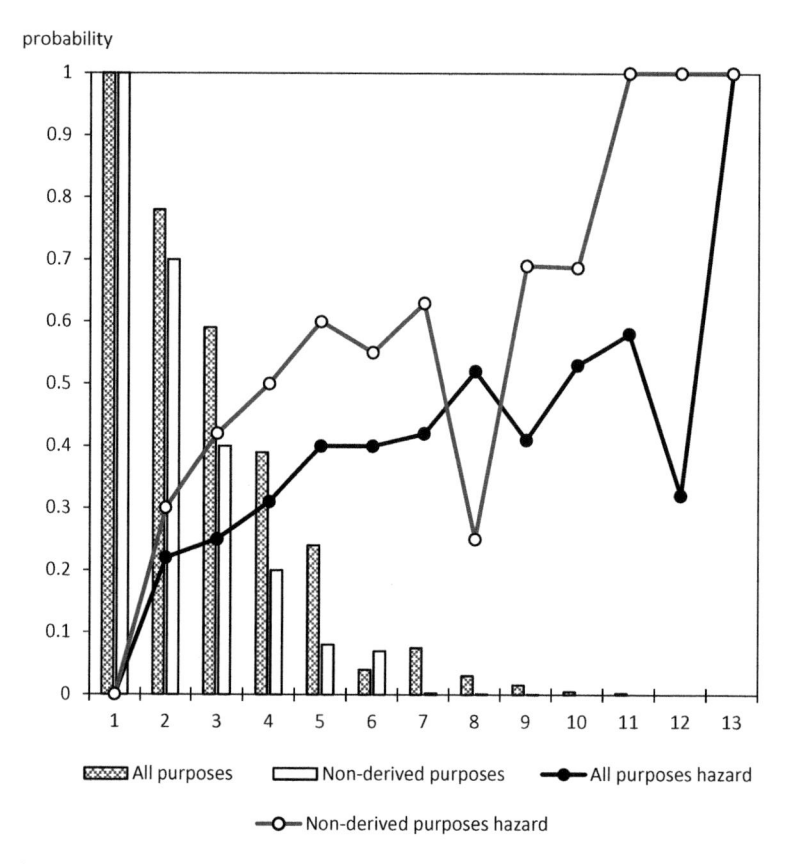

Fig. 6.2 Shop-around step distribution and hazards

around was a holiday, annual leisure expenses were high, and the period of residence was long. For non-derived purposes, significant differences are recognized for age and for the period of residence. In both cases of all and non-derived purposes, there are few shop-around steps from those with less than 3 years of residence, perhaps due to familiarity with the geography of the area.

4 Nonparametric Estimation of Shop-Around (*Kaiyu*) Distance Distribution Function

4.1 Method

We would like to explore the distance of walk-around between districts within the city center district, so we restrict our analysis to the Markov-type one which excludes staying at the same district. Then we perform the Kaplan-Meier nonparametric

Table 6.2 Tests of differences in shop-around step distributions

Groups	Events	
	Distribution of shop-around steps	
	All purposes (parking = entry district)	Non-derived purposes (parking ≠ entry district)
All groups	3.20	2.43
Age (years old)	0.3893[a]	0.0142
20–30/40–50/60	3.26/3.06/3.36	2.36/2.35/2.79
Gender	0.7347	0.3722
Male/female	3.22/3.19	2.44/2.43
Residence years	0.0066	0.0035
0–3/3–5/5–10 10–20/20 more	2.48/3.32/3.54 3.06/3.26	1.92/2.50/2.54 2.37/2.49
Elementary school district	0.1233[b]	0.6035
	–	–
Annual leisure expense (unit: 1000 yen)	0.0473	0.1499
0–50/50–100/100–300 300–500/500 more	3.00/3.14/3.39 3.54/3.62	2.30/2.37/2.56 2.73/2.85
Arrival time	0.6040	0.0748
am/pm	3.24/3.17	2.55/2.34
Travel mode	0.0007	0.0812
Car bus/foot others	3.36/2.94	2.35/2.56
Day of week	0.0184	0.5168
Weekdays/holidays	3.08/3.38	2.40/2.50
Number of travelers	0.0172	0.1320
One/plural	3.04/3.30	2.38/2.46
Original purpose	0.7983	0.9034
Shopping/others	3.18/3.24	2.43/2.46

[a] The upper figure in each cell of the table is p-value by Wilcoxon test; the lower figures are the average shop-around steps (times) by groups. The average number of shop-around steps for all purposes is greater than that for non-derived purposes because of counting the stop at parking

[b] The averages of shop-around steps for 18 elementary school districts were omitted due to space limitations

estimation of total distance and shop-around distance distribution functions. We estimate the shop-around distance distribution functions for two cases, all purposes and non-derived purposes, and conducted Wilcoxson tests in a similar way as before, to determine whether distribution functions differ between different categories in each factor of individual's attributes and trip characteristics.

Table 6.3 Test of differences in shop-around distance distributions

Groups	Events		
	All distance distribution	Shop-around distance distribution	
		All purposes parking=entry district	Non-derived purposes parking ≠ entry district
All groups	1488	1117	911
Age (years old)	0.2255[a]	0.1192	0.0378
20–30/40–50/60	1521/1395/1604	1178/998/1220	917/825/1081
Gender	0.9021	0.4405	0.9447
Male/female	1550/1443	1199/1058	965/872
Residence years	0.2104	0.0669	0.0406
0–3/3–5/5–10 10–20/20 more	1192/1524/1656 1430/1511	855/1209/1158 1063/1148	707/1019/950 860/933
Elementary school district[b]	0.0731	0.0866	0.2026
	–	–	–
Annual leisure expense (unit: 1000 yen)	0.0176	0.0471	0.0710
0–50/50–100/100–300 300–500/500 more	1310/1574/1523 1625/2045	966/1145/1197 1231/1537	765/948/978 1125/1168
Arrival time	0.4891	0.1016	0.0389
am/pm	1501/1477	1168/1064	998/845
Travel mode	0.2546	0.0001	0.8537
Car bus/foot others	1470/1517	1224/924	936/880
Day of week	0.2344	0.0069	0.0717
Weekdays/holidays	1452/1549	1043/1228	860/1010
Number of travelers	0.6176	0.0243	0.0838
One/plural	1479/1494	1036/1168	857/945
Original purpose	0.1746	0.1712	0.3527
Shopping/others	1443/1606	1056/1255	874/1024

[a]The upper figure in each cell of the table is p-value by Wilcoxon test. The lower figures are the average shop-around distances (meters) by groups
[b]The average shop-around distances for 18 elementary school districts were omitted due to space limitations

4.2 Results of the Analysis

Table 6.3 illustrates the Wilcoxon test results. As for the total distance distribution functions, the only statistically significant difference at 5% level is found for the annual leisure expenditure, which is a proxy variable of income. The average total walk-around distance becomes longer as this expenditure increases.

As for the shop-around distance distribution functions, the factors significant at the 5% level for all purposes include annual leisure expenses, modes, and the number of people who went out together. If the cases are such as annual leisure expenses are high, the automobile is the mode used, more than one person went out together, and they went out on a holiday, a longer average shop-around distance occurs for all purposes.

The average shop-around distance for the entire sample for all purposes is 1117 m. For example, it is 1168 m when more than one person went out together and 1036 m when only one person. As it is not possible to account for the differences in hazards only from the differences in average distances, the shop-around distance distribution functions are estimated and displayed in Fig. 6.3 for two categories of the number of people went out together for all purposes. These functions are drawn with the survival probability as the vertical axis and the shop-around distance as the horizontal axis. The entire group's returning home hazard function is also drawn in this figure.

From Fig. 6.3, we see that the shop-around distance distribution function for the category of "more than one person went out together" is situated over that for "only one person" up to 3 km, which means that the probability of continuing shop-around when plural people go out together is higher than that for one person. Regarding the graph of hazard function, we see that there are several points and intervals with 0 hazard value at the part with the larger distance values, but this means that no returning home samples are observed. Thus we have only to focus on the mountain skyline in the graph. From the hazard graph, we observe the following: The returning home hazard at initial step is very high, and then it falls sharply and gradually increases as the shop-around distance increases. The large value of returning home hazard at the initial step means that many people go back home without shop-around.

For non-derived purposes, significant differences were found in age, years of residence, and arrival time. The entire group's average shop-around distance for non-derived purposes was 911 m, but by age, younger people (those in their 20s and 30s), middle-aged people (40s and 50s), and mature people (aged 60 and over) had distances of 917, 825, and 1081 meters, respectively, and middle-aged people had the shortest shop-around distance. Similarly as before, Fig. 6.4 displays the shop-around distance distribution functions by age for non-derived purposes and the hazard function. Even for non-derived purposes, the returning home hazard can be interpreted as having the same trend as for all purposes.

5 Estimation by Proportional Hazard Model

5.1 Method

We take up the total distance and the shop-around distance as dependent variables and try to explain them by the proportional hazard model incorporating individual's

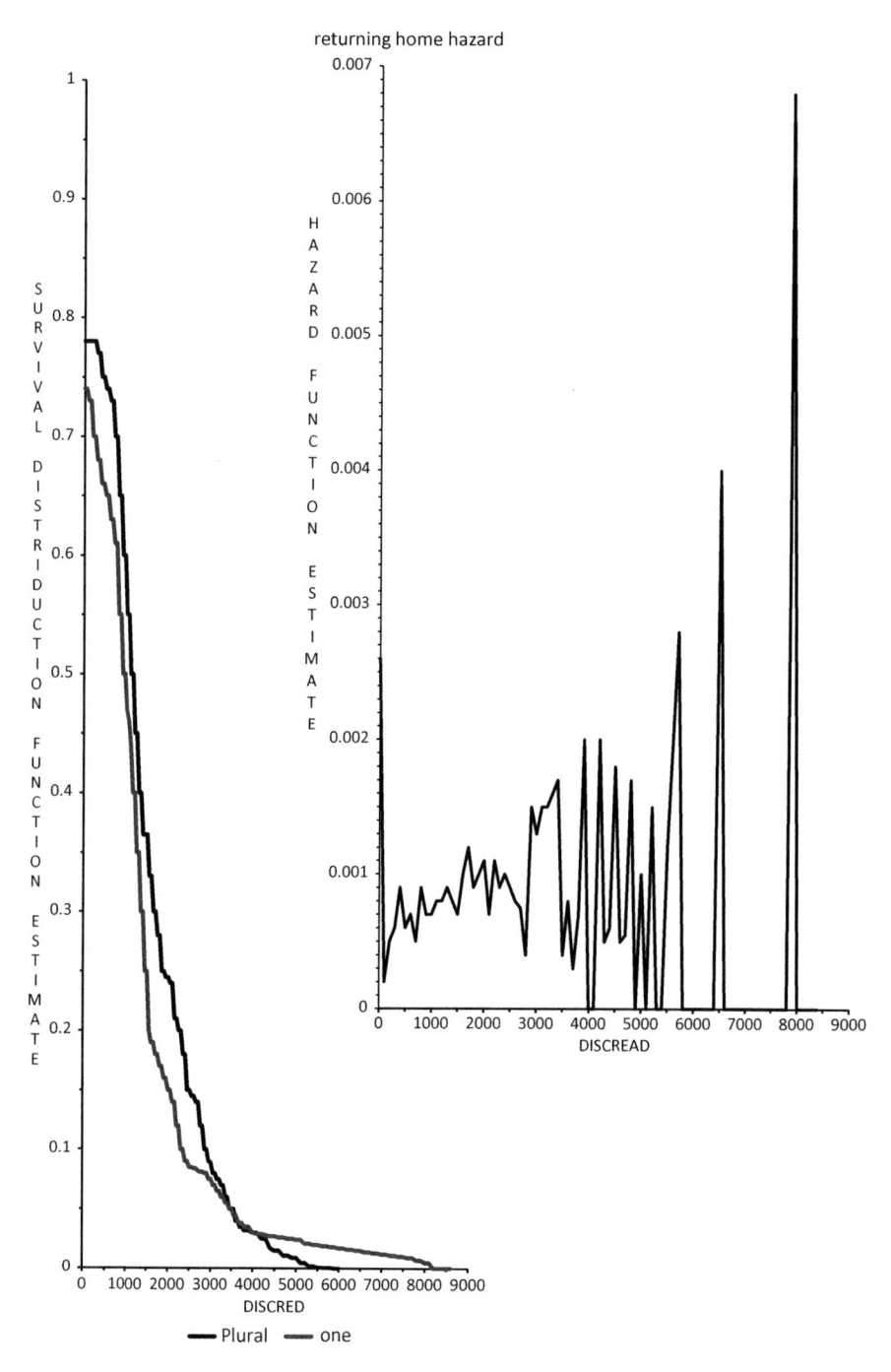

Fig. 6.3 Shop-around distance distribution for all purposes by the number of people who went out together and the returning home hazard for the entire group

6 *Kaiyu* Distance Distribution Function at Downtown Space

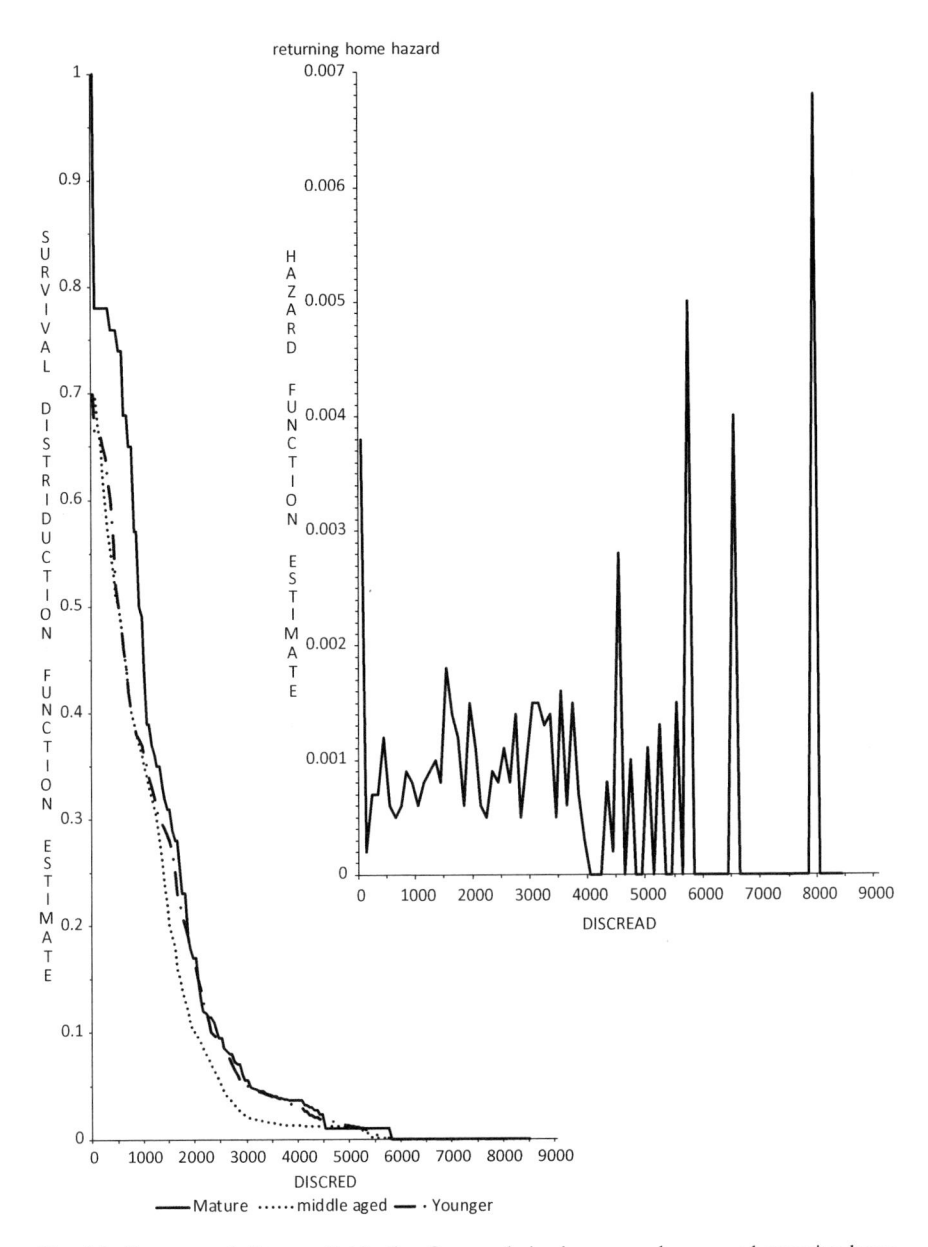

Fig. 6.4 Shop-around distance distribution for non-derived purposes by age and returning home hazard for the entire group

attributes and trip characteristics as explanatory variables. Here, we restrict our analysis to the Markov type one excluding staying and analyze the shop-around distance distribution for the two cases, all purposes and non-derived purposes, in the same way as before. The explanatory variables of personal attributes and trip

Table 6.4 Estimated results of proportional hazard model

		Total distance (km)		Shop-around distance (km)			
				All purposes parking = entry district		Non-derived purpose parking ≠ entry district	
Log-likelihood		−3152.00		−3611.29		−3171.19	
Explanatory variable		Estimate	p-value	Estimate	p-value	Estimate	p-value
Personal attribute	Age 40s–50s	–	–	0.251	0.0019	0.213	0.0128
	Elementary school district[a]	–	–	–	–	−0.184	0.0346
	The second elementary school district						
Trip properties	Travel mode Car, bus	−0.220	0.0100	−0.416	0.0000	−0.224	0.0115
	Day of week Weekday	–	–	0.177	0.0349	0.237	0.0085
	Original purpose Shopping	0.188	0.0478	0.289	0.0012	0.264	0.0053

[a]Elementary school districts are divided concentrically into six groups from the city center district as the first elementary school district to the furthest as the sixth and the corresponding dummy variables were made

characteristics were changed to dummy variables that represent each category of respective explanatory variables. Explanatory variables were selected using a backward-selection method, starting with the model containing all these variables [21].

5.2 Results of the Analysis

Table 6.4 displays the results of the analysis. Regarding how to read the results, when the coefficient β of the explanatory variable is negative, the hazard is decreased by $\exp(\beta)$. Thus, the shop-around distance becomes longer. Regarding the survival function, when the coefficient is negative, the $F(u)$ is shifted upward by exponent of $\exp(\beta)$. According to the result of the total distance distribution function, the two trip characteristics of transportation mode and initial purpose turn out to be statistically significant, as those who use bus or automobile demonstrate a lower returning home hazard and those who have the shopping purpose as the initial purpose show a higher returning home hazard. A likely cause for the result of the decrease in the total distance for the case of shopping might be due to the Markov-type analysis we take here which excludes staying distance. An analysis that includes the staying distance is a topic of future research.

The result of the shop-around distance distribution for all purposes reveals that significant explanatory variables are age, mode, the day of the week going out, and initial purpose. When the transportation modes are bus and automobile, the distribution function shifts upward. In contrast to extending the shop-around distance, the reverse tendency is recognized among those in their 40s and 50s, on weekdays, and with the initial purpose of shopping.

In the shop-around distance distribution for non-derived purposes, statistically significant differences are found in age, residential and elementary school districts, transportation modes, and initial purpose. As in the case of all purposes, the hazard decreases for the transportation modes of bus and automobile, and the hazard increases for those in their 40s and 50s, on weekdays, with the initial purpose of shopping. Even when the residential and elementary school districts are comparatively close to the city center district, the hazard decreases, and the average shop-around distance increases.

6 Estimation by the Exponential Weibull Model

6.1 Method

We conducted a parametric estimation of an exponential Weibull model. Many parametric groups in the distribution function can be used for a survival analysis, but we chose the Weibull distribution because it is a simple extension of the exponential distribution. In addition, depending on whether the shape parameter is greater or less than one, it is possible to determine whether the hazard function increases as the shop-around distance progresses. Here, in the same way as before, the total distance and the shop-around distance excluding staying are used as dependent variables, and the shop-around distance is estimated for the two cases (all purposes and non-derivative purposes).

6.2 Results of the Analysis

Table 6.5 shows the estimated results. Both shape parameters are greater than one, which are the expected results. Equation (6.12) implies that if the explanatory variable's coefficient is negative, the hazard decreases so that the shop-around distance increases. Age, gender, and the number of people who went out together are statistically significant factors in the total distance distribution, and we also see that the initial purpose is added in the shop-around distance distribution for all purposes, and the arrival time furthermore is added for non-derived purposes.

Table 6.5 Parameters estimated by exponential Weibull model

		Total distance (km)		Shop-around distance (km)			
				All purposes parking = entry district		Non-derived purpose parking ≠ entry district	
Log-likelihood		−1405.32		−1312.72		−1171.39	
Explanatory variable		Estimate	p-value	Estimate	p-value	Estimate	p-value
Constant		2.565	0.0001	2.214	0.0001	2.379	0.0001
Personal attributes	Age		0.0000		0.0001		0.0002
	20s–30s	0.000		0.000		0.000	
	40s–50s	0.401	0.0411	0.211	0.2255	−0.604	0.0001
	Over 60	−0.191	0.3346	−0.373	0.0330	−0.339	0.0622
	Gender		0.0001		0.0001		0.0001
	Male	0.000		0.000		0.000	
	Female	0.925	0.0001	0.788	0.0001	0.782	0.0001
Trip characteristics	Arrival time		0.5940		0.1770		0.0068
	pm	−0.079	0.594	0.174	0.1770	0.374	0.0068
	am	0.000		0.000		0.000	
	Travel mode		0.8350		0.0868		0.9822
	Walking/other	−0.034	0.8350	0.237	0.0868	−0.003	0.9822
	Car/bus	0.000		0.000		0.000	
	The day of the week		0.4688		0.7095		0.4652
	Weekday	0.000		0.000		0.000	
	Holidays	−0.118	0.4688	−0.054	0.7095	0.112	0.4652
	Number of travelers		0.0186		0.0052		0.0002
	One	0.000		0.000		0.000	
	Plural	−0.396	0.0186	−0.410	0.0052	−0.589	0.0002
	Original purpose		0.5945		0.0069		0.0359
	Others	−0.086	0.5945	−0.385	0.0069	−0.317	0.0359
	Shopping	0.000		0.000		0.000	
Shape parameters		1.8286		1.6017		1.5972	

7 Conclusion and Future Challenges

Although we will not repeat the above discussions, its results indicate that there are various factors that affect the shop-around distance distribution. In order to give a unified interpretation of influence factors, future research such that the attractiveness of shop-around destinations is incorporated would be needed. However, it is also important to grasp the shop-around events as nonstationary processes, in which the returning home hazard increases as the shop-around progresses, as apparent from the exponential Weibull model's shape parameter. The shop-around distance here is the estimated value, and it would be preferable to obtain high-precision shop-around behavioral data in the future, even with a small number of samples.

References

1. Cox DR, Oakes D (1984) Analysis of survival data. Chapman and Hall, London
2. Fukami T (1974) A study on pedestrian flows in a commercial district part 1. Pap City Plan 43–48. (in Japanese)
3. Hagishima S, Mitsuyoshi K, Kurose S (1987) Estimation of pedestrian shopping trips in a neighborhood by using a spatial interaction model. Environ Plan A 19:1139–1152
4. Ishihara S (1973) Urban social system. The Nikkan Kogyo ShimbunSha. (in Japanese)
5. Kalbfleisch JD, Prentice RL (1980) The statistical analysis of failure time data. Wiley, New York
6. Kondo K (1987) Transport behavior analysis. KoyoShobo. (in Japanese)
7. Kondo K, Kitamura R (1987) Time-space constraints and the formation of trip chains. Reg Sci Urban Econ 17:49–65
8. Kumagai Y (1973) Formation of agglomeration. In: Ishihara S (ed) Urban social system. The Nikkan Kogyo ShimbunSha, p 43. (in Japanese)
9. Lakshmanan T, Hua C-i (1983) A temporal-spatial theory of consumer behavior. Reg Sci Urban Econ 13:341–361
10. Lerman SR (1979) The use of disaggregate choice models in semi-Markov process models of trip chaining behavior. Transp Sci 13:273–291
11. O'Kelly ME (1981) A model of the demand for retail facilities, incorporating multistop, multipurpose trips. Geogr Anal 13:134–148
12. Saito S (1988) Duration and order of purpose transition occurred in the shop-around trip chain at a Midtown District. Pap City Plan 55–60. (in Japanese)
13. Saito S (1988) Assessing the space structure of central shopping district viewed from consumers' shop-around behaviors: a case study of the Midtown District of Saga City. Fukuoka Univ Econ Rev 33:47–108. (in Japanese)
14. Saburo Saito, Analysis report of Saga citizen opinion survey: City attractiveness and policy demand of Saga City: Saga City Government, Saga City, 1986. (in Japanese)
15. Saito S (1984) A disaggregate hierarchical Huff model with considering consumer's shop-around choice among commercial districts: developing SCOPES (Saga Commercial Policy Evaluation System). Plan Publ Manag 13:73–82. (in Japanese)
16. Saito S (1984) A survey report on the modified Huff model: the development of SCOPES (Saga Commercial Policy Evaluation System). Saga City Government, Saga City. (in Japanese)
17. Saito S (1983) Present situation and challenges for the commercial districts in Nobeoka Area. In: Committee for Modernizing Commerce Nobeoka Region Section (ed) The report of regional plan for modernizing commerce Nobeoka area. pp 37–96. (in Japanese)
18. Saito S, Sakamoto T (1988) Analysis of shop-around behavior within a City Center Space. Papers of the 4th annual meeting of JARES (Japan Association for Real Estate Sciences). pp 91–94. (in Japanese)
19. Sakamoto T (1984) An absorbing Markov chain model for estimating consumers' shop-around effect on shopping districts. Pap City Plan:289–294. (in Japanese)
20. SAS Institute (1985) LIFETEST. In: SAS user's guide: statistics
21. Harrel F (1986) The PHGLM procedure. In: SAS Institute Inc (ed) SUGI supplemental library user's guide, 5th edn. SAS Institute, Cary, pp 437–466
22. Takeuchi S (1981) A model for location planning of commercial functions. Plan Publ Manag 25–33. (in Japanese)

23. Kometani E, Kondo K, Kawamoto Y (1982) Absorbing Markov chain model of person trip patterns. Memoirs Fac Eng Fukuyama Univ 4:65–76. (in Japanese)
24. Wrigley N, Dunn R (1984) Stochastic panel-data models of urban shopping behaviour: 1. Purchasing at individual stores in a single city. Environ Plan A 16:629–650
25. Wrigley N, Dunn R (1984) Stochastic panel-data models of urban shopping behaviour: 2. Multistore purchasing patterns and the Dirichlet model. Environ Plan A 16:759–778
26. Yanai H, Takagi H (1986) Handbook of multivariate analysis. Gendai Sugaku Sha, Kyoto. (in Japanese)
27. Saito S, Sakamoto T, Motomura H, Yamaguchi S (1989) Parametric and non-parametric estimation of distribution of consumer's shop-around distance at a Midtown District. Pap City Plan:571–576. (in Japanese)

Chapter 7
The Factors Determining Staying Time of *Kaiyu* in City Center

Saburo Saito, Kosuke Yamashiro, and Mamoru Imanishi

Abstract These days we see many shopping malls all over the world. One of the most critical concerns for management of the malls is how they motivate their visitors to stay longer at their facility. The same holds for management of city center commercial district. The purpose of this study is to answer the question of what factors determine the time length of duration of consumer shop-around at a city center commercial district. Based on the survey of consumer shop-around behavior conducted at the city center of Fukuoka City, first we have analyzed how sojourn time is different among consumers with different individual characteristics. Next we have performed the variable selection procedure in the multiple regression analysis to extract significant factors that affect consumers' staying time. From the analysis, we found the following: Female consumers stay longer than males. The longer the travel time distance to the city center, the shorter the sojourn time. In particular, we found that behavior purposes consumers perform at the city center such as theater, hospital, school, cinema, shopping, etc., have great significant effects on the consumer shop-around staying time.

Keywords Kaiyu · Consumer shop-around behavior · Staying time · Behavior purpose · City center commercial district · Fukuoka City · Sojourn · Shopping mall

---

This chapter is based on the paper, Saburo Saito et.al. [11], "Factors to Determine the Length of Staying Time of Consumer Shop-around (*Kaiyu*) at City Center," *Operations Research and Its Applications* **12**: 443-452, 2010, which is modified for this chapter.

S. Saito (✉)
Faculty of Economics, Fukuoka University, Fukuoka, Japan

Fukuoka University Institute of Quantitative Behavioral Informatics for City and Space Economy (FQBIC), Fukuoka, Japan
e-mail: saito@fukuoka-u.ac.jp

K. Yamashiro · M. Imanishi
Department of Business and Economics, Nippon Bunri University, Oita City, Japan
e-mail: yamashiroks@nbu.ac.jp; imanishimm@nbu.ac.jp

1 Purpose

These days we see many shopping malls all over the world. The most critical concerns for management of the malls are how they attract people and how they motivate their visitors to stay longer at their facility. The same holds for the management of city center commercial district. In order for a city center commercial district to be attractive, it should be a pleasant and entertaining place for people so that people are glad to visit there and want to come again. Here the visitor's satisfaction is a key for making them visit once again and for recruiting new visitors. Thus to facilitate a wide range of visitors' activities, many of city center shopping districts offer not only the shopping establishments but also some amusement functions such as theaters, cinema complexes, restaurants, and so on. Hence several shopping malls in Japan even boast that they are not just a shopping facility only for the shopping purpose but an establishment where the purpose of visitors becomes staying and spending their time there in itself. For the shopping malls, they can expect that their turnover would increase if their visitors extend the length of their staying time. Therefore, the problem for them is to know what kinds of facilities, functions, or events are most effective to make visitors' staying time much longer.

The purpose of this study is to answer this problem of what factors determine the time length of duration of consumers shop-around, called *Kaiyu* in Japanese, at a city center commercial district. Based on data obtained from the survey of consumer shop-around behavior at the city center of Fukuoka City, Japan, we analyze how sojourn time is different among consumers with different individual characteristics. To extract the significant factors, we also perform the explanatory variable selection procedure in the multiple regression analysis in such a way that we can evaluate the size of the effects of significant factors by measuring the dependent variable of staying time in terms of minute and by expressing most of explanatory variables as dummy variables.

As for previous research, it is surprising that there seem to have been no previous studies on the staying time of *Kaiyu*. This study started with Higuchi and Sakaki's thesis study [3]. There are several studies related to the staying time of *Kaiyu*. They include a study on the physical distance of consumers' *Kaiyu* behavior within a city center commercial district [8], studies on the number of steps and expenditure of *Kaiyu* [4, 6, 7], a study on measuring the time value of shoppers [10], and studies on how urban development policies change consumers' *Kaiyu* behaviors and their economic effects [5, 9].

2 Framework of the Analysis

2.1 Definition of the Length of Staying Time of Shop-Around

Our survey of consumer shop-around behavior includes four question items concerning time which can be utilized to define the length of staying time at the city center. They are (1) the leaving time when respondents left home, (2) the arrival time when respondents arrived at the city center, (3) the survey interview time when respondents were asked by interviewers to answer survey questions, and (4) the estimated departure time when respondents will depart from the city center.

To be precise, the time length of duration of consumer shop-around at the city center should be the difference between the estimated departure time and the arrival time. However the estimated departure time from the city center respondents answered is just a schedule that has yet to finish. Respondents also may well often change their departure time. Moreover, we would like to find some significant relations between the staying time and the history of shop-around behaviors which already have finished. From these considerations, we define the length of staying time as the difference between the survey interview time and the arrival time.

Generally speaking, the staying time defined as above is called censored data. While there are statistical methods to deal with censored data, we will not treat our data as censored data and would like to leave it for further study since this is the first empirical analysis to analyze the staying time of consumer shop-around behavior.

2.2 Data Used

We use data obtained from the on-site survey of consumer shop-around behavior (The 7th Fukuoka Shop-Around Survey) conducted from June 28 to 30 in 2002. The data contains 945 samples in total.

The 7th Fukuoka Shop-Around Survey is an on-site questionnaire interview survey which takes about 15 min and asks the respondents who visit the city center about the history of their shop-around behavior, that is to say, the sequence of the triple decisions made by the respondents, places visited, purposes done, and expenditure spent there, which is recorded in the order of their occurrence. We set up nine shopping facilities as sampling points, and the respondents are randomly sampled at those sampling points. Other items we asked the respondents are their individual characteristics such as age, gender, and place of residence, travel mode, travel time, and transport fare to city center, and the frequency of visits at city center and so on.

2.3 Multiple Regression Analysis Employed

We wish to clarify what factors affect the time length of shop-around and to determine the size of their influences. For the purpose, while we simply employ multiple regression analysis, most of explanatory variables are provided as dummy variables since the sizes of effects of explanatory variables can be expressed as the multiple regression coefficients.

In Table 7.1 we give the list of explanatory variables. From Table 7.1 we see explanatory variables include many behavior purposes performed at places visited. The Fukuoka Shop-Around Survey distinguishes behavior purposes at the city

Table 7.1 List of the explanatory variables

Main purpose	Shopping	1	Men's clothes	Affairs	50	Travel agency
		2	Women's street clothes		51	Ticket office
		3	Women's home clothes		52	Culture lesson
		4	Children's wearing		53	Beauty salon, haircut
		5	Underwear		54	Hospital
		6	Traditional costuming		55	Part-time job
		7	Shoes		56	Job application
		8	Handbag		57	Government office
		9	Jewelry, accessories		58	Errands to financial institutions
		10	Fresh foods		59	Post office
		11	Furniture		60	Meeting someone by appointment
		12	Bedding		61	Visit
		13	Home electric appliances		62	Lodgment
		14	Audio video		63	Wedding party
		15	Computer		64	School
		16	Household utensils		65	Work
		17	Cosmetics	Day of the survey	66	*June 28th, Friday, 2002*
		18	Books		67	*June 29th, Saturday, 2002*
		19	CDs		68	*June 30th, Sunday, 2002*
		20	Cameras	Gender	69	*Male*
		21	Toys		70	Female
		22	Medicine	Job	71	Fixed job
		23	Sports appliances		72	*Others*
		24	Souvenirs	Age	73	10s
		25	Automobile		74	20s
		26	Motorcycle, bicycle		75	30s
		27	Artwork		76	40s
		28	Gifts		77	50s
		29	Window-shopping		78	60s
		30	Foods		79	*Over 70*
		31	Others	Marriage	80	Married
	Leisure	32	Movie		81	*Unmarried*
		33	Game arcade	Car	82	Having a car
		34	Karaoke		83	*Having no car*
		35	Event	Accompanying person	84	Family
		36	Exhibition		85	Same-gender friends and acquaintances
		37	Concert		86	Opposite-gender friends and acquaintances
		38	Theatergoing		87	Business associate
		39	Watching a sports game		88	*Others*
		40	Gamble	Commuting to work, school to the city center	89	Yes
		41	Taking up sports		90	*No*
		42	Walking, jogging	Continuous variables	91	Time distance to the city center (unit: minute)
		43	Sightseeing		92	Travel cost to the city center (unit: 100 yen)
		44	Festival		93	Visit of frequency (unit: times per month)
		45	Praying		94	Budget (unit: thousand yen)
		46	Killing time		95	Number of accompanying person (unit: persons)
	Drinking and eating	47	Tea drinking			
		48	Taking a meal			
		49	Drinking			

Note 1: No. 1–No. 90 are dummy variables
Note 2: Italicized variables are categories not used in the analysis

center as about 90 kinds of purposes. They are broadly classified into five classes: shopping, eating and drinking, leisure, work, and transportation. The class of shopping includes 35 kinds of shopping purposes distinguished mainly by the kinds of commodities to purchase. Similarly we have 42 kinds of behavioral purposes concerning eating and drinking, leisure, and work. We also have 16 kinds of transportation purposes such as get-on and get-off public transport, and parking. From all of these purposes except 16 transportation purposes, we employed 77 kinds of purposes of shopping, eating and drinking, and leisure as explanatory variables in the form of dummy variable. These dummy variables take 1 if the respondent has finished those purposes before the respondent takes the survey interview and 0 otherwise. In addition to behavior purposes, we include the following explanatory variables: personal attributes, travel cost to city center (unit: yen), time distance to city center (unit: minute), frequency of visits to city center (unit: times per month), and the weekend dummy, i.e., 1 if the survey day is Saturday or Sunday and 0 otherwise. The dependent variable is the length of staying time of shop-around at city center (unit: minute).

In this analysis we only use the sample whose main purpose to visit the city center commercial district was shopping, eating and drinking, or leisure. Two kinds of multiple regression models are applied: The one is the multiple regression model using all explanatory variables and the other the multiple regression model using backward variable selection method.

3 The Length of Shop-Around Staying Time Differs Among People with Different Individual Characteristics?

3.1 Average Length of Shop-Around Staying Time

Table 7.2 shows average shop-around time by all samples. The average shop-around time is 173 min, approximately 3 h.

3.2 Average Shop-Around Staying Time by Gender

Table 7.3 shows the averages of shop-around staying time at city center by gender. From the table, we see that the average shop-around staying time, about 175 min, for female visitors is longer than that of male visitors, about 167 min, by 8 min.

Table 7.2 Average shop-around staying time	N	Average (unit: minute)	SD
	851	173.04	162.38

Table 7.3 Average shop-around staying time by gender

	Average (unit: minute)	N	SD
Male	167.2	222	165.98
Female	174.5	623	160.13
All	172.6	845	161.62

Table 7.4 Average shop-around staying time by age groups

	Average (unit: minute)	N	SD
10s	178.3	192	170.43
20–24	154.4	296	145.50
25–29	206.5	109	180.41
30s	179.7	89	165.93
40s	168.9	55	178.26
50s	183.3	54	160.80
Over 60	175.3	53	157.92
All	173.3	848	162.46

Table 7.5 Average shop-around staying time by main purposes

	Average (unit: minute)	N	SD
Shopping	159.1	501	143.86
Leisure	159.1	46	167.82
Eating and drinking	115.1	63	126.43
Work	223.8	108	192.53
Others	239.4	91	190.60
All	173.3	809	160.60

3.3 Average Shop-Around Staying Time by Age

Table 7.4 gives the averages of shop-around staying time at city center by age. We classify ages into seven age groups. From the table we see that the length of shop-around staying time assumes the longest of 207 min for the age group of 25–29, which is followed by 183 min for the age group of 50s and 180 min for the age group of 30s.

3.4 Average Shop-Around Staying Time by Main Purposes

In Table 7.5 we give the average lengths of shop-around staying time by main purposes the respondents plan to do at the city center.

One of the questionnaire items of this survey asks the respondents to choose which one is their main purpose to visit the city center among the five: shopping, leisure, eating and drinking, work, and others. The purpose classified as others assumes approximately 239 min, the longest among all. The shortest length of shop-around staying time becomes 115 min for eating and drinking. The purposes of shopping and leisure have the same length of staying time of 159 min.

3.5 Average Shop-Around Staying Time by Travel Time Distances

Table 7.6 shows the averages of shop-around staying time by travel time distance to the city center. Here the travel time distance is defined as the response to the question of how many minutes it takes from your home to the city center today. We classify the travel time distance into the following four classes: less than 30 min, 30 min to less than 1 h, 1–2 h, and more than 2 h.

Table 7.6 clearly shows a very interesting fact that the longer the travel time distance the shorter the shop-around staying time. Considering the constraint of total amount of time given to visitors, it can be said that if the travel time from home to city center increases, the remaining amount of time that can be allocated to shop-around decreases so that the length of shop-around staying time may inevitably be shortened.

3.6 Average Shop-Around Staying Time by Travel Costs

Similarly we show in Table 7.7 the averages of shop-around staying time by travel costs. Here the travel cost is the response to the question of how much you have paid for transportation from home to the city center today. We classify travel costs

Table 7.6 Average shop-around time by travel time distances

	Average (unit: minute)	N	SD
Less than 30 min	182.7	385	159.29
Less than 1 h	181.2	239	158.78
Less than 2 h	177.6	122	178.15
Over 2 h	114.6	104	152.46
All	173.2	850	162.39

Table 7.7 Average shop-around time by travel cost to the city center

	Average (unit: minute)	N	SD
0–200 yen	170.9	213	166.78
201–500 yen	171.4	281	153.47
501–1000 yen	180.3	95	157.97
1001 yen or more	165.7	59	152.72
All	172.0	648	158.23

into four classes: less than 200 yen, 201–500 yen, 501–1000 yen, and more than 1001 yen.

We see from the table that visitors who spend the transportation expense of 500–1000 yen show in average the longest length of shop-around staying time of about 180 min. In contrast, visitors who spend more than 1000 yen for transportation spend in average the shortest shop-around staying time. This result can be said to parallel the above fact that the longer the travel distance the shorter the staying time.

4 Multiple Regression Analysis for Exploring Factors to Determine the Length of Shop-Around (*Kaiyu*) Staying Time

4.1 The Model Using All the Explanatory Variables

Table 7.8 reports estimated results of the multiple regression model using all the explanatory variables. In the table, we arrange the estimated multiple regression coefficients in the descending order of their values.

It should be noticed that most of factors that have large effects on shop-around staying time are related to specific objectives visitors plan to do at the city center. The factor that has the largest effect on shop-around staying time turns out the purpose of "Theatergoing" and the next largest one is "Work," followed by "Hospital," "School," and "Karaoke."

4.2 The Model Estimated by Backward Variable Selection Method

We have tried to obtain the optimal model all of which explanatory variables become significant by employing backward variable selection method (Table 7.9).

The backward variable selection method is an automatic model selection method in which at each step one explanatory variable that has the least contribution is deleted and the step is repeated until all remaining explanatory variables become significant. We set the significant probability level as 10%.

In Table 7.9 we give the estimated model obtained by backward variable selection method. From the estimated result, it turns out that the purpose of "Theatergoing" has the largest significant effect on shop-around staying time of 290 min, which is followed by "School" (223 min), "Work" (219 min), and "Hospital" (216 min).

If we restrict our attention only to the purposes of shopping, leisure, and eating and drinking, the next largest to "Theatergoing" is the purpose of "Movie" whose

7 The Factors Determining Staying Time of *Kaiyu* in City Center

Table 7.8 Estimated model with all explanatory variables

Variables	Estimate	SD	t-value	p-value
38. Theatergoing	291.16	119.91	2.43	0.0155
65. Work	229.80	53.17	4.32	<0.0001
54. Hospital	218.99	118.35	1.85	0.0649
64. School	212.85	46.55	4.57	<0.0001
34. Karaoke	176.52	104.78	1.68	0.0927
56. Job application	130.75	60.16	2.17	0.0302
25. Automobile	117.69	178.98	0.66	0.5111
52. Culture lesson	115.29	121.87	0.95	0.3447
59. Post office	112.90	70.66	1.60	0.1108
10. Fresh foods	112.05	46.57	2.41	0.0165
53. Beauty salon, haircut	109.95	145.84	0.75	0.4512
32. Movie	100.25	87.18	1.15	0.2508
20. Cameras	99.34	55.39	1.79	0.0735
28. Gifts	95.56	69.73	1.37	0.1712
23. Sports appliances	91.32	85.59	1.07	0.2865
2. Women's street clothes	88.55	15.00	5.90	<0.0001
22. Medicine	86.48	42.59	2.03	0.0428
14. Audio video	83.46	122.50	0.68	0.4960
29. Window-shopping	78.90	12.11	6.51	<0.0001
1. Men's clothes	78.06	23.25	3.36	0.0008
43. Sightseeing	76.43	85.76	0.89	0.3733
3. Women's home clothes	75.52	19.47	3.88	0.0001
9. Jewelry, accessories	71.05	39.16	1.81	0.0702
31. Shopping others	69.22	34.07	2.03	0.0427
18. Books	68.45	36.01	1.90	0.0579
87. Business associate (accompanying person)	62.82	43.44	1.45	0.1488
40. Gamble	61.35	118.18	0.52	0.6039
7. Shoes	60.91	23.94	2.54	0.0113
27. Artwork	60.90	75.17	0.81	0.4183
73. 10s (age)	60.28	21.64	2.79	0.0056
15. Computer	58.17	125.79	0.46	0.6440
74. 20s (age)	55.79	20.69	2.70	0.0073
55. Part-time job	52.64	83.98	0.63	0.5311
80. Married	51.06	19.90	2.57	0.0106
84. Family (accompanying person)	47.60	19.45	2.45	0.0148
48. Taking a meal	46.19	14.77	3.13	0.0019
5. Underwear	44.47	61.54	0.72	0.4703
19. CDs	37.93	69.99	0.54	0.5881
85. Same-gender friends and acquaintances (accompanying person)	36.02	13.60	2.65	0.0083
42. Walking, jogging	30.98	120.33	0.26	0.7970
75. 30s (age)	29.95	26.67	1.12	0.2620
24. Souvenirs	25.90	129.99	0.20	0.8422

(continued)

Table 7.8 (continued)

Variables	Estimate	SD	t-value	p-value
86. Opposite-gender friends and acquaintances (accompanying person)	20.70	18.11	1.14	0.2535
8. Handbag	18.24	38.05	0.48	0.6320
16. Household utensils	14.14	27.73	0.51	0.6102
12. Bedding	12.52	92.58	0.14	0.8925
17. Cosmetics	11.70	32.16	0.36	0.7162
46. Killing time	11.57	18.77	0.62	0.5381
47. Tea drinking	8.87	26.96	0.33	0.7421
33. Game arcade	4.56	49.75	0.09	0.9270
89. Commuting to work, school to the city center	4.54	13.38	0.34	0.7347
78. 60s (age)	1.31	38.53	0.03	0.9728
93. Visit of frequency (unit: times per month)	0.25	0.98	0.25	0.7991
91. Time distance to the city center (unit: minute)	0.23	0.18	1.26	0.2089
94. Budget (unit: thousand yen)	−0.07	0.32	−0.21	0.8309
92. Travel cost to the city center (unit: hundred yen)	−0.33	1.01	−0.89	0.3730
95. Number of accompanying person (unit: person)	−0.36	0.41	−0.33	0.7450
60. Meeting someone by appointment	−2.54	13.05	−0.19	0.8456
68. June 30th Sunday (day of the survey)	−2.99	13.57	−0.22	0.8259
77. 50s (age)	−3.18	35.35	−0.09	0.9284
71. Fixed job (job)	−4.82	13.54	−0.36	0.7219
37. Concert	−6.87	122.16	−0.06	0.9552
67. June 29th Saturday (day of the survey)	−7.77	13.48	−0.58	0.5650
82. Having a car (car)	−7.84	11.57	−0.68	0.4985
70. Female (sex)	−15.51	14.76	−1.05	0.2940
30. Foods	−20.28	31.53	−0.64	0.5204
4. Children's wearing	−20.36	86.10	−0.24	0.8132
76. 40s (age)	−30.85	34.31	−0.90	0.3690
11. Furniture	−42.13	85.26	−0.49	0.6214
58. Errands to financial institutions	−80.96	70.39	−1.15	0.2507
13. Home electric appliances	−122.16	99.67	−1.23	0.2209
Adj R^2	0.6656			

effect is 139 min, and the purposes of "Fresh foods" (106 min), "Cameras" (100 min), and "Women's Street Clothes" (93 min) follow.

If you want to go to the theater, you need more than 3 or 4 h at least. Similarly if you want to see movies, you need more than 2 h at least. From these considerations, the above estimated results quite conform to our intuition.

Table 7.9 Estimated model obtained by backward variable selection method

Variables	Estimate	SD	t-value	p-value
38. Theatergoing	290.50	115.05	6.38	0.0119
64. School	223.60	41.91	28.46	<0.0001
65. Work	219.58	50.48	18.92	<0.0001
54. Hospital	216.58	114.45	3.58	0.0590
32. Movie	139.92	67.41	4.31	0.0384
56. Job application	126.19	58.36	4.68	0.0310
10. Fresh foods	105.89	44.25	5.73	0.0170
20. Cameras	99.86	52.42	3.63	0.0573
2. Women's street clothes	93.15	12.99	51.43	<0.0001
1. Men's clothes	90.18	19.93	20.47	<0.0001
22. Medicine	90.09	37.51	5.77	0.0167
29. Window-shopping	76.50	10.88	49.46	<0.0001
3. Women's home clothes	71.92	18.14	15.72	<0.0001
31. Shopping others	65.29	32.54	4.03	0.0453
7. Shoes	64.55	22.62	8.14	0.0045
87. Business associate (accompanying person)	64.20	37.80	2.88	0.0900
48. Taking a meal	60.21	13.07	21.23	<0.0001
18. Books	59.27	32.82	3.26	0.0715
73. 10s (age)	53.17	13.91	14.62	0.0001
80. Married	46.75	14.52	10.36	0.0014
74. 20s (age)	43.42	11.18	15.09	0.0001
84. Family (accompanying person)	39.72	16.68	5.67	0.0176
86. Opposite-gender friends and acquaintances (accompanying person)	34.50	16.22	4.52	0.0339
85. Same-gender friends and acquaintances (accompanying person)	34.06	12.00	8.05	0.0047
76. 40s (age)	−42.91	23.94	3.21	0.0736
Adj R^2	0.6936			

5 Concluding Remarks

While there has been few empirical research to measure actual consumer shop-around staying time at city center commercial district, we have clarified what factors determine the length of shop-around staying time and what size of effect they have based on the survey of actual consumer shop-around behavior at the city center of Fukuoka. We have the following findings.

1. Most important factors that affect the length of shop-around staying time are related to the behavior purpose visitors plan to do at city center. In particular, the purpose of theatergoing has the largest effect on the staying time. The purpose movie also has the large effect on staying time.

2. Female visitors stay longer than male visitors. The age group of 25–29 shows the longest staying time among all age groups.
3. The longer the travel time distance the shorter the staying time.

It is quite interesting to note that the first findings are consistent with the recent fads to attach cinema complexes with multi-screens to most large shopping malls. As for "Theatergoing," in the city center of Fukuoka City, we have a theater, *Hakataza*, where Japanese traditional drama *Kabuki* is performed every year, which takes 3 or 4 h. Thus *Hakataza* seems to play an important role for extending visitors staying time at Fukuoka City. It would be an interesting topic to investigate to what extent *Hakataza* contributes to extending visitors staying time at the city center of Fukuoka City. Also important is to compare visitors staying time among different cities and to explore its relation to the composition and functions of their facilities provided at their city centers.

As for the economic theory of allocation of time, there is vast literature. (See [1, 2] among others.) The results we obtained here quite coincide with the so-called household production approach, which says consumers must use input such as time and money to get the final consumption services or commodities as if they are producing the final consumption services and commodities using various input resources. For example, to get the service of movies and theater, consumers must spend some amount of time. Our results seem to reflect these simple facts.

The mechanism behind how consumers decide the length of shop-around (*Kaiyu*) staying time should be explored further.

References

1. Becker GS (1965) A theory of the allocation of time. Econ J 75:493–517
2. DeSerpa AC (1971) A theory of economics of time. Econ J 81:828–846
3. Higuchi R, Sakaki T (2004) The study on expenditure and factors determining visitor's sojourn time at the city center of Fukuoka City, Japan. Graduation thesis of Faculty of Economics, Fukuoka University. (in Japanese)
4. Saito S (1988) Duration and order of purpose transition occurred in the shop-around trip chain at a Midtown District. Pap City Plan (23):55–60. (in Japanese)
5. Saito S, Nakashima T, Kakoi M (1999) Identifying the effect of city center retail development on consumer's shop-around behavior: an empirical study on structural changes of city center at Fukuoka City. Stud Reg Sci 29:107–130. (in Japanese)
6. Saito S, Nakashima T, Iwami M, Kiguchi T (2001) The position of maximal spending on the consumer's shop-around steps. Collected papers for presentation in the 38th annual meeting of the Japan Section of Regional Science Association International (JSRSAI). pp 197–204. (in Japanese)
7. Saito S, Nakashima T, Kakoi M, Iwami M Kiguchi T (2002) On the position of maximal spending in the course of consumer's shop-around revisited. Collected papers for presentation in the 39th annual meeting of the Japan Section of Regional Science Association International (JSRSAI). pp 425–432. (in Japanese)
8. Saito S, Sakamoto T, Motomura H, Yamaguchi S (1989) Parametric and non-parametric estimations of distribution of consumer's shop-around distance at a Midtown District. Pap City Plan (24):571–576. (in Japanese)

9. Saito S, Yamashiro K (2001) Economic impacts of the downtown one-dollar circuit bus estimated from consumer's shop-around behavior: a case of the downtown one-dollar bus at Fukuoka City. Stud Reg Sci 31(1):57–75. (in Japanese)
10. Saito S, Yamashiro K, Kakoi M, Nakashima T (2003) Measuring time value of shoppers at city center retail environment and its application to forecast modal choice. Stud Reg Sci 33 (3):269–286. (in Japanese)
11. Saito S, Yamashiro K, Imanishi M, Sakaki T, Igarashi Y, Kakoi M (2010) Factors to determine the length of staying time of consumer shop-around (*Kaiyu*) at city center. Oper Res Appl 12:443–452. (in Japanese)

Chapter 8
Little's Formula and Parking Behaviors

Saburo Saito, Kosuke Yamashiro, Masakuni Iwami, and Mamoru Imanishi

Abstract Many local cities in Japan now are facing the critical problem of managing car parking to alleviate traffic congestion occurred chronically at weekends in their city center commercial area. The crucial point, we believe, to make matters worse for solving the problem is that planners at local cities seem to have no means to assess whether or not parking lots are insufficient and if so, how many parking lots they actually need. Taking up the actual instance of city center retail environment of Fukuoka City and conducting the interview survey of parking behaviors of consumers who visit there by automobile, the purpose of this paper is to address the problem to show a simple method to determine how much capacity the parking space must need at the city center by using Little's formula with taking into account the distinctive feature of consumers' parking behaviors at city center retail district.

Keywords Little's formula · Parking space · City center retail district · Consumer behavior

---

This chapter is based on the paper, Saburo Saito, Kosuke Yamashiro, Masakuni Iwami, and Mamoru Imanishi [8], "Parking space policy for midtown commercial district and consumers' parking and shop-around behaviors: Applying Little's formula to the analysis of demand-supply balances for parking capacity in Tenjin area, the midtown of Fukuoka, Japan," *Fukuoka University Review of Economics,* vol. 58, pp.75–98, 2014, which is modified for this chapter.

S. Saito (✉)
Faculty of Economics, Fukuoka University, Fukuoka, Japan

Fukuoka University Institute of Quantitative Behavioral Informatics for City and Space Economy (FQBIC), Fukuoka, Japan
e-mail: saito@fukuoka-u.ac.jp

K. Yamashiro · M. Imanishi
Department of Business and Economics, Nippon Bunri University, Oita City, Japan
e-mail: yamashiroks@nbu.ac.jp; imanishimm@nbu.ac.jp

M. Iwami
Fukuoka University Institute of Quantitative Behavioral Informatics for City and Space Economy (FQBIC), Fukuoka, Japan
e-mail: miwami@econ.fukuoka-u.ac.jp

1 Purpose

Many local cities in Japan now are facing the critical problem of managing car parking to alleviate traffic congestion occurred chronically at weekends in their city center commercial area. The crucial point, we believe, to make matters worse for solving the problem is that planners at local cities seem to have no means to assess whether or not parking lots are insufficient and if so, how many parking lots they actually need.

The basic idea behind this paper is that while we have been conducting the on-site survey of consumer shop-around (*Kaiyu*) behavior at the city center retail environment of Fukuoka city every year since 1996, we have just noticed that if we utilize respondents' responses in this survey about their arrival time at and scheduled departure time from the city center, a simple application of Little's formula can get around the above problem (Cf. [6, 7]). However, since our on-site survey is designed to sample the respondents from visitors who are on the way of their shop-around, that is, have yet to finish their shop-around trip, their responses about the departure time are their self-estimated time they roughly plan to leave from the city center. Thus to implement our idea, we have carried out another on-site interview survey of parking behaviors of consumers who visit the city center by automobile to get their accurate departure time. We have selected several parking lots. The survey was designed to sample the respondents from the visitors who parked at these parking lots and just were going to leave from there so that the samples for this survey can be considered to have finished their shop-around trips.

On the other hand, we think there are distinctive features in parking behaviors of consumers who visit city center retail facilities. For example, in Fukuoka, the retail establishments are located at the middle of the city center area, and the accesses to the middle by car can be divided into four directions: from north, south, east, and west. Note that we have no circular roads surrounding the middle of city center area. If there are no circular roads, the access direction to the middle greatly matters. The destination facilities are located only at the middle so that it becomes the most congested area. Hence if you are accessing the middle from some direction and would like to change the direction to another one, you must pass through the middle, the most congested area. Thus the visitors by car who are accessing to the middle from some direction would not change the direction and like to find the parking space on the way of the direction to the middle. We must take into account these peculiar characteristics of drivers' incentives to avoid the congestion to enhance their trip utility.

While many previous studies on chronic traffic congestion at city center area (Cf. [1]) suggest that main factor and the large part of the congestion are due to driver's cruising behavior for searching the vacancy of parking lots, there have been few empirical studies to firmly support this suggestion. Moreover, there have been no previous studies taking into account the above features of drivers' behaviors.

With these in mind, taking up the actual instance of city center retail environment of Fukuoka City and conducting the interview survey of parking behaviors of consumers who visit there by automobile, the purpose of this paper is to show a simple method to determine how much capacity the parking space must need at the city center by using Little's formula while taking into account the distinctive feature of consumers' parking behaviors at city center retail district.

The remaining parts of this chapter are composed of as follows. Next we review Little's formula and data used. In Sect. 3 we use Little's formula to analyze the needed capacity of parking space dealing with the city center as one area. Section 4 discusses distinctive features of parking behaviors of consumers. Section 5 makes the parking capacity analysis by access directions. Section 6 ends with conclusion.

2 Little's Formula and Data Used

2.1 Little's Formula

As is well known, Little's formula, $L = \lambda W$, prescribes the relationship between L, the length of the queue or waiting line; λ, the arrival rate; and W, the waiting time (Cf. [2, 3]).

Here we explain how to interpret and apply Little's formula to the parking capacity analysis. The Little's formula is concerned with the queuing theory. Suppose some system provides some service for customers. As a concrete example, suppose a ticketing device at railway station where travelers line to get tickets. The length of waiting line L is the number of travelers to wait in front of the ticketing device. The arrival rate λ is the number of travelers to come to the ticketing device per unit of time. The waiting time W is the length of time from starting to join the waiting line to leaving the line after getting the ticket. Here we interpret a whole city center retail environment as a system, which provides shopping services for consumers who visit the city center by car.

Look at Fig. 8.1. The vertical axis indicates the number of cars, and the horizontal axis is the time. In the figure the dark curve expresses the cumulative number of car arrivals and the light line the cumulative number of car departures. Let S be the area surrounded by two curves, let N be the total number of car arrivals (departures), and let T be the total time length from the beginning to the end.

Notice that the vertical difference between the two curves means the number of parking cars, which corresponds to the length of waiting line, and that the horizontal difference between the two means the staying time, which corresponds to the waiting time. It also should be noticed that the unit of the area S must be the multiplication of time and the number of cars.

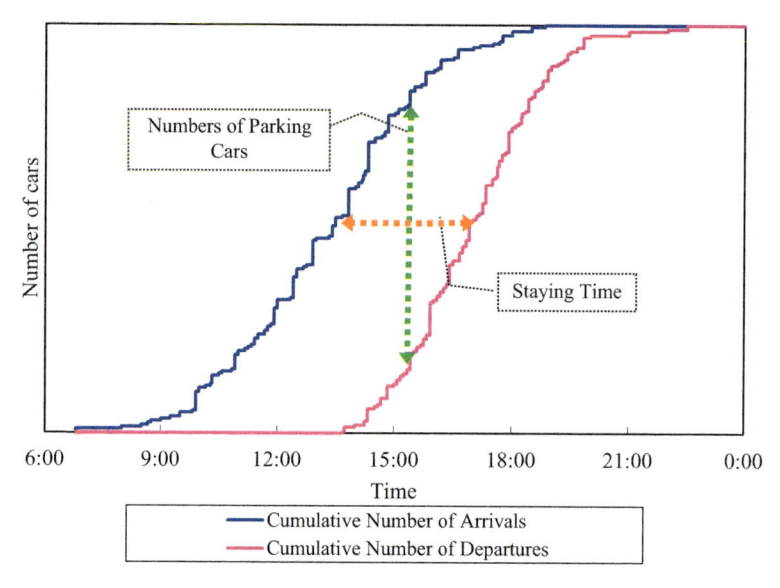

Fig. 8.1 Exposition of Little's formula

The average waiting time can be expressed by $W = S/N$, and the average length of waiting line can be represented by $L = S/T$. Noting that the average arrival rate can be formulated by $\lambda = N/T$, Little's formula is derived as follows (Cf. [4]).

$$L = \frac{S}{T} = \left(\frac{N}{T}\right)\frac{S}{N} = \lambda W$$

2.2 Data Used

In this paper we use the data obtained from the following two surveys. The first one is the 12th survey of consumer shop-around behavior at city center of Fukuoka City, and the second is the 1st survey of consumer parking behavior at city center of Fukuoka. Table 8.1 gives the outline of the 12th survey of consumer shop-around behavior at city center of Fukuoka City.

Similarly, Table 8.2 gives the outline of the 1st survey of consumer parking behavior at city center of Fukuoka. In this survey, we picked up 11 parking facilities as sampling sites. We conducted on-site sampling in which the respondents were sampled at random from visitors who parked their cars at these parking facilities and were going to leave from there. We implemented our on-site interview survey for the consumers who can be thought to have finished their shop-around trip. As shown in Table 8.2, in this survey we asked the respondents from which direction they entered

8 Little's Formula and Parking Behaviors

Table 8.1 Outline of 12th survey of consumer shop-around behavior

Name of survey	The 12th survey of consumer shop-around behavior at city center of Fukuoka City
Date of survey	2007.06.30 (Sat), 2007.07.01 (Sun)
Survey time	12:00–18:00
Sampling points	8 shopping facilities
	Solaria Plaza, JR Hakata Station, Canal City Hakata
	Shoppers *Daiei*, *Iwataya*, *Daimaru*, *Mitsukoshi*, Hakata Riverain
Number of samples	686 samples
Survey method	1. Samples drawn at random from visitors at the sampling points
	2. Interview with questionnaire for 15–20 min
Main questionnaire items	1. Personal profiles (residence, age, gender, occupation, etc.)
	2. Shop-around history (places visited, purposes done there, and expenditure there if any)
	3. Travel time to the city center of Fukuoka
	4. Travel means to the city center of Fukuoka
	5. Frequency of visits to the city center of Fukuoka
	6. Frequency of visits to various shops at the city center of Fukuoka

Table 8.2 Outline of the 1st survey of consumer parking behavior

Name of Survey	The 1st survey of consumer parking behavior
Data of survey	2008.5.24 (Sat) 15:00–20:00, 2008.5.25 (Sun) 14:00–19:00
Sampling points	11 parking facilities
	Chikudo Parking, *Ankoku* Parking, F-Parking *Kitatenjin*, Tenjin *Chuokoen* Parking
	Ayasugi Parking, Hakata Riverain Parking, *Kamiyo* Parking, Trust Park *Kego*
	Daiyoshi Park Tenjin Big Tower, Solaria Terminal Parking, N-Parking Tenjin
Number of samples	204 samples
Survey method	1. Samples drawn at random from visitors to each parking of survey points
	2. Interview with questionnaire for 10–15 min
Questionnaire items	1. Personal profiles (residence, age, sex, occupation, etc.)
	2. Shop-around history (places visited, purposes done there, and expenditure there if any)
	3. Travel time to the city center of Fukuoka
	4. Frequency of visits to the city center of Fukuoka
	5. The respondents from which direction they entered the city center area of Fukuoka
	6. How many minutes they had cruised before they found the vacancy of the parking space

the city center area of Fukuoka, how many minutes they had cruised before they found the vacancy of the parking space, and how many minutes they spent from starting to join the waiting line for the parking lots to finally parking their car and leaving there for starting their shop-around trip.

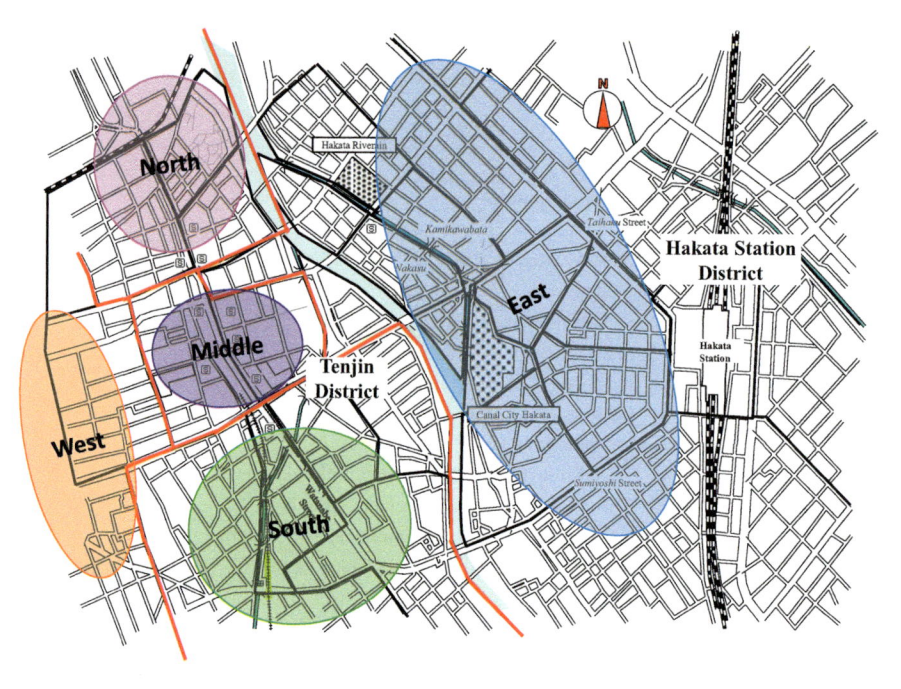

Fig. 8.2 Division of city center retail district, Tenjin

2.3 Dividing the City Center Retail District by Access Directions

The city center retail district of Fukuoka City is called Tenjin. To characterize consumers' parking behaviors by accessing directions, we divide the Tenjin area into five blocks as shown in Fig. 8.2. The five blocks are the north, the south, the east, the west, and the middle blocks. We distinguish the access directions of consumers to the middle by checking what block they entered first among the four blocks: north, south, east, and west.

2.4 Total Number of Visitors Who Visit the City Center by Car

For the parking capacity analysis, we need the number of actual incoming visitors at the city center of Fukuoka. From our previous study, the number of incoming visitors at the city center of Fukuoka for the purposes of shopping, leisure, and eating out was estimated to be 150,000 persons per day in average over the year as of 2000 (Cf. [5]). According to the 12th survey of consumer shop-around behavior at city center of Fukuoka City in 2007, the percentage of the visitors who visit the city center by car turns out to be 14.1%.

8 Little's Formula and Parking Behaviors

Table 8.3 Parking capacity by five blocks of Tenjin, 2008

	Parking capacity	Number of parking lots
North	4,506	100
East	998	53
South	2,732	121
West	2,692	91
Middle	2,399	21
Total	13,327	386

Hence the number of visitors who come to the city center by car is estimated to be 21,150 persons per day in average over the year.

2.5   Parking Capacity for Each Block of Tenjin

Other information we need is the parking capacity of the city center retail environment. We have carried out the field survey of parking space at the city center retail district of Tenjin. First we checked the location of parking lots by using the city map and counted the number of parking lots by the visual check on site while visiting the spot. As for the parking space like high-storied parking facilities, whose number of parking lots cannot be counted by the visual check on site, we asked the management of the parking space how many cars it can accommodate.

From these field survey efforts, the numbers of parking capacity are obtained. They are shown in Table 8.3.

3   Little's Formula and Parking Capacity Analysis

In this section we deal with the city center retail district as one area and analyze whether or not the capacity of parking space is sufficient by using Little's formula. We regard the length of waiting line in Little's formula as the demand for parking capacity. For the purpose, we need the arrival rate and the average waiting time.

3.1   Arrival Rate

Figure 8.3 shows the distribution of arrival time of consumers who visit Tenjin by car. We see that the two thirds of the arrival time are between 10:00 and 15:00. Here we assumed the time period from 6:00 to 24:00 as the operating time of parking lots a day.

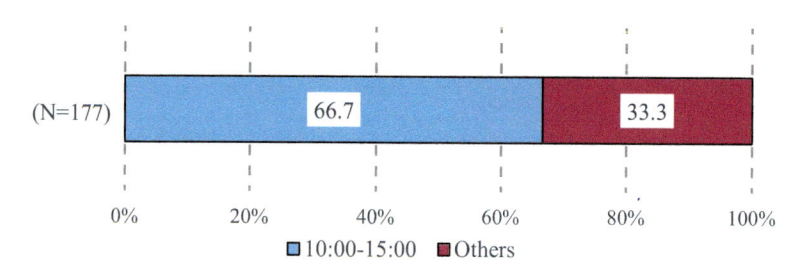

Fig. 8.3 Distribution of arrival time

Table 8.4 Number of arrivals by time zone

Time zone	Total	Percentage	Number of arrivals
10:00–15:00	21,150	66.7	14,101
Others	21,150	33.3	7,049

Table 8.5 Arrival rate (unit: person per minute)

Time zone	Number of arrivals	Time (minutes)	Arrival rate
10:00–15:00	14,101	300	47.00
All time	21,150	1080	19.58

We have assumed that the total number of visitors who visit the city center by car is 21,150 persons per day in average over the whole year.

Thus the number of arrivals during the time period of 10:00–15:00 becomes 14,101 people per day. During other time period, the number of arrivals is 7,049 people per day. They are shown in Table 8.4.

From these results, we can calculate the arrival rate as shown in Table 8.5.

3.2 Average Staying Time for Visitors by Car

According to the result of the 1st survey of consumer parking behavior, the average length of staying time for visitors by car turns out 224.6 min.

3.3 Demand for Parking Lots Calculated by Little's Formula

Since we have the arrival rate and waiting time, we can estimate the demand for parking lots using Little's formula. The results are shown in Table 8.6.

8 Little's Formula and Parking Behaviors   153

Table 8.6 Demand for parking lots calculated by Little's formula

Time zone	Arrival rate (λ)	Average waiting time (W)	Demand for parking lots (number of cars) (L)
10:00–15:00	47.00	224.6	10,557
All time	19.58	224.6	4,398

Table 8.7 Demand and supply of parking capacity

Time zone	(a) Demand for parking capacity	(b) Supply of parking capacity	(a)–(b) Excess demand
10:00–15:00	10,557	13,327	−2770
All time	4,398	13,327	−8929

3.4 Parking Capacity Analysis

Dealing with the city center retail district as one area, we can now compare the demand and supply of parking lots to make the parking capacity analysis. Table 8.7 gives the result. From the table we see that the number of parking lots turns out to be oversupplied. While the number of parking lots would be likely to be expected to be insufficient from the chronic congestion, the opposite becomes true.

It should be noticed that in this analysis, we have excluded the demand for the parking lots by business use so that accurately speaking, we must estimate the size of demand by business use. We would like to save this issue for a further study.

4 Analysis of Parking Behavior by Access Directions

Now we analyze data obtained from the 1st survey of consumer parking behavior to investigate from which access direction they have entered the city center, how long they have cruised to find the vacant parking lot, and at which parking block they have parked their cars.

4.1 Access Directions by Car to the City Center

Look at Fig. 8.4. The figure gives percentages of which access directions visitors by car have taken for entering the city center among the four access directions (Fig. 8.4).

Expanding these percentages by the number of total incoming visitors, 21,150, we have the numbers of actual incoming visitors by four accessing directions. They are given in Table 8.8 below.

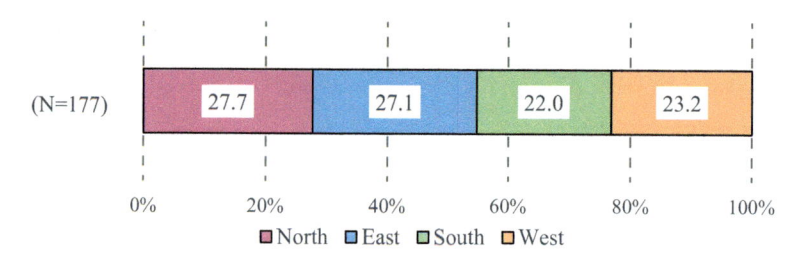

Fig. 8.4 Access directions by car

(N=177) | North 27.7 | East 27.1 | South 22.0 | West 23.2

Table 8.8 Numbers of incoming visitors by access directions

	North	East	South	West
The number of cars	5,854	5,736	4,659	4,898
%	27.68	27.12	22.03	23.16

Table 8.9 Choices of parking blocks by access direction

			Parking block					Total
			North	East	South	West	Middle	
Access direction	North	Frequency	35	1	0	9	4	49
		%	71.4	2.0	0.0	18.4	8.2	100.0
	East	Frequency	8	16	5	6	13	48
		%	16.7	33.3	10.4	12.5	27.1	100.0
	South	Frequency	5	4	14	7	9	39
		%	12.8	10.3	35.9	18.0	23.1	100.0
	West	Frequency	11	7	2	10	11	41
		%	26.8	17.1	4.9	24.4	26.8	100.0
	Total	Frequency	59	28	21	32	37	177
		%	33.3	15.8	11.9	18.1	20.9	100.0

4.2 Choices of Parking Blocks

Table 8.9 shows the results of choices of parking blocks by access directions. As was mentioned before, the largest choice of parking block for each access direction becomes the parking block that coincides with that access direction.

Similarly using the numbers of incoming visitors by access directions in Table 8.8, we can expand the choice probabilities of Table 8.9 into the actual number of cars parked at five parking blocks by access directions (Table 8.10).

Table 8.10 Numbers of cars parked at parking blocks by access directions

			Parking block					
			North	East	South	West	Middle	Total
Access direction	North	Number of people	4,182	119	0	1,075	478	5,854
		%	71.4	2.0	0.0	18.4	8.2	100.0
	East	Number of people	956	1,912	598	717	1,553	5,736
		%	16.7	33.3	10.4	12.5	27.1	100.0
	South	Number of people	597	478	1,673	836	1,075	4,659
		%	12.8	10.3	35.9	18.0	23.1	100.0
	West	Number of people	1,314	836	239	1,195	1314	4,898
		%	26.8	17.1	4.9	24.4	26.8	100.0
Total		Number of people	7,049	3,345	2,509	3,823	4,420	21,150
		%	33.3	15.8	11.9	18.1	20.9	100.0

Table 8.11 Cruising time to find parking lots by parking blocks

	N	Mean (unit: minute)	SD	Min	Max
North	59	8.3	6.1	1	30
East	27	15.4	12.8	2	60
South	21	6.4	5.4	1.5	25
West	32	8.5	5.4	1	20
Middle	37	11.0	7.0	5	40
Total	176	9.8	7.9	1	60

4.3 Cruising Time to Find Parking Lot by Access Directions

In Table 8.11, we provide the cruising time visitors spend to find the vacancy of parking lots from entering the city center for each parking block.

While it is natural that the middle block, the center of city center retail environment, shows the longer time the visitors take to find the vacancy of parking lots, the east block attains the longest cruising time.

5 Parking Capacity Analysis by Parking Blocks

5.1 Arrival Time and Departure Time by Parking Blocks

First we see the distributions of arrival time and departure time of visitors by car for each of five parking blocks. Figure 8.5 illustrates the distributions of arrival time and departure time for the whole sample. How to read the figure is the same as Fig. 8.1, where the bigger or smaller of the area surrounded by two cumulative distribution functions in the figure indicates the longer or shorter of the staying time at the city center.

Figure 8.6a–e depicts cumulative distributions of arrivals and departures for car visitors who utilize the north, east, south, west, and middle parking blocks, respectively. It seems that the average staying time for the car visitors who utilize the middle parking bock is shorter than those who utilize other parking blocks.

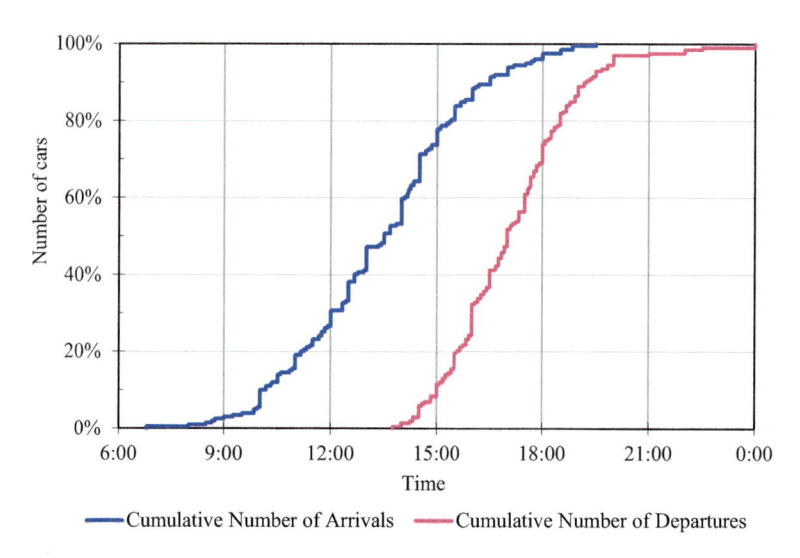

Fig. 8.5 Cumulative distributions of arrivals and departures (whole samples)

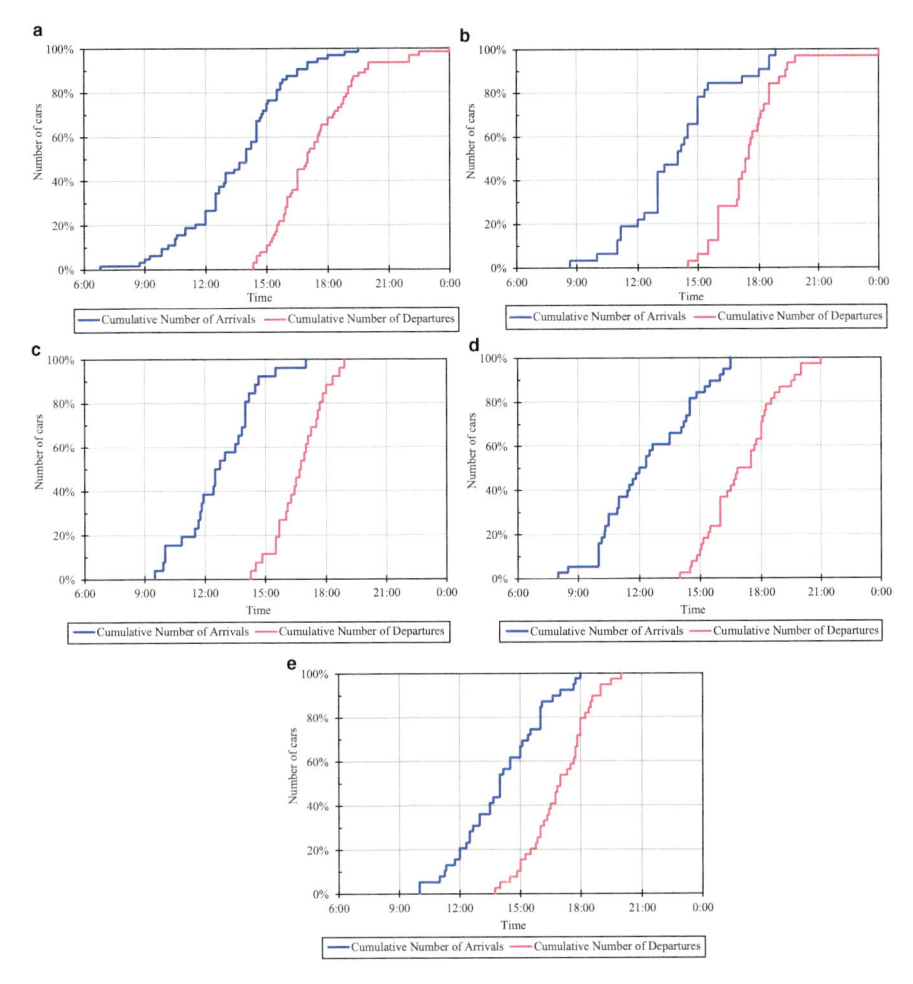

Fig. 8.6 Cumulative distributions of arrivals and departures by five parking blocks (**a**) North parking block (**b**) East parking block (**c**) South parking block (**d**) West parking block (**e**) Middle parking block

5.2 Arrival Rates for Parking Blocks

Now we will make parking capacity analysis by parking blocks. As in Sect. 3, we need the arrival rates and average staying time for five parking blocks.

First, in Fig. 8.7 we show the distributions of arrival time by parking blocks.

Next in Table 8.12 we give the actual numbers of arrivals by time zones and parking blocks. Here we used the actual number of incoming visitors by parking blocks shown in the bottom row of Table 8.10 and the distributions of arrival time by parking blocks shown in Fig. 8.7.

Last, from Table 8.12 we calculate the arrival rates for five parking blocks as shown in Table 8.13.

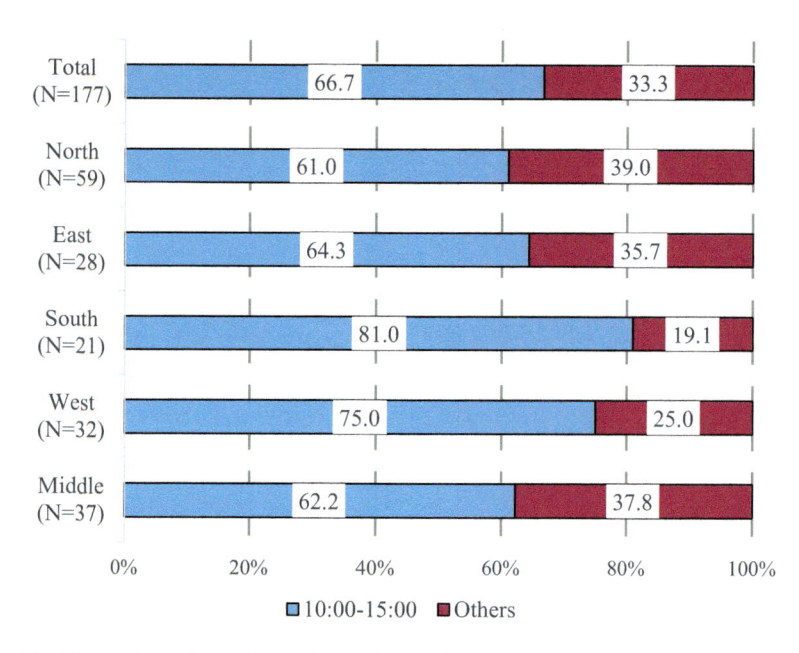

Fig. 8.7 Distributions of arrival time by parking blocks

Table 8.12 Numbers of arrivals by parking block and time zone

	Time zone	Number of arrivals	Percentage	Total
North	10:00–15:00	4,301	61.0	7,049
	Other time	2,748	39.0	
East	10:00–15:00	2,151	64.3	3,346
	Other time	1,195	35.7	
South	10:00–15:00	2,030	81.0	2,508
	Other time	478	19.1	
West	10:00–15:00	2,868	75.0	3,824
	Other time	956	25.0	
Middle	10:00–15:00	2,747	62.2	4,420
	Other time	1,673	37.8	

Table 8.13 Arrival rate by parking block

	Time zone	Number of arrivals	Time (minutes)	Arrival rate
North	10:00–15:00	4,301	300	14.34
	All time	7,049	1,080	6.53
East	10:00–15:00	2,151	300	7.17
	All time	3,346	1,080	3.10
South	10:00–15:00	2,030	300	6.77
	All time	2,508	1,080	2.32
West	10:00–15:00	2,868	300	9.56
	All time	3,824	1,080	3.54
Middle	10:00–15:00	2,747	300	9.16
	All time	4,420	1,080	4.09

Table 8.14 Parking lots demand for five parking blocks by Little's formula

	Time zone	Arrival rate (λ)	Average staying time (W)	Demand for parking lots (number of cars) (L)
North	10:00–15:00	14.34	224.6	3,220
	All time	6.53	224.6	1,466
East	10:00–15:00	7.17	224.6	1,610
	All time	3.10	224.6	696
South	10:00–15:00	6.77	224.6	1,520
	All time	2.32	224.6	522
West	10:00–15:00	9.56	224.6	2,147
	All time	3.54	224.6	795
Middle	10:00–15:00	9.16	224.6	2,057
	All time	4.09	224.6	919

5.3 Average Staying Time for Visitors by Car

We use the same data of average staying time (waiting time) for all parking blocks. Hence the staying time is 224.6 min as in Sect. 3.

5.4 Demand for Parking Lots for Five Parking Blocks Calculated by Little's Formula

Now we can estimate the demand for parking lots for each parking block using Little's formula. Table 8.14 gives these results.

5.5 Parking Capacity Analysis for Five Parking Blocks

Now we can carry out the parking capacity analysis for each parking block. Table 8.15 shows the result of the analysis.

From the table, contrary to our intuition, four parking blocks out of the five have enough parking capacity. All parking blocks except the east block are excess supplied by parking lots. The numbers of over supplied parking lots for each of five parking blocks seem to be roughly proportional to the cruising time for vacancy for each parking block given in Table 8.11.

There might be several reasons for this result. One is that as mentioned before, since we have ignored the demand for business use, the demand for parking capacity calculated here might be underestimated. But our 1st survey of consumer parking behavior was conducted on Saturday and Sunday so that the effect of business use can be small. Other factors might be due to the choice of time period to calculate arrival rates and the number of total incoming visitors by car. It is apparent that if we choose the peak arrival rate, the parking capacity needed would become large. Thus the issue of how to choose the time period for calculating arrival rates becomes important, but we would like to leave it for a further study.

Here we would like to elaborate on the latter issue of the number of incoming visitors by car. We have employed 21,150 as the number of visitors by car per day in average over the year. This estimated number is obtained by averaging the numbers of weekdays and weekends. Hence it does not represent the number of weekends.

Now assume that the numbers of Saturday's and Sunday's visitors are, respectively, 1.5 times and 2 times as large as that of weekday's visitors. Thus let us assume that the average number of Saturday's and Sunday's visitors by car is 1.75 times 21,150, that is, 37,013 persons per day. Table 8.16 gives the result of parking capacity analysis under this assumption.

Table 8.15 Demand and supply of parking capacity by parking blocks

	Time zone	(a) Demand for parking capacity (unit: number of cars)	(b) Supply of parking capacity (unit: number of cars)	(c)= (a)–(b)	(d)= (c)/(b)
North	10:00–15:00	3,220	4,506	−1286	−0.29
	Others	1,466	4,506	−3040	−0.67
East	10:00–15:00	1,610	998	612	0.61
	Others	696	998	−302	−0.30
South	10:00–15:00	1,520	2,732	−1212	−0.44
	Others	522	2,732	−2210	−0.81
West	10:00–15:00	2,147	2,692	−545	−0.20
	Others	795	2,692	−1897	−0.70
Middle	10:00–15:00	2,057	2,399	−342	−0.14
	Others	919	2,399	−1480	−0.62

Table 8.16 Parking capacity analysis under another assumption

	Time zone	(a) Demand for parking capacity (unit: number of cars)	(b) Supply of parking capacity (unit: number of cars)	(c)= (a)–(b)	(d)= (c)/(b)
North	10:00–15:00	5,635	4,506	1,129	0.25
East	10:00–15:00	2,818	998	1,820	1.82
South	10:00–15:00	2,660	2,732	−72	−0.03
West	10:00–15:00	3,758	2,692	1,066	0.40
Middle	10:00–15:00	3,600	2,399	1,201	0.50

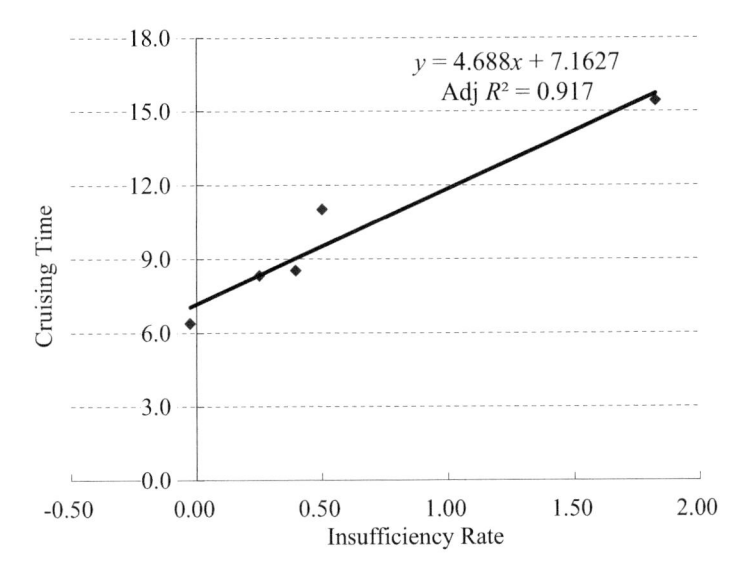

Fig. 8.8 Relation of insufficiency rates and cruising time

We see that all parking blocks except the south now are insufficient for parking lots. From Fig. 8.8, we have known that those insufficiency rates for parking blocks are closely related to the cruising time for each parking block.

6 Conclusion

We have shown a simple method to investigate whether or not parking capacity is sufficient at the actual city center retail environment based on the survey of consumer parking behavior using Little's formula. By a concrete example of Fukuoka City, we have demonstrated that we can determine numerically how much parking capacity is needed at the actual city center retail district using Little's formula when we have the following information: (1) the total number of incoming visitors at city center by car, (2) the length of staying time of visitors who visit city center by car for shopping, and (3) the arrival time distribution of visitors by car. Also we have indicated that as shown in the analysis of access directions, for the analysis of parking space policy, the access structure to the city center of a specific city and the driver's behavioral mechanism such as avoiding congestion may greatly affect the effectiveness of the policy measures for the specific city.

We also have noted that sufficiency and insufficiency of parking capacity critically depend on the number of incoming visitors and the choice of time period for calculating arrival rates. Thus in addition to the parking capacity analysis for business demand, further studies should be needed for exploring how we should decide to choose the time period of peak arrival rates and how we should deal with various variations of the numbers of incoming visitors such as seasonal, day to day, within-a-day, between weekdays and weekends, and so on.

References

1. Arnott R, Rave T, Schob R (2005) Alleviating urban traffic congestion. MIT Press, Cambridge, MA
2. Little JDC (1961) A proof for the queuing formula: $L = \lambda W$. Oper Res 9(3):383–387
3. Ramalhoto MF, Amaral JA, Teresa Cochito M (1983) A survey of J. Little's formula. Int Stat Rev 51(3):255–278
4. Takahashi Y (1991) Queuing model. In: Hironaka H (ed) Part XII or of encyclopedia of modern mathematical sciences. Osaka Books, pp 674–675. (in Japanese)
5. Saito S (2000) Report on the survey of consumer shop-around behavior at city center commercial district of Fukuoka City with focusing on underground space and comparison of attractiveness of Japanese and Korean cities. Institute of Urban Science, City Government of Fukuoka. (in Japanese)
6. Saito S, Sato T, Yamashiro K (2008) An analysis of consumers' parking behaviors at city center commercial district. Paper presented at the 45th annual meeting of the Japan Section of Regional Science Association International (JSRSAI). (in Japanese)
7. Saito S, Sato T, Yamashiro K (2010) Little's formula and parking space policy viewed from consumers' parking behaviors at city center retail environment. Paper presented at The Ninth International Symposium on Operations Research and Its Applications held at Chengdu-Jiuzhaigou, China. Oper Res Appl 12:500–511
8. Saito S, Yamashiro K, Iwami M, Imanishi M (2014) Parking space policy for midtown commercial district and consumers' parking and shop-around behaviors: applying Little's formula to the analysis of demand-supply balances for parking capacity in Tenjin area, the midtown of Fukuoka, Japan. Fukuoka Univ Rev Econ 58:75–98. (in Japanese)

Part III
Economic Effects by Accelerating *Kaiyu*

Chapter 9
The Economic Effects of City Center 100-Yen Circuit Bus

Saburo Saito and Kosuke Yamashiro

Abstract NNR (Nishi-Nippon Railroad Co., Ltd.) introduced the city center 100-yen (1-dollar) circuit bus in Fukuoka City, Japan, in July 1999. NNR first announced that it was an experiment until March 2000, while it could be continued as a business after the experiment period if its operation record would pass the given criteria of profitability. Surprisingly, during the past 8 months, the city center 100-yen bus has attracted over 1.8 times as many customers as before. NNR has decided to keep on running the city center 100-yen bus as a business from April 2000. The purpose of this paper is to estimate the economic effect of the city center 100-yen bus at city center retail environment of Fukuoka City based on data obtained from surveys on consumer's shop-around behavior conducted in March 2000.

Keywords 100-yen bus · Circuit bus · *Kaiyu* · Shop-around behavior · Economic effect · Fukuoka City · NNR · City center · One dollar

---

This chapter is based on the paper, Saburo Saito and Kosuke Yamashiro [49], "Economic Impacts of the Downtown One-Dollar Circuit Bus Estimated from Consumer's Shop-Around Behavior: A Case of the Downtown One-Dollar Bus at Fukuoka City," *Studies in Regional Science*, vol. 31, pp. 57–75, 2001 (in Japanese).

S. Saito (✉)
Faculty of Economics, Fukuoka University, Fukuoka, Japan

Fukuoka University Institute of Quantitative Behavioral Informatics for City and Space Economy (FQBIC), Fukuoka, Japan
e-mail: saito@fukuoka-u.ac.jp

K. Yamashiro
Department of Business and Economics, Nippon Bunri University, Oita City, Japan
e-mail: yamashiroks@nbu.ac.jp

© Springer Nature Singapore Pte Ltd. 2018
S. Saito, K. Yamashiro (eds.), *Advances in Kaiyu Studies*, New Frontiers in Regional Science: Asian Perspectives 19, https://doi.org/10.1007/978-981-13-1739-2_9

1 Purpose of Research

In recent years, circulating bus services that travel around major facilities in city center urban areas are operated on trial basis throughout the country. The main objectives of these pilot circulating bus services are (1) to alleviate traffic congestion by suppressing the inflow of private cars into central urban areas, (2) to encourage use even for short distances (by setting a low fare), and (3) to revitalize central shopping streets and central urban areas, etc. To estimate the economic impacts of this service, some have proposed using traditional methods based on the measurement of consumer surplus based on aspects such as the shortening of traffic time [46]. However, obtaining an actual measurement has proven to be difficult, and these services have remained as social experiments mainly because they are small-scale experiments and, in some cases, they are being implemented free of charge [43].

Nishi-Nippon Railroad Co., Ltd. (NNR) introduced the city center 100-yen circuit bus around the main consumer attraction facilities in the city center of Fukuoka City, Japan in July 1999. It was a revolutionary attempt in the following sense. First, although city center circuit buses in other parts of the country were introduced on an experimental basis, the Fukuoka 100-yen circuit bus was introduced as a fully commercial service from the beginning with an end-date of March 2000. Second, a 100-yen fare system was introduced like the *Musashino* City Community Bus (Mu-bus), and most importantly, with the introduction of the 100-yen bus, a 100-yen zone was established within the city center district since the fares of all scheduled buses running through it were reduced and unified to 100 yen from the then-prevailing fare of 180 yen. The introduction of the city center 100-yen bus gained much attention, the number of bus ridership in the city center district increased by a substantial 1.8 times, and since April 2000, it has been fully operational.

Saito and others have studied *Kaiyu*, which is the Japanese term for the consumer shop-around behavior within a city center retail environment, as they believe that is an important evaluation framework for city center commercial environments as seen from the viewpoint of urban development [8–10, 12, 19–39]. The most recent review of these studies is in [9].

Here, we outline these studies focusing on their relationship with the present study.

The three research topics that have served as background for the above studies on *Kaiyu* in city center commercial districts led by Saito are as follows: (1) research on pedestrian flow (Fukami [2, 3] and Hagishima et al. [4]), (2) research on the trip chain (Hanson [5–7], Kondo [13], Kondo and Kitamura [14], Lerman [15], O'Kelly [18], Saito [19, 22, 23], Saito et al. [40], Sakamoto [41], Sasaki [42], and Wrigley and Dun [45]), and (3) research on modification and extension of the Huff model (Takeuchi [44] and Saito [20, 21]).

Borger and Timmermans [1] includes a detailed review of existing studies published in English on pedestrian flow and trip chains in commercial areas. In addition, Saito and Ishibashi [31] contains a detailed review with emphasis on existing research published in Japanese. Ishibashi and Saito [9] is a review that includes recent developments.

Fukami [2, 3] focused on the fact that the pedestrian flow in commercial areas has a large influence on sales of individual stores, including the traffic volume in front of the store, and that its forecast is indispensable for the assessment of commercial area development plans. Thus, to address this, he developed a Monte Carlo simulation model for pedestrian flow forecast. This issue was similarly addressed in Hanson [5–7]. We find it interesting that Western authors were also tackling this same issue around the same time. Inspired by Fukami [2, 3], Saito [19] considered that shop-around behavior is a concrete manifestation of the agglomeration effect of commercial facilities and developed a shop-around Markov model of infinite number of shop-around as its measurement model. This model is reported in Sakamoto [41].

Meanwhile, Takeuchi [44] introduced the idea of simultaneous behaviors in the Huff model, whereby while consumers perform their main purpose such as commuting, they may engage in secondary purposes such as eating and drinking. This resulted in increasing the prediction accuracy of the sales amount in the retail sector of eating and drinking whose prediction accuracy was previously quite low. Motivated by the research of Takeuchi [44], Saito [20, 21] considered shopping trips in city center commercial districts as a form of *Kaiyu*, where consumers walked around stores and commercial sites within a city center district, and developed a disaggregate hierarchical choice Huff model considering *Kaiyu* among commercial districts. The model of Saito [20, 21] is the first disaggregate Huff model in Japan, which employed a disaggregate logit model. It also modeled consumers' multistage hierarchical decision-making such as the destination city choice, large-scale store choice, and choice of commercial district to shop-around as a recursive causal relationship, formulating it as a multivariate logit model that expressed the relationship by a recursive system of structural equations, which was an unprecedented model globally at that time.

However, although the model of Saito [20, 21] considered *Kaiyu*, it still belonged to the class of traditional probability-based models in the sense that it attempted to explain the choice probability of the Huff model. Thus, the study that actually triggered the recent *Kaiyu* research led by Saito was the one carried out by Saito and Ishibashi [30], which sparked the turning from a probability-based approach to the exploration of a frequency-based approach, such as how many people come to town.

Saito and Ishibashi [30] introduced three new aspects. It might be said that the subsequent research has been more or less related to the refinement and extension of these three aspects.

The first was the method of collecting *Kaiyu* data in the city center district. Previous studies collected *Kaiyu* data by using shopping behavior surveys conducted where consumers reside, that is, home-based surveys. In contrast, Saito and Ishibashi [30] collected *Kaiyu* data through on-site *Kaiyu* surveys conducted at the city center district, in which respondents are randomly sampled from the visitors there. The second was the introduction of explanatory variables into the consumer *Kaiyu* Markov model, which had been limited to describing the current state, making it possible to predict how the pattern of *Kaiyu* would be affected by redevelopment. The third was to forecast and estimate based on "real numbers" how many people actually come to the city center district and how many people move from where to where in the city center area. In fact, in their model, the number of visitors incoming the city center district at present is estimated separately for each type of transportation modes, and this estimated number at present is fixed as given, and under this given present number of incoming visitors, the changes in the *Kaiyu* pattern after the city center retail redevelopment are predicted on a real number basis.

Recently, Saito [24], Saito et al. [33], and Kakoi [12] developed and refined a Poisson model that can predict the increase in the number of incoming visitors due to redevelopment only with the data obtained from on-site surveys of *Kaiyu* behaviors. This model has been applied to the construction of a consumer *Kaiyu* Markov model [36, 37] that considers the increase in the number of incoming visitors due to redevelopment and the construction of a consumer *Kaiyu* Markov model on a monetary basis [8, 11, 37]. In addition, Saito et al. [35] have also presented some theoretical issues related to the on-site surveys of *Kaiyu* behaviors.

In the city center of Fukuoka City, there are two core business districts, the Tenjin district, where the NNR line and the subway connect, and the district around the Japan Railways (JR) Hakata Station where the JR line and the subway connect. In the city center of Fukuoka City, there was a series of large-scale commercial redevelopment projects that were carried out from 1996 to 1998 led by the opening of Canal City Hakata in April 1996, and the Tenjin district also changed significantly, for example, the sales floor area in department stores increased 2.7 times with the opening of two new department stores and the expansion of the sales floor areas in existing stores.

Saito et al. [38] conducted on-site *Kaiyu* surveys three times during the period that these redevelopment projects took place and investigated how *Kaiyu* patterns, which are movements of consumers among commercial facilities in the city center district measured by actual numbers, had changed as redevelopment projects proceeded. They verified that the center of gravity of the Tenjin district, from the perspective of flow of visitors, moved 105 m southward. This finding was featured widely in the local media [47].

Under these circumstances, the city center 100-yen circuit bus for the city center district of Fukuoka City was originally planned with the aim to increase *Kaiyu* between the Hakata Station district and the Tenjin district [48].

Saito et al. [38] showed that the redevelopment had an effect on the number of incoming visitors and *Kaiyu*. Thus, an additional point of interest is to measure how much *Kaiyu* was promoted by the introduction of the city center 100-yen bus, how it contributed to the revitalization of commerce in the city center district, and how much economic effect it created.

The aim of this research exactly is to measure the economic effect the city center 100-yen bus brought on the city center district.

Limited research has been conducted to measure and estimate the effectiveness of city center circuit buses for revitalization, even though revitalization of city center commercial districts is often the reason for its introduction. The reason includes those that have already been mentioned, as well as others such as the following: (1) conventional methods based on traditional transportation research require "person trip" OD (origin destination), but OD estimation in small areas, such as city center districts, has rarely been attempted; (2) in addition to estimating in a small area, it is necessary to estimate the OD trips with shopping and leisure purposes; and (3) merely changing the means of transportation from walking to riding the city center circuit buses does not create any new final demand.

On March 2000, there was an opportunity to conduct the fifth *Kaiyu* survey at the city center of Fukuoka City. The purpose of the present study is to analyze the *Kaiyu* data of the samples in this survey who utilized the city center 100-yen bus and to estimate the economic effect of the city center 100-yen bus using the following criteria.

Passengers of the city center 100-yen buses are expected to increase the number of stops in the city center district for shopping and leisure, that is, the number of consumer *Kaiyu* steps, due to shortening traveling time between facilities. On the other hand, from the *Kaiyu* survey data, we have obtained the average expenditure per consumer *Kaiyu* step. Therefore, the expected increase in expenditure in the city center commercial district due to the use of the city center 100-yen bus can be calculated by multiplying the average expenditure amount per consumer *Kaiyu* step with the increase in the number of consumer *Kaiyu* steps expected by the use of the city center 100-yen bus. The idea is that this increase in the expenditure at the city center commercial district can be thought of as the economic effect of the city center 100-yen bus since this expenditure increase is accrued to the use of the city center 100-yen buses.

Following this approach, the purpose of this study is to estimate the economic effect of introducing the city center 100-yen bus on the retail sector of the city center commercial district of Fukuoka City by utilizing data published by NNR and data from our above survey of *Kaiyu* behaviors.

2 Features of the Fukuoka City Center 100-Yen Bus

2.1 Operating Area and Routes of the Fukuoka City Center 100-Yen Bus

In this study, we call the city center 100-yen circuit bus that NNR introduced in the city center district of Fukuoka City as the 100-yen circuit bus and the city center 100-yen route bus as the 100-yen route bus. When we do not make a distinction between these two, we call them both as the city center 100-yen bus or simply the 100-yen bus.

Figure 9.1 shows the city center district of Fukuoka City, the 100-yen fare zone, and the operating routes of the 100-yen circuit buses.

Following previous studies [27–29, 38], in this study we define the city center district of Fukuoka City as the area that includes the Tenjin district and the Hakata Station district surrounded by the following 50-m-wide roads, *Watanabe* street, *Sumiyoshi* street, *Taihaku* street, and *Showa* street. The 100-yen fare zone introduced

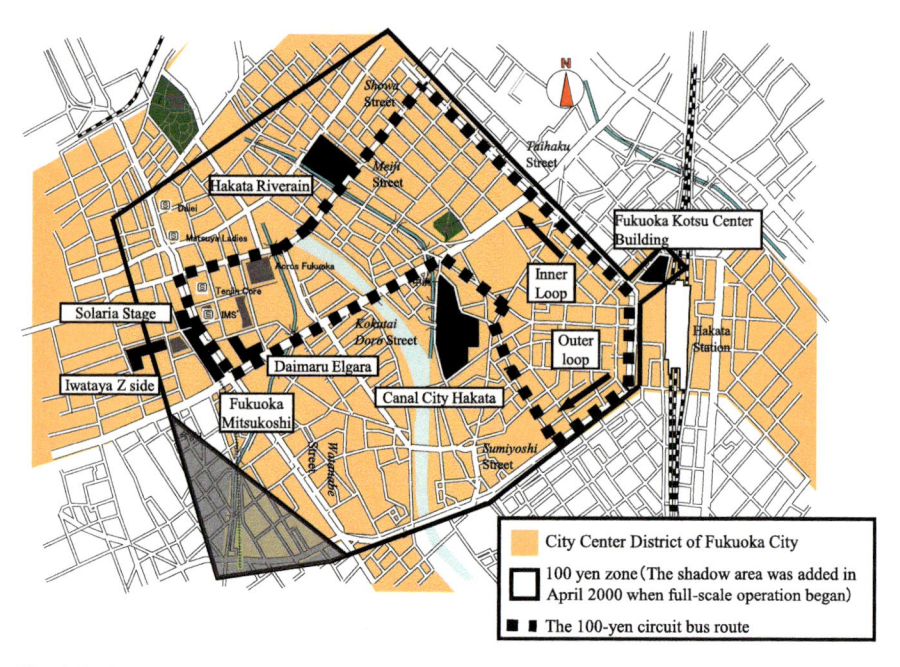

Fig. 9.1 City center of Fukuoka and the 100-yen bus zone

by NNR is set to be almost the same as the city center district of Fukuoka City as defined above.

The operating routes of the 100-yen circuit bus shown in Fig. 9.1 have two directions, the outer and inner routes. The outer route starts at JR Hakata Station, the entrance to the Kyushu region, and passes by Canal City Hakata, which opened in 1996, the commercial establishments in the Tenjin district, and the Hakata Riverain, which opened in 1999, before looping back to Hakata Station. The inner route circulates in the opposite direction.

2.2 Number of Units in Operation and Frequencies

The total number of the city center 100-yen circuit buses that are operating is 255 on weekdays, weekends, and national holidays. The breakdown is 126 operating on the outer route and 129 on the inner route, both with Hakata Station as the starting point. Regarding the frequencies, there are buses about every 5 min with the first one departing at 9:15 a.m. and the last one departing at 7:47 p.m. from Hakata Station. The meaning that the city center 100-yen circuit bus was introduced on a commercial basis from the start is reflected in the operation interval of 5 min. The number of people who feel that it is too far to walk sharply increases when the walking distance exceeds 400 m, and at 800 m, almost everyone feels that it is too far (Saito et al. [40]

and Nagoya Urban Institute [46], p. 9, Figure 2–20). In the city center Fukuoka, it is a 10-min walk from Tenjin to Canal City Hakata. Therefore, if the operation interval exceeds 10 min, people may prefer to walk even if they feel it is too far, rather than wait for the 100-yen circuit bus. Also, the route connecting Tenjin, Riverain, and Hakata Station partly overlaps with the Fukuoka City subway line, so one cannot miss the aspect of price competition with the subway's minimum fare which is 200 yen. Therefore, the introduction of the city center 100-yen bus was not a one-off or experimental introduction like in other cities. Rather, it was introduced in response to the demand for a short-distance transportation for business, commuting, shopping, and leisure in the city center district of Fukuoka City.

2.3 Results After the Introduction of the City Center 100-Yen Bus

Tables 9.1 and 9.2 show the trends in the number of monthly passengers of the city center 100-yen bus released by NNR and the ridership of the 100-yen circuit bus and the ordinary route bus within the zone.

Initially, the city center 100-yen bus was introduced for a limited period from July 1999 to March 2000. One hundred-yen buses have been introduced in various cities as policy experiments, but this was the first case of introducing it on a large scale in a city center district. It was also the first time that a 100-yen bus zone had been established by setting a uniform fare of 100 yen for all buses operating in it. According to NNR, the number of the city center 100-yen bus passengers in March 2000 was about 51,000 per day [17] and increased by 80% compared to the period before its introduction. With the surprisingly high popularity, the 100-yen zone was partially expanded, and full-scale operation began from April 2000.

3 User Characteristics of the City Center 100-Yen Bus According to the Survey of *Kaiyu* Behaviors at the City Center of Fukuoka City

3.1 Survey of Kaiyu Behaviors at the City Center of Fukuoka City

In this study, we use survey data from the fifth survey of *Kaiyu* behaviors at the city center of Fukuoka City conducted by the Saito Laboratory at Fukuoka University on the following 3 days, March 17 (Friday), 18 (Saturday), and 19 (Sunday), 2000. The survey was conducted for the first time in September 1996, and the second, third, and fourth surveys were conducted in July 1997, January and February 1998, and June 1999, respectively.

Table 9.1 Changes in the number of passengers of the city center 100-yen bus (number of passengers per day, unit: people, %)

	Weekday			Saturday			Sunday and holidays			Average		
	This year	Previous year	YoY	This year	Previous year	YoY	This year	Previous year	YoY	This year	Previous year	YoY
July	55,861	35,596	156.9	45,470	26,112	174.1	39,433	19,178	205.6	51,535	31,418	164.0
August	57,733	32,877	175.6	45,977	22,168	207.4	39,740	17,921	221.8	53,314	29,083	183.3
September	55,631	33,382	166.6	43,414	24,277	178.8	38,410	18,820	204.1	50,558	29,256	172.8
October	56,760	34,315	165.4	43,629	23,820	183.2	37,028	18,392	201.3	50,823	29,879	170.1
November	55,439	33,207	166.9	42,153	24,037	175.4	37,989	18,085	210.1	50,178	28,960	173.3
December	55,225	31,569	174.9	48,842	24,656	198.1	37,636	18,196	206.8	51,564	28,069	183.7
January	53,498	30,660	174.5	42,744	18,893	226.2	32,768	14,599	224.5	47,430	25,136	188.7
February	56,616	31,217	181.4	42,962	22,218	193.4	35,562	16,486	215.7	51,102	27,436	186.3
March	58,862	32,661	180.2	43,789	23,593	185.6	38,810	17,543	221.2	53,683	29,052	184.8
July – March 31	56,214	32,851	171.1	44,343	23,266	190.6	37,317	17,612	211.9	51,145	28,756	177.9

Increase in number of passengers per day: 22,389 people
Created by authors from "Changes in the number of passengers of the Fukuoka city center 100-yen bus (As of the end of March 2000)" by NNR [17]

9 The Economic Effects of City Center 100-Yen Circuit Bus

Table 9.2 Ridership of the 100-yen circuit bus and scheduled route buses within the zone (average number of passengers per day, unit: people, %)

	The 100-yen circuit bus			The scheduled route bus		
	Passengers (people)	Share (%)	Riders per bus (people)	Passengers (people)	Share (%)	Total (people)
July	9784	19.0	36	41,751	81.0	51,535
August	10,682	20.0	39	42,632	80.0	53,314
September	8926	17.7	37	41,632	82.3	50,558
October	9061	17.8	37	41,762	82.2	50,823
November	8575	17.1	34	41,603	82.9	50,178
December	8568	16.6	34	42,996	83.4	51,564
January	8447	17.8	33	38,983	82.2	47,430
February	8376	16.4	33	42,726	83.6	51,102
March	8701	16.2	34	44,982	83.8	53,683
Average	9013	17.6	35	42,119	82.4	51,132

Average number of passengers of the 100-yen circuit bus: 9020 people
Created by authors from "Utilization of 100-yen circuit buses and scheduled route buses within the zone" by NNR [17]

The on-site survey of *Kaiyu* behaviors at city center Fukuoka is about 15-min interview survey with questionnaire sheets, in which multiple survey points are set up in the city center district, the respondents are randomly sampled from the visitors visiting survey points, and the respondents are asked about their *Kaiyu* patterns or paths concerning how they walked around the city center district. The main survey items, besides the *Kaiyu* pattern in the city center on the day of the survey, are respondents' attributes, frequency of visits to the city center of Fukuoka City, shopping behaviors, and so on. Regarding *Kaiyu* patterns, respondents were asked which place they visited, what purpose done there, and how much they spent there in the order of their occurrence. Also, the respondents were asked about their planned *Kaiyu* behavior after leaving the survey point where they were interviewed.

In the fifth survey of *Kaiyu* at the city center of Fukuoka City, 12 survey points were set up, and valid 1328 samples were collected. As for the survey points, to cover the core of the city center district, which is serviced by the city center 100-yen bus, and also the Tenjin district, Hakata Station district, Canal City Hakata, and Hakata Riverain, 7 points were set up at major stores and the underground shopping mall in the Tenjin district, 2 points in the station concourse and underground shopping mall in the Hakata Station district, 2 points in Hakata Riverain, and 1 in Canal City Hakata, for a total of 12 points. The survey was administered for 8 h from noon to 8 p.m. The characteristics of the 1328 samples that were collected in the survey are shown in Tables 9.3, 9.4, 9.5, and 9.6.

Table 9.3 shows the number of samples by day of the week, Table 9.4 shows the number of samples by survey point, Table 9.5 shows the ratio by gender, and Table 9.6 shows the ratio by age. Tables 9.5 and 9.6 show characteristics of visitors to the city center commercial district of Fukuoka City, for example, the high proportion of women, and high rate of people in their 20s.

Table 9.3 Number of samples by survey day

	Samples	%
March 17, 2000 (Fri)	446	33.6
March 18, 2000 (Sat)	456	34.3
March 19, 2000 (Sun)	426	32.1
Total	1328	100.0

Table 9.4 Number of samples by survey point

		Samples	%
Tenjin district	Solaria Plaza	107	8.1
	Shoppers Daiei	111	8.4
	Iwataya Z side	41	3.1
	Daimaru Elgala	148	11.1
	Fukuoka Mitsukoshi	139	10.5
	Tenjin Underground Shopping Mall	154	11.6
	Shintencho Shopping Street	95	7.2
Hakata Station district	Hakata Station Concourse	141	10.6
	Hakata Station Underground Shopping Mall	90	6.8
Hakata Riverain	Hakata Riverain	101	7.6
	Atrium Garden	55	4.1
Canal City Hakata		146	11.0
Total		1328	100.0

Table 9.5 Number of samples by gender

	Samples	%
Male	330	24.9
Female	993	75.1
Total	1323	100.0
Unknown	5	

Table 9.6 Number of samples by age

	Samples	%
16–19 years old	285	21.5
20–24 years old	460	34.7
25–29 years old	150	11.3
30–34 years old	72	5.4
35–39 years old	44	3.3
40s	105	7.9
50s	105	7.9
60s	67	5.1
Above 70	36	2.7
Total	1324	100.0
Unknown	4	

9 The Economic Effects of City Center 100-Yen Circuit Bus

Table 9.7 Number of samples by occupation

	Samples	%
High school student	119	9.1
College/University student	349	26.6
Specialty school student	65	5.0
Housewife	135	10.3
Office worker	176	13.4
Technical worker	68	5.2
Sales/service industry worker	101	7.7
Labor employee	16	1.2
Company executive	43	3.3
Self-employed	28	2.1
Liberal professions	17	1.3
Part-time worker	81	6.2
Agriculture/fishery worker	1	0.1
Unemployed	60	4.6
Others	54	4.1
Total	1313	100.0
Unknown	15	

Table 9.8 Number of samples by place of residence

	Samples	%
Higashi-ku	135	10.4
Hakata-ku	126	9.7
Chuo-ku	130	10.0
Minami-ku	103	8.0
Nishi-ku	41	3.2
Jonan-ku	112	8.6
Sawara-ku	96	7.4
Fukuoka City total	743	57.4
The Fukuoka daily living area excluding Fukuoka City	211	16.3
The Fukuoka daily living area total	954	73.7
The Fukuoka metropolitan area excluding Fukuoka living area	126	9.7
The Fukuoka metropolitan area total	1080	83.4
Other	215	16.6
Total	1295	100.0
Unknown	33	

Table 9.7 is a summary by occupation. Junior college and university students accounted for 26.6%, followed by office workers that accounted for 13.4%.

Table 9.8 summarizes the number of samples according to the respondents' place of residence sorted into the following four categories: the seven wards of Fukuoka City, Fukuoka daily living area, Fukuoka metropolitan area, and other areas. The

Fukuoka daily living area is an area where 20% or more of the employed permanent residents work or attend school in Fukuoka City.[1] As for the Fukuoka metropolitan area, following the definition by the national census for the Kitakyushu and Fukuoka metropolitan area, it is defined as the area of continuous municipalities where more than 1.5% of their permanent population commute to Fukuoka City. Approximately 57% of the visitors were residents of Fukuoka City, approximately 74% were from the Fukuoka daily living area, and approximately 83% were from the Fukuoka metropolitan area.

3.2 Characteristics of the City Center 100-Yen Bus Users

In this section, we verify whether the utilization of the city center 100-yen bus increased *Kaiyu* by comparing the *Kaiyu* behaviors between users and nonusers of the city center 100-yen bus based on the data obtained from the fifth survey of *Kaiyu* behaviors at the city center of Fukuoka City. But first, we examine features such as the number of stops, while the respondents were doing *Kaiyu* to understand the concept of consumer *Kaiyu* steps, which is important in estimating the economic effect of the city center 100-yen bus.

Number of Steps During *Kaiyu*

In this study, *Kaiyu* is defined as the behavior of walking around in the city center district. "Walking around" behavior is considered to have occurred, when the respondents either changed their visit place or changed their purpose, while they are at the same place[2].

In the fifth survey of *Kaiyu* behaviors at the city center of Fukuoka City, the purposes of the visits were classified into 42 types, including 35 types of shopping, dining, business, leisure, etc. In addition, there were 16 types of purposes related to travel behaviors such as getting on and off transportation, parking, and use of transportation facilities. The entire city center district was divided into 27 blocks, and numbers were assigned to major stores such as retail, leisure, services, as well as transportation nodes such as parking lots, bus stops, entrances to subways, and trains. In other words, we classified nodes into two types, attraction nodes, which

---

[1]This includes 9 cities and 13 towns: Fukuoka City, *Ogori* City, *Chikushino* City, *Kasuga* City, *Onojo* City, *Munakata* City, *Dazaifu* City, *Maebaru* City, *Koga* City, *Nakagawa* Town, *Umi* Town, *Sasaguri* Town, *Shime* Town, *Sue* Town, *Shingu* Town, *Kasuya* Town, *Fukuma* Town, *Genkai* Town, *Nijo* Town, *Hisayama* Town, *Tsuyazaki* Town, and *Shima* Town as of 2000.

[2]We consider as "walking around" occurred or *Kaiyu* occurred the cases where the purpose changed, while a consumer stays at the same destination. For example, the case when a consumer purchased a different item from the previous ones within the same store, such as a department store is recorded as *Kaiyu*.

Table 9.9 Total number of stops including attraction and transportation nodes

Step	Frequency	%
2	5	0.4
3	111	8.4
4	163	12.3
5	173	13.0
6	170	12.8
7	156	11.7
8	124	9.3
9	103	7.8
10	81	6.1
11	74	5.6
12	51	3.8
13	30	2.3
14	27	2.0
15	19	1.4
16 over	41	3.1
Total	1328	100.0

are visited for purposes such as shopping, leisure, dining, errands, and services including stores and hotels, and transportation nodes, which are visited for the purpose of using transportation facilities, such as railroad entrances, bus stops, and parking lots.

The main survey items on *Kaiyu* pattern focus primarily on how people walked around the attraction nodes in the city center district after leaving their homes. However, with regard to movements between attraction nodes, in cases when people traveled by other means than by foot, such as using their own cars or public transportation, their use was recorded in terms of getting on and off at transportation nodes, such as getting on and off buses/subways/trains or parking, in order to understand the details of their movement.

Table 9.9 shows the frequency distribution of the total number of stops including attraction and transportation nodes for all samples. In addition, Table 9.10 shows the frequency distribution of the number of stops at attraction nodes.[3] For example, in the case of *Kaiyu* pattern of walking around in the following manner, the total number of stops is six, and the number of stops that are attraction nodes is two: home → subway station exit → store one → getting on at bus stop → getting off at bus stop → store two → subway entrance → home. Hereafter, the number of stops at attraction nodes is called the number of consumer *Kaiyu* steps.

---

[3]From the definition of *Kaiyu*, both of the total number of destinations and the number of consumer *Kaiyu* steps include the changes in purposes at the same place (i.e., staying).

Table 9.10 Number of consumer *Kaiyu* steps (number of attraction node stops)

Step	Frequency	%
0	13	1.0
1	126	9.5
2	208	15.7
3	208	15.7
4	189	14.2
5	165	12.4
6	129	9.7
7	78	5.9
8	64	4.8
9	56	4.2
10	35	2.6
11 over	57	4.3
Total	1328	100.0

Table 9.11 Average number of consumer *Kaiyu* steps of users and nonusers of the city center 100-yen bus

	Samples	Average number of *Kaiyu* steps	Standard deviation
Users of 100-yen bus	150	5.26	2.84
Nonusers of 100-yen bus[a]	1178	4.56	2.98
Total	1328	4.65	2.97

[a] This sample is used for the *t*-test performed in Table 9.13.

Average Number of Consumer *Kaiyu* Steps of the City Center 100-Yen Bus Users

In this section, we compare the average number of consumer *Kaiyu* steps between users and nonusers of the city center 100-yen bus. As mentioned earlier, *Kaiyu* is defined as the behavior of walking around in the city center district, and "walking around" occurs either when consumers change the commercial facility they visit or change the purpose of visit, while they stay at the same place. Thus, the number of consumer *Kaiyu* steps is the number of commercial facilities and purposes which consumers visited and performed, while they walked around the city center district.

Also, to compare the average number of consumer *Kaiyu* steps between users and nonusers of the city center 100-yen bus, it is necessary to distinguish these two groups. To do this, considering how we recorded *Kaiyu* patterns as described above, we decided to define as users of the city center 100-yen bus those samples who consecutively stopped at two transportation nodes of bus stops. As a result, the number of people who used the city center 100-yen buses was 150 out of 1328 samples.

Table 9.11 compares the average number of consumer *Kaiyu* steps of users and nonusers of the city center 100-yen bus. The average number of consumer *Kaiyu* steps for the users is 5.26, and for nonusers it is 4.56; users of the city center 100-yen bus walk around 0.7 steps more.

9 The Economic Effects of City Center 100-Yen Circuit Bus

Table 9.12 Average number of consumer *Kaiyu* steps after using the city center 100-yen bus

	Samples	Average number of *Kaiyu* steps	Standard deviation
City center 100-yen bus users	150	3.56	2.53

Table 9.13 Average number of consumer *Kaiyu* steps according to when the city center 100-yen bus was used during the *Kaiyu* process

Stage of 100-yen bus ride during *Kaiyu* process	Samples	Average number of *Kaiyu* steps	Standard deviation
Before *Kaiyu* process	69	4.64	2.42
On the way of *Kaiyu* process[b]	67	6.28***	2.99
After *Kaiyu* process	14	3.43	2.28

[b] The *t*-test is performed to test the difference of averages of *Kaiyu* steps between the sample of nousers of 100-yen bus ("a" in Table 9.11) and the sample of users of 100-yen bus on the way of *Kaiyu* ("b" in this table). It is statistically significant at less than 0.1% level.
*** $p < 0.001$

Table 9.14 Average number of consumer *Kaiyu* steps after the 100-yen bus ride for users who utilized the 100-yen bus on the way of their *Kaiyu* process

	Samples	Average *Kaiyu* steps after bus ride	Standard deviation
Users who utilized 100-yen bus on the way of *Kaiyu* process	67	3.19	2.11

In addition, Table 9.12 shows the average number of consumer *Kaiyu* steps that bus users do *Kaiyu* or shop-around after using the city center 100-yen bus. The table shows that the average number of consumer *Kaiyu* steps after using the city center 100-yen bus is 3.56.

In order to analyze in further detail the effect of the utilization of the city center 100-yen bus on *Kaiyu*, we examined the average number of consumer *Kaiyu* steps of the city center 100-yen bus users depending on whether the city center 100-yen bus was used at the start of, on the way of, or at the end of the *Kaiyu* process. Table 9.13 shows the result. Among the 150 users of the city center 100-yen bus, (1) 69 people used it before visiting the first commercial facility, (2) 67 people used it on the way of *Kaiyu*, and (3) 14 people used it after visiting the last commercial facility. The average number of consumer *Kaiyu* steps is known to vary greatly depending on when the 100-yen bus was used during the *Kaiyu* process, which turns out to be 4.64, 6.28, and 3.43, respectively, for the three categories mentioned above.

The difference of the number of *Kaiyu* steps between 100-yen bus nonusers (4.56 steps) and 100-yen bus users on the way of their *Kaiyu* (6.28 step) is statistically significant at less than 0.1% level by performing *t* test.

In addition, Table 9.14 shows the average number of *Kaiyu* steps after the bus ride for those who rode the city center100-yen bus on the way of *Kaiyu*. The average number of *Kaiyu* steps after the bus ride turns out to be 3.19 for those 100-yen bus users who took the city center 100-yen bus on the way of their *Kaiyu*.

Table 9.15 Average expenditure per *Kaiyu* step for users and nonusers of the city center 100-yen bus (unit: yen)

	Kaiyu steps	Average expenditure per *Kaiyu* step	Standard deviation
Users of 100-yen bus	789	2537.3	22946.5
Nonusers of 100-yen bus	5383	1509.4	18846.6
Total	6172	1640.8	19419.9

Table 9.16 Average expenditure amount per consumer *Kaiyu* step excluding outliers (unit: yen)

	Kaiyu steps	Average expenditure per *Kaiyu* step	Standard deviation
Users of 100-yen bus	786	1781.8	8558.4
Nonusers of 100-yen bus	5373	1279.4	8101.9
Total	6159	1343.5	8162.5

Table 9.17 Average expenditure per consumer *Kaiyu* step according to when they used the city center 100-yen bus (excluding outliers) (unit: yen)

Stage of 100-yen bus ride during *Kaiyu* process	*Kaiyu* steps	Average expenditure per *Kaiyu* step	Standard deviation
Before *Kaiyu* process	317	1883.2	5734.1
On the way of *Kaiyu* process	421	1545.7	8507.9
After *Kaiyu* process	48	3182.7	18785.5

Average Expenditure for City Center 100-Yen Bus Users

Table 9.15 shows the average expenditure per step by visitors who used the city center 100-yen bus and those who did not use it. The average expenditure per *Kaiyu* step by the city center 100-yen bus users was 2537 yen, and for nonusers it was 1509 yen. However, samples of extremely large expenditure amounts are included (a spending of 1,250,000 yen among bus users, and 2 samples of 600,000 yen each among nonusers). Therefore, Table 9.16 shows the average expenditure excluding these outliers. Then, the average expenditure becomes 1782 yen for the city center 100-yen bus users and 1279 yen for nonusers.

Table 9.17 summarizes the average expenditure depending on which stage of their *Kaiyu* process consumers used the city center 100-yen bus to analyze in more detail the effect of the city center 100-yen bus utilization. The average expenditure is 1883 yen for users who got on the bus before visiting the first commercial facility, 1546 yen for users who used the bus on the way of *Kaiyu*, and 3183 yen for users who used it after visiting the last commercial location. The average expenditure by those who got on the bus after visiting the last commercial location was larger than the others, but this is not conclusive, as the number of samples was small.

Table 9.18 Average expenditure per consumer *Kaiyu* step after the bus ride for the city center 100-yen bus users on the way of their *Kaiyu* process

Stage of 100-yen bus ride during *Kaiyu* process	*Kaiyu* steps	Average expenditure per *Kaiyu* step	Standard deviation
On the way of *Kaiyu* process	214	1395.2	10533.6

In addition, Table 9.18 shows the average expenditure per *Kaiyu* step after utilizing the 100-yen bus for users who rode the 100-yen bus on the way of their *Kaiyu* process. The table shows that the average expenditure for this group was 1395 yen.

4 Estimation of the Economic Effect of the Fukuoka City Center 100-Yen Bus

Using data published by NNR and results of the analysis of the fifth survey of *Kaiyu* behaviors at the city center of Fukuoka City, let us estimate the economic effect of introducing the city center 100-yen bus in the city center district of Fukuoka City.

4.1 Estimating the Economic Effect

City center 100-yen bus users are mainly visitors who switched from walking, driving their cars, or riding the subway. Therefore, if visitors, who had been visiting the city center district, merely changed their travel means and did not change their actions other than travel means within the city center district, then it is apparent that there is no economic effect. Based on the results of the previous analysis, we consider the economic effect in the following way.

(a) Among the city center 100-yen bus users, users who utilized it for shopping, dining, and leisure purposes increased the number of their *Kaiyu* steps in the city center district by the utilization of the city center 100-yen bus.

(b) Among people who used the city center 100-yen bus for shopping, dining, and leisure purposes, it is the visitors who used the city center 100-yen bus on the way of their *Kaiyu* process that increased the number of consumer *Kaiyu* steps. In other words, this refers to visitors who used the city center 100-yen bus in the middle of their *Kaiyu* among commercial facilities in the city center district excluding those who used the city center 100-yen bus at the beginning or at the end of their *Kaiyu* in the city center district.

(c) We define the economic effect that the city center 100-yen bus brings to the city center as the increased amount of money dropped to the city center district contributed by the expenditure that visitors who used the city center 100-yen bus on the way of their *Kaiyu* to extend their *Kaiyu* steps spend at their extended stops at the commercial facilities.

Based on the above assumptions, the following four measurements are necessary to estimate the economic effect of the city center 100-yen bus: (1) among the city center 100-yen bus passengers, the number of people who use it for the purpose of shopping, dining, or leisure, (2) the number of people who use the city center 100-yen bus on the way of their *Kaiyu*, (3) the number of consumer *Kaiyu* steps the city center 100-yen bus users increase, and (4) the amount of expenditure spent on purchases per consumer *Kaiyu* step.

Data on (1) was obtained from the data published by NNR. Data on (2), (3), and (4) were obtained from the *Kaiyu* survey data.

4.2 Results of Estimation

1. *The number of the city center 100-yen bus passengers who used it for shopping, dining, and leisure purposes*

According to data published by NNR, the annual average increase per person per day for the city center 100-yen bus was 22,389 people/day, of which 9020 people/day was for the 100-yen circuit bus [17]. According to a survey conducted by NNR, 70% of 100-yen circuit bus passengers stated that their purpose of use was shopping and leisure [16].

$$6314 \text{ people/day} = 0.7 \times 9020 \text{ people/day}$$

In general, the passengers using commuting regular ticket among route bus passengers on weekdays are 50%, so let us consider the purpose of shopping and leisure as the passengers not using regular ticket. Also, let us consider all utilization on the weekends as being for shopping and leisure purposes. According to the weekday and weekend data on the number of 100-yen route bus riders released by NNR, approximately 61.26% of 100-yen route bus users (13,369 people/day = 22,389 people/day − 9020 people/day) are assumed to use the bus for shopping and leisure purposes[4]. (See Tables 9.1 and 9.2.)

$$8190 \text{ people/day} = 0.6126 \times 13,369 \text{ people/day}$$

Together, the number of city center 100-yen bus passengers that use the bus for the purpose of shopping, dining, and leisure is as follows.

---

[4]From the bottom row in Table 9.1, the percentage of 61.26 is obtained by (0.5*56,214*5 + 44,343 + 37,317)/(56,214*5 + 44,343 + 37,317).

$$14,504 \text{ people/day} = 6314 \text{ people/day} + 8190 \text{ people/day}$$

2. *Ratio of the city center 100-yen bus users on the way of Kaiyu*

Among 150 city center 100-yen bus users, 67 people used the city center 100-yen bus during *Kaiyu*.

$$0.447 = 67/150$$

3. *Increase in the number of consumer Kaiyu steps*

We consider the number of consumer *Kaiyu* steps that increased due to the use of the city center 100-yen bus as the number of consumer *Kaiyu* steps that are taken after using the bus by visitors who used the city center 100-yen bus in the middle of their *Kaiyu*, which is their moving around commercial facilities in the city center district. From the results of the above analysis, the number of consumer *Kaiyu* step increased was 3.19 steps.

4. *Amount of expenditure spent on purchases per consumer Kaiyu step*

We consider the following two ways. First, the average expenditure per consumer *Kaiyu* step after the use of the 100-yen bus was 1395 yen for the passengers who used the bus during *Kaiyu*, and the average expenditure per consumer *Kaiyu* step for all *Kaiyu* steps was 1546 yen for the same passengers who used the city center 100-yen bus during *Kaiyu*.

5. *Economic effect of introducing the city center 100-yen bus*

Based on the above, when the average expenditure per consumer *Kaiyu* step is assumed to be 1395 yen, the economic effect of introducing the city center 100-yen bus is:

$$\begin{aligned} 10,530,600,000 \text{ yen} = &\ 14,504 \text{ (people)} \times 0.447 \times 3.19 \text{ (step/person)} \\ &\times 1395 \text{ (yen/step)} \times 365 \text{ (days)} \end{aligned}$$

In addition, when the average expenditure per consumer *Kaiyu* step is assumed to be 1546 yen, then:

$$\begin{aligned} 11,670,047,000 \text{ yen} = &\ 14,504 \text{ (people)} \times 0.447 \times 3.19 \text{ (step/person)} \\ &\times 1546 \text{ (yen/step)} \times 365 \text{ (days)}. \end{aligned}$$

From the above, we find that the economic effect of introducing the city center 100-yen bus is between 10.5 billion and 11.7 billion yen.

5 Conclusion and Future Challenges

The significance of this research centers on the estimation of the economic effect of the city center 100-yen bus from the perspective of *Kaiyu* of the passengers. The method of estimating the increase in consumption expenditure due to the use of the city center 100-yen bus based on the increase in the number of consumer *Kaiyu* steps has not been seen in previous studies. We believe that this study shows that it is effective to evaluate city center commercial policies and urban transportation policies from the viewpoint of *Kaiyu*, and that this approach can become a new universal analytical framework based on the fundamental reference frame of "behavior."

The following can be considered as future challenges. First, the economic effect estimated in this paper is just the amount of the increased consumption expenditure caused by an increase in consumer *Kaiyu* steps due to the use of the city center 100-yen bus. Thus, it is similar to the estimation of the amount of the increased final demand or final consumption expenditure in an input-output analysis. In this sense, the economic effect estimated in this paper is equivalent to estimating the direct effect in an input-output analysis. Therefore, it would be an interesting task to estimate the indirect effect that the increase in final consumption expenditure has brought to the entire Fukuoka City through local inter-industry linkages.

However, this extension may be a major challenge that may involve the development of a new estimation method given complications such as how the wider economic effect including this indirect effect should be attributed to a small area, the city center district of Fukuoka City, which is the area of interest of this research, or in estimating the economic effect for each district within the city center district, which would be an indispensable piece of information in determining how each district should pay for the operating cost of the city center 100-yen bus.

The second challenge is related to prediction and applicability to other geographical areas. This study is an analytical framework based on the universal reference frame of *Kaiyu*; thus, there is no doubt about its applicability to other areas. However, if we think carefully, we see that there remain many questions such as under what kind of conditions and to what extent the introduction of a city center 100-yen bus can increase the number of consumer *Kaiyu* steps; further, how many visitors will use it; and how much they will spend at their extended *Kaiyu* steps. These questions are related to "prediction," whereas the purpose of this study was "measurement" of the economic effect due to the introduction of the city center 100-yen bus. It is an important future challenge to construct a model for these individual prediction aspects and to develop a predictive model that can determine whether the introduction of a city center 100-yen bus can be introduced on a commercial basis or if it can be significant in terms of its economic effect.

References

1. Borgers A, Timmermans H (1986) City centre entry points, store location patterns and pedestrian route choice behaviour: a microlevel simulation model. Socio Econ Plan Sci 20:25–31
2. Fukami T (1974) A study on pedestrian flows in a commercial district part 1. Pap City Plan:43–48. (in Japanese)
3. Fukami T (1977) A study on pedestrian flows in a commercial district part 2. Pap City Plan:61–66. (in Japanese)
4. Hagishima S, Mitsuyoshi K, Kurose S (1987) Estimation of pedestrian shopping trips in a neighborhood by using a spatial interaction model. Environ Plan A 19:1139–1152
5. Hanson S (1979) Urban travel linkages: a review. In: Hensher D, Stopher P (eds) Behavioral travel modeling. Croom-Helm, London, pp 81–100
6. Hanson S (1980) The importance of the multi-purpose journey to work in urban travel behavior. Transportation 9:229–248
7. Hanson S (1980) Spatial diversification and multipurpose travel: implications for choice theory. Geogr Anal 12:245–257
8. Ishibashi K (1998) A study of an evaluation model for development program in central commercial district using consumer shop-around behavior. Doctoral Dissertation, The Graduate School of Science and Engineering, Tokyo Institute of Technology. (in Japanese)
9. Ishibashi K, Saito S (2000) Evaluation of space of city center viewed from shop-around model. In: Kumata Y (ed) Planning theory of public system. Gihodo, pp 177–193. (in Japanese)
10. Ishibashi K, Saito S, Kumata Y (1998) A disaggregate Markov shop-around model to forecast sales of retail establishments based on the frequency of shoppers' visits: its application to city center retail environment at Kitakyushu City. Pap City Plan 349–354. (in Japanese)
11. Ishibashi K, Saito S, Kumata Y (1998) Forecasting sales of shopping sites by use of the frequency-based disaggregate Markov shop-around model of consumers: its application to central commercial district at Kitakyushu City. Presented at the 5th Summer Institute of the Pacific Regional Science Conference Organization (PRSCO), Nagoya, Japan
12. Kakoi M (2000) Theory and application of statistical inverse problem from choice-based samples: constructing simultaneous inverse estimation method of entrance frequency and shop-around pattern in downtown shopping area and its evaluation. Doctoral Dissertation, No.779 (Economics), Fukuoka University. (in Japanese)
13. Kondo K (1987) Transport behavior analysis. Koyo Shobo. (in Japanese)
14. Kondo K, Kitamura R (1987) Time-space constraints and the formation of trip chains. Reg Sci Urban Econ 17:49–65
15. Lerman SR (1979) The use of disaggregate choice models in semi-Markov process models of trip chaining behavior. Transp Sci 13:273–291
16. Nishi-Nippon Railroad (NNR) Co., Ltd. News release 99-18. http://www.nnr.co.jp/nnr/inf/release/release99_18.htm (in Japanese)
17. Nishi-Nippon Railroad (NNR) Co., Ltd. News release 99-66. http://www.nnr.co.jp/nnr/inf/release/release99_66.htm (in Japanese)
18. O'Kelly ME (1981) A model of the demand for retail facilities, incorporating multistop, multipurpose trips. Geogr Anal 13:134–148
19. Saito S (1983) Present situation and challenges for the commercial districts in Nobeoka Area. In: Committee for Modernizing Commerce Nobeoka Region Section (ed) The report of regional plan for modernizing commerce Nobeoka Area. pp 37–96. (in Japanese)
20. Saito S (1984) A disaggregate hierarchical Huff model with considering consumer's shop-around choice among commercial districts: developing SCOPES (Saga Commercial Policy Evaluation System). Plan Public Manag 13:73–82. (in Japanese)
21. Saito S (1984) A survey report on the modified Huff model: the development of SCOPES (Saga Commercial Policy Evaluation System): development of SCOPE (Saga Commercial Policy Evaluation System). Saga City Government, Saga City. (in Japanese)

22. Saito S (1988) Duration and order of purpose transition occurred in the shop-around trip chain at 551 a Midtown District. Pap City Plan:55–60. (in Japanese)
23. Saito S (1988) Assessing the space structure of central shopping district viewed from consumers' shop-around behaviors: a case study of the Midtown District of Saga City. Fukuoka Univ Econ Rev 33:47–108. (in Japanese)
24. Saito S (1993) Report on evaluating urban development plans at the area around the Murasaki River from the viewpoint of Pedestrian *Kaiyu* Behavior. (in Japanese)
25. Saito S (1997) Analysis of the structure of city center space viewed from consumer shop-around behavior. In: Abstracts of The 1997 Spring National Conference of ORSJ. pp 20–21. (in Japanese)
26. Saito S (1997) Urban development viewed from consumer shop-around behavior. In: Abstracts of the 25th annual meeting of the behaviormetric society, Sendai. pp 22–25. (in Japanese)
27. Saito S (1998) Report on surveys of consumer *Kaiyu* behavior at Midtown of Fukuoka City 1996–1998. (in Japanese)
28. Saito S (1999) Report of survey on consumer *Kaiyu* behavior at city center of Fukuoka City 1999: analyzing the changes in the city center structure over years due to retail redevelopment at city center based on consumer *Kaiyu* behavior with emphasis on the effects by the opening of Hakata Riverain. (in Japanese)
29. Saito S (2000) Report on the survey of *Kaiyu* behaviors at city center of Fukuoka, Japan 2000: with focusing on underground space and comparison of city attractiveness between Japan and Korea, Fukuoka Asia Urban Research Center. (in Japanese)
30. Saito S, Ishibashi K (1992) Forecasting consumer's shop-around behaviors within a city center retail environment after its redevelopments using Markov chain model with covariates. Pap City Plann 27:439–444. (in Japanese)
31. Saito, Ishibashi K (1992) A Markov chain model with covariates to forecast consumer's shopping trip chains within a central commercial district. Presented at the the fourth world congress of the regional science association international, Palma de Mallorca, Spain
32. Saito S, Ishibashi K, Kumata Y (2001) An opportunity cost approach to valuation of the river in a city center retail environment: an application of consumer's shop-around Markov model to the Murasaki River at Kitakyushu City. Stud Reg Sci 31:323–337. (in Japanese)
33. Saito S, Kakoi M, Nakashima T (1999) On-site Poisson regression modeling for forecasting the number of visitors to city center retail environment and its evaluation. Stud Reg Sci 29:55–74. (in Japanese)
34. Saito S, Kakoi M, Nakashima T (1999) Inverse estimation of entry frequency from the number of visitors observed at shopping sites under consumer's shop-around. In: Paper presented at 16th PRSCO (Pacific Regional Science Conference Organization) held at Seoul Korea Abstracted in Proceedings
35. Saito S, Kakoi M, Nakashima T (2000) Joint inverse estimation of consumer's entry and shop-around pattern among shopping sites in a city center retail environment. Stud Reg Sci 30:213–229. (in Japanese)
36. Saito S, Kumata Y, Ishibashi K (1995) A choice-based Poisson regression model to forecast the number of shoppers: its application to evaluating changes of the number and shop-around pattern of shoppers after city center redevelopment at Kitakyushu City. Pap City Plan:523–528. (in Japanese)
37. Saito S, Kumata Y, Ishibashi K (1996) A choice-based Poisson regression model: its integrated use with Markov shop-around model to evaluate city center retail redevelopment. Presented at the 3rd international conference on recent advances in retailing and service science, Telfs/Buchen, Austria
38. Saito S, Nakashima T, Kakoi M (1999) Identifying the effect of city center retail redevelopment on consumer's shop-around behavior: an empirical study on structural changes of city center at Fukuoka City. Stud Reg Sci 29:107–130. (in Japanese)
39. Saito S, Sakamoto T (1992) Extraction and analysis of long sightseeing travel routes in Kyusyu Island. Pap City Plan:523–528. (in Japanese)

40. Saito S, Sakamoto T, Motomura H, Yamaguchi S (1989) Parametric and non-parametric estimation of distribution of consumer's shop-around distance at a Midtown District. Pap City Plan:571–576. (in Japanese)
41. Sakamoto T (1984) An absorbing Markov chain model for estimating consumers' shop-around effect on shopping districts. Pap City Plan:289–294. (in Japanese)
42. Sasaki T (1971) Estimation of person trip patterns through Markov chains. In: Newell GF (ed) Traffic flow and transportation. Elsevier, New York
43. Takano N, Kagaya S, Akazawa Y (1999) An experiment of city loop bus for the traffic of the midtown area. Jpn J Transp Econ 1998:21–30. (in Japanese)
44. Takeuchi S (1981) A model for location planning of commercial functions. Plan Public Manag:25–33. (in Japanese)
45. Wrigley N, Dunn R (1984) Stochastic panel-data models of urban shopping behaviour 1, 2. Environ Plan A 16:629–650., pp. 759–778
46. Nagoya Urban Institute (1998) Study survey on city center transportation policy at Nagoya: to increase visitor mobility. Research Report, No.12. (in Japanese)
47. Asahi Shimbun Morning edition, August 7, 1998, 10th page; Mainichi Shimbun Morning edition, August 7, 1998, 8th page; Nihon Keizai Shimbun Morning edition, August 7, 1998, 13th page, Nikkei Distribution Newspaper September 3, 1998, 17th page; Nishi Nippon Shimbun Morning edition, August 7, 1998, 9th page; Yomiuri Shimbun Morning edition, August 7, 1998, 8th page
48. Personal communications with persons involved
49. Saito S, Yamashiro K (2001) Economic impacts of the downtown one-dollar circuit bus estimated from consumer's shop-around behavior: a case of the downtown one-dollar bus at Fukuoka City. Stud Reg Sci 31:57–75. (in Japanese)

Chapter 10
Time Value of Shopping

Saburo Saito and Kosuke Yamashiro

Abstract In general, travel time and travel fares are different by travel means. With the travel time transformed into money terms by time value of travelers, which can be estimated by data of modal choices, the generalized travel cost is defined as the sum of travel fares and the money term travel time. It enables one to compare different travel modes in terms of money. While many efforts have been made to measure the generalized travel costs and the time value, they are all for the purpose of transportation planning for a wide area such as the metropolitan area. As a result, they rarely have measured the time value for travelers other than commuters. In particular, few research has been done to measure the time value for shoppers who shop around a relatively small area such as the city center retail district. In part, this is because there are no such data as the consumer shop-around OD (origin-destination) flows that include their modal choices at city center retail district. The purpose of this paper is to measure the time value of shopping. Based on the actual data obtained from surveys of consumer shop-around behaviors at the city center retail district of Fukuoka City, we have measured the time value of shopping and analyzed how the time value differs by attributes of shoppers and types of shop-around behaviors. We also provide the application in which the estimated time value of shopping is utilized to forecast modal choices when the 100-yen (one-dollar) circuit bus is introduced at the city center retail district.

This chapter is based on the paper, Saburo Saito, Kosuke Yamashiro, Masakuni Kakoi, and Takaaki Nakashima [22], "Measuring time value of shoppers at city center retail environment and its application to forecast modal choice," *Studies in Regional Science*, vol. 33, pp. 269–286, 2003 (in Japanese), which is modified for this chapter.

S. Saito (✉)
Faculty of Economics, Fukuoka University, Fukuoka, Japan

Fukuoka University Institute of Quantitative Behavioral Informatics for City and Space Economy (FQBIC), Fukuoka, Japan
e-mail: saito@fukuoka-u.ac.jp

K. Yamashiro
Department of Business and Economics, Nippon Bunri University, Oita City, Japan
e-mail: yamashiroks@nbu.ac.jp

Keywords Time value · Shopping · Modal choice · *Kaiyu* · Shop-around behavior · Subway · Bus · City center · 100-yen bus · One-dollar circuit bus · Retail district

1 Purpose

In general, travel time and travel fares vary with travel means. With the travel time transformed into money terms by time value of travelers, which can be estimated by actual data of modal choices, the generalized travel cost is defined as the sum of travel fares and the money term travel time. It enables one to compare different travel modes in terms of money. This method for measuring the time value is based on the behavioral result of choosing a travel mode. Therefore, it is called a measurement method that uses RP (revealed preference) data obtained from actual choices among travel means or travel routes.

Until now, while many efforts have been made to measure the generalized travel costs and the time value, they are all for the purpose of transportation planning for a wide area such as the metropolitan area or for the forecast of transport modal choices for commuting. As a result, they rarely have measured the time value for travelers other than commuters. In particular, few research has been done to measure the time value for shoppers who shop around a relatively small area such as the city center retail district. In part, this is because there are no such data as the consumers' shop-around OD (origin-destination) flows that include their modal choices at city center retail district.

However, the importance of measuring time value is not limited to the analysis of transportation behavior. From the perspective of the "consumer production" approach of Becker [1], which is considered the basis of the economic theory of time value, time is regarded as an essential input factor for consumers to produce more "basic commodities." Therefore, in order to understand how consumers allocate income and time to activities such as shopping, leisure, and eating out, it is important to understand how they evaluate the time value of these activities and how these vary among different segments of consumers.

Especially in the city center district of a city where various types of consumers with different preferences and motives come together, it is important to know the differences in time values among these different segments of consumers as well as the differences in time values among different kinds of activities for creating an attractive city center district and introducing efficient downtown transportation systems. In fact, the fact has been pointed out that the utilization rate of bicycle posts varies widely depending on where they are installed in the city center district of Fukuoka City. For example, the utilization rate is low for bicycle posts in front of a large discount commercial facility, which are mainly for shoppers buying food at the supermarket on the basement floors, but is high for posts located in front of a department store. (Cf. [21]) This is considered to reflect the differences in time values between discount shopping for food items and shopping for items in department stores. Furthermore, regarding the judgment as to whether to charge a fee for parking lot information and town information provision services, if the time value for staying in the city center is different, the utilization rate should differ greatly. Thus, measuring the time value by consumer segment will become increasingly important.

As a first step toward solving this problem, Saito et al. [15] used data from annual surveys of consumers' *Kaiyu* behaviors in downtown Fukuoka to (1) measure and estimate the time value and generalized travel cost of visitors who do not have a specific purpose (e.g., shopping, leisure, etc.) for small areas and (2) analyze how these differ depending on individual attributes, such as age, gender, and the distance from home to the area in question. This is possible because the consumer behavior microdata from these surveys contain RP (revealed preference) data on consumers' modal choices in these areas. In this sense, the attempt made by Saito et al. [15] is a new attempt to measure generalized travel costs, and time value based on the microdata of consumer behaviors should be highly appreciated. However, several issues such as model estimation methods have been pointed out.

The objective of this research is to resolve these issues and to (1) measure the time value of visitors who visit the city center district for shopping and leisure purposes, (2) analyze how this time value varies depending on visitor's attributes and purposes for being there, and (3) apply the time value measurement to the problem of forecasting how the choices of travel means changed after the introduction of the Fukuoka city center "100-yen (one-dollar) bus."[1] Then, we evaluate the accuracy of the time value measurement and the forecast.

With regard to forecasting the change in modal share, we specifically forecast the change in modal share due to the introduction of the 100-yen bus in the city center district of Fukuoka City in July 1999. In this research, it becomes possible to utilize the data from surveys of *Kaiyu* behavior at the city center of Fukuoka City conducted before and after the introduction of the 100-yen bus. Utilizing this advantage, we carry out ex ante and ex post forecast evaluations. In other words, we conduct an ex ante forecast by estimating the generalized travel cost using data before the introduction of the bus and then forecast the modal share after the introduction of the 100-yen bus using the estimated results. Then, we compare these ex ante forecasts with the actual modal share obtained from post-introduction data. Conversely, we estimate the generalized travel costs after the introduction of the 100-yen bus based on actual modal choice data, forecast from the backward the modal share before the introduction of the 100-yen bus, and compare these ex post forecasts of modal share with the modal share obtained from actual data before the introduction of the 100-yen bus.

The structure of this chapter is as follows. In Sect. 2, we outline previous studies related to time value measurement and clarify the position of this study. In Sect. 3,

---

[1] Saito and Yamashiro [14] first utilized the modal choice data of the city center 100-yen bus by the visitors at the city center of Fukuoka City to estimate economic effects of the city center 100-yen bus. Also see Chap. 9 of this book.

we explain the time value measurement method and the data used. Section 4 presents the measurement results of time value. In Sect. 5, we analyze how time value varies depending on factors such as consumers' personal attributes and shopping purposes. In Sect. 6, we apply the measured time value to ex ante and ex post forecasts of the change in modal share due to the introduction of the 100-yen bus. Section 7 concludes the chapter.

2 Previous Studies

The importance of measuring time value has long been pointed out since the benefits due to time savings among the benefits of transportation facility development reaches as much as 80%. In fact, the Time Value Study Group [8], organized by transportation researchers, noted its importance as early as at the time when the availability of SP (stated preference) data had attracted attention in addition to non-aggregated models and RP data. Recently, Kono and Morisugi [11] indicated again the importance of time value and reconsidered the economic theory of time value.

The economic theory of time value measurement started from the time allocation model of Becker [1]. The characteristics of his model are to consider that consumers, according to his consumer production approach, invest both of market goods and time as inputs to produce basic commodities or activities. As is the same as in other consumer models, consumers are assumed to maximize their utility by combining these basic activities or commodities under income and time constraints. More specifically, the model is considered a two-stage optimization model. In the first stage, consumers produce these basic activities using time and market goods as input factors. In the second stage, they determine the level of these basic activities that will maximize their utility [7, 10, 17, 20]

Subsequently, as to the model of Becker [1], a theoretical problem was raised that because the time constraint had been considered only as a resource constraint, the time value as an input factor becomes the same for all activities or commodities. Thus, time becomes homogeneous as a resource. To avoid this problem, the model was expanded to be able to deal with the heterogeneity of time where the time value differs depending on the kinds of activities. They include the following. The model of DeDonnea [3] takes into account the environment of time when activities are being carried out, presuming that the time value should differ depending on whether people travel by private car or by bus even for the same commuting time. The model of DeSerpa [2] introduces the constraints of the consuming time required for each activity, assuming that the minimum amount of time to spend is necessary for performing each activity. In particular, DeSerpa's model derives the formulation of "time value as a product" = "time value as a resource" – "value of time saving." From this formula, the model contributed to clarifying several concepts, such as defining leisure activities as those that consume more than the minimum necessary time, where the value of time saving becomes zero. The model of DeSerpa [2] provided a theoretical foundation for subsequent studies on the measurement of time value [7–9, 11, 17].

On the other hand, because transportation behavior does not occur in isolation, there are some transportation behavior models which consider transportation behavior in relation to various other behaviors. Fujii et al. [4] integrated the conventional activity-based approach into Becker's time allocation model and constructed a time allocation model under income and time constraints. In the latter model, the marginal utility of time allocated to each activity is not the same between individuals and activities. Fukuda et al. [5] focused on this issue and constructed an activity choice model where consumers choose activities with the highest marginal utility of time when the time constraints are relaxed. Then, they integrated this into the time allocation model of Fujii et al. [4] and proposed a method for estimating the time value of the activity itself, not the movement, and attempted to estimate it using virtual data.

Although it is a measurement of time value, many studies have measured the time value of transportation behavior (see [16] for a list of examples of time value measurement in Japan).

Time value measurement methods can be classified from two viewpoints: the type of data used and the method used. Then, the type of data can be divided into those cases that use RP (revealed preference) data, which is the result of actual activity choices, and those that use SP data, which is the result of the experiment where the research subject chooses the activity under hypothetical situations. The method used is classified as an income method, production cost method, incidence method to housing and land prices, and transportation behavior-based method, among others [8]. Among these, the methods that use transportation behavior are most popular. Representative methods here are those based on a conversion price [18, 19] and methods that use means of travel/route choice. Currently, the majority of studies are based on means of travel/route choice methods, which analyze the results of the choices using logit or probit models and SP and RP data.

In the travel/route choice method, the utility function v_k of travel means k is expressed as $v_k = \alpha t_k + \beta c_k$ or $v_k = \alpha_k t_k + \beta c_k$ in the logit model. According to Truong and Hensher [17], the former form corresponds to Becker's model and the latter form, in which the coefficient of time for means of travel means k changes as α_k, is referred to as the indirect utility function that corresponds to the model of DeSerpa. (Cf. [7]).

Time value measurement research in Japan mainly is directed to the measurement of the time saving benefits in the cost-benefit analysis for road planning and the modal choice forecast for work and school commuting. However, as far as we know, we cannot find examples where a detailed time value has been measured for the activities with free purposes other than commuting. In particular, there are no cases where the time value of consumer's shopping behavior is measured by shopping purpose or by individual attributes using RP data for small urban areas. The characteristic and significance of this study are that we use RP data on modal choices for shopping behavior and estimate the time value of the shopping behavior for each segment of consumers, based on Becker-type and DeDerpa-type indirect utility functions.

3 Framework of the Analysis

3.1 Time Value Measurement Method

In this study, time value is measured using a modal choice model. The measurement method is as follows.

We assume the utility U_{ij}^m obtained by consumers who move by means of travel m to go from the commercial area i to commercial area j is given as follows:

$$U_{ij}^m = V_{ij}^m + \varepsilon_{ij}^m. \tag{10.1}$$

Here, V_{ij}^m represents the nonstochastic utility of the consumer obtained by moving from commercial area i to commercial area j using travel means m, and ε_{ij}^m represents the error term.

In addition, as described above, according to Truong and Hensher [17], the nonstochastic (indirect) utility function V_{ij}^m of each travel means m is expressed as a Becker-type linear function with respect to the required time t and the required cost (transport fare) c, as follows.

$$V_{ij}^m = \alpha t_{ij}^m + \beta c_{ij}^m = \beta \left(\frac{\alpha}{\beta} t_{ij}^m + c_{ij}^m \right). \tag{10.2}$$

Here, α and β are unknown parameters.

Thus, considering three travel means (bus, subway, and walking), the utility function V_{ij}^m is expressed as follows:

$$V_{ij}^{BUS} = \alpha t_{ij}^{BUS} + \beta c_{ij}^{BUS} \tag{10.3}$$

$$V_{ij}^{SUB} = \alpha t_{ij}^{SUB} + \beta c_{ij}^{SUB} \tag{10.4}$$

$$V_{ij}^{WLK} = \alpha t_{ij}^{WLK} + \beta c_{ij}^{WLK}. \tag{10.5}$$

In addition, denoting by $P_{ij}^{BUS}, P_{ij}^{SUB}$, and P_{ij}^{WLK} the choice probabilities of using the bus, the subway, and walking, respectively, and assuming the independently and identically distributed Gumbel distribution as an error term, the following relationship holds:

$$P_{ij}^{\text{BUS}} = \frac{\exp\left[V_{ij}^{\text{BUS}}\right]}{\exp\left[V_{ij}^{\text{BUS}}\right] + \exp\left[V_{ij}^{\text{SUB}}\right] + \exp\left[V_{ij}^{\text{WLK}}\right]} \tag{10.6}$$

$$P_{ij}^{\text{SUB}} = \frac{\exp\left[V_{ij}^{\text{SUB}}\right]}{\exp\left[V_{ij}^{\text{BUS}}\right] + \exp\left[V_{ij}^{\text{SUB}}\right] + \exp\left[V_{ij}^{\text{WLK}}\right]} \tag{10.7}$$

$$P_{ij}^{\text{WLK}} = \frac{\exp\left[V_{ij}^{\text{WLK}}\right]}{\exp\left[V_{ij}^{\text{BUS}}\right] + \exp\left[V_{ij}^{\text{SUB}}\right] + \exp\left[V_{ij}^{\text{WLK}}\right]} \tag{10.8}$$

$$P_{ij}^{\text{BUS}} + P_{ij}^{\text{SUB}} + P_{ij}^{\text{WLK}} = 1. \tag{10.9}$$

When choosing a means of travel, it is reasonable to assume there is a substituting relationship between the travel fare and the required time.

The condition of keeping the nonstochastic utility V_{ij}^m constant for the modal choice can be expressed as follows: (Cf. [12])

$$\Delta V = -\alpha(-\Delta t) - \beta(\Delta c) = 0. \tag{10.10}$$

From this, the following relation holds between t and c:

$$\left.\frac{\Delta c}{\Delta t}\right|_{V=\text{const}} = \frac{\alpha}{\beta}.$$

This is a cost that may be paid to reduce the time by one unit, while keeping utility constant, which represents the time value:

Time value:

$$\left.\frac{\Delta c}{\Delta t}\right|_{V=\text{const}} = \frac{\alpha}{\beta}. \tag{10.11}$$

In this study, we measure the time value by estimating parameters α, β using the maximum likelihood estimation method based on the ordinary logit model.

Furthermore, to see how the time value varies among travel means, we try to measure the time value for each travel means. To do this, the different parameters α_1, α_2, and α_3 are set to the coefficients of the required time of three travel means. The nonstochastic utilities for three travel modes become as follows:

$$V_{ij}^{\text{BUS}} = \alpha_1 t_{ij}^{\text{BUS}} + \beta c_{ij}^{\text{BUS}} \tag{10.12}$$

$$V_{ij}^{\text{SUB}} = \alpha_2 t_{ij}^{\text{SUB}} + \beta c_{ij}^{\text{SUB}} \qquad (10.13)$$

$$V_{ij}^{\text{WLK}} = \alpha_3 t_{ij}^{\text{WLK}} + \beta c_{ij}^{\text{WLK}}. \qquad (10.14)$$

The time value of each travel means is obtained by substituting the parameter of the required time of each travel means into α in formula (10.11). As described above, in previous studies, this is a representation of the DeSerpa-type indirect utility function, according to the classification by Truong and Hensher [17].

3.2 Forecasting Method of Transport Mode Share

Based on the estimated result of the modal choice model, it is possible to predict the change in transport mode share under various fare schemes. We forecast the changes in transport mode share under a change of fare structure as follows. In this study, here we exclude the mode of walking and forecast transport share between remaining two travel modes, i.e., bus and subway. As for the forecasted bus share ratio, forecasted values are obtained by substituting the estimated values of parameters α, β and the required time and required cost (transport fare), which are explanatory variables, into formula (10.6). Similarly, the forecasted values for using the subway and walking are obtained using formulas (10.7) and (10.8).

3.3 Data

Survey of *Kaiyu* Behaviors at the City Center of Fukuoka City: Data on Transport Modal Choice

We use the fourth on-site survey of *Kaiyu* behaviors at the city center of Fukuoka City conducted by the Saito Laboratory at Fukuoka University on June 11 (Friday), 12 (Saturday), and 13 (Sunday), 1999 (hereafter, abbreviated as "1999 data"), and the fifth on-site survey of *Kaiyu* behaviors at the city center of Fukuoka City conducted on March 17 (Friday), 18 (Saturday), and 19 (Sunday), 2000 (hereafter, abbreviated as "2000 data").

The on-site survey of *Kaiyu* behaviors is about 15-min interview survey with questionnaire sheets conducted as follows. Multiple survey sampling points are set up within the city center district. The respondents are randomly sampled from the visitors at those sampling points. The respondents are asked about their *Kaiyu* behaviors on the day sampled. The *Kaiyu* behavior is defined as the shop-around behavior which is walking around or moving around commercial facilities within a city center district for shopping. The respondents are asked about the history of their *Kaiyu* behaviors concerning which places they visited, for what purposes, and how much they spent at those places in the order of occurrence from the first place they

10 Time Value of Shopping

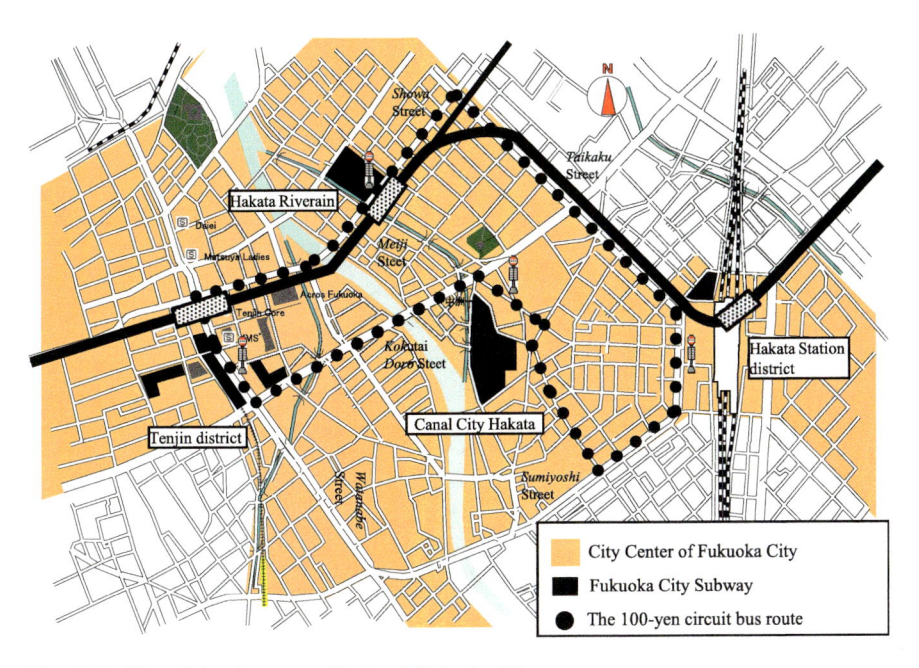

Fig. 10.1 Map of the city center district of Fukuoka City

visited up to the point of time they are interviewed, as well as their planned *Kaiyu* behaviors after the point of time interviewed until the last place they planned to exit from. Major question items other than the history of *Kaiyu* behaviors are personal attributes of respondents, frequency of visits to the city center district and the sampling points, and purchasing attitude, and so on.

For the fourth on-site survey, ten survey points were set up in major commercial facilities in Tenjin, Hakata Riverain, Canal City Hakata, and Hakata station. Valid samples of 2373 questionnaires were collected. For the fifth on-site survey, 12 survey points were set up in Tenjin, Hakata Riverain, Canal City Hakata, and Hakata station, yielding valid samples of 1328 questionnaires.

Figure 10.1 shows the city center district of Fukuoka City, the route of Fukuoka Municipal Subway, and the route of Fukuoka city center 100-yen (one-dollar) circuit bus. The city center district of Fukuoka City is surrounded by 50-m-wide roads, which are *Taihaku* Street, *Showa* Street, *Watanabe* Street, and *Sumiyoshi* Street.

Here, we take up the three particular districts in the city center district of Fukuoka City, Hakata Station district, Tenjin district, and Hakata Riverain district, where the subway and bus compete with each other and deal with modal choices of the OD (origin-destination) movements between these three districts.

The *Kaiyu* behavior survey data takes the same form as the person-trip survey data in that both data record the trip chains of the respondents. However, the focus of the *Kaiyu* behavior survey is on the *Kaiyu* behaviors of consumers within the city center district so that the survey does not ask directly about the choice of travel mode. Therefore, in order to extract the respondent's choice of travel mode from the

Kaiyu survey data, ingenuity was required. In order to understand this, we now look into the items of *Kaiyu* history obtained from the *Kaiyu* surveys.

Kaiyu History Recorded

In order to identify the choice of travel mode, it is necessary to specify the travel means used along with the information on the origin and destination of each OD. Places the respondents visited which are recorded in their *Kaiyu* behavior history include transportation nodes such as the entrances and exits of subway and railway stations and bus stops, in addition to attraction nodes such as stores and commercial facilities. Thus, it was possible to identify which travel mode the respondents used for moving among places they visited. In addition to subway and bus, walking also is considered one of travel means in this study. As for the mode of walking, making use of the information of places visited and purposes done there enables to specify the mode of walking. In the next section, we examine how to identify the choice of travel mode using *Kaiyu* behavior survey data.

Criteria for Modal Choice

Here, we state how to identify the choice of travel mode (subway, bus, and walking). Note that as stated above, we take up the three specific districts, Tenjin district, Hakata Station district, and Hakata Riverain district, and restrict our attention to the respondents' modal choices for their movements between two different districts among these three districts.

1. Subway: When two nodes of subway stations located at different districts are included in the *Kaiyu* history as consecutive two nodes.
    Example (origin) subway Hakata Station, (destination) subway Tenjin Station.
2. Bus: When two nodes of bus stops located at different districts are included in the *Kaiyu* history as consecutive two nodes. Note that here, the movements by bus within the same district are excluded.
    Example (origin) bus stop in Tenjin district, (destination) bus stop in Hakata Station district.
3. Walking: When two attraction nodes located at different districts are included in the *Kaiyu* history as consecutive two nodes of origin and destination nodes whose purposes are not transport behavior.

Choice Result by Mode

Tables 10.1 and 10.2 show the results of choices of travel means and modal share between each OD pair obtained from the 1999 data and 2000 data, respectively.

10 Time Value of Shopping

Table 10.1 Modal choice and share among travel means for each OD pair (1999)

OD	Section	Samples				Share		
		Bus	Subway	Walk	Total	Bus	Subway	Walk
1	Hakata Station -> Hakata Riverain	6	22	7	35	17.1%	62.9%	20.0%
2	Hakata Riverain -> Hakata Station	7	12	4	23	30.4%	52.2%	17.4%
3	Hakata Station -> Tenjin	19	123	10	152	12.5%	80.9%	6.6%
4	Tenjin -> Hakata Station	21	103	8	132	15.9%	78.0%	6.1%
5	Hakata Riverain -> Tenjin	6	11	75	92	6.5%	12.0%	81.5%
6	Tenjin -> Hakata Riverain	5	18	72	95	5.3%	18.9%	75.8%

Table 10.2 Modal choice and share among travel means for each OD pair (2000)

OD	Section	Samples				Modal share		
		Bus	Subway	Walk	Total	Bus	Subway	Walk
1	Hakata Station -> Hakata Riverain	11	5	5	21	52.4%	23.8%	23.8%
2	Hakata Riverain -> Hakata Station	6	3	1	10	60.0%	30.0%	10.0%
3	Hakata Station -> Tenjin	45	56	21	122	36.9%	45.9%	17.2%
4	Tenjin -> Hakata Station	41	37	7	85	48.2%	43.5%	8.2%
5	Hakata Riverain -> Tenjin	9	4	51	64	14.1%	6.3%	79.7%
6	Tenjin -> Hakata Riverain	7	4	41	52	13.5%	7.7%	78.8%

Required Time and Cost by Mode

In order to measure the time value, the time required to move between each OD and the required costs (transport fares) are necessary. These data are explained below.

Data on Required Time

(i) Subway: We obtained the required time between stations from the Direct Train Standard Required Time Table, which is published on the website of the Fukuoka City Subway [6].
(ii) Bus: From the NNR Group's website [13], the required time between each OD was calculated from the timetable of the city center 100-yen (one-dollar) circuit bus. Since the bus route is a circuit, there is an inner route and an outer route, and the required time differs depending on the directions, even though the OD pairs connect the same districts.
(iii) Walking: The shortest route between each OD pair was determined using an electronic map, and the walking speed of pedestrians was set to 4 km per hour:

Table 10.3 Required time by travel means between each OD pair (unit: minute)

OD	Section	Required time (minute)		
		Bus	Subway	Walk
1	Hakata Station -> Hakata Riverain	7	3	22
2	Hakata Riverain -> Hakata Station	8	3	22
3	Hakata Station -> Tenjin	12	5	32
4	Tenjin -> Hakata Station	11	5	32
5	Hakata Riverain -> Tenjin	5	1	16
6	Tenjin -> Hakata Riverain	3	1	16

Table 10.4 Required costs by travel means between each OD pair (2000) (unit: yen)

OD	Section	Fare (yen)		
		Bus	Subway	Walk
1	Hakata Station -> Hakata Riverain	100	200	0
2	Hakata Riverain -> Hakata Station	100	200	0
3	Hakata Station -> Tenjin	100	200	0
4	Tenjin -> Hakata Station	100	200	0
5	Hakata Riverain -> Tenjin	100	200	0
6	Tenjin -> Hakata Riverain	100	200	0

$$\text{Walking Time (min)} = \text{shortest route between each OD (km)} \div 4(\text{km/h}) \times 60 \text{ (min)}$$

Table 10.3 shows the result of required time for each OD by three modes.

Data on Required Cost

As of 1999, the required costs for the subway, bus, and walking in the city center district were 200 yen, 180 yen, and 0 yen, respectively, at a flat rate. For 2000, the required costs were 200 yen, 100 yen, and 0 yen, respectively, for each district as shown in Table 10.4.

4 Results of Measuring Time Value

4.1 Measurement Results from Two Modes by Subway and Bus

Here, we measure the time value drawing upon the concepts discussed thus far. From the 1999 data and 2000 data, we pick up subway users and bus users for five sections between subway stations at two different districts from among Hakata Station, Hakata Riverain, and Tenjin districts, excluding the section from Tenjin to Hakata

Riverain district,[2] and conducted a parameter estimation for modal choice between the subway and bus.

In addition, the time value was calculated from the estimated parameters using formula (10.11) and converted to values per hour because the data on required time is given in minutes. Table 10.5 shows the estimated results of parameters and measurement results of the time value. The *t*-values show that all except the required cost in 1999 are significant. In addition, the hit ratio in 1999 was very high, but the hit ratio in 2000 was lower. The time value measurement results were 854.1 yen per hour in 1999 and 948.3 yen in 2000.

These measurement results of the time values were close to the part-time hourly wage, which is quite interesting. In both 1999 and 2000, there was no large differences between the coefficients of required time and required cost, which can be said that they were stable. These measurement results are used for the ex ante and ex post forecasts of the modal share in Sect. 6.

4.2 Results of Measurement from Two Modes by Day of Week

In general, time value is considered to be different between weekdays and weekends. Therefore, we divide the data between visits on a weekday (Friday) and those on a weekend and measure the respective time values for weekday and for weekend from the modal choice data between subway and bus, as in the same way as the previous section. The estimated results are shown in Table 10.6.

On weekdays, both required time and required cost were significant at the 5% level. Neither were significant in the case of weekends. The coefficients on weekends were not significant, so the implications of the results cannot be said to be definitive. However, when we consider the estimated results, on weekends, the estimates of the absolute values for both α and β coefficients are smaller than on weekdays. Thus, it appears that the disutility (cost consciousness) to the additional increase in the required cost and the required time is smaller than on weekdays. However, on weekends, compared to weekdays, the cost consciousness for transport fare decreases considerably, much more than that for time so that the time value becomes 849.0 yen on weekdays and 1062.8 yen on weekends. Thus, on weekends it is approximately 200 yen per hour higher.

---

[2]The result of measuring time value when the samples moved from Tenjin to Hakata Riverain are included is 1920.3 yen per hour, which is about 1000 yen higher than when those samples are excluded. This is because there is a large range of required times for buses due to traffic jams in this section, and it is possible for visitors to get off at a bus, stop and walk, or use the subway for a short distance. We leave this problem as a topic for future research.

Table 10.5 Estimated results of parameters and time values from modal choices between subway and bus (1999, 2000)

	1999 ($n = 330$)			2000 ($n = 217$)		
	Estimate	t value	SD	Estimate	t value	SD
Time (minute) α	−0.3325	−2.5067**	0.1327	−0.3493	−2.4588**	0.1421
Cost (yen) β	−0.0234	−0.5887	0.0397	−0.0221	−2.4804**	0.0089
Hit ratio	82.12%			56.68%		
Time value (yen per hour)	854.1			948.3		
Loglikelihood with all parameters set to zeros $L(0)$	−228.739			−150.413		
Loglikelihood with estimated values of parameters $L(\hat{\beta})$	−151.908			−147.117		
$-2 \times \text{Loglikelihood} \rho = -2\left[L(0) - L(\hat{\beta})\right]$	153.66			6.5914		
McFadden's R-square s $\rho^2 = 1 - L(\hat{\beta})/L(0)$	0.3359			0.0219		
Adjusted R-square $\rho^{-2} = 1 - \left(L(\hat{\beta}) - K\right)/L(0)$	0.3271			0.0086		

* significant at 10%, ** significant at 5%, *** significant at 1% K: the number of parameters

10 Time Value of Shopping

Table 10.6 Estimated results of parameters and time values by day of week from modal choice between subway and bus (2000 data)

	Weekdays ($n = 61$)			Weekends/holiday ($n = 156$)		
	Estimate	t value	SD	Estimate	t value	SD
Time (minute) α	−0.7231	−2.1075**	0.3431	−0.2533	−1.4991	0.1215
Cost (yen) β	−0.0511	−2.3018**	0.0222	−0.0143	−1.1252	0.1609
Hit ratio	58.53%					
Time value (yen/hour)	849.0			1062.8		

*significant at 10%, **significant at 5%, ***significant at 1%
$L(0) = -150.413$, $L(\hat{\beta}) = -143.608$, $\rho = 13.611$, $\rho^2 = 0.0452$

Table 10.7 Estimated results of parameters and time values from modal choice among subway, bus, and walking (2000 data)

	Estimate	t value	SD
Time (minute) α	−0.1186	−6.6828***	0.0178
Cost (yen) β	−0.0124	−5.8863***	0.0021
Hit ratio	58.06%		
Time value (yen/hour)	573.0		

*significant at 10%, **significant at 5%, ***significant at 1% ($n = 302$)
$L(0) = -331.78$, $L(\hat{\beta}) = -306.446$, $\rho = 50.75$, $\rho^2 = 0.00765$, $\rho^{-2} = 0.0703$

4.3 Results of Measurement from Three Modes Including Walking

Next, people that walk were added to subway users and bus users, and the parameters were estimated using data on all three modes.

The estimated results of parameters and time values are shown in Table 10.7.

The estimated results of the parameters become significant at the 1% level for both required time and required cost. However, the hit rate is not so high. The estimated result for time value is 573.0 yen per hour, which is about 350 yen lower than when measured using data on the subway and bus only. This is because pedestrians' time value when walking is low.

In the next section, we measure the time value for all three modes of travel.

4.4 Measurement Results by Modal Choices of Subway, Bus, and Walking

First, we estimate the parameter of required cost and the parameters of required time for three travel modes according to the formulae (10.12)–(10.14). The estimated results are shown in Table 10.8.

Table 10.8 Estimated results of parameters and time values for subway, bus, and walking (2000 data)

	Estimate	t value	SD
Bus time α_1	−0.4296	−1.7982*	0.2389
Subway time α_2	−0.2681	−0.5902	0.4543
Walk time α_3	−0.3009	−2.8812***	0.1044
Cost β	−0.0351	−5.5323***	0.0064
Hit ratio	68.65%		
Bus time value (yen/hour)	733.7		
Subway time value (yen/hour)	457.9		
Walk time value (yen/hour)	513.8		

*significant at 10%, **significant at 5%, ***significant at 1% ($n = 302$)
$L(0) = -331.78$, $L(\hat{\beta}) = -306.446$, $\rho = 98.155$, $\rho^2 = 0.1418$, $\rho^{-2} = 0.0643$

Although the time required for the subway is not statistically significant, the times required for using the bus and walking are significant at the 10% and 1% levels, respectively. Therefore, the hit rate is also high.

With regard to the time value measurement for each travel mode, the highest time value was 733.7 yen per hour for bus, the second highest was 513.8 yen per hour for walking, and the lowest was for subway at 457.8 yen per hour.

This is a surprising result. We expected the time value to be in the order of subway, bus, and walking.

The reason for the result is given by the coefficient of required time: the absolute value of subway is the smallest, which means the disutility is smallest with respect to the additional increase in required time. We consider that this represents the utility and comfort of taking the subway, compared to going by bus or walking. However, since the coefficient of the time required for the subway is not statistically significant, we would like to examine this further in future research.

5 Measurement of Time Value by Attribute

In this section, we examine how the time value changes depending on the attributes of consumers visiting the city center district.

The estimation is carried out using the following procedure. First, as in Sect. 3, we use the data on respondents who chose bus or subway in the five sections between subway stations located at different two districts from among Hakata Station, Hakata Riverain, and Tenjin districts, excluding the section from Tenjin to Hakata Riverain district. The parameter estimates of required time and required cost for each attribute class are obtained using the maximum likelihood estimation method as the coefficients of the interaction terms of the dummy variables of each attribute class with the required time and required cost. The time value per hour for each attribute class is calculated using the estimated results of the required time and required cost parameters, as before.

5.1 Measurement of Time Value by Respondent's Travel Cost and Its Characteristics

First, we investigate whether or not there are the differences in their time values between the visitors who had paid high travel fares when they came to the city center district and those who had not. Here, we classify visitors into two groups: the one is a group of consumers whose travel fares from their residence to the city center of Fukuoka were less than 1500 yen, which roughly corresponds to the residents in the Fukuoka metropolitan area[3] and its surrounding area, and the other is a group of people whose travel fares were 1500 yen or more, which roughly corresponds to the residents in the Kyushu area and other areas than the previous group. Then, we conducted an estimation for each parameter. Table 10.9 shows the estimated results of the parameters and the time value measurement.

The estimation results show that the parameters for both groups are significant. When the travel fares were less than 1500 yen, the time value was 816.6 yen, and when the travel fares were 1500 yen or more, the time value was 1261.8 yen. Thus, the time value was higher for people who had paid more to get to the city center district.

To examine why this is so, we compare the coefficients of required time and required cost between the two groups. Interestingly, the group with high travel fares has a disutility of -0.0228, even when the cost increases, the absolute value of which is smaller than that of -0.0272 for the group with low travel fares. In contrast, their additional disutility when time increases is -0.4799, whose absolute value is larger than that of -0.3695 for the group with low travel fares. In other words, consumers coming from farther away and who have paid higher travel fares are less

Table 10.9 Estimated results of parameters and time values by travel fare (2000 data)

	Less than 1500 yen ($n = 117$)			1500 yen or more ($n = 59$)		
	Estimate	t value	SD	Estimate	t value	SD
Time (minute)α	−0.3695	−1.9319*	0.1913	−0.4799	−1.4991**	0.3201
Cost (yen) β	−0.0272	−2.2796**	0.0119	−0.0228	−1.1252**	0.0203
Hit ratio	63.64%					
Time value (yen/hour)	816.6			1261.8		

*significant at 10%, **significant at 5%, ***significant at 1%
$L(0) = -121.994$, $L(\hat{\beta}) = -111.868$, $\rho = 20.251$, $\rho^2 = 0.083$

---

[3]We adopt the definition of the national census in which the Fukuoka metropolitan area is define as the area of Kitakyushu and Fukuoka City. More than 1.5% of the resident population commutes to Fukuoka City from these municipalities.

conscious about additional costs. In contrast, the cost consciousness for additional time is high when there are time constraints. To put it another way, the utility of saving time for consumers from farther away is higher than those from nearer. We interpret these as both being convincing results.

5.2 Measurement of Time Value by Individual Attributes

In this section, we divide consumers into segments according to individual attributes, their purpose of visiting the city center district, and differences in purchasing attitudes. Then, we measure their time values.

Gender

Table 10.10 shows the parameter estimated results by gender and the measurement results of time value. The t-value is significant for females at the 5% level for both parameters, but both parameters for males are not statistically significant.

The measurement results for time value are 1038.9 yen for males and 919.0 yen for females; for males it is about 120 yen higher. Comparing the parameters of required time and required cost among males and females, the disutility for additional costs is lower for females than it is for males. In addition, the disutility for additional time is lower for females than it is for males. These results suggest that females are less conscious about additional costs and additional time when shopping than males are. However, because the estimated result for males is not significant, it is not a definitive conclusion, but rather a subject for future research.

Age

For age, we classify consumers into three groups: young people (10–20 years old), middle-aged people (30–40 years old), and senior adults (50 years or older). Then, we conduct a parameter estimation for each group. Table 10.11 shows the parameter estimated results by age and the measurements for time value.

Table 10.10 Estimated results of parameters and time values by gender (2000 data)

	Male ($n = 33$)			Female ($n = 183$)		
	Estimate	t value	SD	Estimate	t value	SD
Time (minute) α	−0.5848	−1.4647	0.3992	−0.3134	−2.0476**	0.1531
Cost (yen) β	−0.0338	−1.3238	0.0255	−0.0205	−2.1402**	0.0096
Hit ratio	57.41%					
Time value (yen/time)	1038.9			919.0		

*significant at 10% **significant at 5% ***significant at 1%
$L(0) = -149.72$, $L(\widehat{\beta}) = -145.618$, $\rho = 8.204$, $\rho^2 = 0.027$

Table 10.11 Estimated results of parameters and time values by age (2000 data)

	10–20 years old ($n = 161$)			30–40 years old ($n = 21$)			50 years or older ($n = 34$)		
	Estimate	t value	SD	Estimate	t value	SD	Estimate	t value	SD
Time (minute) α	−0.4099	−2.1996**	0.1863	−0.3253	−0.7626	0.4266	−0.2327	−0.8268	0.2814
Cost (yen) β	−0.0256	−2.1605**	0.0118	−0.0267	−0.9989	0.0268	−0.0134	−0.8075	0.0166
Hit ratio	57.41%								
Time value (yen/hour)	961.4			730.4			1040.2		

*significant at 10%, **significant at 5%, ***significant at 1%
$L(0) = -149.72$, $L(\hat{\beta}) = -145.285$, $\rho = 8.871$, $\rho^2 = 0.027$

According to the estimated results, the senior group is highest (1040.2 yen), followed by the young group (961.4 yen), and then the middle-aged group (730.4 yen). It is reasonable that the senior has the highest time value, but the younger group having a higher value than the middle-aged group is an unexpected result.

Looking into the estimated results of the required cost coefficients, the disutility for the additional cost is the smallest for the senior group and highest for the middle-aged group. In addition, the estimated result of the required time coefficient shows that the disutility for additional time is smallest in senior group, followed by the middle-aged group, and then largest for the young group. This may explain the highest consciousness for additional cost among the middle-aged group, because their disposable income tends to be small. On the other hand, the cost consciousness with respect to time is highest in the young group, followed by the middle-aged group, and then the senior group is unexpected. If there are activities with high utility other than shopping, obviously, the cost for additional time and expenses will increase. This may suggest that there are activities with high utility, other than shopping, in the city center district, for the young group.

5.3 Measurement of Time Value by Purpose and Purchasing Attitude in the City Center District

Main Purpose in the Downtown Area

The on-site survey of *Kaiyu* behaviors at the city center of Fukuoka City asked the respondents who visited the city center about their main purpose to visit there. Here, we examine how the time value differs depending on the purpose for their visits. Here, we classify consumers into three groups: shopping, leisure and eating/drinking, and errands and others. Then, we conducted the parameter estimation.

Table 10.12 shows the estimated results of the parameters and time value measurements. The time value for shopping was 942.0 yen, for leisure was 820.7 yen, and for errands was 1351.8 yen. Thus, the value for the purpose of errands is highest.

Comparing the coefficients of required time and required cost, interestingly, the absolute values are in ascending order errands and other, leisure and eating/drinking, and shopping. For errands, the awareness of additional costs was very small and was highest for shopping. For additional time, shopping had the highest cost consciousness. However, when the purpose was shopping, the utility of saving time was largest, and in the case of errands, was smallest.

Of the above results, those results for shopping may seem to be surprising. However, here in this analysis, we focus on consumers' movements as derived demand in order to achieve their main purpose to visit the city center district and analyze the result of their choices among travel means for their movements. Thus, the measurement result in the case of shopping can be thought of as representing an opportunity cost of the movements in the process of shopping activities. In other

10 Time Value of Shopping

Table 10.12 Estimated results of parameters and time values by main purpose of visit (2000 data)

	Shopping ($n = 132$)			Leisure and eating/drinking ($n = 32$)			Errands ($n = 34$)		
	Estimate	t value	SD	Estimate	t value	SD	Estimate	t value	SD
Time (minute) α	−0.3980	−2.1210**	0.1876	−0.2713	−0.8224	0.3298	−0.2140	−0.5927	0.3611
Cost (yen) β	−0.0254	−2.1374**	0.0119	−0.0198	−0.9925	0.0200	−0.0095	−0.4264	0.0223
Hit ratio	59.09%								
Time value (yen/hour)	942.0			820.7			1351.8		

* significant at 10%, ** significant at 5%, *** significant at 1%
$L(0) = -137.243$, $L(\widehat{\beta}) = -133.147$, $\rho = 8.192$, $\rho^2 = 0.0298$

words, it might be said that in this analysis of time value, we are measuring consumers' inherent value or utility of their shopping activities that would have been obtained if they could have relocated the time spent for their movements to the inherent shopping activities. Using this interpretation, it should also be clear that the utility of time saving is largest in the case of shopping. However, the theoretical relationship between the time value of the movement and the time value of the activity itself is an important subject for future research (Cf. [5]). In addition, even for the same shopping purpose, the time value should differ between shopping for clothing or high priced goods and that for food. Thus, further examination is necessary in the future.

Purchasing Attitude

In the survey, the respondents who visited for shopping purposes were asked about their attitude toward shopping on that day, based on three categories: (1) coming to preview goods and prices, (2) thinking about buying if there are good items, and (3) a definite intention to buy the items. Here, we analyze how the time value varies depending on the purchasing attitude. Table 10.13 shows the parameter estimated results and the time value measurement results by purchasing attitude.

The "coming to preview goods and prices" was 857.1 yen, "thinking about buying if there are good items" was 1004.0 yen, and "a definite intention to buy the items" was 332.7 yen. Thus, people who responded that they were thinking about buying if there were good items had the highest time value.

The interesting thing is that the coefficient of the required time and required cost of people who responded that they would definitely buy items has an absolute value that is one digit smaller than the other two groups. This can also be interpreted as indicating that when the purpose of the activity has already been determined, even if additional cost or time is added "to produce" that activity, its marginal utility would be small. In fact, in contrast, the marginal utilities of time saving for the groups to preview goods and prices and those thinking about buying if there are good items are very high.

Unlike the case of "errands" which was contrasted with the shopping purpose in the previous section, when limiting to the shopping purpose, rather than conclusive activities such as definitely intending to buy the item, the activities with high uncertainty, such as doing previews and thinking about buying, have higher time values. These results suggest that the theoretical clarification of time value for shopping while regarding shopping as an exploratory search behavior with uncertainty is a major challenge for future research.

Table 10.13 Estimated results of parameters and time values by purchasing attitude (2000 data)

	Coming to preview goods and prices ($n = 15$)			Thinking about buying if there are good items ($n = 75$)			A definite intention to buy the items ($n = 41$)		
	Estimate	t value	SD	Estimate	t value	SD	Estimate	t value	SD
Time (minute) α	−0.9163	−0.8365	1.0955	−0.6668	−2.5208**	0.2645	−0.0374	−0.1278	0.2925
Cost (yen) β	−0.0641	−0.8869	0.0723	−0.0399	−2.3948**	0.0166	−0.0067	−0.3713	0.0182
Hit ratio	61.83%								
Time value (yen/hour)	857.1			1004.0			332.7		

* significant at 10%, ** significant at 5%, *** significant at 1%
$L(0) = -90.802$, $L(\hat{\beta}) = -85.28$, $\rho = 11.045$, $\rho^2 = 0.0608$

6 Application of Time Value to the Ex Ante and Ex Post Forecast of Changes in Travel Mode Choices Caused by the Introduction of City Center 100-Yen Bus

In this section, as an application of the time value measurement, we applied the time value measurement result to the analysis of changes in modal share due to the introduction of the Fukuoka city center 100-yen (one-dollar) bus. The procedure for the analysis is as follows.

In Sect. 6.1 using the time value measured from the result of choice of travel mode in the 1999 data, which was carried out before the introduction of the city center 100-yen bus, we conduct an ex ante forecast of the transport mode share in 2000 after the introduction of the city center 100-yen bus. Then, we compare the forecasted result and the actual measured values of the travel mode share using the 2000 data and evaluate the accuracy of the forecast. Conversely, in Sect. 6.2, using 2000 data after the introduction of the city center 100-yen (one-dollar) bus, we measure the time value and then conduct an ex post forecast of the travel mode share before the introduction of the city center 100-yen bus in 1999. This forecast value is then compared with the actual measured value of the travel mode share using the 1999 data, and the forecast accuracy is evaluated.

The ex ante forecast of modal share and the accuracy evaluation of ex post forecast generally correspond to what is called cross-validation. If we merge 1999 and 2000 data together into one data, we can divide this data into two groups for 1999 and 2000, measure the time value using only one half data set, and predict the modal share of the remaining half of the data using the measurement result. Since the same can be done with the other data, the method here corresponds to an evaluation of the forecast accuracy by repeating this twice, for the ex ante forecast and the ex post forecast. This corresponds to a method called a twofold cross-validation method. If the average prediction accuracy is high, it can be said that the measurement accuracy of the time value is also high.

6.1 Ex Ante Forecast of Modal Share After the Introduction of City Center 100-Yen Bus Using 1999 Data and Its Evaluation

In Sect. 4.1, since the values of the parameters of subway and bus using the 1999 data are obtained, using these values in formulae (10.6) and (10.7), we substituted the required time and the required cost after the introduction of the downtown 100-yen bus (200 yen for subway and 100 yen for bus). Then, we forecast the modal share of subway and bus. The difference between the forecasted values using the 1999 data and the measured value in 2000 was evaluated as an error. Table 10.14 compares the forecasted result of the modal share with the measured value.

Table 10.14 Comparison between ex ante forecasted modal share in 2000 using 1999 data and actual measured modal share in 2000

OD	Section	Forecasted values in 2000 using the data in 1999		The measured value in 2000		The error |a−b|
		Bus (a)	Subway	Bus (b)	Subway	
1	Hakata Station -> Hakata Riverain	73.2%	26.7%	68.8%	31.3%	4.4%
2	Hakata Riverain -> Hakata Station	66.2%	33.8%	66.7%	33.3%	0.5%
3	Hakata Station -> Tenjin	50.2%	49.8%	44.6%	55.4%	5.6%
4	Tenjin -> Hakata Station	58.5%	41.5%	52.6%	47.4%	5.9%
5	Hakata Riverain -> Tenjin	73.2%	26.8%	69.2%	30.8%	4.0%

The error in each section shows that the section from Hakata Riverain to Hakata Station district has the smallest error of 0.5%, and the largest error was 5.9% in the section from Tenjin to Hakata Station district. Every section has a small error (within 6%), which shows that ex ante forecast can predict the modal share after the introduction of the city center 100-yen bus with high accuracy.

6.2 Ex Post Forecast of Modal Share Before the Introduction of City Center 100-Yen Bus Using 2000 Data and Its Evaluation

Similarly, from the 2000 data, the modal share of subway and bus in 1999 was forecasted and compared with the actual measured value of the modal share of subway and bus in 1999. By using the parameters in the 2000 data estimated in Sect. 4.1, we made an ex post forecast of the modal share of subway and bus in 1999 by substituting the required time and required cost for 1999. Table 10.15 compares these forecasted results with the actual measured values of modal share in 1999.

Here, the errors in the actual measurement show the smallest was 0.8% for the section from Tenjin to Hakata Station and then 1.5% for the section from Hakata Station to Tenjin district. In addition, since even the greatest error was 14.0% for the section from Hakata Riverain to Hakata Station district, this forecasted result shows that every interval is within a small error except for the section from Hakata Riverain to Hakata Station district.

The results of these ex ante and ex post forecasts are all within the range of small errors from actual measured values of modal share. Thus, they are considered to indicate a high accuracy of time value measurement.

Table 10.15 Comparison between ex post forecasted modal share in 1999 using 2000 data and actual measured modal share in 1999

OD	Section	The forecasted values using the 2000 data		The measured value in 1999		The error
		Bus (a)	Subway	Bus (b)	Subway	|a−b|
1	Hakata Station -> Hakata Riverain	27.8%	72.2%	21.4%	78.6%	6.4%
2	Hakata Riverain -> Hakata Station	22.8%	77.2%	36.8%	63.2%	14.0%
3	Hakata Station -> Tenjin	11.9%	88.1%	13.4%	86.6%	1.5%
4	Tenjin -> Hakata Station	16.1%	83.9%	16.9%	83.1%	0.8%
5	Hakata Riverain -> Tenjin	27.8%	72.2%	35.3%	64.7%	7.5%

7 Conclusion and Future Challenges

The significance in this study is that while until now the measurement of time value has been limited to the data on modal choices in wide areas for transportation planning, we carry out the measurement of time value for consumers who visited the city center district with free or discretionary purposes such as shopping, leisure, meals, and so on, using data on their travel mode choices for the movements within a small city center district. Second, we measured and clarified the differences in time values depending on the differences in consumer attributes. Third, we conducted an ex ante forecast of modal share from data before the introduction of the city center 100-yen bus and an ex post forecast of modal share from data after the introduction of the city center 100-yen bus by using the estimated results of parameters obtained from the time value measurement process. As a result of comparing the forecasts with actual measured values, it was shown that measured values of modal share can be forecasted with high accuracy. Thus, we conclude that the measurement model of time value for shopping purposes in this study has high accuracy and can be applied to forecasting changes, such as modal share in real transportation projects, in small areas.

In future research, we would like to clarify the consumer behavioral characteristics at the city center district for free purposes by measuring the time value by each segment of consumers. In addition, as mentioned in the above analysis, we wish to explore the theoretical relationship between time value and shopping behavior as an exploratory search behavior, as well as the theoretical relationship between the value of movement and the value of the activity itself. Furthermore, we will refine the measurement model of time value by incorporating convenience of access and waiting time and make a comparative analysis of time value with other cities as future works.

References

1. Becker GS (1965) A theory of the allocation of time. Econ J 75:493–517
2. DeSerpa AC (1971) A theory of the economics of time. Econ J 81:828–846
3. de Donnea FX (1972) Consumer behaviour, transport mode choice and value of time: some micro-economic models. Reg Urban Econ 1:355–382
4. Fujii S, Kitamura R, Kumata Y (1999) A monetary and temporal constrained consumption-behavior model for travel demand analysis. J JSCE:99–112. (in Japanese)
5. Fukuda D, Yoshino H, Yai T, Prasetyo I (2003) Estimating value of activity time based on discretionary activities on weekends. J JSCE 211–221. (in Japanese)
6. Fukuoka City Subway (2003) Fukuoka city subway home page. Available: http://subway.city.fukuoka.jp/ (in Japanese)
7. Hensher D (1997) Behavioral value of travel time savings in personal and commercial automobile travel. In: The full costs and benefits of transportation. Springer, pp 245–278
8. Time Value Study Group (1987) Study on the theory of value of time and the method of its measurement. The Japan Research Center for Transport Policy. (in Japanese)
9. Time Value Study Group (1988) Study on the theory of value of time and the method of its measurement. The Japan Research Center for Transport Policy. (in Japanese)
10. Katayama K (1974) Value of time and travel demand. Transp Econ 34:67–77. (in Japanese)
11. Kono T, Morisugi H (2000) Theoretical examination on value of time for private trips. J JSCE:53–64. (in Japanese)
12. Morisugi H, Miyagi T (1996) Evaluation of urban transportation projects. CoronaSha, Tokyo. (in Japanese)
13. Nishi-Nippon Railroad Co (2003) Nishi-nippon railroad home page. Available: http://www.nnr.co.jp/ (in Japanese)
14. Saito S, Yamashiro K (2001) Economic impacts of the downtown one-dollar circuit bus estimated from consumer's shop-around behavior: a case of the downtown one-dollar bus at Fukuoka City. Stud Reg Sci 31:57–75. (in Japanese)
15. Saito S, Yamashiro K, Kakoi M, Nakashima T (2001) Measuring and analyzing generalized travel cost and value of time for shoppers at city center retail environment. Papers of the 38th annual meeting of JSRSAI (Japan Section of Regional Science Association International). 207–214. (in Japanese)
16. Suzuki S, Harada N, Ota K (1987) A study on the value of time in road planning: an analysis of awareness data by disaggregate behavior model. Expresways Automobiles 30:p28–p36. (in Japanese)
17. Truong TP, Hensher DA (1985) Measurement of travel time values and opportunity cost from a discrete-choice model. Econ J 95:438–451
18. Tsukahara S (1970) Valuation of time for transportation. Transp Econ 30:35–44. (in Japanese)
19. Ueda T, Nitta Y, Mori Y (1993) Coefficients of equivalent time and value of time by transport mode for aged people. The 48th annual meeting of JSCE. pp 488–489. (in Japanese)
20. Yamauchi H, Imahashi T (1988) Economic theoretical background for measuring value of time. Expressways Automobiles 31:p38–p42. (in Japanese)
21. Hearing from officials of Fukuoka City Administration
22. Saito S, Yamashiro K, Kakoi M, Nakashima T (2003) Measuring time value of shoppers at city center retail environment and its application to forecast modal choice. Stud Reg Sci 33:269–286. (in Japanese)

Chapter 11
Roles of City Center Cafés and Their Economic Effects on City Center: A Consumer Behavior Approach Focusing on *Kaiyu*

Saburo Saito, Masakuni Iwami, and Kosuke Yamashiro

Abstract Recently, at the city center of Fukuoka City, Japan, new kinds of coffee shops such as "Doutor" and "Starbucks" and so on, which we call city center cafés, have opened one after another. Unlike the conventional coffee shops, these cafés provide cheap coffee and fashionable shop space with a form of open café with open-air seats.

The attractiveness of city center retail environment clearly consists in the mixed location of different types of shops or establishments such as department stores, restaurants and movie theaters, and so on. Apparently these new cafés add a new element to the attractiveness of city center. However, no research has been done to clarify to what extent each kind of shops, cafés in particular, enhances the attractiveness of city center retail environment.

We have been conducting studies to evaluate city center retail environment from the viewpoint of consumer shop-around behaviors or *Kaiyu*: the more the consumer

---

This chapter is based on the paper by Saburo Saito, Tomoyuki Kiguchi, Masakuni Kakoi, and Takaaki Nakashima [10], "Roles of open cafés and their economic effects on city center: Based on consumer behavior using open cafés located at city center," in *Collected Papers for Presentation in The 39th Annual Meeting of Japan Section of Regional Science Association International (JSRSAI)*, pp. 417–424, 2002 (in Japanese), which is modified for this chapter.

S. Saito (✉)
Faculty of Economics, Fukuoka University, Fukuoka, Japan

Fukuoka University Institute of Quantitative Behavioral Informatics for City and Space Economy (FQBIC), Fukuoka, Japan
e-mail: saito@fukuoka-u.ac.jp

M. Iwami
Fukuoka University Institute of Quantitative Behavioral Informatics for City and Space Economy (FQBIC), Fukuoka, Japan
e-mail: miwami@econ.fukuoka-u.ac.jp

K. Yamashiro
Department of Business and Economics, Nippon Bunri University, Oita City, Japan
e-mail: yamashiroks@nbu.ac.jp

shop-arounds or *Kaiyu*, the better the city center retail environment. From this viewpoint, the function of city center cafés might be evaluated by their effect on the consumer shop-around behaviors (*Kaiyu*). With this in mind, we conducted the sixth survey of consumer shop-around behaviors at city center of Fukuoka City in June 2001.

Based on this survey, the purpose of this paper is to clarify consumer's behaviors of using city center cafés and to estimate the economic effects of the city center café from its effect on increasing the number of shop-around (*Kaiyu*) steps.

Keywords *Kaiyu* · Open café · City center · Shop-around · Attractiveness · Economic effect · Increase in shop-around step · Role of café

1 Purpose

In recent years, coffee shop chains such as Doutor and Starbucks have been opening shops at the city center district of Fukuoka City, Japan. These coffee shops are proving successful and are developing a new business category that differentiates them from traditional coffee shops. In fact, they offer low-priced and original coffee and have a Western open café style with an atmosphere inviting casual visits. In this study, we refer to the coffee shops in this new business category as "city center cafés."

In the city center district, needless to say, various industries such as department stores, movie theaters, street shops, and restaurants are mixedly located. These individual industries are performing a variety of roles for the city center district, and their roles are thought to constitute the unique attractiveness of the city while complicatedly interdepending and interacting with each other collaboratively. The city center cafés are certainly one of these, with many consumers indicating that they visit coffeehouses as a break from shopping, for meetings, and as places to enjoy conversations.

However, until now, none of the studies have clarified concretely what kind of roles each industry and business category plays in the city attractiveness and what kind of effects it brings about to the city center district of the city, including this new business category of city center cafés.

So far, the authors have been conducting studies to evaluate the city center district from the perspective of *Kaiyu* behaviors, taking the view that "a space structure that induces more *Kaiyu* behaviors in the city center space is more desirable" [1–3].

Here, "*Kaiyu*" is a Japanese term which refers to consumer shop-around behavior and is defined as consumers' walking or moving around shopping sites in a city center district which is recorded as the sequence of places visited, purposes done there, and expenses spent there in the order of their occurrence.

In particular, Saito and Yamashiro [5] first established a framework to evaluate urban development policies from the economic effects of extending *Kaiyu* steps brought by those policies. While they took up as an urban policy the city center 100-yen bus introduced in the city center of Fukuoka City, the framework can be applied to any other urban policies.

Thus, if standing on this viewpoint of *Kaiyu* behaviors to consider the recent opening of city center cafés, it is seen to become possible to evaluate the effects of these city center cafés on the city center district since we can make the following hypothesis.

The hypothesis here is that consumers make a longer *Kaiyu* for shopping and visit more shopping sites when they stay at a city center café in their *Kaiyu* than when they do not do so.

The purpose of this study is to focus on a new business category called city center cafés, and to investigate *Kaiyu* behaviors of visitors who stopped at these city center cafés, based on the sixth survey of *Kaiyu* behavior at city center of Fukuoka City, carried out in June 2001. Further, this study also tries to measure whether visitors extend their total stops at shopping sites when they use these city center cafés on their *Kaiyu* process than when they do not do so and estimates the economic effects of these city center cafés on the retail sector at the city center district from their effects on those changes of consumers' *Kaiyu* behaviors.[1]

2 Framework of the Analysis

2.1 Data Used

In this study, we use data from the sixth survey of *Kaiyu* behavior at the city center of Fukuoka City, carried out on June 15 (Friday), 16 (Saturday), and 17 (Sunday), 2001, by the Fukuoka University Institute of Quantitative Behavioral Informatics for City and Space Economy (FQBIC), the Saito Laboratory, and the Kakoi Laboratory at the Faculty of Economics of Fukuoka University.

The on-site survey of *Kaiyu* behaviors at the city center of Fukuoka City has been conducted every year since September 1996 (by the Saito Laboratory at Fukuoka University). The survey is an about 15-min on-site interview survey using questionnaire sheets. In the survey, multiple survey points were set up in major commercial facilities in the district. The respondents are randomly sampled from the visitors at the survey points. The respondents are asked about their *Kaiyu* behaviors on the survey day, that is, which places they visited, for what purposes, and how much they spent there in the order of their occurrence. In addition to the history of their *Kaiyu*, the survey also asks about respondents' personal attributes, the frequency of their visits to the city center district and the survey points, the main purpose to visit the district, their purchasing attitude, and so on.

---

[1]This study started from the thesis research [7, 8]. The research results were further elaborated and reported in [9, 11].

In the questionnaire of the sixth survey, we added city center cafés to the list of places where respondents can stop, as well as new question items about the awareness of city center cafés in the city center district and the frequency with which the respondents visit them.

As for the respondents who used a city center café on the day of the survey, other question items are also added to ask about the reason for stopping at the city center café that day, the time they did so, how long they stayed there, and so on.

The survey was conducted over an 8-h period, from 11:00 a.m. to 7:00 p.m. on each day, with ten survey points, including two department stores, *Iwataya* Z-Side and Fukuoka Tenjin *Daimaru*, and a large commercial complex of Canal City Hakata.

A total of 1191 valid samples was collected.

2.2 Procedure of the Analysis

The analysis of this study proceeds as follows. First, we analyze the individual attributes of the users of city center cafés. Then, we analyze the usage characteristics of city center cafés such as the utilization rate, the purpose of use, the average expenditure, the length of time stayed there, the number of *Kaiyu* steps on the day, and the average expenditure per *Kaiyu* step. Here, the number of *Kaiyu* steps is a concept representing the number of stops in *Kaiyu*. Each time a visitor moves to a different place[2] or changes the purpose, the number of *Kaiyu* steps increases by one.[3] Finally, we define the economic effects of city center cafés based on the number of *Kaiyu* steps and the expenditure per step. Then, we describe the results for city center Fukuoka.

3 Functions of City Center Cafés from the Viewpoint of Consumers' Use Behavior

3.1 Attributes of City Center Café Users

City center café users provided 193 samples. Below, we examine the attributes of these café users by gender and by age.

---

[2]The places to visit during *Kaiyu* consist of two kinds of nodes: the attraction nodes such as commercial facilities and the transportation nodes for getting on and off transport means.

[3]In this study, we only analyze *Kaiyu* behavior between attraction nodes with excluding transportation nodes.

Fig. 11.1 Percentage of city center café users by gender

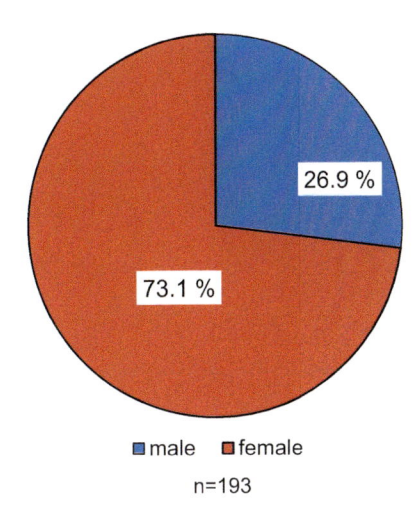

26.9 %
73.1 %
■ male ■ female
n=193

Percentage by Gender

Males account for 26.9% of the responses and females 73.1%. This shows that the number of female users is overwhelmingly larger than that of males (Fig. 11.1).

Percentage by Age

The largest group is the age group of 20–24 years old, which accounts for 43.2%, followed by the 16–19 years group, and then the 25–29 years group. In other words, young people under 29 years of age account for over 70% of visitors (Fig. 11.2).

3.2 Utilization Rate of City Center Cafés

City Center Café Utilization Rate by Gender

With regard to the utilization rate by gender, males account for 17.2% and females for 15.9%, which means there is little difference in utilization rates between males and females (Fig. 11.3).

Utilization Rate of City Center Cafés by Age

The utilization rate of city center cafés by age is 28.1% for the 30–34 years group, followed by 21.4% for the 35–39 years group, and then people in their 40s, indicating that the utilization rate is high for people in their 30s. The utilization

Fig. 11.2 Percentage of city center café users by age

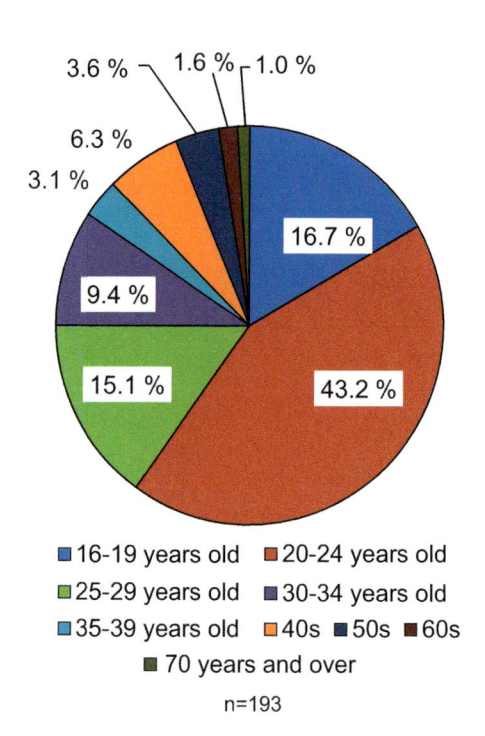

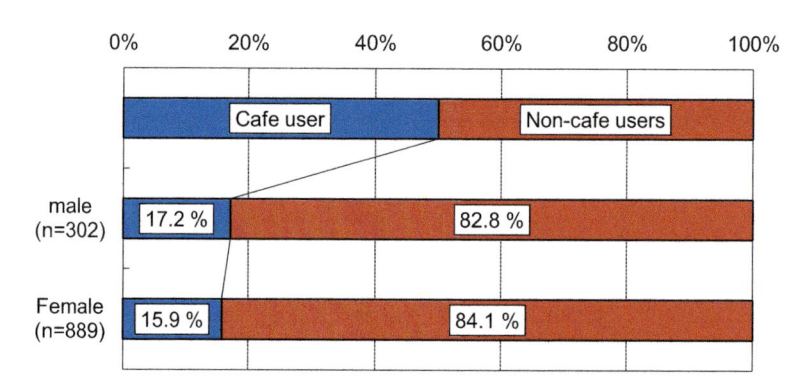

- 16-19 years old: 16.7 %
- 20-24 years old: 43.2 %
- 25-29 years old: 15.1 %
- 30-34 years old: 9.4 %
- 35-39 years old: 3.1 %
- 40s: 6.3 %
- 50s: 3.6 %
- 60s: 1.6 %
- 70 years and over: 1.0 %

n=193

male (n=302): Cafe user 17.2 %, Non-cafe users 82.8 %
Female (n=889): Cafe user 15.9 %, Non-cafe users 84.1 %

Fig. 11.3 Utilization rate of city center cafés by gender

rate for the 16–19 years group (the second largest number of samples in Fig. 11.4) is relatively low at 10.4%.

Utilization Rate of City Center Cafés by Occupation

The breakdown by occupation shows that the main visitors are, in descending order, general labor employee (30.8%), technical workers (24.7%), and sales/service industry workers (21.3%) (Fig. 11.5).

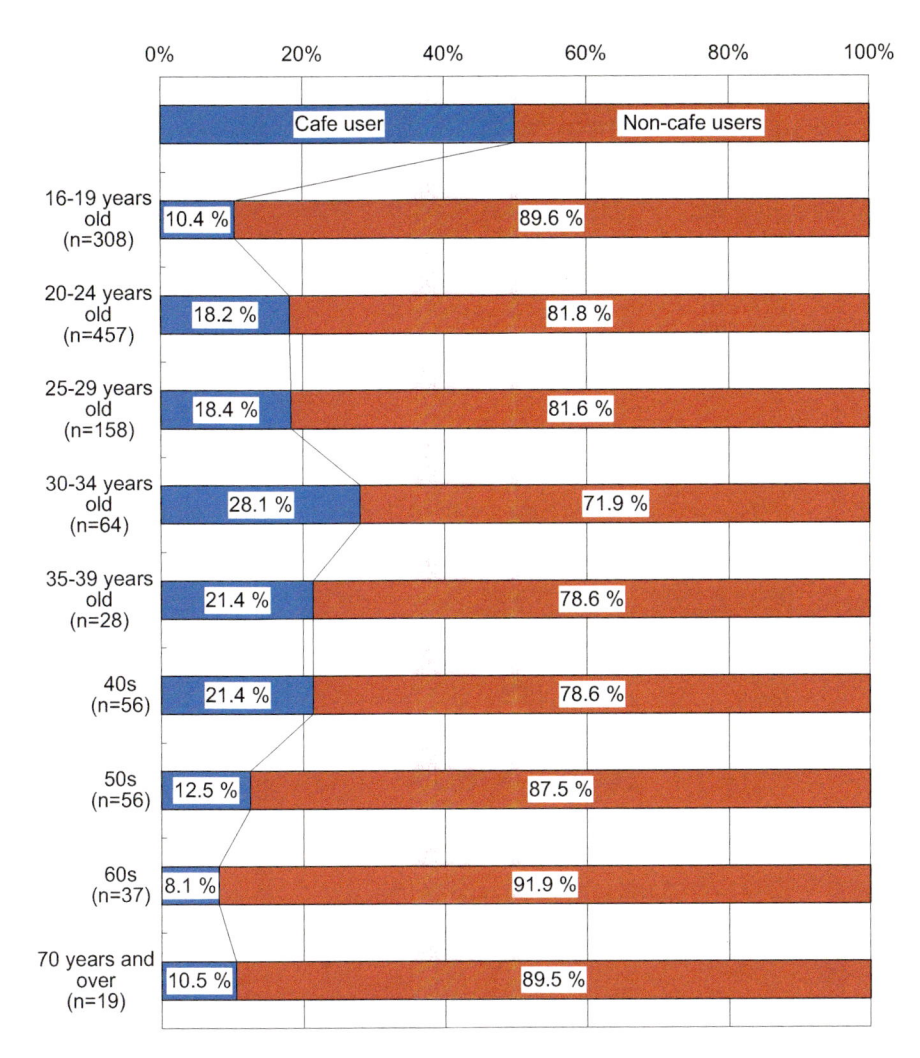

11 Roles of City Center Cafés and Their Economic Effects on City... 223

	Cafe user	Non-cafe users
16-19 years old (n=308)	10.4 %	89.6 %
20-24 years old (n=457)	18.2 %	81.8 %
25-29 years old (n=158)	18.4 %	81.6 %
30-34 years old (n=64)	28.1 %	71.9 %
35-39 years old (n=28)	21.4 %	78.6 %
40s (n=56)	21.4 %	78.6 %
50s (n=56)	12.5 %	87.5 %
60s (n=37)	8.1 %	91.9 %
70 years and over (n=19)	10.5 %	89.5 %

Fig. 11.4 Utilization rate of city center cafés by age

Utilization Rate of City Center Cafés by the Number and Types of Companions

Figures 11.6 and 11.7 summarize the city center café utilization rate by the number and types of accompanying persons, respectively. As for the number of companions, the utilization rate is higher for visitors who visited the city center district with plural companions than visitors visiting the district alone. With regard to the types of companions, the highest result (26.7%) is for friends/acquaintances of the opposite gender, while the other categories were much the same as each other (around 20%).

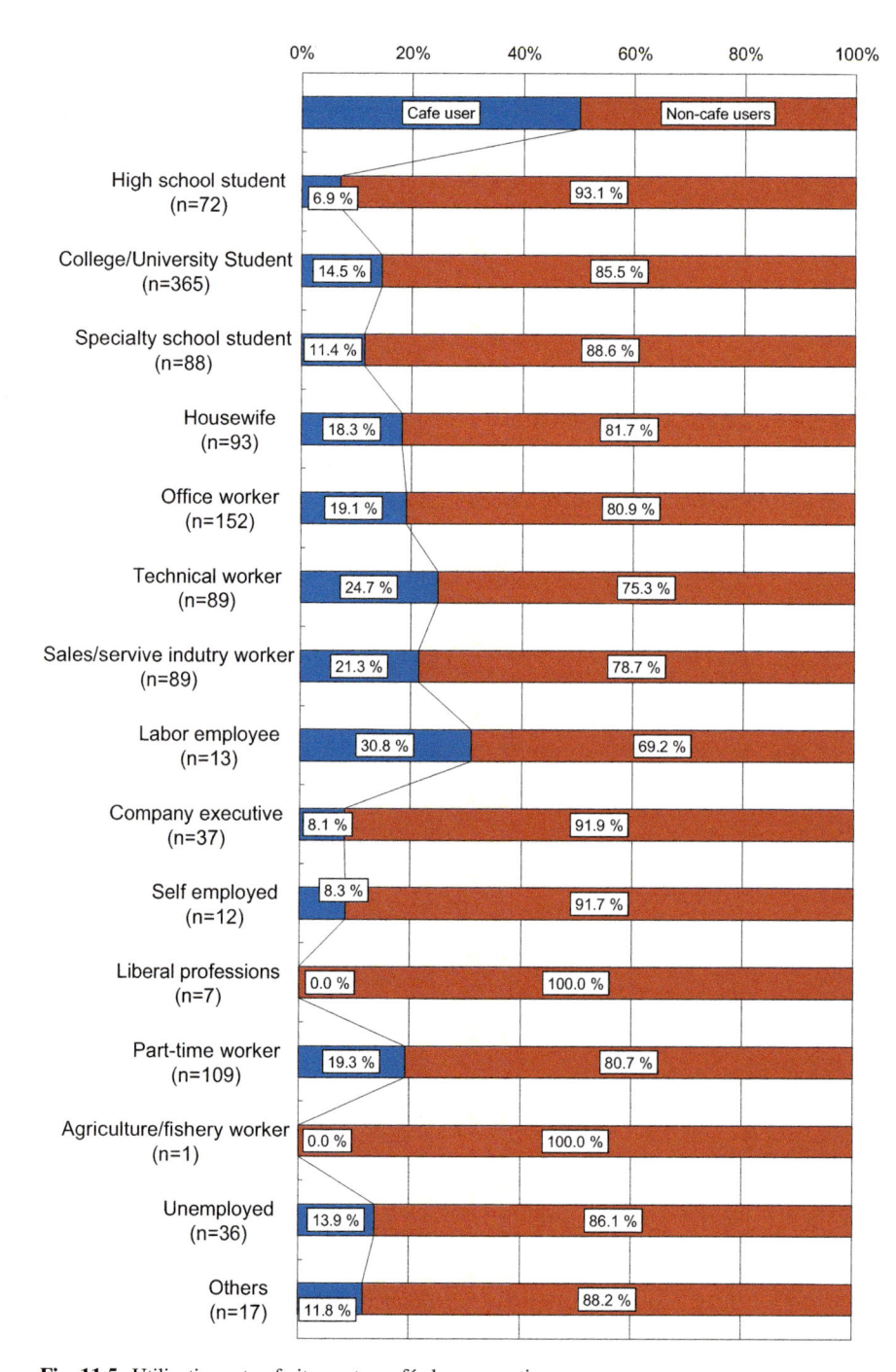

Fig. 11.5 Utilization rate of city center cafés by occupation

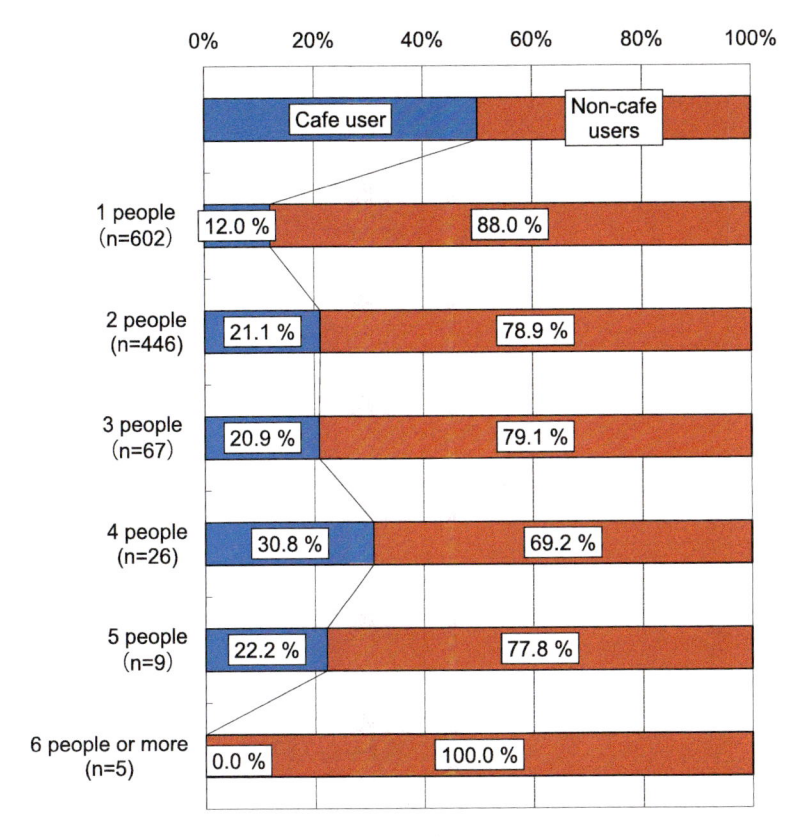

Fig. 11.6 Utilization rate by the number of companions

3.3 Characteristics of the Utilization of City Center Cafés

In order to grasp the characteristics of city center cafés, we examine why visitors to the city center district used city center cafés, as well as their average expenditure when there. Then, we analyze the frequency of visits to and the length of time staying at city center cafés.

Reasons for Using City Center Cafés

We received multiple answers on the reasons for using city center cafés. "Because I was tired" was the most frequent, followed by "because I was thirsty" and "because I wanted to drink coffee at that shop." Many people use city center cafés as a way to take a break (Fig. 11.8).

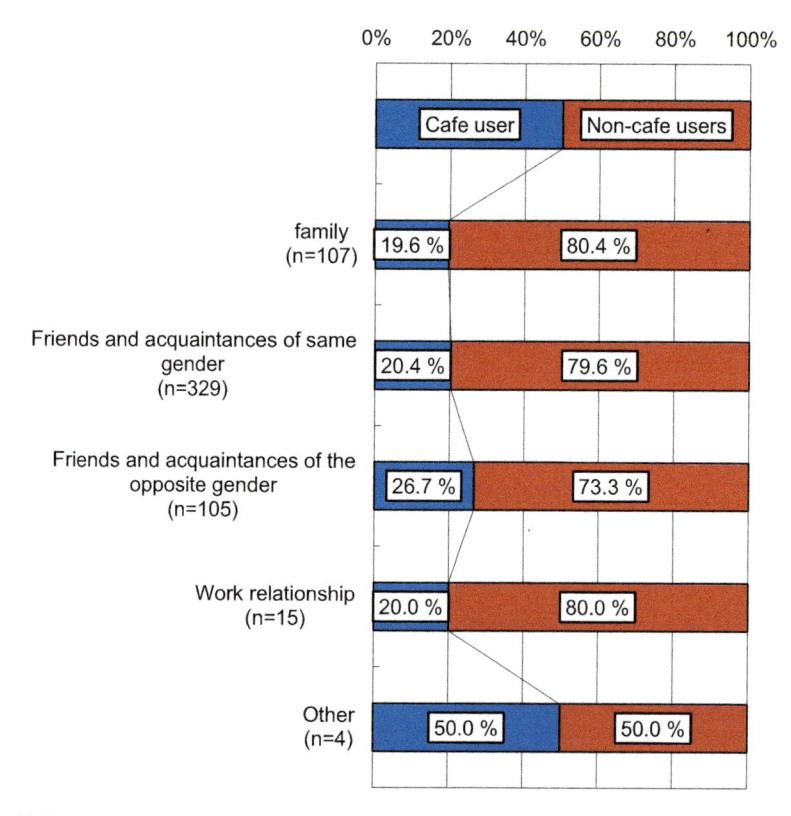

Fig. 11.7 Utilization rate by types of accompanying person

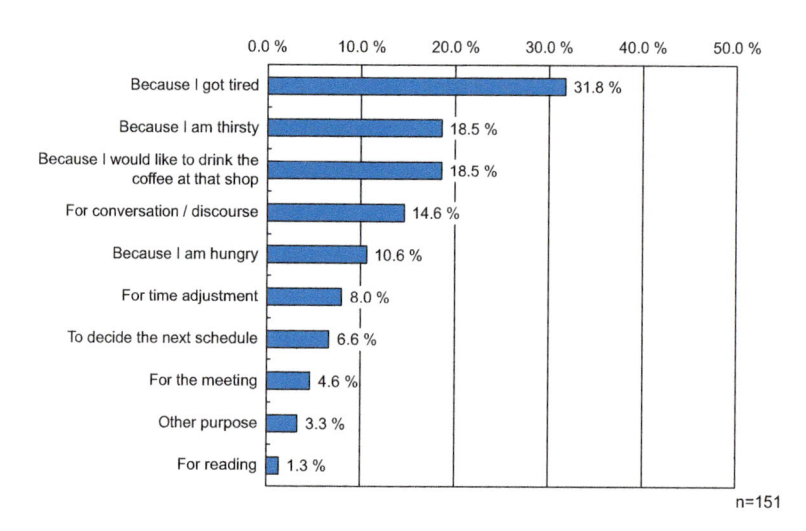

Fig. 11.8 Purposes for using city center cafés

Table 11.1 Average expenditure at city center cafés per visit

	Average (yen)	Number of samples	Standard deviation
Expenditure at café	533	172	376.01

Table 11.2 Average frequency of visits to city center cafés per ten visits to the city center district

	Average (times)	Number of samples	Standard deviation
Café users	5.8	193	3.19
Non-café users	3.2	964	3.06
All	3.6	1157	3.23

The probability to use city center cafés when visiting the city center

Average Expenditure at City Center Cafés Per Visit

The average expenditure at a city center café per visit is about 533 yen (Table 11.1).

Frequency of Visits to City Center Cafés

We asked respondents how many times they visit city center cafés per ten visits to the city center district, determining the probability that respondents use the city center cafés when they are in the district. The average is 5.8 times in ten visits for the respondents who used a city center café on the day of the survey and for the respondents who did not do so 3.2 times in ten visits. Overall, the average is 3.6 times per ten visits (Table 11.2).

Length of Time Staying at City Center Cafés

As for the respondents who used city center cafés on the day of the survey, we asked how much time they had spent in city center cafés on that day. Those spending less than 30 min accounted for 62.2%, and those who spent less than an hour were 94.6% (Fig. 11.9).

City Center Cafés Most Frequented in the City Center District

The survey asked which city center cafés respondents like to use (e.g., Starbucks and Doutor). The available options also included "the nearest city center café." The results are shown in Fig. 11.10. Starbucks has the largest support (50.7%), followed by Doutor (18.5%). The option of the nearest (most convenient) city center café was chosen by 12.2% of respondents.

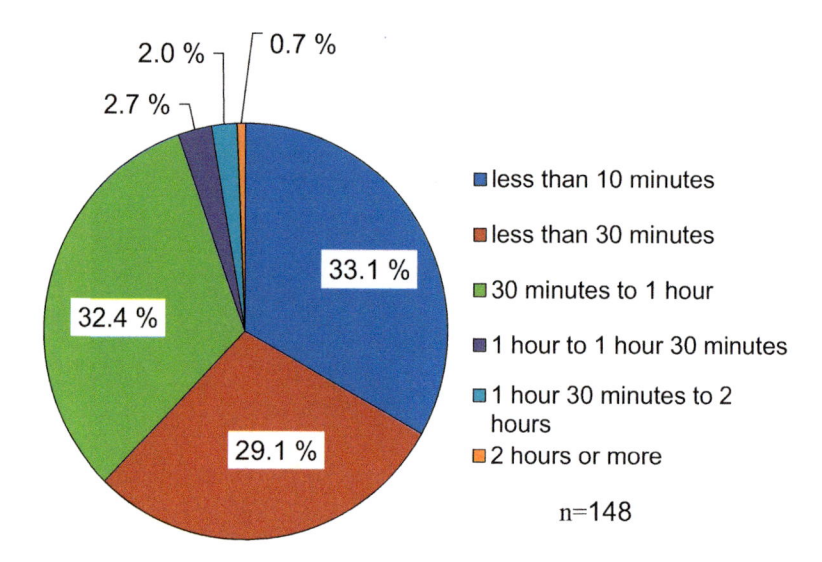

Fig. 11.9 Length of time staying at city center cafés

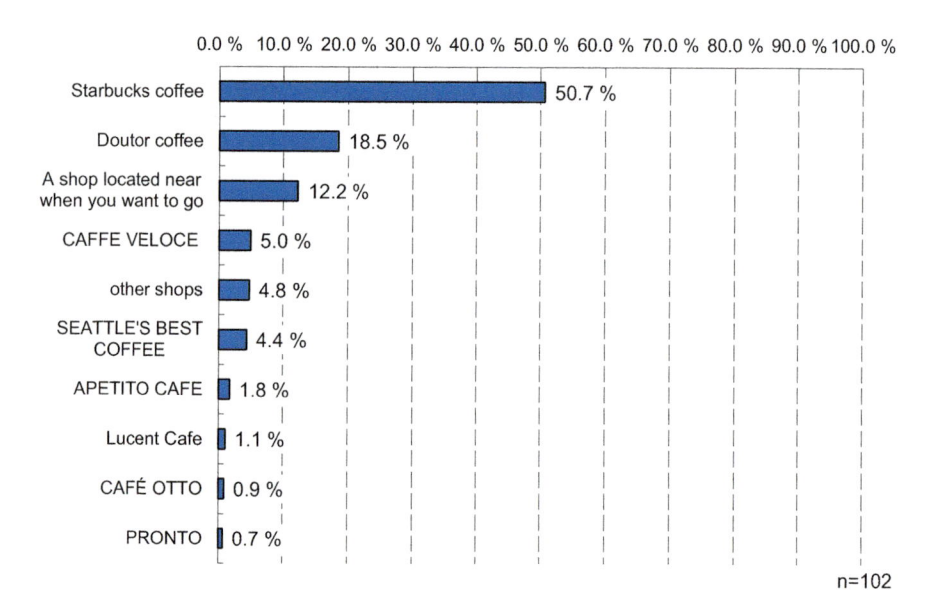

Pie chart values:
- less than 10 minutes: 33.1 %
- less than 30 minutes: 29.1 %
- 30 minutes to 1 hour: 32.4 %
- 1 hour to 1 hour 30 minutes: 2.7 %
- 1 hour 30 minutes to 2 hours: 2.0 %
- 2 hours or more: 0.7 %

n=148

Fig. 11.10 City center cafés most frequented

- Starbucks coffee: 50.7 %
- Doutor coffee: 18.5 %
- A shop located near when you want to go: 12.2 %
- CAFFE VELOCE: 5.0 %
- other shops: 4.8 %
- SEATTLE'S BEST COFFEE: 4.4 %
- APETITO CAFE: 1.8 %
- Lucent Cafe: 1.1 %
- CAFÉ OTTO: 0.9 %
- PRONTO: 0.7 %

n=102

3.4 Comparison of the Utilization of Starbucks, Doutor, and Other City Center Cafés

Here, we classified the city center cafés used by the respondents on the day of the survey into Starbucks and Doutor (the two most popular options) and other city center cafés and then compare the results.

City Center Cafés Used on the Day of the Survey

The ratio of city center cafés used on the survey day was 47.9% for Starbucks, 14.5% for Doutor, and 37.6% for others (Fig. 11.11).

Reasons for Stopping at a City Center Café

The reasons given for using Starbucks were "because I was tired" (35.1%), "because I was thirsty" (26.0%), and "because I wanted to drink the coffee at that shop" (24.7%). For Doutor, the reasons given were "because I was tired" (23.3%) and "because I was hungry" (20.0%). In addition, the reasons given for using other coffeehouses were "because I was tired" (30.0%), "a conversation/meeting" (26.0%), and "because I was thirsty" (18.0%). This shows the difference between the attractiveness of Starbucks' unique coffee and that of Doutor, where people can have light meals (Fig. 11.12).

Fig. 11.11 City center cafés used on the survey day

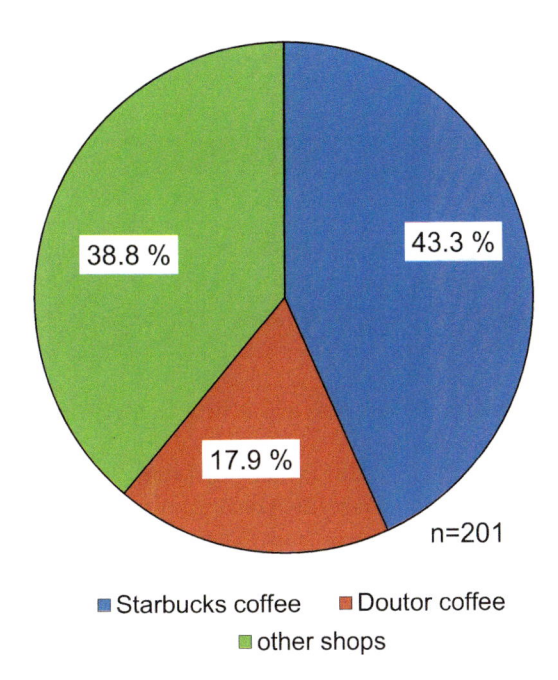

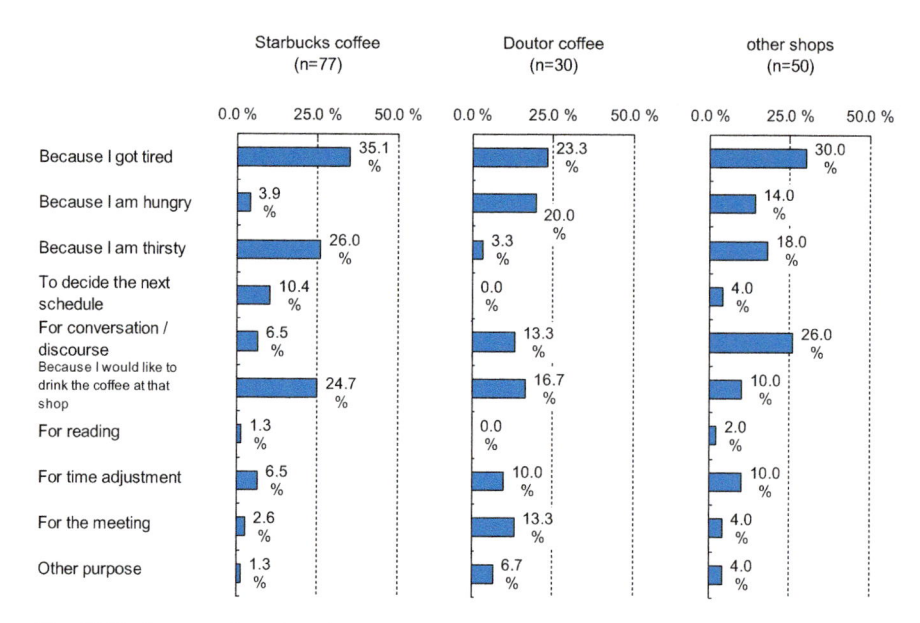

Fig. 11.12 Purposes for using city center cafés by Starbucks, Doutor, and others visited on the survey day

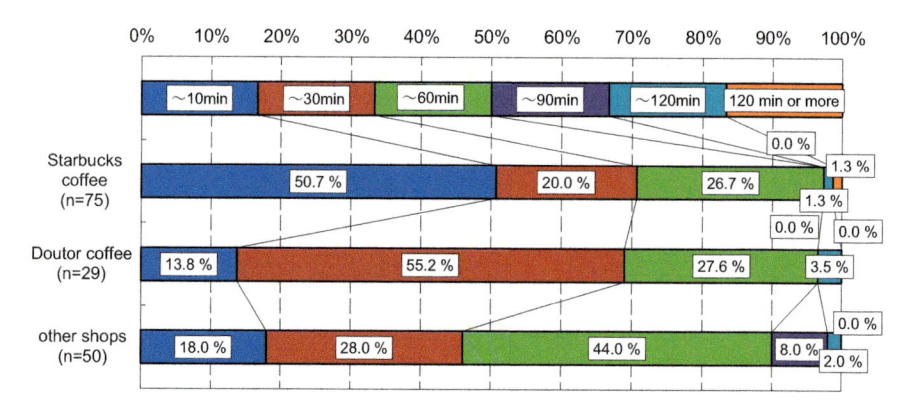

Fig. 11.13 Length of time staying at city center café by Starbucks, Doutor, and others visited on the survey day

Length of Time Staying at City Center Cafés

Figure 11.13 shows the length of time staying at the city center cafés visited on the survey day. The length of time spent at Starbucks was "less than 10 min" (50.7%) for about half users, while for Doutor it was for more than half users "10 to 30 minutes" (55.2%). For other city center cafés, most people responded with "30 to 60 minutes" (44.0%) (Fig. 11.14).

11 Roles of City Center Cafés and Their Economic Effects on City...   231

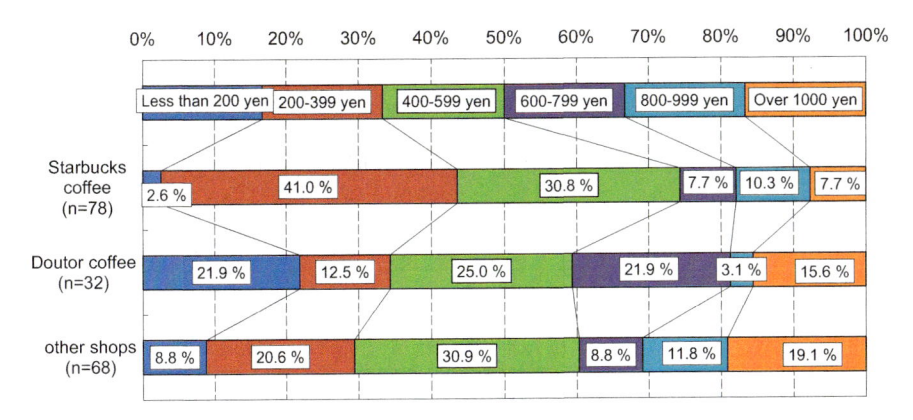

Fig. 11.14 Expenditure spent at city center cafés by Starbucks, Doutor, and others visited on the survey day

Table 11.3 Average expenditure at city center cafés by Starbucks, Doutor, and others visited on the survey day (per visit)

	Average (yen)	Number of samples	Standard deviation
Starbucks coffee	508	78	290.41
Doutor coffee	548	32	419.63
Other shops	604	68	461.54
All	552	178	387.39

Expenditure Spent at City Center Cafés

Figure 11.14 depicts the expense spent at the city center cafés used on the day of the survey. For Starbucks, expenditure of "200 yen to 399 yen" accounted for 41.0%. In the case of Doutor, the figures varied from 21.9% for "up to 200 yen," 25.0% for "400 yen to 599 yen," and 15.6% for "1000 yen or more."

Table 11.3 gives the average expenditure at the city center cafés visited on the survey day. The average expenditure at Starbucks was 508 yen, while that at Doutor was 548 yen, and other coffeehouses was 604 yen.

3.5 *Average Number of* Kaiyu *Steps of Visitors Who Used City Center Cafés*

Here, we investigate the number of *Kaiyu* steps (i.e., the number of commercial facilities that consumers stopped at) in order to determine whether the use of city center cafés is a cause to induce *Kaiyu* behaviors in the city center district.

Table 11.4 Comparison between average numbers of *Kaiyu* steps for users and nonusers of city center cafés

	Average (place)	Number of samples	Standard deviation	Difference in average numbers of steps to stop (a-b)	Significance probability
a Café user	3.10	193	1.98	0.26	0.085
b Non-café user	2.84	998	1.75		
All	2.88	1191	1.79		

Table 11.5 Average number of *Kaiyu* steps after stopping at city center cafés

Average (place)	Number of samples	Standard deviation
1.44	193	1.51

Comparison Between Average Numbers of *Kaiyu* Steps for Users and Nonusers of City Center Cafés

First, we examine the average number of *Kaiyu* steps for the entire sample. The average number of *Kaiyu* steps for the overall sample was 2.88 steps. Next, we compared the average numbers of *Kaiyu* steps for users and nonusers of city center cafés. Table 11.4 gives the result. The number of *Kaiyu* steps for city center café users and nonusers was 3.10 steps and 2.84 steps, respectively. A city center café user has about 0.26 steps more than a nonuser,[4] which suggests that the use of city center cafés increases the number of *Kaiyu* steps.[5]

Next, we analyze how the number of *Kaiyu* steps changed after stopping at a city center café. Table 11.5 shows that the average number of *Kaiyu* steps of city center café users after visiting a city center café is about 1.4 steps. This indicates that city center cafés may induce a *Kaiyu* effect, which is the effect on an increase of *Kaiyu* steps.

Average Number of *Kaiyu* Steps of City Center Café Users by Time of Use

Next, in order to analyze the induced *Kaiyu* effect of city center cafés in more detail, we classify the sample of city center café users into three groups based on the time when they used the city center cafés: the group who used the city center cafés at the start of *Kaiyu*, in the middle of *Kaiyu*, and at the end of *Kaiyu*. Then, we calculate the average number of *Kaiyu* steps for each group. Table 11.6 shows the result. From the table, we see that the group who used the city center cafés at the start of *Kaiyu*

---

[4] In this case, the *p*-value is 0.085, which means statistically significant at the 10% level.

[5] Here, the number of *Kaiyu* steps does not include the stops at city center cafes visited and the stops at commercial facilities for the purpose of passing or transit, either.

Table 11.6 Average number of *Kaiyu* steps by time of use

	Average (place)	Number of samples	Standard deviation
At the start of *Kaiyu*	1.73	56	1.73
In the middle of *Kaiyu*	4.26	87	1.73
The end of *Kaiyu*	2.62	50	1.41
All	3.10	193	1.98

Table 11.7 Test for the difference between the average numbers of *Kaiyu* steps for city center café users in the middle of *Kaiyu* and for nonusers

	Average (place)	Number of samples	Standard deviation	Difference in average numbers of steps to stop (a-b)	Significance probability
a Café user (in the middle of *Kaiyu*)	4.26	87	1.73	1.42	0.000
b Non-café user	2.84	998	1.75		

Table 11.8 Average number of *Kaiyu* steps after stopping at a city center café for users using city center cafés in the middle of *Kaiyu*

Average (place)	Number of samples	Standard deviation
2.07	87	1.23

comprised 56 samples, the group in the middle of *Kaiyu* consists of 87 samples, and 50 samples composed the group at the end of *Kaiyu*. The most frequent use of the city center cafés is known to be in the middle of *Kaiyu*. The average number of *Kaiyu* steps for the group who used the city center cafés in the middle of *Kaiyu* is 4.26 steps, which is the longest among all groups.

Furthermore, we perform a statistical test for checking whether the average number of *Kaiyu* steps for the group who used city center cafés in the middle of *Kaiyu* is statistically different from that for the group who did not use the city center cafés. Table 11.7 shows the result. The difference of the number of *Kaiyu* steps between these two groups of users and nonusers is 1.42 steps, which is statistically significant at 0.1% level.

In addition, focusing on the group of city center café users in the middle of *Kaiyu*, we examine the average number of *Kaiyu* steps after using city center cafés. Table 11.8 gives the result. The average number of *Kaiyu* steps after using city center cafés was 2.07 steps for this group.

3.6 Average Expenditure of City Center Café Users

We calculated the average expenditure per *Kaiyu* step for visitors who used city center cafés and for those who did not do so. Note that in this calculation of

Table 11.9 Average expenditure per *Kaiyu* step for users and nonusers of city center cafés

	Average (yen)	Number of samples	Standard deviation
Café user	1457	617	4191.97
Non-café user	1441	2828	6920.62
All	1444	3445	6515.95

A sample with high expenditure 500,000 yen is excluded from non-café users

Table 11.10 Average expenditure per *Kaiyu* step by time of use

When the drop to the café	Average (yen)	Number of samples	Standard deviation
At the start of *Kaiyu*	616	107	1818.58
In the middle of *Kaiyu*	1480	376	4342.99
The end of *Kaiyu*	2065	134	4962.75
All	1457	617	4191.97

expenditure, the amount of money spent in the city center cafés visited is not included in the amount of expenditure calculated here.

Average Expenditure Per *Kaiyu* Step of City Center Café User

Table 11.9 shows that the average expenditure per *Kaiyu* step of city center café users and nonusers is about 1400 yen, with little difference between the two groups.[6]

Average Expenditure by Time of Use

Next, in order to analyze the effect of city center cafés on the consumers' expenditure, we calculate the average expenditure per *Kaiyu* step for each group classified by the time when the city center cafés were used, at the start, in the middle, and at the end of *Kaiyu*. The result is shown in Table 11.10.

The average expenditures of visitors who used city center cafés in the middle and at the end of *Kaiyu* are 1480 yen and 2065 yen, respectively, whereas those who used city center cafés at the start of *Kaiyu* spent approximately 620 yen, which is significantly lower than the other two.

---

[6]The data include high expenditure (nonuser of city center cafes: 500,000 yen). The average expenditure calculated here excluded this sample.

4 Estimation of the Economic Effect of City Center Cafés

4.1 Conceptualizing the Economic Effect of City Center Cafés

In the previous section, we have analyzed consumer *Kaiyu* behaviors from the viewpoint of how the use of city center cafés affects the *Kaiyu* behaviors focusing on the number of *Kaiyu* steps and expenditure. As a result, we have found the fact that the use of city center cafés in the middle of *Kaiyu* has an effect to increase the number of *Kaiyu* steps in comparison with the case when visitors did not use the city center cafés. This effect can be interpreted as a refreshing effect or taking a break effect, when consumers who are tired of shop-around or moving around for shopping take a break to stay at city center cafés and continue their shopping by extending their *Kaiyu* steps. In this sense, the city center cafés can be thought of as inducing additional *Kaiyu* behaviors.

If we further advance this thinking, consumers who extend their *Kaiyu* behaviors due to the use of city center cafés not only should increase the number of their *Kaiyu* steps but also must increase their expense by the amount spent at commercial facilities they visited as their extended *Kaiyu* steps. The increase of the expense spent at the extended *Kaiyu* steps would not have occurred if consumers had not used the city center cafés. Thus the increase of the expenditure spent at the extended *Kaiyu* steps can be considered as accruing to the city center cafés. Therefore, we regard the economic effect of city center cafés as the amount of money spent at the extended *Kaiyu* steps due to the use of city center cafés.

More specifically, we define the economic effect of city center cafés as the increase of turnover of the city center district which is brought about by the increased amount of money the consumers who used the city center cafés spend at the commercial facilities they visited at their extended *Kaiyu* steps.

Based on this definition, we estimate the economic effect of city center cafés as the product of the following four values: (1) the number of city center café users, (2) the percentage of city center café users in the middle of *Kaiyu*, (3) the increase of *Kaiyu* steps for visitors who use city center cafés in the middle of *Kaiyu*, and (4) the amount of money spent per *Kaiyu* step.

4.2 Estimated Results

Number of City Center Café Users

Here, as the number of people entering the city center district of Fukuoka City, we use the estimate of 231,265 (people/day), obtained using the on-site Poisson regression model by Saito et al. [4] (this result is included in Saito [6]). The total number of samples is 1191, with 193 being city center café users[7] so that the rate using city

---

[7]See Sect. 3.1.

center cafés is 16.2%. Based on these, the number of the city center café users per day is estimated as follows:

$$37,476 \text{ (person/day)} = 231,265 \text{ (person/day)} \times (193 \div 1191).$$

Percentage of City Center Café Users in the Middle of *Kaiyu*

Among all of 193 city center café users, 87 users visited the city center cafés in the middle of their *Kaiyu*. Therefore, the percentage of city center café users in the middle of *Kaiyu* is as follows:

$$0.451 = 87 \text{ people} \div 193 \text{ people}.$$

Increase of the Number of *Kaiyu* Steps

We presume that the number of *Kaiyu* steps after using city center cafés represents the increase resulting from stopping at city center cafés.

The average number of *Kaiyu* steps after using city center cafés in the middle of *Kaiyu* was 2.07 steps, as shown in Table 11.8.

Average Expenditure per *Kaiyu* Step

We employ two values for the estimate of the average expenditure per *Kaiyu* step: the average expenditure of all city center café users and the average expenditure of those who use city center cafés in the middle of *Kaiyu*. The former is 1457 yen and the latter is 1480 yen (see Table 11.10).

Economic Effect of City Center Cafés

The economic effect of city center cafés is estimated as follows.
In the case where the average expenditure per *Kaiyu* step is 1457 yen:

$$\begin{aligned}18,590,650,000 \text{ (yen/year)} &\fallingdotseq 37,476 \text{ (person/day)} \\ &\times 0.451 \text{ (percentage of use in the middle)} \times 2.07 \text{ (step/person)} \\ &\times 1,457 \text{ (yen/step)} \times 365 \text{ (day)}.\end{aligned}$$

In the case where the average expenditure per *Kaiyu* step is 1480 yen:

$18{,}881{,}130{,}000 \text{ (yen/year)} \fallingdotseq 37{,}476 \text{ (person/day)}$
$\times\, 0.451 \text{ (percentage of use in the middle)} \times 2.07 \text{ (step/person)}$
$\times\, 1{,}480 \text{ (yen/step)} \times 365 \text{ (day)}.$

5 Conclusion and Future Challenges

In the city center district, various industries are located mixedly, and individual industries complicatedly interact with each other to create the unique attractiveness of the city. However, up to now few studies have revealed concretely the effects of individual industries in the city center district. The significance of this study is that taking up city center cafés as one of the components of the attractiveness of the city, it analyzes their function from the viewpoint of consumers' *Kaiyu* behaviors.

This study makes two major contributions. First, we found the several empirical facts. The number of users of city center cafés is particularly large for people younger than 30. However, the utilization rate is higher for people in their 30s. In addition, there are differences in the utilization purpose and expenditure per shop. The most important empirical fact we found is that the number of *Kaiyu* steps for those who use the city center cafés is larger than that for those who do not use the city center cafés.

Second, we defined the economic effect of city center cafés as the increase of the turnover of the retail sector at the city center district which is brought about by the increased amount of money the consumers who used the city center cafés spend at their extended *Kaiyu* steps. Further, based on this definition, we actually calculate the estimate of the economic effect of the city center cafés at the city center district of Fukuoka City, Japan.

Future research topics include expanding above analysis to the industries other than city center cafés and clarifying the structure of the attractiveness of the city by exploring concretely the effects and roles that each individual industry plays at the city center district.

References

1. Saito S (1983) Present situation and challenges for the commercial districts in Nobeoka area. In: The report of regional plan for modernizing commerce: Nobeoka area. Committee for Modernizing Commerce Nobeoka Region Section, pp 37–96. (in Japanese)
2. Saito S (1984) A disaggregate hierarchical Huff model with considering consumer's shop-around choice among commercial districts: developing SCOPES (Saga Commercial Policy Evaluation System). Plan Public Manag 13:73–82. (in Japanese)
3. Saito S, Ishibashi K (1992) Forecasting consumer's shop-around behaviors within a city center retail environment after its redevelopments using Markov chain model with covariates. Pap City Plan 27:439–444. (in Japanese)

4. Saito S, Kakoi M, Nakashima T (1999) On-site Poisson regression modeling for forecasting the number of visitors to city center retail environment and its evaluation. Stud Reg Sci 29:55–74. (in Japanese)
5. Saito S, Yamashiro K (2001) Economic impacts of the downtown one-dollar circuit bus estimated from consumer's shop-around behavior: a case of the downtown one-dollar bus at Fukuoka city. Stud Reg Sci 31:57–75. (in Japanese)
6. Saito S (2000) Report on the survey of *Kaiyu* behaviors at city center of Fukuoka, Japan 2000: with focusing on underground space and comparison of city attractiveness between Japan and Korea. Fukuoka Asia Urban Research Center, Fukuoka. (in Japanese)
7. Maruki T, Kusaba K, Yoshiya Y, Morishita Y, Ushijima K, Kimura S (2001) The economic effects of midtown cafes based on behaviors of cafe users. In: Proceedings for the 2nd survey research conference on marketing for urban development at city center of Fukuoka, Japan. Fukuoka, pp 56–59. (in Japanese)
8. Yoshiya Y, Morishita Y, Ushijima K, Maruki T, Kusaba K, Kimura S (2001) Some features of location dynamics of midtown cafes and behaviors of cafe users. In: Proceedings for the 2nd survey research conference on marketing for urban development at city center of Fukuoka, Fukuoka, pp 49–55. (in Japanese)
9. Saito S, Kakoi M, Nakashima T, Igarashi Y, Kiguchi T (2008) A consumer behavior approach to estimating the economic effects of open cafes at city center retail district: how further do those open cafes accelerate the shop-around behavior of their customers? Fukuoka Univ Rev Econ 52:435–458. (in Japanese)
10. Saito S, Kiguchi T, Kakoi M, Nakashima T (2002) Roles of open cafés and their economic effects on city center: based on consumer behavior using open cafés located at city center. In: Collected papers for presentation in the 39th annual meeting of Japan Section of Regional Science Association International (JSRSAI), pp 417–424. (in Japanese)
11. Saburo Saito, Tomoyuki Kiguchi, Masakuni Kakoi, Takaaki Nakashima (2003) The economic effect and function of city center café from the viewpoint of consumer shop-around behavior. 2003 Fall Natl Conf Opera Res Soc Japan Abstr:218–219. (in Japanese)

Part IV
Economic Effects by Increasing Visitors

Chapter 12
The Economic Effects of Opening a New Subway Line on City Center Commercial District

Saburo Saito and Kosuke Yamashiro

Abstract This study forecasted the economic effects that the new Fukuoka City Subway Nanakuma Line would induce on the city center retail sector. The subway will connect the suburban residential area to the city center. Suburban residents can drastically reduce their travel time to the city center from 40 or 50 to 20 min by changing from bus to subway. This improvement in accessibility should increase the frequency of visits to the city center so that the number of visitors and the turnover at the city center would increase. This increase of the turnover was defined and estimated as the economic effects of the new subway on the city center retail sector. We called this procedure a consumer behavior approach because all estimations depend on behavioral changes in consumers after the subway is introduced. First we estimated the modal choice and visit frequency models from the data obtained from a survey of consumer travel behavior conducted at the city center for randomly sampled visitors. We predicted modal choice, frequency of visits, and expenditure at the city center by residents for each 278 residential divisions along the subway line and summed the results. The economic effects were estimated to be 17.7 billion yen per year.

Keywords Consumer behavior approach · City center district · New subway line · Economic effect · Increase in visit frequency · Retail sector · Turnover · *Kaiyu*

This chapter is based on the paper by Saburo Saito, Kosuke Yamashiro, Takaaki Nakashima, and Yasufumi Igarashi [1], "Predicting economic impacts of a new subway line on a city center retail sector: A case study of Fukuoka City based on a consumer behavior approach," *Studies in Regional Science*, vol. 37, pp. 841–854, 2007, which is modified for this chapter.

S. Saito (✉)
Faculty of Economics, Fukuoka University, Fukuoka, Japan

Fukuoka University Institute of Quantitative Behavioral Informatics for City and Space Economy (FQBIC), Fukuoka, Japan
e-mail: saito@fukuoka-u.ac.jp

K. Yamashiro
Department of Business and Economics, Nippon Bunri University, Oita City, Japan
e-mail: yamashiroks@nbu.ac.jp

1 Purpose

In Fukuoka City, Japan, the opening of Fukuoka City Subway Line 3 (nickname: Nanakuma Line) connecting Tenjin, the city center commercial district of Fukuoka City, to the southwest part of Fukuoka City is scheduled to be held in February 2005. The area along the Fukuoka City Subway Nanakuma Line is a residential area with many population as well as with four universities located; it has characteristics as a region that plays a role of consumption at the city center commercial district of Fukuoka City. Also, as the past development of the city center of Fukuoka City had been triggered by the development of transport systems such as transportation infrastructure and terminal station building development, people involved with retail sector at the city center are greatly concerned with the change and the effects caused by the opening of the Fukuoka City Subway Nanakuma Line. In particular, from the viewpoint of revitalizing the city center business, it can be said that there is a long-awaited need for prediction of effects such as an increase in the number of visitors to the city center district due to the opening of the subway and an increase in sales volume.

So far, we have predicted and measured the effect of city center commercial redevelopment mainly at the city center of Fukuoka City from the increase in the number of visitors and the change in visitors' *Kaiyu* behaviors [2–7]. A major feature of these studies is that they predict and measure the effect focusing on how consumer behaviors change before and after the redevelopment project (with or without). In particular, they do not rely on probability-based models such as shopping destination choice models but try to measure the economic effect of the commercial redevelopment project concretely in terms of the consumer's frequency of visits, the number of visitors, and the amount of turnover at the city center district. We called this procedure as the consumer behavior approach.

Following this consumer behavior approach, the purpose of this study is to measure economic effects of the new Fukuoka City Subway Nanakuma Line on the city center commercial district. The new Subway Nanakuma Line connects the southwest part of Fukuoka City with the city center district of Fukuoka City. Because Subway Nanakuma Line brings a drastic shortening to the travel time to the city center of Fukuoka City for the residents along Subway Nanakuma Line, they would increase their frequency of visits to the city center commercial district. This increase of visit frequency can be thought of as the increase of travel demand to the city center district since, for the residents along the subway line, the relative price of city center goods declines as the generalized travel cost falls. Since the residents increase their frequency of visits, the number of visitors at the city center district increases. Thus the turnover of city center retail sector would be increased. We define the increase of turnover as the economic effects on city center commercial district induced by the new subway line. We call this method as the consumer behavior approach because all the impacts are captured by the predictive behavioral changes of consumers after the new subway line is introduced. This way of thinking is a different concept from the traditional concept of the benefit assessment of transportation facilities.

In existing studies on the benefit assessment of transport facilities, various calculation methods have been considered [8, 9]. A representative example is the development of partial equilibrium theory focusing on user benefits such as time reduction and transportation expenses including Foster and Beesley [10]. There are also studies such as Small and Rosen [11] which theoretically showed consumer surplus of society as a whole under disaggregated logit model. These are studies based on applied welfare economics that seek to measure changes in utility levels of consumers as a benefit. The characteristic of these studies is that they show how to measure changes in the ordinal utility level of consumers with changes in prices and incomes as the benefit of money terms by the concept of EV (equivalent variation) or CV (compensation variation) strictly following the concept of traditional consumer surplus. In this trend, Kanemoto and Mera [12] developed consumer surplus analysis from partial equilibrium analysis to general equilibrium analysis, and also there is a development of the shortcut theory [8] which evaluates the benefits of transportation facilities focusing only on the information related to the transport market while relying on the framework of general equilibrium analysis. In particular, the user benefit method based on the shortcut theory is commonly used as a method of evaluating benefits [13, 14]. However, the user benefit method has a problem that it is difficult to objectively measure whether or not the benefit actually was realized as a money term.

As another method of objectively measuring the realized or to be realized benefits on a cash basis, there is a concept of a capitalization hypothesis that the generated benefits ultimately result in asset value [15, 16]. In this trend there are studies such as Hidano et al. [17] based on the land value reduction method that the effect of the development of the new railway line will be attributed to land prices. Although the capitalization method can verify the magnitude of the realized benefit, there is a problem that it takes some time before the benefit reaches the asset price.

On the other hand, based on the framework of a general equilibrium analysis with a single-city model, Wheaton [18] theoretically derived that social net benefits due to reduced transportation expenses can be measured as the product of initial traffic demand and transportation cost reduction. Pines and Weiss [19] show that Wheaton's method cannot be applied because it changes the traffic demand itself when the traffic time is shortened. Although these studies mainly focus on what can be evaluated by market value, the method of capitalization has been applied to the evaluation of nonmarket goods such as parks and the environment as a hedonic approach, together with CVM and travel cost methods [15].

We have looked at previous research. Regarding the measurement of the economic effect by developing the railway new line, as seen in the manual [14] by the Railway Bureau of Ministry of Transportation, the main effect they are concerned with is the measurement of the direct effect of the user's benefits obtained from the reduction of transportation costs and the shortening of the travel time. However, the problem with this method is that the benefits generated for users include virtual things such as the travel time shortening so that it is almost impossible to prove that such benefits actually had occurred as flows on a monetary basis. Also, even with the money-based capitalization hypothesis such as the land value reduction method, what is measured is a surplus attributable to society as a whole, which is a concept of long-term effects on stock, and what is concerned with is how much economic net benefit is generated for society as a whole. Both of these studies are not paying attention to concrete monetary flows of specific industries at specific areas such as commercial turnover at the city center district of Fukuoka City, which we are concerned with. In this way, most traditional transport studies to measure economic effects have been employing the user benefit approach that measures user benefits induced by decrease of the travel time due to the introduction of new railroad. While these previous studies are appropriate for estimating the net economic benefits for society as a whole, it must not be so for our present purpose to estimate the economic effects for the retail sector at the restricted area as city center.

In contrast, the feature of this study is that the effect of reducing the travel time due to the opening of the subway is regarded as the effect of increasing the frequency of trips to the city center district by the residents along the subway line, and the resulting increase in expenditures at the city center district is predicted as the economic effect, the magnitude of which can be verified ex post facto. As will be described later, forecasts are conducted for the areas along the subway line by town-*chome* unit, which is the smallest area division for the municipal population statistics. Thus our method makes possible such a measurement as what amount of effects has been brought about from which district to the city center district of Fukuoka City, enabling more detailed analyses, which is also the feature of this study.

The aim of this research is to show that the economic effect that the opening of the new subway line brings to the city center can be measured concretely by the consumer behavior approach. According to that aim, this study defines the economic effect to the city center by opening the new subway line based on the consumer behavior approach as follows.

The opening of the new subway line will shorten the travel time required for residents along the subway line to visit the city center and improve the accessibility to the city center. As the accessibility to the city center improves, the residents along the subway line will change their transportation means from existing means to the subway. Also the improvement of the accessibility to the city center will increase their frequency of visits to the city center. Under the assumption that the average expenditure at the city center per visit remains the same before and after the opening of the subway, the retail sales at the city center will increase.

In this study, this increase of retail sales at the city center is defined as the economic effect due to the opening of the new subway line. According to this definition, this economic effect represents the effect the retailers located at the city center would obtain so it does not include the effect on the retailers located in the area along the subway line other than the city center. Thus its concept is different from the benefit society as a whole would receive. Also note that the effect just indicates the increase of the turnover not the increase of the profits.

This definition of the economic effect presumes that the opening of the subway line would induce the residents in the area along the subway line to increase their frequency of visits to the city center. This presumption is derived from the following behavioral hypothesis of consumers.

Consumers who live along the subway line have local goods purchased locally and city goods purchased at the city center. It is assumed that generalized transport costs are included in the prices of these goods. By opening the subway line, the travel time required from the residential area to the city center area is shortened, and the generalized transport cost is decreased, so that the price of city good decreases. Thus the relative price of the city good to the local good decreases than before the opening of the subway line. Therefore, the demand for the city good is considered to increase because its relative price falls after the opening of the subway line.

In this study, the increase of the demand for the city good is assumed to correspond to the increase of visits to the city center. Thus the number of visitors who visit the city center increases more than before the opening of the subway line.

The purpose of this study is to estimate the economic effect caused by the opening of Fukuoka City Subway Nanakuma Line based on the above consumer behavioral approach. To do this, first we constructed two models: modal change and visit frequency models. Next, the parameters of these two models are estimated from data obtained from the survey of consumer travel behavior that was conducted at the city center for the respondents sampled from visitors there. Before the prediction, we delineate the border of the residential tract whose inhabitants have the possibility to use the new subway line. The resulted tract contains divisions of 278 residential addresses or town-*chome* divisions. Then using the population data of Fukuoka City, and complementing the data about the required travel time and travel costs by each transportation means for all 278 area divisions by using GIS, we forecast how many people change their modal choice from existing travel means to the subway and predict the increased number of visitors to the city center. Finally, we forecast the amount of expenditure the increased number of visitors spends at the city center.[1]

The remaining parts of this paper are organized as follows. In the next section, we show the route of Subway Nanakuma Line and the study area. In Sect. 3, we outline our procedure of analysis. In Sect. 4, we formulate the two models of modal choice and visit frequency and present their estimated results. In Sect. 5, we forecast how many people would change their modal choice from bus into subway and how they would increase the frequency of visits to city center of Fukuoka. We also forecast the economic effects of the new subway line on the city center commercial district. We conclude with Sect. 6.

---

[1]This study started from the thesis research [20].

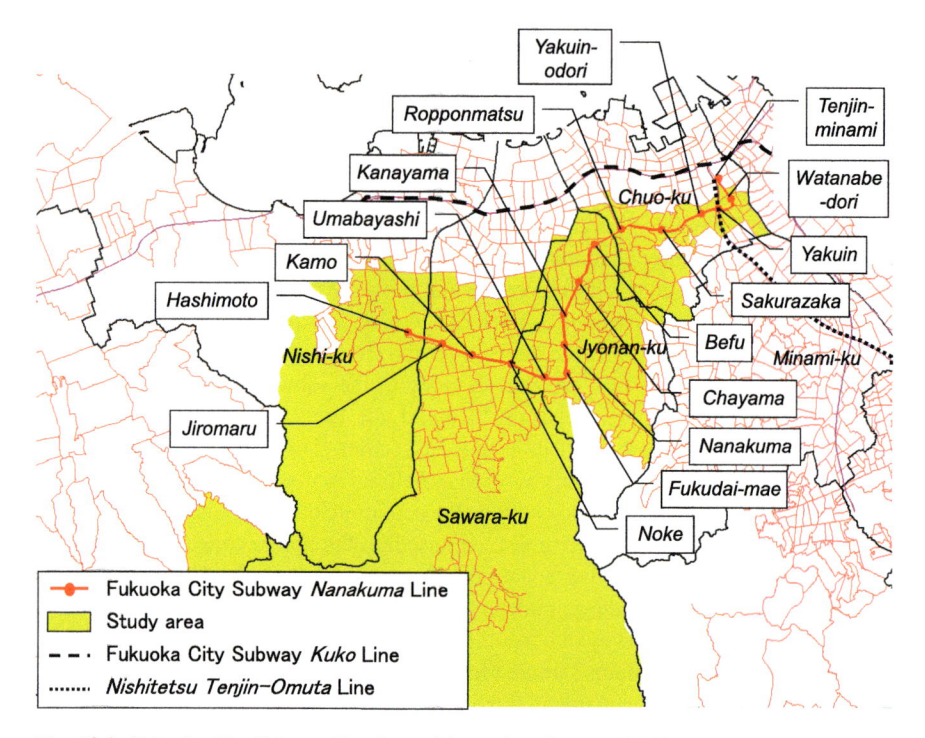

Fig. 12.1 Fukuoka City Subway Nanakuma Line and study area (Cf. [21])

2 Route of Fukuoka City Subway Nanakuma Line and Study Area

Figure 12.1 gives the map of Fukuoka City Subway Nanakuma Line and study area. The map depicts five wards out of seven in Fukuoka City: *Chuo-ku*, *Minami-ku*, *Nishi-ku*, *Sawara-ku*, and *Jonan-ku*.

The Fukuoka City Subway Nanakuma Line has 16 stations from *Tenjin-Minami* Station in *Chuo-ku* to *Hashimoto* Station in *Nishi-ku* connecting two terminal stations in 24 min. The southwest part of Fukuoka City where the subway runs has been rapidly becoming a bedroom town after the mid-1960s, and now it is an area where about 500,000 people, about 40% of population in Fukuoka City, live. Nonetheless, public transport has only buses and taxis, and on the main roads, chronic traffic jams are still going on. The opening of the new subway is expected to alleviate congestion in the area along the subway line.

As the details are shown in Table 12.1, among 672 town-*chome* divisions in the above 5 wards, the study area is composed of 278 town-*chome* divisions whose inhabitants have the possibility to use the new subway line. The remaining town-*chome* divisions are close to the existing subway line 1 and *Nishitetsu Tenjin Omuta* Line, and they are excluded from the study area because they are town-*chome* ·

Table 12.1 Numbers of town-chome divisions in five wards of Fukuoka City and those along the new subway

	Numbers of town-chome divisions in the ward	Numbers of town-chome divisions along Nanakuma Line
Chuo-ku	122	51
Minami-ku	161	0
Nishi-ku	129	29
Jonan-ku	93	84
Sawara-ku	167	114
Total	672	278

divisions whose travel times are not expected to be shortened even if using the new subway line after the opening.[2]

3 Procedure to Estimate Economic Effect

In this research, we predict the economic effect on the city center commercial district of Fukuoka City by the following procedure:

1. The parameters of transport modal choice model are estimated using two kinds of data: One is the data obtained from two on-site surveys conducted at Tenjin by FQBIC (Fukuoka University Institute of Quantitative Behavioral Informatics for City and Space Economy) which provide respondents' travel behaviors to Tenjin concerning their choice of transport means on the day to Tenjin, the travel expense, and the travel time to Tenjin. The other is the supplementary data calculated by GIS which provides attributes of alternative travel means if respondents took other alternative travel means than that used on the day concerning their travel expenses and travel times to Tenjin where the alternative travel means are the ones the respondents listed in the surveys.

---

[2]Because Fukuoka City Subway Nanakuma Line and *Nishitetsu Tenjin Omuta* Line are connected at *Yakuin* Station and run in parallel from there to Tenjin, the lines compete with each other between the Nanakuma Line section of *Yakuin–Watanabe-dori–Tenjin-Minami* station and the Nishitetsu Tenjin Omuta Line section of *Yakuin–Fukuoka-Tenjin* station. In this study, the residents along the subway line are assumed to make a trip to Tenjin city center of Fukuoka to get on from the bus stop or subway station with the shortest distance. As for the town-*chome* divisions where *Yakuin* Station is the nearest station, the alternative set of transport means are assumed to be composed of subway, bus, and automobile. The railway of the *Nishitetsu Tenjin Omuta* Line is excluded from the options as well as on foot and bicycle. This topic should be left as a future subject.

2. Using the same data as in (1), we estimate the parameters of the Poisson model to predict the frequency of visits to Tenjin with explanatory variables of transport expense and travel time to Tenjin.
3. For each town-*chome* division, the transport expense and travel time to Tenjin by transport means are measured by GIS, and by using the transport modal choice model estimated in (1), we estimate for each town-*chome* division the probability of modal choice in current two modes of bus and automobile and the probability of modal choice in three modes with adding the new subway after its opening.
4. Using transport expense and travel time to Tenjin by town-*chome* division by transport mode measured by GIS in (3), we estimate the frequency of visits to Tenjin by town-*chome* division by transport mode in two cases of current and after the opening of the subway.
5. Using population data by town-*chome* division and the results of (3) and (4), we forecast the number of visitors to the city center of Fukuoka in two cases of current and after the opening and calculate the increased number of visitors after the opening of the new subway line.
6. Using the data on the average expenditure at the city center district of Fukuoka City per visit and multiplying this with (5), we obtain the increase of the turnover at the city center district of Fukuoka City as the economic effect on the city center commercial district of Fukuoka City due to the opening of the new subway line.

In the following, procedures (1) and (2) are explained in Sect. 4 and procedures (3) to (6) in Sect. 5.

4 Estimating Models of Modal Choice and Visit Frequency

4.1 Data Used

Data mainly used is data obtained from on-site consumer travel behavior surveys conducted by FQBIC, which are the on-site interview surveys with questionnaire sheets asking the respondents randomly sampled from the visitors at the city center district of Fukuoka City about their travel behavior on the survey day.

In the surveys, in order to reduce the burden on the survey cooperation of respondents, we asked them about the travel time and the transport expense only for the transport means they used on the day of the survey.

Regarding alternative transport means, we asked only the means and have not asked about their transport expenses and travel times. For the estimation of the transport modal choice model, the transport expenses and the travel times for the alternative transport means are required. Therefore, GIS is used to prepare for each sample the complementary data on the transport expenses and travel times for alternative transport means.

Consumer Travel Behavior Survey at the City Center of Fukuoka City

In this study, we use data obtained from two surveys conducted by FQBIC: the first Fukuoka Travel Behavior and *Kaiyu* Compound Survey from June 28 to 30 in 2002 and the first Fukuoka Travel Behavior Survey from September 6 to 8 in 2002.

The first Fukuoka Travel Behavior and *Kaiyu* Compound Survey is designed to ask the residents who live along the new subway line about their travel behavior from their home to Tenjin district, the city center district of Fukuoka City. The items asked are the transport means they used on the day of the survey from home to Tenjin, transport expense, travel time, other alternative transport means they did not use on the day, and their *Kaiyu* behavior: the places visited, the purposes done there, and the expenditure spent there in the order of their occurrence. We set six survey points from shopping sites at Tenjin district: three department stores, *Daimaru*-Elgala, Fukuoka *Mitsukoshi*, and *Iwataya* Z-side, and three commercial complexes, Solaria Plaza, Solaria Stage, and Shoppers *Daiei*. The survey was carried out from 12:00 to 20:00 as an on-site interview survey with questionnaire sheets which takes about 15 min where the respondents are sampled at random from the visitors at the survey points. We collected 59 valid samples.

In order to obtain a sufficient number of samples, in September 2002, the question items were narrowed down to items of travel behavior, and the additional survey was conducted as the first Fukuoka Travel Behavior Survey. We set three survey points of shopping sites at Tenjin, *Daimaru*-Elgala, Fukuoka *Mitsukoshi*, and Solaria Plaza. As in the previous survey, the survey time is 8 h from 12:00 to 20:00. We collected valid 240 samples.

In both surveys, in order to measure the distance, which is necessary for model estimation, from the respondent's home to the city center district of Fukuoka City by using GIS, we asked respondents about the nearest bus stop and the walking time from home to the nearest bus stop.

Complementary Data by GIS

In both of the first Fukuoka Travel Behavior and *Kaiyu* Compound Survey and the first Fukuoka Travel Behavior Survey, we asked the respondents only about the transport means they chose on the day, the travel expense, and the travel time from home to Tenjin, but have not asked about the travel expenses and travel times for their alternative transport means.

Therefore, the complementary data was separately prepared by using the distance measured by GIS and the travel time and transport expenses provided by each transport agency home page [22, 23]. The method of creating complementary data is as follows.

Narrowing down to samples who choose at least two alternative transport means among the four transport means to Tenjin, bus, subway line 1, *Nishitetsu* train, and automobile, for each of these samples, we measure the distance from home to the city center by GIS, and by using this, the transport expense and the travel time by transport means are calculated.

Distance Measurement by GIS Software

In order to measure the distance from home to the city center, Zenrin Digital Map Z6 [21] was used as GIS. Ultimately, the physical distance from home to the city center is converted to the travel time and the transport cost by transport means. Therefore, in the case of railway, the distance from the nearest bus stop to the nearest station and, in the case of automobile, the distance from the nearest bus stop to the Tenjin district are measured. The measured distance is the shortest road distance from the nearest bus stop to the destination.

Calculation of Travel Time and Travel Cost

We calculate travel time and travel cost by transport means. For automobile, parking waiting time is 20 min, parking fee is 800 yen, hourly speed is 18 km, fuel consumption is 6 km/liter, and fuel price is 100 yen/liter. Also, the waiting time for railway users for *Nishitetsu Tenjin Omuta* Line and subway line 1 is 15 min. For railways and buses, we estimated the travel time and travel cost from each website. For motorcycles, the speed is 18 km/h, and the bicycle is supposed to be 8 km/h.

The method of measuring the travel time and the transport expense by transport means is summarized as follows:

(1) Bus

Travel time: travel time from home to the nearest bus stop (survey item) + travel time from the nearest bus stop to the Tenjin district bus stop (travel time on Nishitetsu website [23])

Transport cost: fare from the nearest bus stop to the Tenjin district bus stop (fare described in the same website [23])

(2) Subway Line 1

Travel time: travel time from home to the nearest bus stop (survey item) + measured distance (GIS) from the nearest bus stop to the nearest station ÷ 4 km/h + travel time from the nearest station to the Tenjin subway station (stated on the Fukuoka City Subway website about time required [22]) + waiting time 15 min

Travel cost: fare from nearest station to Tenjin subway station (fare described in the same website [22])

(3) Nishitetsu Train

Travel time: travel time from home to the nearest bus stop (survey item) + measured distance (GIS) from the nearest bus stop to the nearest station ÷

4 km/h + travel time from the nearest station to *Nishitetsu Fukuoka-Tenjin* station (travel time on *Nishitetsu* website [23]) + waiting time 15 min

Travel cost: fare from the nearest station to *Nishitetsu Fukuoka-Tenjin* station (fare described in the same website [23])

(4) Automobile (Private Car)

Travel time: distance from the nearest bus stop to the Tenjin area (GIS) ÷ hourly speed 18 km + parking lot waiting time 20 min

Travel cost: distance from nearest bus stop to Tenjin area (GIS) ÷ fuel consumption 6 km/liter × 100 yen/liter + parking fee 800 yen

4.2 Modal Choice Model: Formulation and Estimated Results

Formulation

Let $P_m^{i,k,s}$ denote the modal choice probabilities that sample i with age k and gender s chooses travel mode m. The probabilities $P_m^{i,k,s}$ are formulated as a logit model (12.1).

$$P_m^{i,k,s} = \frac{\exp(V_m^{i,k,s})}{\sum_{j=1}^{n}\exp(V_j^{i,k,s})}, \quad m = 1\ldots n \tag{12.1}$$

where $V_m^{i,k,s}$ are the non-stochastic utilities that represent the utility level of mode m for each individual with characteristics i, k, and s.

We formulate the non-stochastic utility as a linear function of time distance t and cost c as follows:

$$V_m^{i,k,s} = \alpha t_m^i + \beta c_m^i \tag{12.2}$$

Here t_m^i and c_m^i are, respectively, travel time and travel cost for consumer i when taking travel mode m. The parameters α and β are estimated by maximum likelihood estimation method.

Estimated Results of Parameters for Modal Choice Model

Table 12.2 shows estimated results of parameters of the modal choice model. While the signs of two parameters are in accord with the hypothesis, the t value is not significant at 10% in either case. Regarding the hit rate, it is not so high at 56.1%, and the degree of freedom adjusted coefficient is 0.0277, so it cannot be said that the fitness of the model is very good. Looking at the $-2 \times$ log likelihood ratio, it is the

Table 12.2 Estimated results of parameters for modal choice model

	Estimate	t value	std
Time (minute) α	−0.008313	−0.88	0.00948
Cost(yen) β	−0.000760	−1.31	0.000582
Hit ratio	56.1%		
$L(0)$	−109.53		
$L(\beta)$	−106.45		
$\rho = -2[L(0)-L(\beta)]$	6.1609**		
$\rho^2 = 1-L(\beta)/L(0)$	0.0281		
$\bar{\rho}^2 = 1-(L(\beta)-K)/L(0)$	0.0099		

$L(0)$: Log likelihood with all parameters set to zeros
$L(\beta)$: Log likelihood with estimated values of parameters
**Significant at 5%, K the number of parameters, $n = 242$

result that the model as a whole with the travel time and the travel cost as explanatory variables is significant at 5%, but the improvement of accuracy of the model is a subject in the future research.

4.3 Visit Frequency Model: Formulation and Estimated Results

Formulation of Visit Frequency Model

To forecast the frequency of visits to city center commercial district, we formulate Poisson model, in which the logarithm of the parameter λ is explained by linear function as follows:

$$\log \lambda_i = \beta_0 + \beta_1 t_m^* + \beta_2 c_m^* \tag{12.3}$$

Here, t_m^* and c_m^* are the travel time and the travel cost of the travel means sample i chose on the day of the survey. Note that the frequency of visits here is the frequency by the trips with purposes of shopping, leisure, and eating and does not include the trips with purposes of business and commuting for work and school.

Estimated Results of Parameters

To estimate the visit frequency model, we use the weighted Poisson model which takes into consideration that the data used for the estimation are obtained from the on-site samples. Since in the present case the data obtained are collected at the city

Table 12.3 Estimated results of parameters of the weighted Poisson model for predicting visit frequency to the city center of Fukuoka City

	Estimate	t value	std
Constant	1.375437	7.09***	0.1939
Time (minute)	−0.015105	−3.18**	0.0047
Cost (yen)	−0.000622	−2.00**	0.0003

$L(0) = -325.56, L(\hat{\beta}) = -282.60, \rho = 85.9314, \rho^2 = 0.1320, \rho^{-2} = 0.1227$
**Significant at 5% ***Significant at 1%, n = 242

center, the visit frequency to which the visit frequency model tries to explain, the choice-based sampling biases necessarily occur because of endogenous sampling. Elsewhere we developed the weighted Poisson model to remove these choice-based sampling biases (for details, refer to Saito et al. [2, 3]).

Table 12.3 shows estimated results of parameters of the weighted Poisson model for predicting visit frequency to the city center of Fukuoka City. In the estimated results, the *t* values are large for all parameters, and all coefficients are significant at 1% or 5% level, and the signs are in accord with the hypothesis. However, the adjusted coefficient of determination is 0.1258, which is not high. Using this estimated model, we forecast the frequency of visits to the city center by transport means by town-*chome* division.

5 Forecasting Economic Effects on the City Center Commercial District of Fukuoka City

In this section, we forecast the economic effect of the opening of the new subway line on the city center commercial district of Fukuoka City. We estimate the economic effect in the following two cases:

Case 1: For the use of Fukuoka City Subway Nanakuma Line, users from areas that are less than 15 min on foot from each station are assumed to access the station on foot, and users from areas that are more than 15 min on foot are assumed to access the station by 25 km/h. This is a case assumed to take a policy to expand the attraction area of station such as park and ride and connecting with bus. This is a case with park and ride.

Case 2: This is a case without taking the policy of park and ride. A case where all users, including users far away 15 min on foot from each station, are supposed to access the station on foot. This is a case without park and ride.

Table 12.4 Population of study area by gender by age group (unit: person)

	15–29	30–49	50–79	Total
Male	35,296	41,025	47,060	123,381
Female	36,533	44,464	56,515	137,512
Total	71,829	85,489	103,575	260,893

5.1 Data Required for Forecasting

Two types of data are necessary. One is population data of Fukuoka City by town-*chome* division (by age group by gender). This is published on the homepage of Fukuoka City administration [24]. The other is data on travel time and travel cost by town-*chome* division by transport means when using buses, automobiles, and the new subway line. Since this data does not exist, GIS is used to prepare complementary data on the travel time and the travel cost to the city center by town-*chome* division by transport means.

Population Data of Study Area

We use the population data of study area published on website by Fukuoka City administration. Data include the number of population in every town-*chome* division by age and sex. We use the data as of September 30, 2003. The age category is a 5-year-old division from 0 to 4 years old and 5–9 years old in the order up to the class of 100 years old and over, totaling 21 classes in the age category. However, in this study, we use the age group from 15 to 79 years old for prediction, excluding the group of less than 14 years old and that of over 80 years old.

Table 12.4 shows the population of the study area by gender by three age groups. The total population along the subway line is about 260,000 people.

Calculating Travel Time and Travel Cost by Transport Means by Town-*Chome* Division Using GIS

As will be stated later, in the forecast, we will take up three transport modes, bus, automobile, and the new subway line. Therefore, data on travel time and travel cost to the city center Tenjin district by these three transport means are required for all 278 town-*chome* divisions in the study area. However, since there is no published data on these, data is created using GIS. The method is similar to the preparation of complementary data by GIS used for estimation of the modal choice model in the previous section.

The differences are as follows. First, we set a reference point at each town-*chome* division, and in the case of the automobile, we measure the road distance from there to the Tenjin district and convert it to the travel time and travel cost as before. Regarding the bus, the nearest bus stop closest to the reference point is obtained, the road distance is measured, and the travel time by walking from the reference point to the nearest bus stop is sought. The calculation of the travel time from the nearest bus stop to the Tenjin district bus stop and its fare is based on the *Nishitetsu* website [23] as before. For Fukuoka City Subway Nanakuma Line as well, we find the nearest station from the reference point, measure its road distance, and calculate the travel time by walking from the reference point to the nearest station. However, as for this new subway line, the travel time from the nearest station to the *Tenjin-Minami* subway terminal station and its fare were not disclosed yet, so we got these data by hearing from the Fukuoka City Transport Bureau [25].

5.2 Predicted Results for Modal Choice After the Subway Opens

Assumptions of Predicting Modal Choice for Town-*Chome* Division

Based on the estimated result of modal choice model shown in Sect. 4.2, we predict the share of transport means the residents choose when traveling from home to the city center Tenjin district for each of 278 town-chome divisions. To do this, we make the following two assumptions:

1. Under the current circumstances, there are two transport means to the city center, bus and automobile, and the new subway line will be added as the third transport means after the opening of the new subway line.
2. We do not assume that the traffic congestion would be decreased by the opening of the new subway line so that the travel time by bus and car would also decrease. In other words, the travel times by bus and by car to the city center shall be the same both at present and after the opening of the new subway line.

Predicted Results for Modal Shift by Town-*Chome* Division

We will look at the current situation and the change in the number of people using each of transport means after the opening of the new subway line from population data and the forecasted modal share predicted by the transport modal choice model.

Case with Park and Ride

As for the case with park and ride, in Table 12.5, we give the predicted number of users by each transport means before and after the opening of the new subway line. As a result, it is shown that about 125,000 people would shift to the new subway line after the opening. In other words, it is the result of forecasting which transport means the residents along the new subway line of 260,000 people would choose when they go to the city center Tenjin district after the opening of the new subway line. Currently, about 160,000 people are choosing buses, and about 100,000 people are

Table 12.5 Predicted numbers of users by travel means before and after the opening of the new subway line (case with park and ride) (unit: people)

	Before		After		
	Bus	Car	Bus	Car	Subway
All residence	163,465	97,428	84,759	50,617	125,517

Table 12.6 Predicted numbers of users by travel means before and after the opening of the new subway line (case without park and ride) (unit: people)

	Before		After		
	Bus	Car	Bus	Car	Subway
All residence	163,465	97,428	89,737	53,513	117,643

choosing automobiles. But after the opening of the new subway line, about 75,000 people from the bus and about 50,000 from the automobile would change their transport means into the new subway line.

Case Without Park and Ride

On the other hand, as for the case without park and ride, in Table 12.6, we show the predicted number of users by each transport means before and after the opening of the new subway line. In this case, about 120,000 people will shift to the new subway line, which is about 10,000 less than the case with park and ride.

5.3 *Forecast Result of the Number of Visitors to the City Center*

Next, by each transport means of bus, automobile, and the subway, we predict how many visits one consumer in each town-*chome* division will make to the city center district per month before and after the opening of the new subway line by using the estimated result of the visit frequency model. Combining this result with the number of the residents who use each transport means we predicted in the previous section by the modal choice model, we obtain the predicted number of visitors to the city center by each transport means by each town-*chome* division before and after the opening of the new subway line. By adding all these, we forecast to what extent the number of visitors to the city center will increase after the opening of the new subway line.

Table 12.7 Case with park and ride: changes in numbers of visitors to the city center before and after the subway opens (unit: people per month)

	Before		After		
	Bus	Car	Bus	Car	Subway
All residence	255,205	96,865	134,222	51,321	379,490
Total	352,071		565,033		
After-before	212,962				

Table 12.8 Case without park and ride: changes in numbers of visitors to the city center before and after the subway opens (unit: people per month)

	Before		After		
	Bus	Car	Bus	Car	Subway
All residence	255,205	96,865	140,617	53,523	299,348
Total	352,071		493,488		
After-before	141,417				

Case with Park and Ride

For the case with park and ride, in Table 12.7, we show the predicted number of visitors to the city center district before and after the opening of the new subway line. Currently, about 350,000 visitors per month increased to about 570,000 visitors after the opening, resulting in the increase of about 210,000 visitors per month.

Case Without Park and Ride

As for the case without park and ride, Table 12.8 gives the current situation and the change in the number of visitors to the city center district per month after the opening of the new subway line. After the opening, it has an increase of about 140,000 people, about 70,000 less than the case with park and ride. This result means that in the absence of policies such as park and ride and feeder transportation, the number of subway users decreases, and the attraction area of station becomes smaller.

5.4 Average Expenditure at the City Center District per Visit

The average of expenditure at the city center district was 6925 yen per visit per person when visitors visited the city center district of Fukuoka City. This is the result of our on-site survey of consumer *Kaiyu* behaviors that we carried out in July 2003. The expenditure is the one obtained when a main purpose of the visit to the city center district was limited to shopping, leisure, and meal. We use this result.

5.5 Result of Economic Effects

Therefore, the amount of economic effects on the city center district of Fukuoka City turns out as follows.

Case with Park and Ride

17,697 million yen = 212,962 (people/month) × 6925 (yen/person) × 12 (months)

Case Without Park and Ride

11,752 million yen = 141,417 (people/month) × 6925 (yen/person) × 12 (months)

6 Conclusion and Future Challenges

We believe that the novelty of this study is to propose the consumer behavior approach to estimate economic effects of the opening of a new subway line on the specific small area such as a city center commercial district and actually has carried out the approach to show the concrete figure of the effects based on how consumers increase their frequency of visits to city center district. While our approach is different from the traditional one, its conceptual framework is quite natural since for consumers who reside in suburbs the improvement of accessibility to the city center implies the decline of the relative price of city center composite goods so that the increase of the visit frequency corresponds to the income effect for them.

As for the prediction of the economic effect this time, we gave a prediction result in the case where, for the residents in the areas outer than the distance of 15 min on foot to each station, the access by bus to the station and the access by car with park and ride are functioning. We also gave a prediction result when there was no such policy. As a result, we obtained the prediction that the number of users of the new subway will be reduced by approximately 70,000 people per month, and the economic effect on the city center district will be 6 billion yen less than that when park and ride is functioning. In the future, it is important to take the policy of expanding the attraction area of station as much as possible, such as park and ride and get on the bus at the suburban station, to increase the number of users of the subway.

The assessment may be divided whether the predicted figure of 17.7 billion yen is small or large. We thought the economic effects would be much larger. However, mulling over this result, we should have known that since Fukuoka Subway Nanakuma Line has just 260 thousand inhabitants along the line, is only 12 km long, and does not connect to any other networks, its effects must be limited.

According to our forecast, it is very hard to attract many users for Fukuoka Subway Nanakuma Line. This is because if we do not provide some means to improve the accessibility of inhabitants along the line to each station, the number of inhabitants who access to each station on foot is predicted quite small. Hence further study is needed to investigate how such polices as feeder transport, park and ride, and so on would increase the number of users for each station.

References

1. Saito S, Yamashiro K, Nakashima T, Igarashi Y (2007) Predicting economic impacts of a new subway line on a city center retail sector: a case study of Fukuoka City based on a consumer behavior approach. Stud Reg Sci 37:841–854. (in Japanese)
2. Saito S, Kakoi M, Nakashima T (1999) On-site Poisson regression modeling for forecasting the number of visitors to city center retail environment and its evaluation. Stud Reg Sci 29:55–74. (in Japanese)
3. Saito S, Kumata Y, Ishibashi K (1995) A choice-based Poisson regression model to forecast the number of shoppers. Pap City Plan 30:523–528. (in Japanese)
4. Saito S, Nakashima T, Kakoi M (1999) Identifying the effect of city center retail development on consumer's shop-around behavior: an empirical study on structural changes of city center at Fukuoka City. Stud Reg Sci 29:107–130. (in Japanese)
5. Saito S, Yamashiro K (2001) Economic impacts of the city center one-dollar circuit bus estimated from consumer's shop-around behavior: a case of the city center one-dollar bus at Fukuoka City. Stud Reg Sci 31(1):57–75. (in Japanese)
6. Saito S, Yamashiro K, Kakoi M, Nakashima T (2003) Measuring time value of shoppers at city center retail environment and its application to forecast modal choice. Stud Reg Sci 33(3):269–286. (in Japanese)
7. Tamura M, Saito S, Nakashima T, Yamashiro K, Iwami M (2005) Forecasting the impacts of a new monorail line from the survey on consumers' stated prospective behavior changes: a case study in Naha City, JAPAN. Stud Reg Sci 35(1):125–142. (in Japanese)
8. Morisugi H (1997) Benefit evaluation of environmental infrastructure improvement: an approach from general equilibrium theory, Keisoshobo. (in Japanese)
9. Morisugi Y, Miyagi T (1996) The evaluation of urban transportation project: examples and exercises, corona publishing. (in Japanese)
10. Forster CD, Beesley ME (1963) Estimating the social benefit of constructing and underground railway in London. J R Stat Soc 126(1):46–93
11. Small KA, Rosen HS (1981) Applied welfare economics with discrete choice models. Econometrica 49:105–129
12. Kanemoto Y, Mera K (1985) General equilibrium analysis of the benefits of large transportation developments. Reg Sci Urban Econ 15:343–363
13. Nomura T, Aoyama Y, Nakagawa D, Matsunaka R, Shirayanagi H (2001) The evaluation of the benefit of the intercity high-speed train project using the EVGC. Infrastruct Plan Rev 18:627–636. (in Japanese)
14. Ministry of Transport (1999) Infrastructure and Transport, A manual on cost effectiveness analysis of railroad projects '99, Institution for Transport Policy Studies. (in Japanese)
15. Hidano N (2002) The economic valuation of the environment and public policy: a hedonic approach: E. Elgar. (Japanese edition 1997)
16. Kanemoto Y (1988) Hedonic prices and the benefits of public projects. Econometrica 56:981–989

17. Hidano N, Nakamura H, Aratsu Y, Nagasawa K (1986) The estimation of capital gains of property value for equitable cost breaking of urban railway improvement. J JSCE 1986:135–144. (in Japanese)
18. Wheaton WC (1977) Residential decentralization, land rends, and the benefits of urban transportation investment. Am Econ Rev 67(2):136–143
19. Pines D, Weiss Y (1982) Land improvement projects and values: an appendum. J Urban Econ 11:199–204
20. Yamazaki D, Takemoto A, Kawaguchi H (2003) Fundamental study on prediction of impact on city center of Fukuoka after opening subway line 3. Graduation thesis, Faculty of Economics, Fukuoka University. (in Japanese)
21. Zenrin Co (2003) Digital Map Z6, ZENRIN CO., LTD.
22. Fukuoka City Subway Homepage. http://subway.city.fukuoka.jp/ (in Japanese)
23. Nishi-Nippon Railroad Co. Ltd. Homepage. http://www.nishitetsu.co.jp/ (in Japanese)
24. Fukuoka City Home Page. http://www.city.fukuoka.jp/index.html (in Japanese)
25. Hearing from the Fukuoka City Transport Bureau

Chapter 13
Did an Introduction of a New Subway Line Increase the Frequency of Visits to City Center?

Saburo Saito and Kosuke Yamashiro

Abstract This study verified the increase in the frequency of visits to the city center commercial district of Fukuoka City caused by the introduction of the new subway line for the residents who live in the area along the subway line. To do this, the survey was conducted targeting university students after the opening of the subway to obtain the retrospective panel data in which the respondents answered retrospectively their behaviors before the opening of the subway as well as the present behaviors after the opening. The comparison of the frequency of visits before and after the opening becomes the analysis of difference in difference which verified the increase in the visit frequency.

Keywords Frequency of visits · Increase · City center · Introduction of subway · Retrospective panel data · Difference in difference · Fukuoka City Subway

---

This chapter is based on the paper, Saburo Saito, Kosuke Yamashiro, and Takaaki Nakashima [1], "An Empirical Verification of the Increase of Visit Frequency to City Center caused by an Introduction of a New Subway Line: A comparative analysis before and after opening based on retrospective panel data," Paper presented at The 43rd Annual Meeting of the Japan Section of Regional Science Association International (JSRSAI), 2006, which is modified for this chapter.

S. Saito (✉)
Faculty of Economics, Fukuoka University, Fukuoka, Japan

Fukuoka University Institute of Quantitative Behavioral Informatics for City and Space Economy (FQBIC), Fukuoka, Japan
e-mail: saito@fukuoka-u.ac.jp

K. Yamashiro
Department of Business and Economics, Nippon Bunri University, Oita City, Japan
e-mail: yamashiroks@nbu.ac.jp

1 Purpose

In Fukuoka City, Japan, the new subway line called Fukuoka City Subway Nanakuma Line opened in February 2005, which connects the city center district of Fukuoka City and the southwestern suburban part of Fukuoka City.

Saito et al. (2005) had already constructed forecast models of modal choices and visit frequency by using data from a questionnaire survey conducted before the opening of the subway, published data, and measured data by GIS, and predicted the changes in modal choices from bus to subway and the increases in the visit frequency to the city center district of Fukuoka City after the opening of the subway line. Based on these results, they forecasted the economic effect that the opening of the new subway line would bring about on the Fukuoka city center district before the opening of the subway line. However, the ex post evaluation of the above forecast after the opening of the subway line has remained as a challenge to be addressed.

As the new Subway Nanakuma Line had opened in 2005, this study aims to address this challenge. It should be noted that the key point of their forecast lies in the thinking that the new subway line would cause the increase in the frequency of visits to the city center of Fukuoka City by the residents who live in the areas along the subway line.

Traditionally, the above way of thinking has been criticized by the proponents of four step models because the causal direction is opposite. The traditional four-step method first determines the frequency of origin and destination trips for each zone as the trip generation step. While this amount of trip demand is estimated by the land use and demographic pattern of each zone, it is set as fixed in the following three steps, trip distribution, mode choice, and route assignment step. Thus, the introduction of the subway line does not change the frequency of visits to the city center even if the travel time to the city center is changed by the changes of modal choice and route assignment.

In this study, we draw on the conceptual framework of Saito et al. [2] about why and how the increase in the frequency of visits to the city center commercial district would occur. Using data from a questionnaire survey conducted after the subway opening, we try to verify whether the frequency of visits to the city center has increased.

The conceptual framework of the increases in the frequency of visits to the city center commercial district due to the new subway line is based on the following assumption of consumer behaviors.

Consumers who live along the subway line have local goods purchased locally and city goods purchased at the city center. It is assumed that generalized transport costs are included in the prices of these goods. By the introduction of the subway line, the travel time required from the residential area to the city center district is shortened, and the generalized transport cost is decreased, so that the price of city good decreases. Thus the relative price of the city goods to the local goods decreases than before the introduction of the subway line. Therefore, the demand for the city good is considered to increase because its relative price falls after the introduction of the subway line. Hence the frequency of visits to the city center commercial district increases because the increase in the demand for the city goods implies the increase in the visit frequency to the city center.

The purpose of this study is to verify whether the introduction of a new subway line has had an effect on the increase in the frequency of visits to the city center commercial district by residents along the subway line based on the retrospective panel data obtained from a questionnaire survey of use of the subway for university students living along the subway line conducted after the opening of the subway in which the respondents answered retrospectively their behavior before the opening of the subway as well as the present behavior after the opening.

While we have carried out several studies to identify the effects of urban development policies, they are all based on the cross-section data [3–5]. The salient feature of this study is to address the problem of verifying the effect of the new subway line on the increase in the frequency of visits to the city center by the residents along the subway line by using the retrospective panel data. (For the advantages of the retrospective panel data, refer to Chap. 16 of this book.)

2 Framework of the Analysis

2.1 Overview of Fukuoka City Subway Nanakuma *Line*

The Fukuoka City Subway *Nanakuma* Line opened in February 2005 and goes from *Tenjin-Minami* Station located at the city center commercial district of Fukuoka City in *Chuo-ku* ward, Fukuoka City, passing through *Jonan-ku* ward and *Sawara-ku* ward, up to *Hashimoto* Station in *Nishi-ku* ward, Fukuoka City, in a total time of 24 min and a total length of 12 km. The route map of the Fukuoka City Subway Nanakuma Line is shown in Fig. 13.1.

In the southwest area of Fukuoka City, despite being home to 500,000 people or about 40% of the population of Fukuoka City, there was no railway system before the opening of the Nanakuma Line, and there was no choice but to rely on buses or private vehicles as transport means to the city center. Therefore, the opening of the Nanakuma Line greatly shortened the time required to reach the city center of Fukuoka.

For example, from Tenjin in the city center district to Fukuoka University, a journey taking 40 min by bus was shortened to 16 min with the opening of the subway. At the same time, some areas along the subway line have cheaper fares than buses. In the area around Fukuoka University, the bus used to cost 370 yen, but the subway charged 290 yen, which means the subway is superior in both travel time and cost.

There are four universities along the Nanakuma Line, *Kyushu* University, *Nakamura Gakuen* University, Fukuoka University, and Fukuoka Dental College, and many students use the subway line.

Fig. 13.1 Route map of Fukuoka City Subway Nanakuma Line (Cf. [6])

2.2 Data Used

In this study, we used data from the second survey on the actual usage situation of the Ecole card by Fukuoka University students. The Ecole card is the commuter pass discounted for students. The survey was conducted by the Fukuoka University Institute of Quantitative Behavioral Informatics for City and Space Economy (FQBIC) over the course of 5 days in 2005: November 17, 21, and 30 and December 1 and 5.

The second survey on the actual usage situation of the Ecole card by Fukuoka University students was a 10-min questionnaire survey employing a method to distribute and collect in the classroom, which targeted Fukuoka University students. The survey asked the respondents about the last transport means they utilized to come to the university, changes in the frequency of school attendance, changes in the transport means used to visit the city center, and changes in the visit frequency to the city center, as well as the expenditure per visit the respondents spend when they visit the city center. A survey summary is shown in Table 13.1.

Table 13.1 Overview of the second survey on the actual usage situation of the Ecole card by Fukuoka University students

Survey date	November 17, 2005 (Thursday)
	November 21, 2005 (Monday)
	November 30, 2005 (Monday)
	December 1, 2005 (Thursday)
	December 5, 2005 (Thursday)
Survey location	Carried out in the rooms of six lectures at Fukuoka University
Survey method	Distributed and collected the questionnaires at the lectures
Main question items	Personal attributes; place of residence; nearest bus stop/train station
	Changes in transport means used for coming to the university and the frequency of attendances at the university before and after the opening of the subway
	Changes in the transport means and the frequency of visits to the city center
	Stated preference among Ecole card, commuter ticket, and *chika* passport
Number of samples collected	313 samples

Table 13.2 Number of samples by gender

		Number of obs.	Percent (%)	Effective percent (%)
	Male	237	75.7	76.0
	Female	75	24.0	24.0
	Total	312	99.7	100.0
Missing value		1	0.3	
Total		313	100.0	

The noticeable characteristics of this survey is that having the respondents answer their current behavior, as well as answer retrospectively their behavior before the opening of the subway, made possible an efficient collection of panel data at two points in time from the same sample. We call this kind of panel data as the retrospective panel data.

In this study, we use this retrospective panel data, which contains the frequency of visits to the city center of Fukuoka City before and after the opening of the subway and changes in travel modes to visit the city center, and verify the hypothesis of the increase in the frequency of visits to the city center of Fukuoka City before and after the opening.

The characteristics of the samples obtained from the survey are shown in Tables 13.2, 13.3, and 13.4.

Table 13.2 shows the number of samples by gender, and Table 13.3 shows the number of samples by age. By gender, males accounted for 75%, and by age, the ratio was highest at 19 years of age.

Table 13.4 shows the number of samples by place of residence. As this survey targeted Fukuoka University students, the percentage for *Jonan*-ku ward was the highest.

Table 13.3 Number of samples by age

		Number of obs.	Percent (%)	Effective percent (%)
	18 years old	29	9.3	9.4
	19 years old	91	29.1	29.4
	20 years old	76	24.3	24.6
	21 years old	57	18.2	18.4
	22 years old	33	10.5	10.7
	23 years old and over	13	7.3	7.4
	Total	309	98.7	100.0
Missing value		4	1.3	
Total		313	100.0	

Table 13.4 Number of samples by place of residence

		Number of obs.	Percent (%)	Effective percent (%)
	Higashi-ku ward	16	5.1	6.0
	Hakata-ku ward	10	3.2	3.8
	Chuo-ku ward	17	5.4	6.4
	Minami-ku ward	25	8.0	9.4
	Nishi-ku ward	6	1.9	2.3
	Jonan-ku ward	124	39.6	46.6
	Sawara-ku ward	14	4.5	5.3
	Within Fukuoka City (subtotal)	212	67.7	79.7
	Outside Fukuoka City	72	23.0	20.3
	Total	284	90.7	100.0
Missing value		29	9.3	
Total		313	100.0	

2.3 Method to Verify the Increase in the Frequency of Visits to the City Center District of Fukuoka City

In this study, we verify the increase in the frequency of visits to the city center of Fukuoka City after the opening of the subway in the following way.

1. We compare before and after the opening of the subway to observe whether residents along the subway line has increased their frequency of visits to the city center of Fukuoka City.
2. We divide the residents along the subway line into subway users and nonusers according to their choice of the transport means to the city center and observe whether or not the frequency of visits to the city center increased after the opening of the subway compared to before.

Table 13.5 State of use of the Nanakuma Line to the city center of Fukuoka by the required time to get to the nearest station among samples who responded that the nearest station was a Nanakuma Line station

Walking distance	Nanakuma Line users		Nanakuma Line nonusers	
	Obs.	%	Obs.	%
Less than 5 min	16	80.0	4	20.0
5 min or more and less than 10 min	28	52.8	25	47.2
10 min or more and less than 15 min	23	60.5	15	39.5
15 min or more	5	14.3	30	85.7
Total	72	49.3	74	50.7

In step (1), the goal is to verify whether there is an increase in the frequency of visits for the whole area along Nanakuma Line, and in (2) the goal is to verify whether the subway line causes an increase in the frequency of visits to the city center of Fukuoka City by dividing the residents into two groups of subway users and nonusers to see if there is a difference between two groups. In Sect. 4.1, we examine step (1), and in Sect. 4.2, we examine step (2).

We define the residents along Nanakuma Line as the samples who satisfy the following two conditions.

1. The samples who answered a subway station on Nanakuma Line to the question item asking about their nearest train or subway station
2. The samples that responded that the required time to get to their nearest station was within 15 min on foot

The reason for setting the area within 15 min on foot to define the residents along the subway line is that, in cases where the nearest station is more than 15 min away, people will not use the subway as often. In fact, from Table 13.5 showing the usage situation of Nanakuma subway line to the city center of Fukuoka City by the required time to the nearest subway station, we can see that up to 15 min, the percentage of users of the subway is higher than the nonusers, but when it is more than 15 min, the percentage of nonusers becomes higher.

Hence, of the total 313 samples, we will narrow down to 127 samples of the residents along the Nanakuma Line who live within 15 min to the nearest subway station and conduct verification.

3 Changes in Transport Means to the City Center Before and After the Opening of the Subway Line

Table 13.6 shows the transition of changes in choices of transport means to the city center of Fukuoka City before and after the opening of the subway. When we look at transport means to the city center before the opening of the subway, the bus is the most used, followed by motorcycle, automobile, and bicycle. On the other hand, after the opening, the Nanakuma subway line is the most used, followed by

Table 13.6 Transition of choices of transport means to the city center before and after the opening of the subway (among residents within 15 min walking distance to the nearest Nanakuma Line station)

From			After the opening of the subway							Total
		To	Express bus	Non-express bus	Private car	Motorcycle	Bicycle	On foot	Subway Nanakuma Line	
Before the opening of the subway	JR bullet train	Obs.	0	0	0	0	0	0	1	1
		%	0.0	0.0	0.0	0.0	0.0	0.0	100.0	100.0
	Express bus	Obs.	1	0	0	0	0	0	1	2
		%	50.0	0.0	0.0	0.0	0.0	0.0	50.0	100.0
	Bus other than express bus	Obs.	0	14	0	0	0	0	45	59
		%	0.0	23.7	0.0	0.0	0.0	0.0	76.3	100.0
	Private car	Obs.	0	0	6	0	0	0	1	7
		%	0.0	0.0	85.7	0.0	0.0	0.0	14.3	100.0
	Motorcycle	Obs.	0	0	0	23	0	0	1	24
		%	0.0	0.0	0.0	95.8	0.0	0.0	4.2	100.0
	Bicycle	Obs.	0	0	0	0	4	0	3	7
		%	0.0	0.0	0.0	0.0	57.1	0.0	42.9	100.0
	On foot	Obs.	0	0	0	0	0	1	0	1
		%	0.0	0.0	0.0	0.0	0.0	100.0	0.0	100.0
Total		Obs.	1	14	6	23	4	1	52	101
		%	1.0	13.9	5.9	22.8	4.0	1.0	51.5	100.0

Table 13.7 Changes in the frequency of visits by residents along Nanakuma Line to the city center before and after the opening of the subway

	Before the opening of the subway		After the opening of the subway		
	Average	Obs.	Average	Obs.	Difference
Frequency of visits to the city center (times/month)	3.16	117	4.12	116	0.96

$p < 0.000$

motorcycle, bus, and automobile. It is clear that there are many users of the Nanakuma subway line after its opening.

Next, when looking at users of the Nanakuma subway line after its opening to see how they changed their transport means to the city center before and after the opening of the subway, 76.3% of bus users before the opening changed their travel mode from bus to subway. However, it can be seen that users of motorcycles and automobile have not shifted so much to the Nanakuma subway line, as it is 4.2% and 14.3%, respectively. Simply put, it is clear that many public transport users changed to the subway, whereas automobile and motorcycle users did not change.

4 Verification of the Increase in the Frequency of Visits to the City Center Before and After the Opening of the Subway

4.1 Changes in Visit Frequency Pre and Post the Subway Opening

Now we shall see how the frequency of visits to the city center commercial district of Fukuoka City has increased for the entire area along the Nanakuma subway line after the opening of the subway. Table 13.7 shows the change in the frequency of visits before and after the opening. From the table, we can see that this was 3.16 times per month before the opening and changed into 4.12 times after the opening, showing an increase of 0.96 times. The result of the test on the difference between the averages of visit frequency before and after the opening becomes statistically significant at a significant level at 1% or less.

4.2 Comparison of Changes in the Frequency of Visits to the City Center Before and After the Opening by Nanakuma Line Users and Nonusers

Recall that our objective of this study is to verify the increase in the frequency of visits to the city center due to the introduction of the new subway line. To do this, we

Table 13.8 Difference in difference of changes in the frequency of visits to the city center before and after the opening by the Nanakuma Line users and nonusers

	Before the opening of the subway		After the opening of the subway		
	Average (times/month)	Obs.	Average (times/month)	Obs.	Difference
Nanakuma Line users	2.44	64	4.11	64	1.67
Nanakuma Line nonusers	4.03	53	4.13	52	0.10
Difference in difference					1.57***

***$p < 0.000$

must examine whether or not the use of the new subway line has actually the effect to increase the frequency of visits to the city center compared with the case of the not-use of the new subway line. For the purpose, next we examine whether or not there is a difference in the frequency of visits to the city center between subway user and nonusers while classifying the residents along the Nanakuma subway line into users and nonusers of the Nanakuma subway line.

Table 13.8 shows the result of comparing the changes in the frequency of visits to the city center of Fukuoka City before and after the opening of the subway among users and nonusers of the Nanakuma subway line. From this result, it can be seen that, for users of the Nanakuma subway line, it was 2.44 times per month before the opening and became 4.11 times after the opening, so it increased by 1.67 times. On the other hand, for nonusers, it was 4.03 times before opening and became 4.13 times after the opening, for a difference 0.1 times, a barely negligible increase. When we tested the average difference between users of the Nanakuma Line and nonusers, it is statistically significant at a significant level at 1% or less.

The frequency of visits to the city center district of Fukuoka City by the residents along the subway line who use the Nanakuma subway line has largely increased. Moreover, an interesting result is that the residents along the subway line who use the subway to get to the city center did not visit the city center district so often before the opening of the subway, but they have increased their visit frequency to the extent that it becomes the same as nonusers after the opening of the subway. This increase can be considered an effect of the improvement of the accessibility to the city center district due to the shortening of the travel time after the opening of the subway. Also note that the difference 1.57 obtained by subtracting 0.10 from 1.67 can be thought of as the result of the analysis by the difference in difference, which implies that the difference 0.10 expresses all other changes except the introduction of the subway so that the net effect of the increase due to the new subway line should be the difference in difference of 1.57 obtained by subtracting this effect of 0.10 from the gross effect 1.67 due to the subway.

From the above result, we have verified the increase of the frequency of visits to the city center commercial district of Fukuoka City due to the introduction of the new subway line.

5 Conclusion and Future Challenges

The significance of this study is that we have verified that the introduction of the new subway line has increased the frequency of visits to the city center district of Fukuoka City for the all residents along the subway line based on a questionnaire survey conducted after the opening of the subway. More specifically, we have extracted the net effect of the increase in the frequency of visits to the city center due to the introduction of the new subway.

While we derived an explicit result, we have conducted the verification using a questionnaire survey targeting Fukuoka University students, so it is hard to say that this explicit result can apply to the whole residents along the Fukuoka City Subway Nanakuma Line. Therefore, a future task is to verify whether the frequency of visits has also increased among people in other age groups and whether there is an overall increase.

References

1. Saito S, Yamashiro K, Nakashima T (2006) An empirical verification of the increase of visit frequency to city center caused by the opening of a new subway line: a comparative analysis before and after opening based on retrospective panel data. Paper presented at 43rd annual meeting of the Japan Section of Regional Science Association International (JSRSAI). (in Japanese)
2. Saito S, Yamashiro K, Nakashima T, Igarashi Y (2007) Predicting economic impacts of a new subway line on a city center retail sector: a case study of Fukuoka City based on a consumer behavior approach. Stud Reg Sci 37:841–854. (in Japanese)
3. Saito S, Nakashima T, Kakoi M (1999) Identifying the effect of city center retail redevelopment on consumer's shop-around behavior: an empirical study on structural changes of city center at Fukuoka City. Stud Reg Sci 29:107–130. (in Japanese)
4. Saito S, Yamashiro K (2001) Economic impacts of the downtown one-dollar circuit bus estimated from consumer's shop-around behavior: a case of the downtown one-dollar bus at Fukuoka City. Stud Reg Sci 31:57–75. (in Japanese)
5. Saito S, Yamashiro K, Kakoi M, Nakashima T (2003) Measuring time value of shoppers at city center retail environment and its application to forecast modal choice. Stud Reg Sci 33:269–286. (in Japanese)
6. Zenrin Co (2003) Digital map Z6. ZENRIN CO., LTD

Chapter 14
To What Extent Did the Woodworks Festival Attract People?

Saburo Saito, Kosuke Yamashiro, Masakuni Iwami, and Mamoru Imanishi

Abstract The Okawa Woodworks Festival, held in spring and autumn every year, is the biggest event in Okawa City. Okawa City, Japan, is famous for its clustered agglomeration of furniture industry. In recent years, Okawa City has been forced to retreat from the industrial scale at the time of its peak by imported furniture made in China and low price large retailers. In Okawa City, various policies have been attempted to revitalize furniture and interior industries. Okawa Woodworks Festival is one of such policies. The purpose of this study is to measure how many visitors usually visit Okawa City and to estimate to what extent visitors are attracted by the Woodworks Festival. Specifically, we devise the method to explicitly estimate the visitor attraction effect of a large event such as the Woodworks Festival, which has never been clearly measured so far, at an individual level, and measure its size actually. To that end, we conducted the on-site questionnaire survey of the festival visitors to ask whether they visited Okawa City with the purpose of attending the festival or with other purposes. First, by restricting the samples to those from the Okawa metropolitan area, we forecasted the number of visitors from that area who visit the city center district of Okawa City with the purpose for attending the

---

This chapter is based on the paper of Saburo Saito, Jie Cai, Kosuke Yamashiro, Masakuni Iwami, and Mamoru Imanishi [1], "Measurement of the customer attraction effect of large events: The case of Okawa Woodworks Festival," presented at the 48th Annual Meeting of Japan Section of the Regional Science Association International (JSRSAI), 2011, which is modified for this chapter.

S. Saito (✉)
Faculty of Economics, Fukuoka University, Fukuoka, Japan

Fukuoka University Institute of Quantitative Behavioral Informatics for City and Space Economy (FQBIC), Fukuoka, Japan
e-mail: saito@fukuoka-u.ac.jp

K. Yamashiro · M. Imanishi
Department of Business and Economics, Nippon Bunri University, Oita City, Japan
e-mail: yamashiroks@nbu.ac.jp; imanishimm@nbu.ac.jp

M. Iwami
Fukuoka University Institute of Quantitative Behavioral Informatics for City and Space Economy (FQBIC), Fukuoka, Japan
e-mail: miwami@econ.fukuoka-u.ac.jp

Woodworks Festival. This is the estimation of the visitor attraction effect on Okawa metropolitan residents due to the Woodworks Festival, using the on-site weighted visit frequency Poisson model. While this estimation by Poisson model is thought of as based on the comparison of the visitors attracted by the festival purpose with those with usual purposes, we made an ingenuity to create the panel data of the visitors with the festival purpose as though composed of data at two points of time before and after the festival. Using this panel data for the visitors with the festival purpose, we demonstrated that we can determine the visitor attraction spatial area of the Woodworks Festival in terms of travel time distance. By doing this, we show that the area is the circle with the radius of 313.6 min. Furthermore, applying the same method to the analysis of the effect by the TV commercial advertisement, we show that the enlargement effect of TV commercial advertising on the visitor attraction spatial area in terms of travel time distance turns out to increase the area by 42 min.

Keywords On-site Poisson model · Visit frequency · Attraction effect · Radius of attraction spatial area · Artificial panel data · TV commercial effect · Okawa City · Woodworks Festival · Visitors with festival purpose · Furniture industry

1 Purpose

The Okawa Woodworks Festival, held in spring and autumn every year, is the biggest event in Okawa City,[1] Japan, which is attended by many visitors from within and outside the prefecture. It is well known that Okawa City is a major clustered production area in the furniture industry. The woodwork industry in Okawa City has a long tradition, and in particular, many small- and medium-sized companies, which are involved in the production process up to final products of furniture, such as wood as a material, its processing, painting, assembly as well as fittings for the interior industry, are closely located and agglomerated to form a local network of many woodwork-related industries.

Middle-grade furniture has been a specialty of Okawa. In recent years, however, cheap furniture that is mainly made in China is being sold in large furniture stores in Japan, and Okawa's production has been largely reduced from the time the industrial scale was at its peak. Under these circumstances, various attempts have been made in Okawa toward the reconstruction of the furniture, woodwork, and interior industry with a sense of impending crisis. The Okawa Woodworks Festival conducted in spring and autumn is one of such attempts and is led by an initiative of the Okawa Furniture Manufacturers Association and other organizations. Its main participants

---

[1]Okawa city is located about 45 km down south from Fukuoka city with a population of 37,000 people as of 2010, which is known in its agglomeration of small and medium companies involved in the furniture and woodworks industry.

are furniture manufacturers located in Okawa, furniture retail stores, and retail stores selling imported furniture.

It should be noted that traders, such as wholesalers and retailers, often visit furniture manufacturers looking for opportunities to purchase products in Okawa to conduct their business, whereas, in the Okawa Woodworks Festival, products are mainly sold directly to consumers, such as in exhibitions and spot sale. The Woodworks Festival is advertised on television, and giveaways equivalent to 3 million yen are distributed to attendees, which is the selling point of the event. Thus the Woodworks Festival is said to be tailored as an event that can be enjoyed by families including adults and children. The reason why such an attempt was made is considered to be based on the following policy issues Okawa City currently faces.

First is the viewpoint that the major customers for Okawa's furniture industry should be shift from traders such as furniture wholesalers and retailers to end users. Second, all furniture manufacturers have a showroom of furniture in addition to their factory, which they mainly use for business negotiations with traders. However, it may be necessary to have a furniture exhibition facility that integrates individual separated showrooms, such as the spring and autumn Woodworks Festival, which mainly targets end consumers. Third, for attracting customers by having a furniture exhibition facility, it is desirable to have a permanent facility, rather than a temporary event. Fourth, as the target of attracting customers is to be shifted from traders to final consumers, it is important at the same time to have final consumers or visitors appreciate the value of Okawa City as a multifaceted network interwoven with Okawa's food, history, and culture surrounding Okawa's furniture industry. Fifth, it is necessary to cooperate with tourist spots, such as *Yanagawa* and *Yoshinogari* in the neighboring area, to attract customers.

Although the main motivation of this study is Okawa's furniture industry, it also focuses on identifying what kind of policy would be desirable to increase the number of customers to Okawa while appealing to consumers the value of Okawa City as a network interwoven with its food, culture, and history in collaboration with the neighboring tourist spots.

Therefore, the aim of this research is to obtain fundamental knowledge regarding to what extent Okawa City has the potential to attract people on a daily basis at present, to evaluate the degree of the impact of events such as the spring and autumn Woodworks Festival, and to estimate the number of customers that can be expected to attend these events from the neighboring tourist spots.

More specifically, the purpose of this research is to devise a method to explicitly estimate the customer attraction effect of the Okawa Woodworks Festival in spring and autumn, which has not been clearly formulated so far, and to obtain the actual size of the effect.

2 Framework of the Analysis

2.1 Overview of the Okawa Woodworks Festival

The "Okawa Woodworks Festival" is the biggest event of Okawa City, which is the town of production and sales of furniture and woodwork. It is a festival that widely presents Okawa's industry, tourism, and culture. It is held every year since October 1949, and since 2010, the "Spring Okawa Woodworks Festival" is also held in April. Thus, now, it is held in spring and autumn of each year. Because the scale and the organizers of the festival are different in spring and autumn, these events do not exactly have the same content. However, the same elements are present at various venues, such as the Okawa Industry Hall, including an exhibition and the spot sale of furniture, a village of food stalls, local gourmet, a flea market, and stage events. The event witnesses about 20,000 visitors in spring and about 45,000 visitors in autumn.

2.2 The Method to Measure the Customer Attraction Effect of the Woodworks Festival

In this research, to measure the customer attraction effect of the Okawa Woodworks Festival, we devised question items in the questionnaire survey to ask respondents, "would you come to Okawa city center if there was no Okawa Woodworks Festival today?" If the respondents are residents in Okawa city, we ask instead the respondents whether they would visit the festival venue if there was no Okawa Woodworks Festival today. This allows us to distinguish between "a person who came for the purpose of attending the Okawa Woodworks Festival" and "a person who came for a purpose other than the Okawa Woodworks Festival."

Previously, Saito and Yamashiro [2] estimated the economic effects due to city center one-dollar bus based on the comparison between visitors who used one-dollar city center bus and those who did not (cf. Chap. 9 of this book) The same framework was applied to the estimation of the economic effect due to the city center café [3] based on the comparison between the visitors who used the city center café and those who did not (cf. Chap. 11 of this book). This research differs from these previous studies because the groups compared are formed not by the differences in the respondents' behaviors but by the differences in their visit purposes obtained from the questionnaire survey. This method for the first time has been applied to the estimation of the event effects [4] (cf. Chap. 15 of this book).

In this research, we forecast the number of visitors coming to Okawa City for shopping, leisure, and meal purposes using the Poisson regression model. We assume that the number of visitors who come to Okawa City for the purpose of attending the Okawa Woodworks Festival is the increase in the number of visitors due to the Okawa Woodworks Festival. The customer attraction effect of the Woodworks Festival is defined as this increase in the number of visitors to Okawa

City due to the Woodworks Festival. Using the Poisson regression model for forecasting visit frequency to Okawa City, we estimate the increase in the number of visitors to Okawa City due to the Okawa Woodworks Festival to measure the size of the customer attraction effect caused by the Okawa Woodworks Festival.

2.3 Data Used

In this study, we use the data from the questionnaire survey of the 61st Okawa Woodworks Festival conducted from October 9 to 11, 2010, and the questionnaire survey of the Spring Okawa Woodworks Festival conducted on April 9 and 10, 2011. Each questionnaire survey is a placement survey targeting visitors to the Okawa Woodworks Festival that asks respondents about individual attributes, their transportation mode to Okawa City, places that they stopped by on that day, and the amount of money spent. The survey summary of each is shown in Tables 14.1 and 14.2.

Table 14.1 Overview of the 61st Okawa Woodworks Festival questionnaire survey

Survey date	October 9, 10, 11, 2010
Survey time	11:30–17:00
Survey method	Questinnaire sheets offered to the 61st Okawa Woodworks Festival participants and collected at venue
Main questions	1. Individual attributes
	2. Their transportation mode to Okawa city
	3. Places that they stopped by on that day
	4. The amount of money spent
The number of samples	7463 samples

Table 14.2 Overview of the Spring Okawa Woodworks Festival 2011 questionnaire survey

Survey date	April 9, 10, 2011
Survey time	11:30–17:00
Survey method	Questinnaire sheets offered to the Spring Okawa Woodworks Festival participants and collected at venue
Main questions	1. Individual attributes
	2. Their transportation mode to Okawa city
	3. Places that they stopped by on that day
	4. The amount of money spent
The number of samples	2842 samples

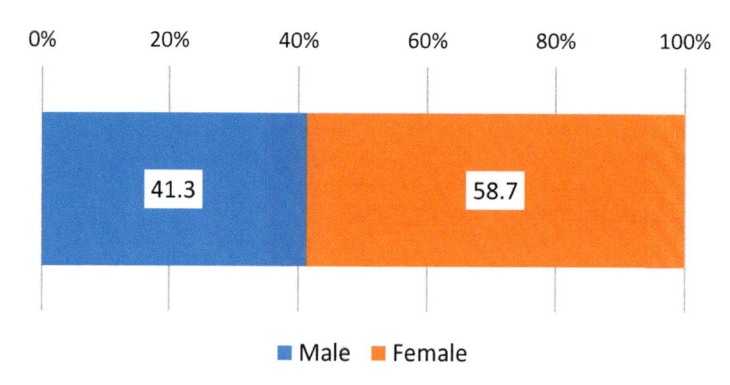

Fig. 14.1 Gender

2.4 Sample Profile

Here, we look at the gender, age, and place of residence of the respondent as a sample profile from the data of the Spring Okawa Woodworks Festival in 2011.

Gender

As for the gender of the visitors, we find from Fig. 14.1 that 41.3% are male and 58.7% are female, which means there were more female visitors.

Age

As for the age of the visitors, we see from Fig. 14.2 that people in their 30s are the largest group, followed by people in their 60s. Overall, the age of the visitors is concentrated in the range of 30–60 years.

Visit Purpose

Figure 14.3 shows the percentage of whether the visitors visited Okawa with the purpose for joining the 2011 Spring Okawa Woodworks Festival or with other purposes.

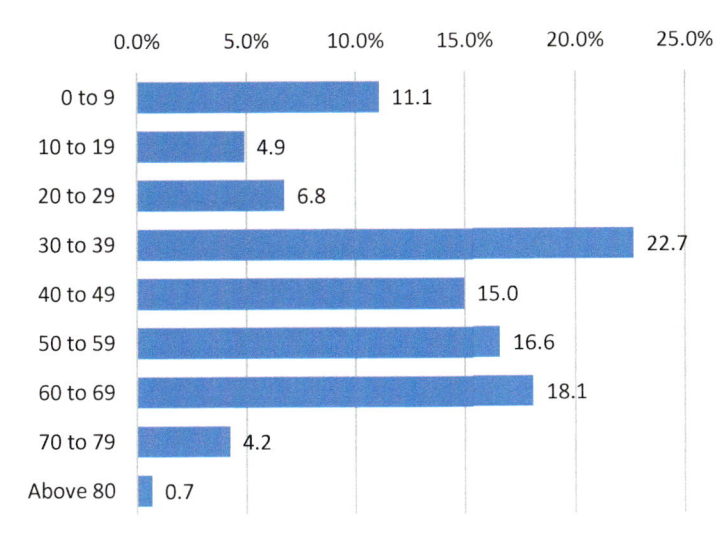

Fig. 14.2 Age

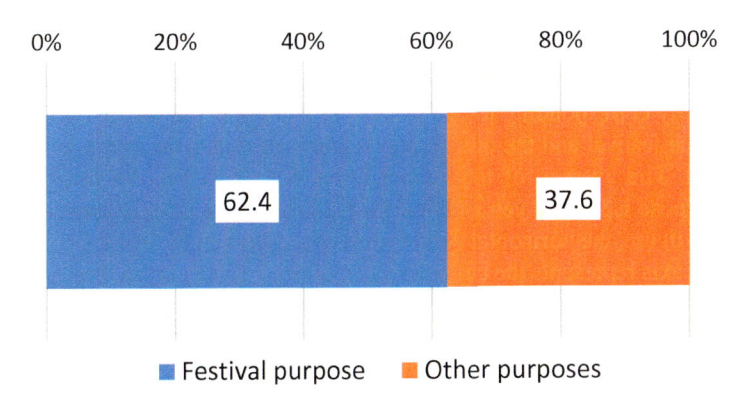

Fig. 14.3 Visit purpose to Okawa

3 Comparison Between Visitors with Festival Purpose and with Other Purposes

In this section, we compare the characteristics of visitors who came to the city for the purpose to attend the Spring Okawa Woodworks Festival 2011 and those visitors who came to the city for purposes other than the festival to investigate whether or not there are differences in their travel time, their expenditure, and their visited places on the festival day.

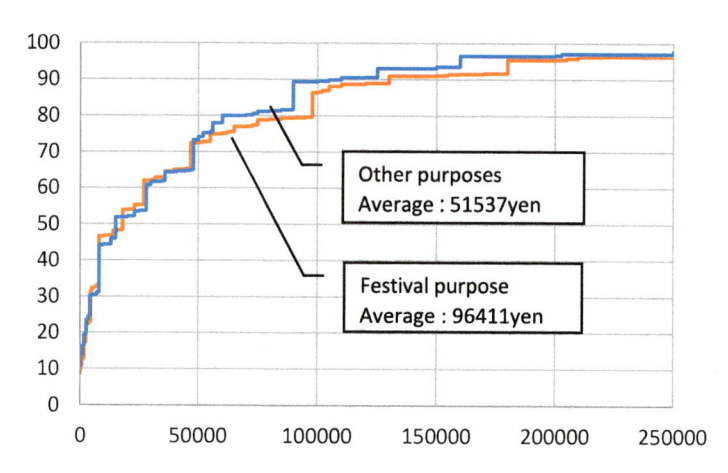

Fig. 14.4 Total expenditure in Okawa city

3.1 Expenditure at the "Spring Okawa Woodworks Festival"

Total Expenditure in Okawa City

Figure 14.4 shows the two cumulative distribution functions of the total expenditure in Okawa city corresponding to two groups: one group is the visitors who came for attending the festival and the other is the visitors who came for purposes other than the festival. The average of total expenditure in Okawa city for the group with festival purpose is 96,411 yen and that for the group with other purposes is 51,538 yen. In this figure, the horizontal axis represents the amount of total expenditure, and the vertical axis represents the cumulative percentage. How to read these cumulative distribution functions is as follows.[2] Focus on the cumulative distribution function (CDF) for the group with festival purpose. Look at 60,000 yen. We see that the value of CDF for the group with festival purpose at 60,000 yen is 75%, which means that among the group with festival purpose, the share of people who spent less than or equal to 60,000 yen is 75%. Similarly, we see that the value of CDF for the group with other purposes at 60,000 yen is 80%. This implies that the percentage of people who spent more than 60,000 yen is smaller for the group with other purposes than for the group with festival purpose. We also see that this fact holds at almost all values of total expenditure since CDF for the group with other purposes lie above that with festival purpose. Consequently, the average of the total expenditure for the group with other purposes becomes smaller than the one for the group with festival purpose. In general, the more the CDF of some variable shifts to the right the larger the average of that variable becomes.

---

[2]The cumulative distribution function F for a random variable X is defined as $F(x) = \text{Prob}(X \leq x)$.

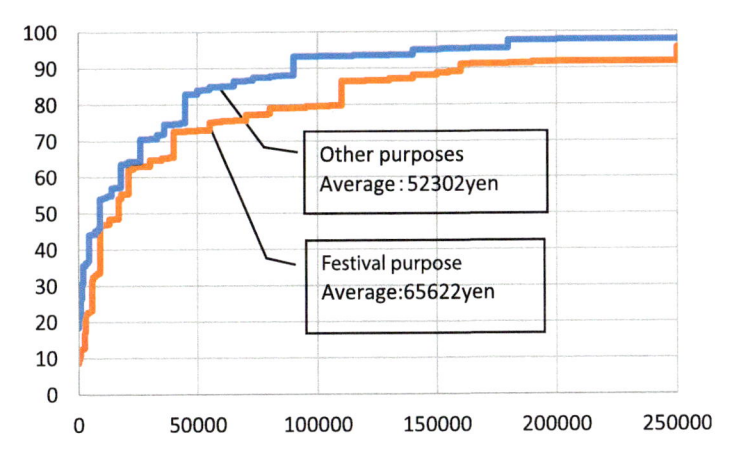

Fig. 14.5 Expenditure at the Okawa Woodworks Festival venue

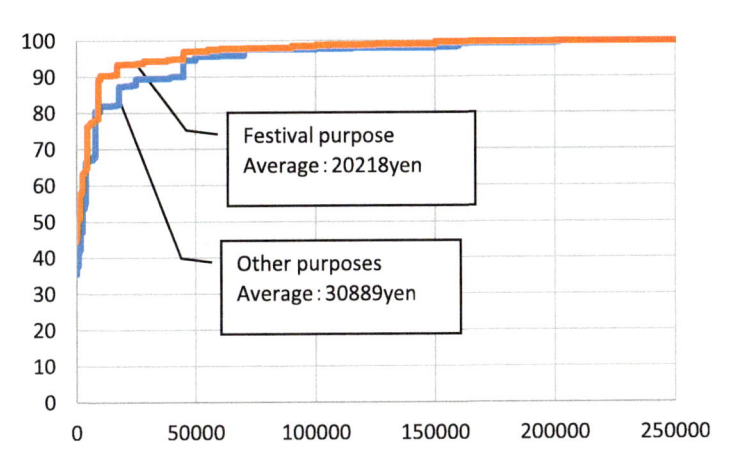

Fig. 14.6 Expenditure outside the Okawa Woodworks Festival venue

Expenditure at Okawa Woodworks Festival Venue

Next, we look at the expenditure at the Okawa Woodworks Festival venue in the same way as above. From Fig. 14.5, we see that the expenditure at the festival venue is higher for those who came for the purpose of attending the Okawa Woodworks Festival than those who came for purposes other than the festival. The average expenditure was 65,621 yen for the group with festival purpose and 52,302 yen for the group with other purposes. This is because most people who came to the city for the purpose of attending the Woodworks Festival have the intention to purchase furniture.

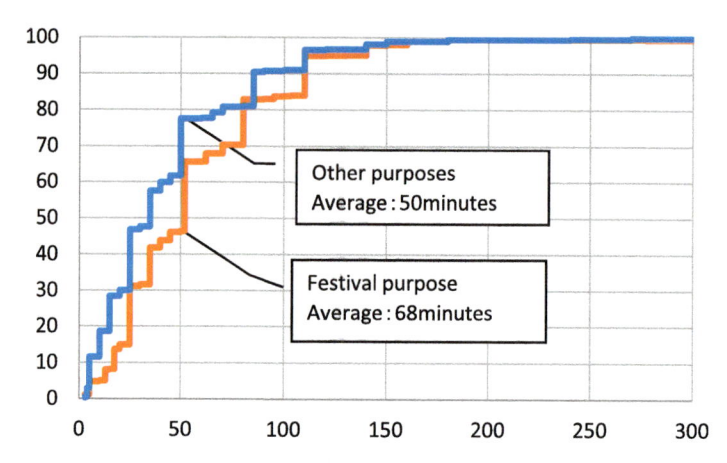

Fig. 14.7 The required time to visit Okawa for visitors with the festival purpose and for those with other purposes

Expenditure Outside the Okawa Woodworks Festival Venue

In the same way, in Fig. 14.6 we depict the cumulative distribution functions of the expenditure outside the Okawa Woodworks Festival venue for the above two groups. From the figure we see that unlike the expenditure inside the venue, the CDF for the group with other purposes lies on the right side of the CDF for the group with festival purpose. This means that the average of the expenditure outside the festival venue for the group with other purposes is larger than that for the group with festival purpose. In fact, the average expenditure was 20,217 yen for the group with festival purpose and 30,889 yen for the group with other purposes. The group with other purposes did not come to the festival as its main purpose. Therefore, they seem to spend more outside the venue than inside.

3.2 Comparison of Travel Time

Here we compare the travel time to Okawa between the visitors with the festival purpose and with other purposes. In the same way as above, we compare the two cumulative distribution functions of the required time to Okawa for visitors with the festival purpose and other purposes. Figure 14.7 illustrates the result.

14 To What Extent Did the Woodworks Festival Attract People?

Fig. 14.8 Places people visited before arriving at the Okawa Woodworks Festival venue

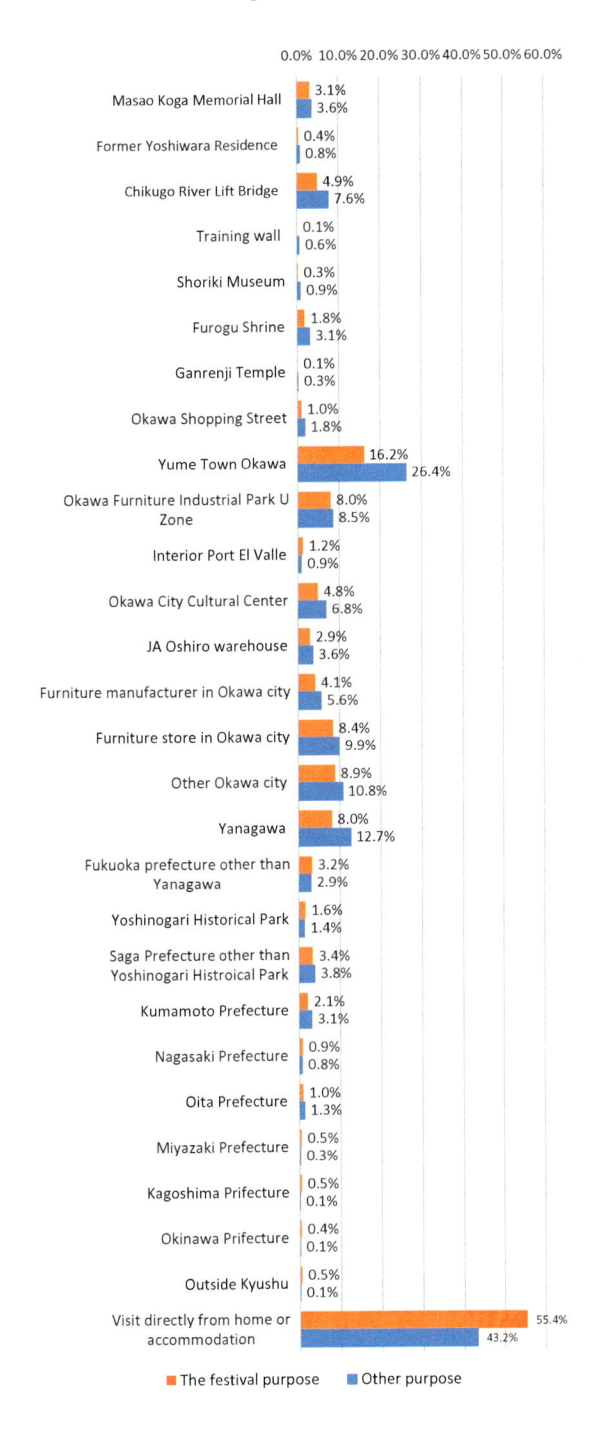

Place	The festival purpose	Other purpose
Masao Koga Memorial Hall	3.1%	3.6%
Former Yoshiwara Residence	0.4%	0.8%
Chikugo River Lift Bridge	4.9%	7.6%
Training wall	0.1%	0.6%
Shoriki Museum	0.3%	0.9%
Furogu Shrine	1.8%	3.1%
Ganrenji Temple	0.1%	0.3%
Okawa Shopping Street	1.0%	1.8%
Yume Town Okawa	16.2%	26.4%
Okawa Furniture Industrial Park U Zone	8.0%	8.5%
Interior Port El Valle	1.2%	0.9%
Okawa City Cultural Center	4.8%	6.8%
JA Oshiro warehouse	2.9%	3.6%
Furniture manufacturer in Okawa city	4.1%	5.6%
Furniture store in Okawa city	8.4%	9.9%
Other Okawa city	8.9%	10.8%
Yanagawa	8.0%	12.7%
Fukuoka prefecture other than Yanagawa	3.2%	2.9%
Yoshinogari Historical Park	1.6%	1.4%
Saga Prefecture other than Yoshinogari Histroical Park	3.4%	3.8%
Kumamoto Prefecture	2.1%	3.1%
Nagasaki Prefecture	0.9%	0.8%
Oita Prefecture	1.0%	1.3%
Miyazaki Prefecture	0.5%	0.3%
Kagoshima Prifecture	0.5%	0.1%
Okinawa Prifecture	0.4%	0.1%
Outside Kyushu	0.5%	0.1%
Visit directly from home or accommodation	55.4%	43.2%

3.3 Comparison of Visited Places on the Day of the Festival

We compare the behavior of those who came for the purpose of attending the Okawa Woodworks Festival and those who came to Okawa for other purposes from the viewpoint of places that they visited on that day. In the questionnaire survey of the "Spring Woodworks Festival," we ask respondents about the places that they visited before coming to the venue on the day and places they were planning to visit after leaving the venue.

In the same way as before, we analyze the behavior of the two groups: one is with festival purpose and the other is with other purposes.

Figure 14.8 shows the visited places before coming to the Okawa Woodworks Festival venue. From the figure, we see that most people came to the venue directly from their home or accommodation facility. This is the same for the two groups. The most frequently visited place other than directly coming from home is *Yume*-Town Okawa, which is a shopping mall located at the city center district. This is also true for the two groups.

However, the percentage of the people who directly visited the festival venue from home is higher for the group with festival purpose than for the group with other purposes, whereas the percentage of the people who visited *Yume*-Town Okawa before coming to the festival venue is higher for the group with other purposes than for the group with festival purpose.

In addition, Fig. 14.9 shows the places that people were planning to visit after leaving the Okawa Woodworks Festival venue. Similar to the case of visited places before coming to the venue, we see from the figure that for the group with festival purpose, the highest percentage is the people who were planning to go back directly to their home or accommodation facility and the second highest is the people who planned to visit *Yume*-Town Okawa. On the other hand, for the group with other purposes, the highest percentage is the people planning to visit *Yume*-Town Okawa and that of directly going back home follows.

To summarize the above observations, it is clear that the people who came for the purpose of attending the Okawa Woodworks Festival tend to spend more inside the venue than outside, go directly from home to the venue, and go back directly home from the venue, whereas people who came for purposes other than the festival tend to spend more outside the venue and plan to visit the places to outside the venue such as *Yume*-Town Okawa on the way back to home.

4 Measuring the Customer Attraction Effect of the Okawa Woodworks Festival

In the previous section, we clarified the characteristics of people who came for the purpose of attending the Okawa Woodworks Festival. We have defined the customer attraction effect due to the Okawa Woodworks Festival as the increase in the number of visitors who came to the city for the purpose to attend the festival. With this

14 To What Extent Did the Woodworks Festival Attract People?

Fig. 14.9 Places people planned to visit after leaving Okawa Woodworks Festival venue

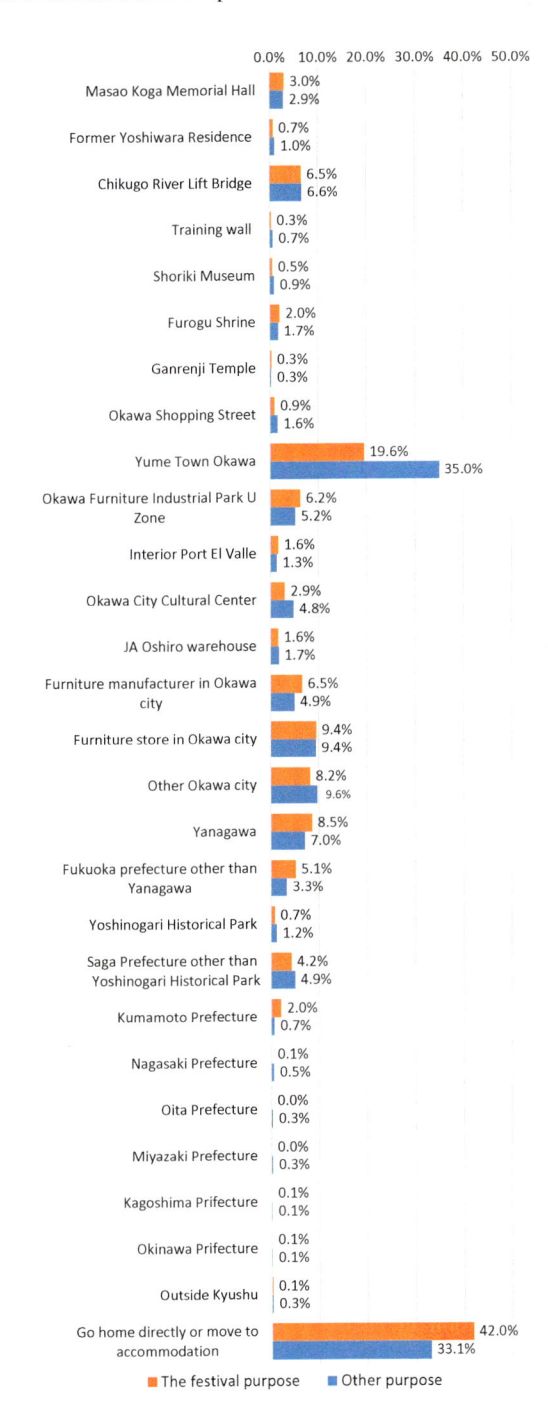

Place	The festival purpose	Other purpose
Masao Koga Memorial Hall	3.0%	2.9%
Former Yoshiwara Residence	0.7%	1.0%
Chikugo River Lift Bridge	6.5%	6.6%
Training wall	0.3%	0.7%
Shoriki Museum	0.5%	0.9%
Furogu Shrine	2.0%	1.7%
Ganrenji Temple	0.3%	0.3%
Okawa Shopping Street	0.9%	1.6%
Yume Town Okawa	19.6%	35.0%
Okawa Furniture Industrial Park U Zone	6.2%	5.2%
Interior Port El Valle	1.6%	1.3%
Okawa City Cultural Center	2.9%	4.8%
JA Oshiro warehouse	1.6%	1.7%
Furniture manufacturer in Okawa city	6.5%	4.9%
Furniture store in Okawa city	9.4%	9.4%
Other Okawa city	8.2%	9.6%
Yanagawa	8.5%	7.0%
Fukuoka prefecture other than Yanagawa	5.1%	3.3%
Yoshinogari Historical Park	0.7%	1.2%
Saga Prefecture other than Yoshinogari Historical Park	4.2%	4.9%
Kumamoto Prefecture	2.0%	0.7%
Nagasaki Prefecture	0.1%	0.5%
Oita Prefecture	0.0%	0.3%
Miyazaki Prefecture	0.0%	0.3%
Kagoshima Prifecture	0.1%	0.1%
Okinawa Prifecture	0.1%	0.1%
Outside Kyushu	0.1%	0.3%
Go home directly or move to accommodation	42.0%	33.1%

definition, we now estimate the degree of the increase in the number of customers attracted to Okawa city by the Okawa Woodworks Festival based on the population of the Okawa metropolitan area, which we consider as composed of Okawa City and neighboring Yanagawa City.[3]

4.1 Procedure of the Analysis and Data

Procedure of the Estimation

We estimate the increase in the number of visitors attracted by the Okawa Woodworks Festival by the following procedure.

1. Among the collapsed data from two surveys of the 61st Okawa Woodworks Festival survey and the Spring Okawa Woodworks Festival 2011 survey, we picked up the samples from the Okawa metropolitan area which is defined as Okawa city and Yanagawa city. Using these samples from Okawa and Yanagawa cities, we estimate the parameters of the visit frequency Poisson model with three explanatory variables: the travel time to Okawa City, the dummy variable whether the visit purpose is the festival purpose or not, and the number of exhibition booths at autumn and spring Okawa woodworks festivals.
2. Using the estimated Poisson model while omitting the festival purpose dummy variable, we forecast the number of usual daily visitors visiting the city center district of Okawa City for shopping, leisure, and having meals from the Okawa metropolitan area under the condition that there are no Woodworks festivals. To obtain each number of visitors from Okawa and Yanagawa City, we used population data for each city, as well as the sample mean of travel time to Okawa City from each city obtained from survey data, and an average number of booths for autumn and spring festivals as the values of explanatory variables of travel time and the number of exhibition booths, respectively.
3. From the estimated coefficient for the dummy variable of whether the visit purpose is the festival purpose, we obtain the ratio of visitors with the festival purpose to those visitors with other usual purposes. From the result, we can estimate the increment number of visitors to Okawa City with the purpose for attending the Okawa Woodworks Festival.

Data

We use two sets of data simultaneously, the 61st Okawa Woodworks Festival questionnaire survey data and the Spring Okawa Woodworks Festival questionnaire

---

[3]Yanagawa City is located about 50 km south from Fukuoka City adjoining to the south border of Okawa City with a population of 71,000 people as of 2010, which is known as a site-seeing spot for tourists.

Table 14.3 Estimated results of the parameters

	Parameter estimate	t value	SD
Travel time (minutes) α	−0.0400	−10.00***	0.0040
Dummy for the festival purpose β	−0.4319	−5.32***	0.0812
Number of booths γ	0.0078	15.60***	0.0005

***: 1% significant

survey data. Further, for forecasting data, we use the population data[4] for the Okawa metropolitan area (Okawa City and Yanagawa City) between 15 and 79 years of age in 2010. The population of Okawa City is 30,659, and that of Yanagawa City is 57,255. The required time to reach Okawa City (downtown area), according to the Spring Okawa Woodworks Festival 2011 questionnaire survey, is 13 min from Okawa city and 22 min from Yanagawa city. The number of exhibition booths was 156 at the 61st Okawa Woodworks Festival and 152 at the 2011 Spring Okawa Woodworks Festival.

4.2 Estimation Model

Visit Frequency Poisson Model

The visit frequency Poisson model considers that sample i's visit frequency Y_i are identically independently distributed according to Poisson distribution with mean λ_i. Then, the logarithm of the mean, $\log(\lambda_i)$, is expressed as the linear function $x_i\beta$ of the explanatory variables x_i and unknown parameters β to be estimated. The unknown parameter β is estimated by the maximum likelihood estimation. In this study, we use the data obtained from on-site random sampling survey so that there occurs the choice-based sampling biases. Elsewhere Saito et al. have developed the weighted Poisson model to remove the choice-based random sampling biases. (For details, see Saito et al. [5, 6]). Their method put to each sample the weight of the inverse of that sample's visit frequency to the site where that sample was sampled. Here we use their weighted Poisson model for the estimation.

Specifically we formulate our Poisson model as follows.

$$\log(\lambda_i) = \alpha t_i + \beta \text{evntd}_i + \gamma \text{booth}_{s(i)} \tag{14.1}$$

Here, λ_i is the mean of sample i's frequency of visits to Okawa city, t_i, is the sample i's travel time to Okawa City. The dummy variable, evntd_i is 1 if sample i's visit purpose is the festival purpose and 0 otherwise. The booth_j is the number of

---

[4] According to population based on the Basic Resident Register, demographics and the number of household survey and Basic Resident Register municipalities by age group population in 2010, Ministry of Internal Affairs and Communications

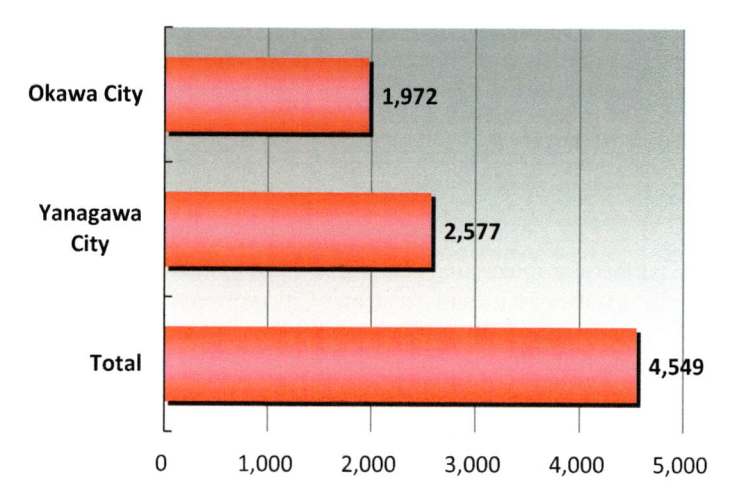

Fig. 14.10 Forecasted number of visitors entering city center of Okawa City from residents of the Okawa metropolitan area (unit: people/day)

exhibition booths for autumn and spring festivals, where $j = 1$ for autumn and $j = 2$ for spring. The index $s(i)$ takes the value 1 if sample i is from autumn festival samples and 2 otherwise. α, β, and γ are unknown parameters.

Estimated Results of the Parameter

Table 14.3 shows the estimated results of the parameters. Looking at the estimated results, the travel time to Okawa City and the dummy for the festival purpose are negative, and the number of exhibition booths is positive. All the parameters are statistically significant at 1% level. The signs for the travel time and the number of booths are convincing results. However the sign for the dummy for the festival purpose might be a little bit contrary to the intuition. It should be noted that we asked the respondent about their frequency of visits to Okawa city as the usual frequency of visits to there. As shown in the previous results, we saw that the visitors with the festival purpose come to the city from farther away than those visitors with other usual purposes. Thus the visitors with the purpose for attending the woodworks festival have lower frequency of visits in comparison with those visitors with other purposes. Hence the result of the sign of the dummy for the festival purpose might become understandable since the visitors with festival purpose are the less frequent visitors.

At any rate, because the parameters of all variables are significant, we use these values to carry out the forecast.

14 To What Extent Did the Woodworks Festival Attract People?

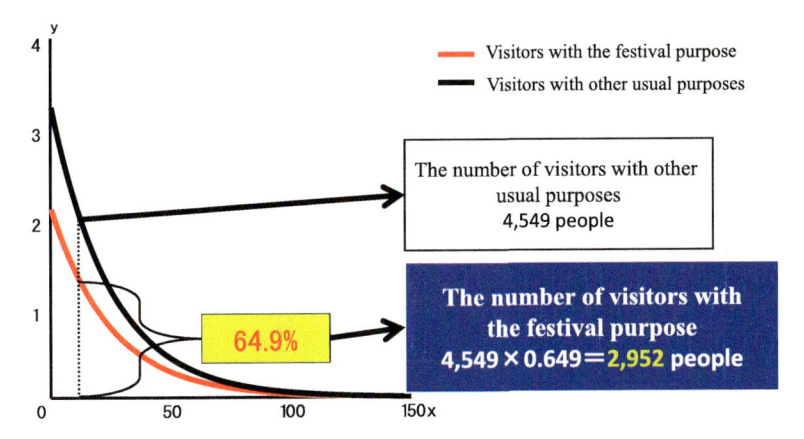

Fig. 14.11 Ratio of the number of visitors to Okawa city with the festival purpose to that of the usual daily visitors (unit: people/day)

4.3 Results of Forecasting the Number of Usual Daily Visitors to Okawa City from the Okawa Metropolitan Area

From the parameter estimated results, we forecast the number of visitors to Okawa city. Figure 14.10 shows the forecasted results of the number of usual daily visitors visiting the city center district of Okawa city from the residents of the Okawa metropolitan area. The number of people who visit from Okawa City is 1,972 people and 2,577 people from Yanagawa City per day, which sums up to 4,549 people visiting the city center district of Okawa city per day from the Okawa metropolitan area.

4.4 Estimating the Number of Visitors Attracted by the Okawa Woodworks Festival from the Okawa Metropolitan Area

Here we focus on the estimated results of the Poisson model for the frequency of visits to the Okawa Woodworks Festival among residents of the Okawa metropolitan area, which is shown in Table 14.3. From the table, the estimated value of the parameter of the dummy variable for whether people came for the purpose to attend the Woodworks Festival is -0.4319. Therefore, the effect by the dummy variable for the festival purpose on the mean of the visit frequency, λ_i, becomes $e^{-0.4319} = 0.649$, which implies that the effect by the festival is measured by the ratio to the usual daily visit frequency with other purposes. Thus we interpret the number of visitors to Okawa City with the purpose for attending the Okawa Woodworks Festival is 64.9% of the number of those visitors with usual daily other purposes.

Therefore, since the 4,549 people was the number of visitors to Okawa city with usual purposes, the number of visitors who came for the purpose of attending the Okawa Woodworks Festival becomes 2,952 people, which is 4,549 people multiplied by 64.9%. From these, the Okawa Woodworks Festival attracted a total of 7501 people, which is 4,549 people plus 2,952 people (Fig. 14.11).

5 How Far the Okawa Woodworks Festival Attracts Visitors From?

5.1 Method for Determining Visitor Attraction Area by the Festival

Here we investigate the method for estimating and determining the spatial area from which the Okawa Woodworks Festival attracts the visitors. In the above analysis to estimate the number of visitors with the festival purpose, the effects by the festival on the visitors with festival purpose are measured by the comparison with the visitors who visit Okawa City if there were no festivals, that is, the usual daily visitors with other purposes than the festival. Thus the higher effects by the festival are represented as the one that attracts more people from the visitors with the lower frequency of usual visits to Okawa city. This is the consequence of the comparison of the visitors with the festival purpose with those with usual purposes.

However, we would like to investigate how the Okawa Woodworks Festival has the influences on among the visitors with the festival purposes. For instance, if certain sample has come to Okawa City with the festival purpose from far away than other samples, that sample can be seen to be affected by the festival much more than other samples.

To do this, we made one trick. We asked the respondents about their frequency of visits to the city center district of Okawa city. This question item is concerned with asking their usual frequency of visits to Okawa City. Thus for the respondents with the festival purpose, we can interpret this question item as asking about the usual frequency of visits excluding the present visit to the festival with the festival purpose. From this interpretation, limiting for the samples with the festival purpose, we created an artificial panel data for these samples with the festival purpose as follows. The first half of this panel data contains the original data with the dummy variable for the festival purpose with the value 0, and the second half contains the original data while modifying the sample's answered value for the frequency of visits to Okawa City into the value obtained by adding the present visit[5] to Okawa City with the festival purpose and setting the value of the dummy variable for the festival purpose to 1.

---

[5]The frequency of visits added is once for 2 or 3 months, which equals the frequency corresponding to the festival period over 2 or 3 days, which is held once per 6 months.

Table 14.4 Estimated results of the model for visitor attraction area of the Okawa Woodworks Festival in terms of travel time distance

	Estimate	t value	SD
Intercept α	−1.2947	−30.25***	0.0428
Travel time β	−0.0085	−12.14***	0.0007
Dummy for festival purpose γ	0.7527	15.30***	0.0492
Travel time × dummy for festival purpose δ	−0.0024	−8.00***	0.0003

***: 1% significant, $n = 2920$

We use this panel data for the analysis. The model is formulated as a visit frequency Poisson model as follows:

$$\log(\lambda_i) = \alpha + \beta t_i + \gamma \text{eventd}_i + \delta(t_i \times \text{eventd}_i) \qquad (14.2)$$

In the same way as before, the sample i's frequency of visits to the city center of Okawa City is assumed to be distributed as independent identical Poisson distributions with mean λ_i. The logarithm of λ_i is expressed as Eq. (14.2), where t_i denotes sample i's travel time to Okawa City and eventd$_i$ sample i's dummy variable for the festival purpose, which takes 1 if sample i's visit purpose is the festival and 0 otherwise.

Further, note that Eq. (14.2) can be transformed as follows:

$$\log(\lambda_i) = \alpha + \beta t_i + (\gamma + \delta t_i)\text{eventd}_i \qquad (14.3)$$

From this, we see that the coefficient of the dummy variable for the festival purpose varies depending on the travel time t_i, and the term eventd$_i$ would disappear if its coefficient $(\gamma + \delta t_i)$ becomes 0.

Under the condition that $\gamma > 0$ and $\delta < 0$, this fact can be interpreted as follows. The larger the travel time becomes the smaller the festival effect becomes and finally it reduces to 0. In this sense, the value of t_i that attains $\gamma + \delta t_i = 0$ turns out to be the farthest travel time distance the festival effect reaches.

Hence if we obtain the estimates of these parameters, we can determine the visitor attraction spatial area the effect of the Okawa Woodworks Festival reaches in terms of the travel time distance.

5.2 Parameter Estimated Results

Table 14.4 shows the estimated results of the parameters. All of the parameters are statistically significant at 1% level.

In contrast to the previous estimated results, the coefficient of the dummy variable for the festival purpose, that is, the indicator of those who visit Okawa city with the purpose for attending the Okawa Woodworks Festival, turns out positive. This is because the comparison was carried out among the visitors coming to Okawa city with the festival purpose while devising ingenuity to deal with the original data as though the sample were from before the festival and the present visit to Okawa city with the festival purpose as though the sample were from after the festival by adding the present visit to the original frequency of visits.

5.3 Visitor Attraction Area of the Okawa Woodworks Festival

Now we determine the visitor attraction spatial area of the Okawa Woodworks Festival in terms of travel time distance.

From the above, we have

$$\gamma + \delta t_i = 0$$
$$t_i = -\frac{\gamma}{\delta} = \frac{0.7527}{0.0024} = 313.6$$

Thus the travel time distance for the visitor attraction area turns out 313.6 min, about 5 h travel time.

6 TV Commercial Is Effective for Enlarging the Visitor Attraction Area?

6.1 Model to Estimate the Effect of TV Commercials

In this section, we investigate how effective the TV commercials contribute to attracting the visitors. As stated in the introduction, the organizers of the Okawa Woodworks Festivals try to attract the consumers to the festival so that the TV commercials to advertise the festival have been aired many times, which appeal the winning of the premium of 3 million yen.

Here we introduce the explanatory variable that represents the number of times TV commercials are aired.[6] By applying the previous method to determine the visitor attraction area, we investigate the effectiveness of TV commercials from the viewpoint of how TV commercials have contributed to enlarging the visitor attraction area of the Okawa Woodworks Festival.

The model is formulated as follows.

---

[6]The number of times TV commercial are aired varies depending on the area. Thus the possibilities of the number of times the respondents view the TV commercials are different between their residences. We use the area-wise values of the numbers of times aired.

Table 14.5 Estimated results of the model for estimating the effect of TV commercials on enlarging the visitor attraction area by the Okawa Woodworks Festival in terms of travel time distance

	Estimate	t value	SD
Intercept α	−1.3954	−27.63***	0.0505
Travel time β	−0.0087	−12.43***	0.0007
CM η	0.0017	4.25***	0.0004
Dummy for festival purpose γ	0.8217	14.17***	0.058
Travel time × dummy for festival purpose δ	−0.0025	−8.33***	0.0003
CM × dummy for festival purpose μ	0.0006	2.00**	0.0003

***: 1% significant, **: 5% significant, n = 2920

$$\log(\lambda_i) = \alpha + \beta t_i + \eta CM_{r(i)} + \gamma \text{eventd}_i + \delta(t_i \times \text{eventd}_i)$$
$$+ \mu\left(CM_{r(i)} \times \text{eventd}_i\right) \quad (14.4)$$

Here $CM_{r(i)}$ denote the number of times TV commercials are aired at the area $r(i)$ where sample i resides. Other explanatory variables are the same as before.

Similarly as before, Eq. (14.4) can be transformed as follows.

$$\log(\lambda_i) = \alpha + \beta t_i + \eta CM_{r(i)} + \left(\gamma + \mu CM_{r(i)} + \delta t_i\right)\text{eventd}_i \quad (14.5)$$

From the equation, the enlargement effect by TV commercials on the visitor attraction area Δt_i can be seen as follows.

$$\gamma + \mu CM_{r(i)} + \delta t_i = 0$$
$$\Delta t_i = -\frac{\mu CM_{r(i)}}{\delta} \quad (14.6)$$

6.2 Parameter Estimated Results and the Enlargement Effect by TV Commercials

Parameters Estimated

Table 14.5 gives the estimated results. All the parameters except the interaction term between CM and the dummy for the festival purpose are statistically significant at 1% level and that interaction term also is significant at 5% level. As expected, the sign of the coefficient μ of that interaction term and that of the coefficient δ of the interaction term between travel time and the dummy for the festival purpose are, respectively, positive and negative so that the enlargement effect becomes positive. Note that since the value of CM is the number of times TV commercials are aired, its coefficient represents the effect of TV commercial when it is aired once.

Enlargement Effect by TV Commercials on Visitor Attraction Area of the Okawa Woodworks Festival

We calculate the enlargement effect by the formula (14.6) which turns out as follows.

$$\Delta t_i = -\frac{\mu CM_{r(i)}}{\delta} = \frac{0.0006 \times 114}{0.0025} = 27.4$$

Here we set the value of the number of times TV commercials are aired to 114, which is the number of times aired covering the area of Fukuoka Prefecture, where 70% of the samples reside. The enlargement effect is known to be 27.4 min.

Now let us determine the visitor attraction area of the Okawa Woodworks Festival with TV commercials aired in terms of travel time distance. The result turns out as follows.

$$\gamma + \mu CM_{r(i)} + \delta t_i = 0$$
$$t_i = -\frac{\gamma + \mu CM_{r(i)}}{\delta}$$
$$= \frac{0.8217 + 0.0006 \times 114}{0.0025} = 356.0$$

The visitor attraction spatial area becomes the circle centered at the city center of Okawa City with the radius of 356.0 min.

Comparing the previous case without TV commercials aired, the area is increased from 313.6 to 356.0 min by 42.4 min.

7 Conclusion and Future Challenges

In this study, based on the questionnaire survey data of the 61st Autumn Okawa Woodworks Festival of 2010 and the Spring Okawa Woodworks Festival of 2011, by restricting the samples to those from the Okawa metropolitan area, we forecasted the number of visitors from that area who visit the city center district of Okawa city with the purpose for attending the Woodworks Festival. This is the estimation of the visitor attraction effect on Okawa metropolitan residents due to the Woodworks Festival, using the on-site weighted visit frequency Poisson model.

While this estimation by Poisson model is thought of as based on the comparison of the visitors attracted by the festival purpose with those with usual purposes, we made an ingenuity to create the panel data of the visitors with the festival purpose as though composed of data at two points of time before and after the festival. Using this panel data for the visitors with the festival purpose, we demonstrate that we can determine the visitor attraction spatial area of the Woodworks Festival in terms of travel time distance.

By doing this, we show that the area is the circle with the radius of 313.6 min. Furthermore, applying the same method to the analysis of the effect by the TV commercial advertisement, we show that the enlargement effect of TV commercial advertising on the visitor attraction spatial area in terms of travel time distance turns out to increase the area by 42 min.

As for the future research, since we found that our method to create the panel data composed of the samples who are directly affected by the festival is effective for scrutinizing the effects by the festival on among the samples affected, we should explore further its potentials for various applications. For example, here we roughly estimate the effect of TV commercials aired over all samples, but we have the varied numbers of times TV commercials are aired depending on different residences so that we can extract much more detailed effect on the different samples residing at different areas.

Another future research area may be the evaluation of the cost of organizing the Okawa Woodworks Festival and its effect on Okawa City. In addition, future research may include Okawa's history, culture, commerce, food, and other aspects, to evaluate how the value of the town can be raised.

References

1. Saito S, Cai J, Yamashiro K, Iwami M, Imanishi M (2011) Measurement of the customer attraction effect of large events: the case of Okawa woodworks festival. Paper presented at the 48th annual meeting of Japan Section of Regional Science Association International (JSRSAI). (in Japanese)
2. Saito S, Yamashiro K (2001) Economic impacts of the downtown one-dollar circuit bus estimated from consumer's shop-around behavior: a case of the downtown one-dollar bus at Fukuoka City. Stud Reg Sci 31:57–75. (in Japanese)
3. Saito S, Kiguchi T, Kakoi M, Nakashima T (2003) The economic effect and function of downtown cafe from the viewpoint of consumer shoparound behavior. The 2003 fall national conference of operations resarch society of Japan, abstracts, pp 218–219. 2003/09/10. (in Japanese)
4. Saito S, Sato T, Yamashiro K (2010) On-site consisitent estimation of shop-around behavior and measuring the effect of events: an application to the measurement of the effect of Kumamoto castle festival, vol 26. In: proceedings of 26th annual meeting of the Japan Association for Real Estate Sciences, pp 175–182. (in Japanese)
5. Saito S, Kumata Y, Ishibashi K (1995) A choice-based poisson regression model to forecast the number of shoppers: its application to evaluating changes of the number and Shop-around pattern of shoppers after city center redevelopment at Kitakyushu City. Pap City Plan 30:523–528. (in Japanese)
6. Saito S, Kakoi M, Nakashima T (1999) On-site poisson regression modeling for forecasting the number of visitors to city center retail environment and its evaluation. Stud Reg Sci 29:55–74. (in Japanese)

Chapter 15
How Did the Effects of the Festival Held on Main Street Spread Over Other Districts Within a City Center?

Saburo Saito, Kosuke Yamashiro, and Masakuni Iwami

Abstract Various tourist attractions and events have been conducted to revitalize towns. As for what kinds of effects tourist attractions and events have on towns, organizers usually announce only the approximate number of visitors. However if a town is regarded as one entity for attaining the goal of its revitalization, the effects of those events cannot be said to be estimated and identified, unless they are clarified in such a way as how many people participated in those events and how their effects are spread over to what extent and to which part of the town. In this study, we propose a method to make it possible to measure the effects of those events on a town in that way. More specifically, we estimate the number of event participants by using the consistent method to estimate consumer shop-around patterns based on the data obtained from the on-site survey of *Kaiyu* behaviors, which enables to estimate the net number of incoming visitors to the town. Using this estimation method, we measure the effect of the Kumamoto Castle Festival on a number of visitors' basis as well as on a monetary basis, by comparing visitors who visited with the purpose for attending the festival and those who did not. From the analysis, we found that the

---

This chapter is based on the paper of Saburo Saito, Takahiro Sato, and Kosuke Yamashiro [1], "Consistent Estimation of Shop-around Behavior and Measurement of the Event Effects Spread Over City Center: An application to the measurement of the effect of Kumamoto Castle Festival," *Papers of the 26th Annual Meeting of The Japan Association for Real Estate Sciences*, vol. 26, pp. 175–182, 2010, which is modified for this chapter.

S. Saito (✉)
Faculty of Economics, Fukuoka University, Fukuoka, Japan

Fukuoka University Institute of Quantitative Behavioral Informatics for City and Space Economy (FQBIC), Fukuoka, Japan
e-mail: saito@fukuoka-u.ac.jp

K. Yamashiro
Department of Business and Economics, Nippon Bunri University, Oita City, Japan
e-mail: yamashiroks@nbu.ac.jp

M. Iwami
Fukuoka University Institute of Quantitative Behavioral Informatics for City and Space Economy (FQBIC), Fukuoka, Japan
e-mail: miwami@econ.fukuoka-u.ac.jp

visitors who came with the purpose for attending the Kumamoto Castle Festival shopped around along the arcades adjoining to the festival venue. Thus the festival effects were spread over these arcades, and the total amount of the effect on a monetary basis was 52,670,000 yen (or 526,700 dollars).

Keyword *Kaiyu* · Event effects · Spread over · Consistent estimation method · Shop-around pattern · Net incoming visitors · Kumamoto · Castle

1 Purpose

Many city administrations and managements of shopping streets often utilize various tourist attractions and events to revitalize towns. However, as pointed out by Saito [2], the net number of incoming visitors to the town, which is obviously necessary and indispensable information when viewing the town as one entity for attaining the goal of its revitalization, has hardly been grasped. If the net number of incoming visitors to the town is unknown, it is difficult to judge whether tourist attractions and events have really been effective for revitalizing the town because we cannot accurately evaluate how these attractions and events have contributed to the increase in the net number of incoming visitors to the town.

Furthermore, it can be said that there are no studies that directly address the question such as how many people participated, how they walked around the town, and to what extent and on which part of the town the effect of the tourist attractions and events was spread.

The aim of this study is to address this issue directly. More specifically, on the day of the event, we carry out an on-site survey of *Kaiyu* behavior, in which a question item is devised to distinguish the visitors who came with the purpose to attend the event from the ordinary visitors. Then we apply the consistent estimation method of *Kaiyu* patterns developed by Saito et al. [3–5] and propose a method to measure how many people participated in the event, how those people shopped around the town, and to what part of the town the effect of the event was spread. Additionally, we also attempt to show the effectiveness of our method with a concrete application example.

While many policies advocating revitalizing towns have been implemented so far, it is surprising that most towns do not know the net number of incoming visitors to their own towns. Some subsidized policies such as vitalization of city center are required to present post-evaluation reports after their implementations, but the situation that policy makers for towns have not known their net number of incoming visitors to their own towns has yet to be improved.

Most of the large retailers such as the department stores and shopping malls have grasped the number of incoming customers to their shops daily. The information is indispensable for their marketing and promoting their shops. The number of incoming customers is usually measured by some people counting system, which automatically counts the number of customers entering and exiting their building in a time interval such as 15 min. As for towns, since it is almost impossible to encircle the whole area of a town, the net number of incoming visitors has not been known. Some areas to implement city center revitalization program employ pedestrian transport volume surveys as a policy evaluation tool, where the sum of the pedestrians counted is used for the substitute for accounting for the number of visitors to their city center.

However, a pedestrian counted at some pedestrian survey point might be counted at another different survey point so that we cannot obtain the net number of incoming visitors to the city center because of this double counting possibility.

Recently, Saito et al. [3] and Saito and Nakashima [4] have developed a unique theoretical method to estimate the net number of incoming visitors to a city center district, which is based on micro-behavioral *Kaiyu* data obtained from the on-site survey of consumer shop-around behaviors. They call their method a consistent estimation method of consumer shop-around patterns for on-site survey data. While Saito et al. [6] also tried to estimate the number of incoming visitors to the city center by using the weighted Poisson model, this consistent estimation method has quite a different theoretical background from their previous efforts.

The aim of this study is to apply this method to the event to estimate the net number of incoming visitors attracted by the event and to evaluate the effect of the event which spreads over other areas than the event venue through tracing *Kaiyu* behaviors of the event participants, which has never been addressed to be analyzed in details.

Thus the purpose of this study is to take up the case of the Kumamoto Castle Festival held at the city center district of Kumamoto City on May 31, 2009, conduct an on-site survey of shop-around behaviors on the day of the event, estimate the net number of incoming visitors to the event as well as to the city center district of Kumamoto City, and evaluate to what extent the Kumamoto Castle Festival has influenced on which part of the city center district of Kumamoto City[1] through the shop-around behaviors of the event participants.[2]

The significant feature of this study is that to estimate the effects of the event we make a comparison between the visitors who visited the city center with the purpose attending the festival and those visitors with other purposes. The respondents are classified into these two groups by their visit purposes obtained from the questionnaire. This is a sharp contrast to previous studies in which the comparison is based on the difference in respondents' behaviors such as use or non-use of one-dollar bus and city center café [8, 9].

---

[1] Kumamoto City is the capital city of Kumamoto Prefecture with a long history and has the Kumamoto Castle near its city center, which is a symbol of the town. It is located in the midst of Kyushu Island with population of 740,822 (census 2015) within its administrative division and population 1,461,794 (census 2010) for its 1.5% metropolitan area. Kumamoto City was designated as one of ordinance-designated cities in Japan in 2011.

[2] This research was also reported in [7].

2 Framework of the Analysis

2.1 Outline of the Kumamoto Castle Festival

The Kumamoto Castle Festival was an event held in accordance with the opening of the Kumamoto Castle *Honmaru* palace to the public on April 20, 2008. It was held during the period of the local expo "Festival of the 400 years since the construction of the Kumamoto Castle," held from December 2006 to May 2008 to commemorate the 400th year of the construction of the Kumamoto Castle in 2007.

The data used in this study was collected from the 2nd survey of shop-around behaviors at the city center of Kumamoto City, which was conducted on May 31, 2009 (Sunday). This survey date was the day when the second Kumamoto Castle Festival was held to celebrate the achievement that Kumamoto Castle became the castle with the highest number of visitors in Japan.

In addition, at the second Kumamoto Castle Festival, trams between *Kumamotojo-Shiyakusho-Mae* Station and *Suidocho* Station were stopped from 10:30 to 16:00, the entry of cars and motorbikes was prohibited, and the area of *Torichosuji* became a pedestrian paradise. The trams in the section other than the pedestrian paradise area were operating halfway, and the fare was free all day. Within the pedestrian paradise, a main stage and substages were installed, where various events were held. The opening time of the Kumamoto Castle Festival was from 11:30 to 15:00.

2.2 Data Used

In this study, we used data from the 2nd survey of consumer shop-around behaviors at the city center of Kumamoto City, which was jointly conducted on May 31, 2009 (Sunday), by the Fukuoka University Institute of Quantitative Behavioral Informatics for City and Space Economy (FQBIC) and the Downtown Atelier of the Faculty of Engineering of Kumamoto University. The 2nd survey of consumer shop-around behaviors at the city center of Kumamoto City was carried out as an on-site interview survey with questionnaire sheets, which takes about 15 min. We set up several survey points. The respondents are sampled at random from the visitors at the sampling points. The respondents are asked about their shop-around behaviors at the city center district of Kumamoto City; the frequency of visits to the city center district of Kumamoto City, the *Kamitori* Arcade, and the *Shimotori* Arcade; the transport means; etc. The survey was carried out for 6 h between 12:00 and 18:00, with survey points located at the *Kamitori* Arcade, the *Shimotori* Arcade, and the *Shinshigai* Arcade. A total of 122 samples were collected. The outline of the survey is shown in Table 15.1.

Table 15.1 Outline of the 2nd survey of consumer shop-around behaviors at the city center of Kumamoto City

Survey date	2009.05.31 (Sun) 12:00–18:00
Survey points	4 points at the city center district of Kumamoto City
	Kamitori Shopping Arcade, *Shimotori* Shopping Arcade (2 points), *Shinshigai* Shopping Arcade
Number of samples	122 samples
Survey method	1. Samples drawn at random from visitors who visit survey points
	2. Interview survey with questionnaire sheets for about 15 min
Main questionnaire items	1. Personal profiles (residence, age, gender, occupation, etc.)
	2. Shop-around history (places visited, purposes done there, and expenditure there if any)
	3. Travel time to the city center of Kumamoto
	4. Transport means to the city center district of Kumamoto City
	5. Frequency of visits to the city center district of Kumamoto City
	6. Frequency of visits to various shops and shopping arcades within the city center district of Kumamoto City

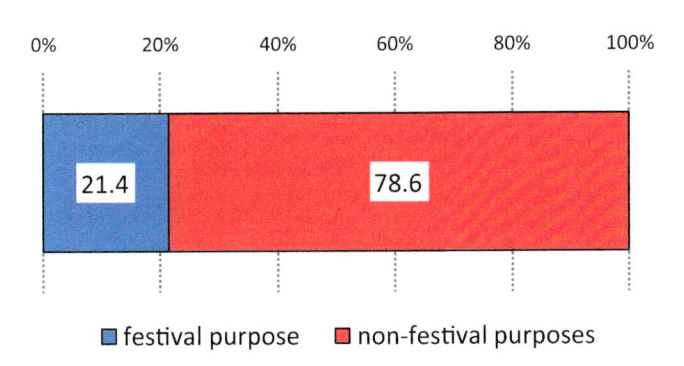

Fig. 15.1 Visitors with the festival purpose and with non-festival purposes

2.3 Method for the Analysis

In this survey, to measure the effect of the Kumamoto Castle Festival, we asked the respondents whether or not they came to the city center district of Kumamoto City with the purpose to attend the festival. Figure 15.1 shows the results of the questionnaire item on whether or not the respondents came to the city center district of Kumamoto City with the festival purpose. We see that 21.4% of the people visited the city center of Kumamoto City with the purpose to attend the Kumamoto Castle Festival.

In this study, we measure the effect of the Kumamoto Castle Festival by the following steps:

1. To estimate the net number of incoming visitors to the city center district of Kumamoto City on the same day of the Kumamoto Castle Festival, first we apply the consistent estimation method of consumer shop-around patterns for on-site survey data to the on-site consumer shop-around data obtained from the 2nd survey of consumer shop-around behaviors at the city center of Kumamoto City and get the consistent estimate of the density of consumer shop-around patterns at the city center district of Kumamoto City, which is defined as the probability distribution over all the observed shop-around routes (paths or trip chains). It should be noted that since we have obtained the probability distribution over all shop-around routes, we can also estimate the density of the origin-destination table by the standard aggregation procedure. Thus we can also obtain the density of the total number of visitors to every commercial site or zone division within the city center district.
2. Next we use the actual number of visitors to a department store on the survey day, which was kindly provided by the department store, divide it by the estimated density of the total number of visitors to that department store obtained in the previous step (1), and get the estimate of the net number of incoming visitors into the city center district of Kumamoto City. Since this is the method of expansion from the density to the actual number, it should also be noticed that the estimates of actual numbers of visitors to various commercial sites or divided zones within the city center district of Kumamoto City can be obtained at the same time.
3. We assume that the percentage of visitors with the festival purpose among the total number of visitors to the city center is equal to the percentage obtained from the shop-around survey data, that is, 21.4%. With this assumption we estimate the respective total numbers of visitors with festival purpose and with non-festival purposes.[3]
4. From the shop-around survey data, we calculate the visit ratios to zones, i.e., the choice probabilities to visit zones, respectively, by samples with the festival purpose and by those with non-festival purposes. From multiplying these respective visit ratios to zones by the total numbers of visitors with the festival purpose and with non-festival purposes estimated in the step (3), we obtain the estimates of the total numbers of visitors to each zone with the festival purpose and with non-festival purposes.
5. Furthermore, in order to measure the effect on the amount of expenditure, we use the following method. We assume that all the amount of expenditure spent in the city center district by the visitors who came to the city center district of Kumamoto City with the festival purpose is included in the effect of the Kumamoto Castle Festival on the expenditure. We estimate this effect by each zone from multiplying the total number of visitors to this zone with the festival purpose estimated in the step (4) by the average amount of expenditure per visitor to this zone obtained from the shop-around survey data. On the other hand, for those visitors who had not come with the festival purpose, we assume that their expenditure in the

---

[3] As this percentage of the festival purpose and the non-festival purposes, it would be more precise to use the percentage obtained from one sampling point. Here we use that from the total samples because of the small number of samples collected.

pedestrian paradise, the venue of the Kumamoto Castle Festival, would not have been spent if it were not for the Kumamoto Castle Festival. Thus we regard this as a part of the effect of the Kumamoto Castle Festival on expenditure and estimate it from multiplying the number of visitors to the venue with non-festival purposes by the amount of expenditure per visitor in the pedestrian paradise. By summing up these two, we measure the effect of the Kumamoto Castle Festival on the amount of expenditure at the city center district of Kumamoto City.

Here we would like to point out two points. First, it should be noted that the effect of this study is limited to the city center district of Kumamoto City, as it does not measure the effect on other areas and it differs from the concept of the net effect on society as a whole. Second, in this study we employed steps (3) and (4), but if we had much more survey samples, we could have applied the consistent estimation method to the shop-around data separately for the samples with the festival purpose and those with non-festival purposes. This is left for a future task.

2.4 Consistent Estimation Method

In this study, we apply the consistent estimation method of consumer shop-around pattern for on-site survey data to the on-site data obtained from the 2nd survey of consumer shop-around behavior at the city center of Kumamoto City. The consistent estimation method was developed by Saito et al. [3]. The method makes it possible to consistently estimate the density of shop-around patterns from the data obtained by the on-site sampled shop-around data while removing the choice-based sampling bias that inherently occurs in the on-site sampled data. The density of shop-around patterns is the probability distribution over all shop-around routes or paths. In other words, it is the distribution over all types of trip chains. The most innovative consequence of the method is that it gives a theoretical foundation to estimate the net number of incoming visitors to the area such as the city center district or the downtown. The reason is that since the method gives us the accurate probability over all shop-around routes, it also gives the density of the amount of visitors to a particular commercial facility when the density of the net number of incoming visitors equals 1. Usually the large retailers such as department stores are equipped with a people counting system, which automatically counts the actual number of people entering and exiting their buildings. Thus the consistent estimation method provides us with a theoretical method to estimate the net number of incoming visitors to the city center if we had the actual number of visitors to at least one commercial facility. In fact, the net number of incoming visitors to the city center can be obtained by using the usual expansion method which divides the actual number of visitors to the facility by the density of its visitors.

In this study, we apply the consistent estimation method to the data obtained from the 2nd survey of consumer shop-around behaviors at the city center of Kumamoto City to estimate the density of shop-around patterns within the city center. We also have the data of the actual number of visitors to one commercial facility. With these data, following the above procedure, we can obtain the net number of incoming visitors to the city center district of Kumamoto City.

The consistent estimation method is formalized as the following formula. This formula shows the algorithm of how to calculate the consistent estimate for the distribution over shop-around routes, $f_c(r)$ from the on-site shop-around behavior history data obtained where the subscript c indicates the distribution f is defined at on-site.[4]

$$\hat{f}_c(r) \propto \frac{\sum_{v>0}\sum_{s \in S}\sum_{t \in T} \delta_t^c(r,v,s)\frac{1}{l_S(r_t)}\frac{\tilde{f}_c(s_t|v_t)\tilde{f}_c(v_t)}{H(s_t)\tilde{f}_c(v_t|s_t)}}{\sum_{r \in R}\sum_{v>0}\sum_{s \in S}\sum_{t \in T} \delta_t^c(r,v,s)\frac{1}{l_S(r_t)}\frac{\tilde{f}_c(s_t|v_t)\tilde{f}_c(v_t)}{H(s_t)\tilde{f}_c(v_t|s_t)}}$$

Here, $f_c(r)$ denotes the distribution over shop-around routes $r \in R$, r denotes a shop-around route, v denotes the visit frequency, $s \in S$ denotes the survey point, $t \in T$ is the sample, and $l_s(r)$ denotes the length of the shop-around route r over the set of survey points S. $\delta_t^c(r,v,s)$ is the random variable that assumes 1 when sample t takes the route r and the visit frequency v, and is sampled at the survey point s, and 0 otherwise. $H(s)$ is the sampling ratio of each sampling point s.

3 The Effect of the Kumamoto Castle Festival from the Viewpoint of *Kaiyu* Behaviors

3.1 Estimating the Net Total Number of Incoming Visitors to the Whole City Center District of Kumamoto City

Here, we estimate the net number of incoming visitors to the city center of Kumamoto City by applying the consistent estimation method of shop-around patterns to the on-site shop-around data obtained from the 2nd survey of consumer shop-around behavior at the Kumamoto city center.

Specifically, first we estimate the density of shop-around patterns in the Kumamoto city center by applying the consistent estimation method of shop-around patterns. The *Tsuruya* Department Store, which is the long-established local department store, kindly provided the data of the actual number of its visitors. Based on the estimated result of probability distribution over shop-around routes and the actual number of visitors at one commercial facility, we can estimate the net number of incoming visitors to the whole city center district of Kumamoto City.

---

[4]Refer to Saito et al. [3] for details including the derivation of the equation.

According to the estimated result of the consistent estimation,[5] the density of the amount of total visitors to the *Tsuruya* Department Store turns out to be 0.153 when the total number of visitors to the city center of Kumamoto City including net incoming visitors and shop-around visitors is set to 1. Also, the number of visitors to *Tsuruya*, according to an interview with people involved with *Tsuruya*, was 26,250 (in this study, we use the data on the number of visitors to *Tsuruya* on September 28, 2008, the previous year when the first Kumamoto Castle Festival was held).

Furthermore, the average number of steps in the shop-around, that is, the average number of all shops visitors visited while they shopped around the city center, which is obtained from the data collected in the 2nd survey of consumer shop-around behaviors at the city center of Kumamoto City, turns out to be 2.78.

Based on the above, the net number of incoming visitors to the whole city center district of Kumamoto City is calculated as follows.[6]

The net number of incoming visitors to the whole city center district = (the actual number of visitors to *Tsuruya* Department Store ÷ the density of the amount of visitors to *Tsuruya*) ÷ the average number of all shops visitors visited.

It becomes as follows:

$$61,675 \text{ (people per day)} = (26,250 \text{ (people per day)} \div 0.153) \div 2.78$$

Also, the following holds.

The total number of visitors to the city center district of Kumamoto City (the net number of incoming visitors + the number of visitors visited by their shop-around) = the net number of incoming visitors × the average number of all shops visitors visited.

It becomes as follows:

$$171,456 \text{ (people per day)} = 61,675 \text{ (people per day)} \times 2.78$$

---

[5]The density calculation here sets the density of the total number of visitors counted at all facilities in the city center to 1, which is equal to the product of the net number of incoming visitors and the average number of all shops visitors visited.

[6]To avoid the explanation to become complicated, we did not go into details of the consistent estimation method we employed here. Actually, here we employed the modified consistent estimation method which ignores the order of sites visitors visited along their shop-around. This modified method is only concerned with the distribution of total number of visitors over all commercial facilities so that the density of the total number of visitors counted at all sites is set to 1.

3.2 Comparing Visit Ratios to Zones by Visit Purposes of Visitors

Figure 15.2 is a map of the city center district of Kumamoto City. The area depicted in the map is used as the area defined as the city center district for this study. Based on the zone divisions shown in the figure, we look at the visit ratios to these zones in the city center district by visitors with the festival purpose and by those with non-festival purposes from the data collected by the 2nd survey of consumer shop-around behaviors. Table 15.2 shows the result of these visit ratios to zones in the city center district for visitors by visit purposes of visitors, that is, by the festival purpose and non-festival purposes. Looking at the top five zones, it can be seen that when comparing visitors who came for the festival purpose with those who did not, the former spreads along the pedestrian paradise and arcades (Zones 3, 5, 8, 10, and 28). In contrast, the latter were concentrated on the pedestrian paradise and around the *Tsuruya* Department Store (Zones 5, 6, and 7).

3.3 Comparing the Number of Visitors to Zones by Visit Purposes of Visitors

Figure 15.3 shows the result of the number of visitors to zones in the city center district of Kumamoto City by visit purposes of visitors. This was calculated by multiplying the total numbers of visitors of respective visit purposes by the corresponding visit ratios to zones within the city center. From the figure, we can compare the differences in shop-around behaviors by visit purposes on a real number basis.

Among visitors who came for the festival purpose, 6900 people stopped by the pedestrian paradise zone, which was the highest number, followed by *Tsuruya* (Zone 5) with 4185 people. Also, *Ginza* Street (Zone 28) had high numbers.

On the other hand, among visitors who came for other purposes, 21,850 people stopped by the *Tsuruya* Department Store which was the most visited, followed by the pedestrian paradise zone with 14,845 people, and PARCO (Zone 6) with 13,579 people. In addition, a lot of visitors stopped by the surroundings of *Shinshigai* Arcade (Zone 12).

4 Measuring the Effect of the Kumamoto Castle Festival on Expenditure

4.1 Average Expenditure per Visit for Each Zone

In this section we estimate the effect of the Kumamoto Castle Festival from the viewpoint of expenditure. Table 15.3 shows the result of calculating the average

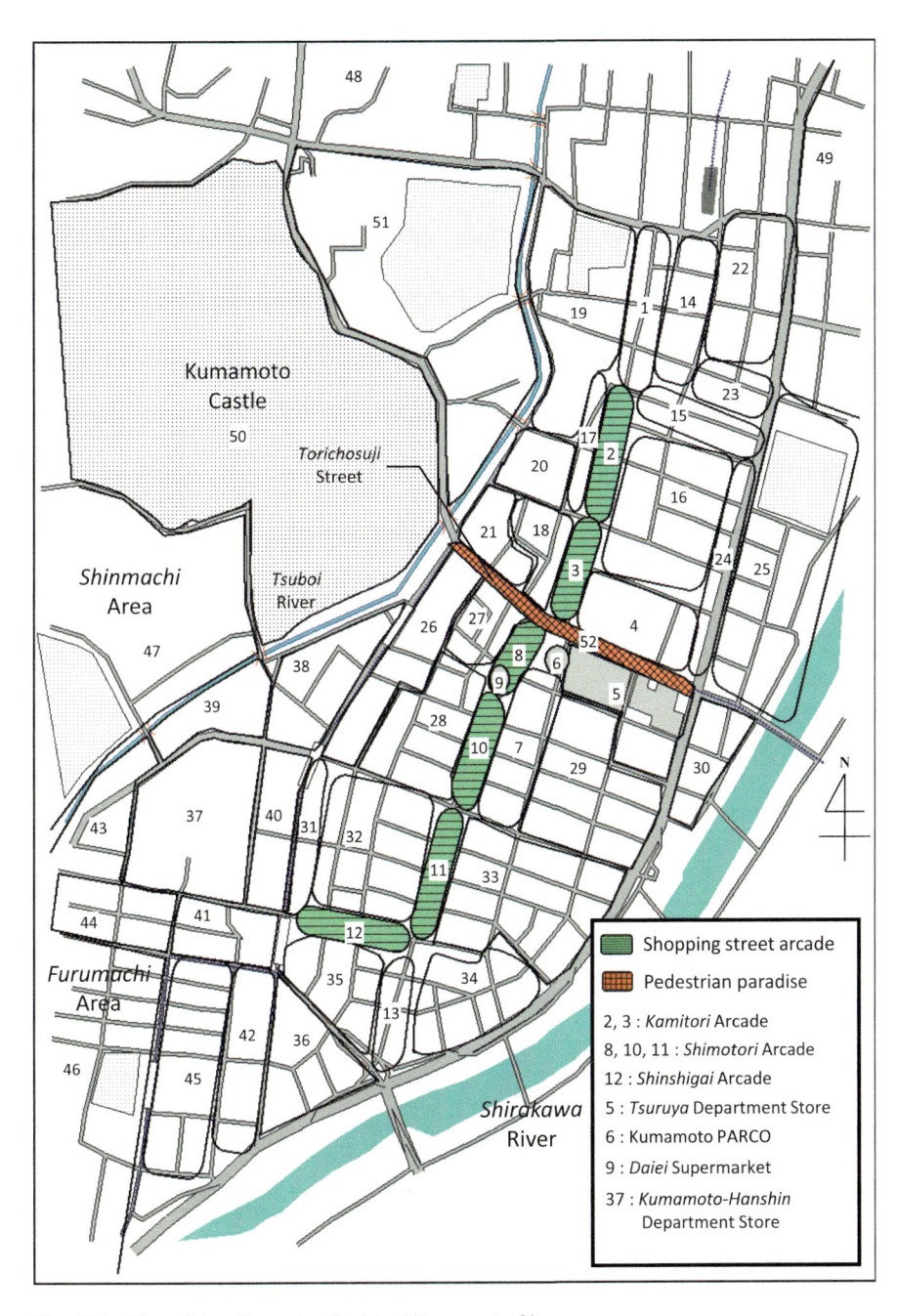

Fig. 15.2 Map of the city center district of Kumamoto City

Table 15.2 Visit ratios to zones for visitors with the festival purpose and with non-festival purposes

Zone no.	Festival purpose	Non-festival purposes
1	0.00	1.51
2 (*Kamitori* Arcade)	6.62	4.63
3 (*Kamitori* Arcade)	7.86	4.98
4	4.61	5.74
5 (*Tsuruya* Department)	11.39	16.22
6 (Kumamoto PARCO)	4.79	10.08
7	0.00	7.98
8 (*Shimotori* Arcade)	7.67	3.72
9 (*Daiei* Supermarket)	0.00	4.29
10 (*Shimotori* Arcade)	7.18	4.73
11 (*Shimotori* Arcade)	5.20	3.34
12 (*Shinshigai* Arcade)	3.89	6.86
13	0.00	0.53
14	0.00	0.70
15	0.00	1.68
16	0.89	0.11
17	0.00	0.16
18	0.00	2.86
19	0.00	0.18
21	0.00	0.42
25	0.00	0.53
27	0.00	0.32
28	8.32	0.00
31	0.00	0.40
32	0.27	1.13
33	5.09	0.00
34	0.18	0.47
35	0.00	1.79
37 (*Kumamoto-Hanshin* Department)	5.05	1.28
38	0.00	0.32
41	0.00	0.13
50	2.22	1.55
51	0.00	0.32
52 (Pedestrian paradise)	18.78	11.02
Total	100.00	100.00

expenditure per visit for each zone within the city center district of Kumamoto City based on the data obtained from the 2nd survey of consumer shop-around behaviors. Note that the notion of the average expenditure per visit is different from the concept of customer-based expenditure. For example, suppose that one customer visited shopping sites, A and B, located in the same zone X and spent 2000 yen and 3000

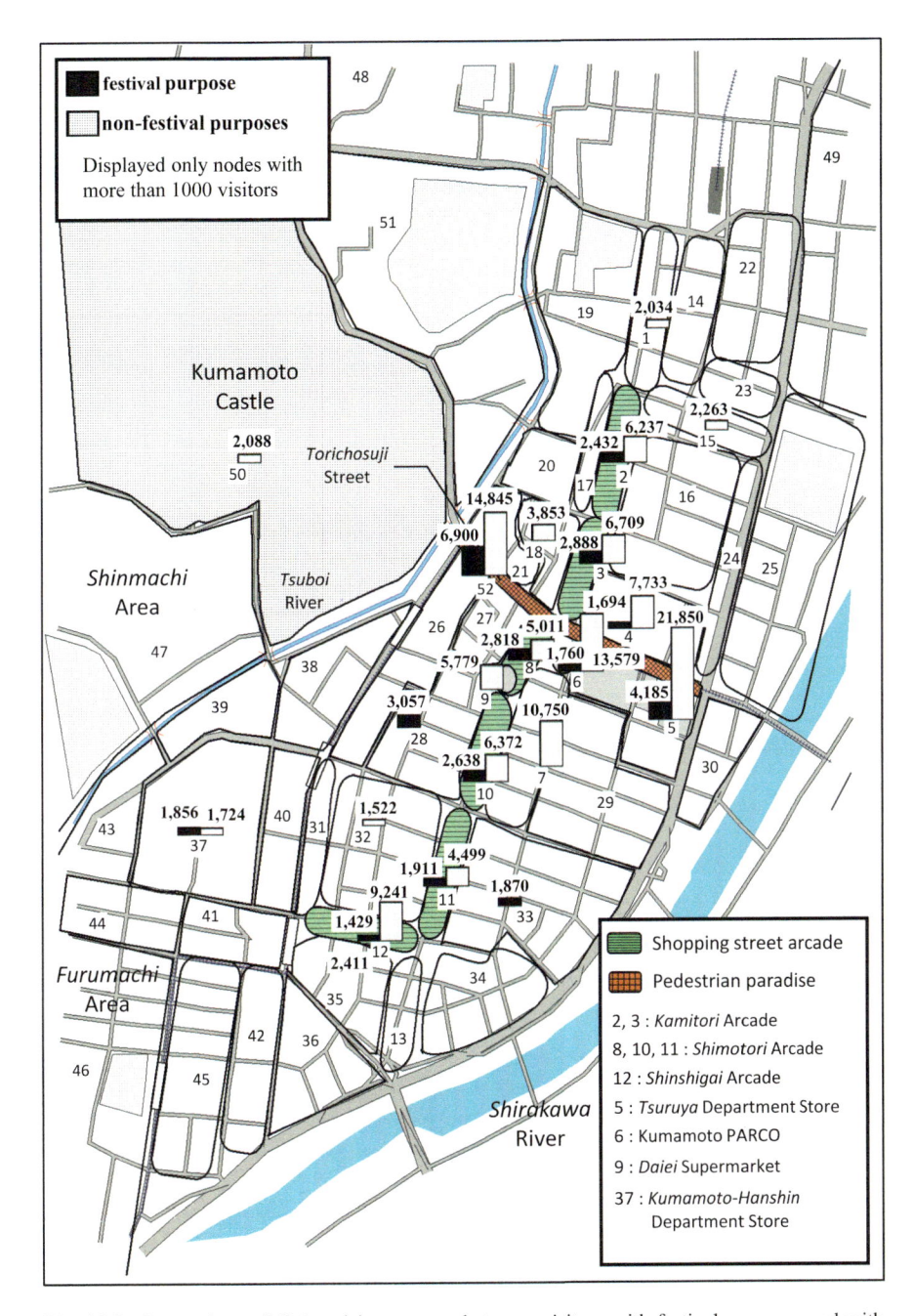

Fig. 15.3 Comparison of *Kaiyu* visits at zones between visitors with festival purposes and with non-festival purposes

Table 15.3 Average expenditure by zone per step

Zone no.	Average expenditure by zone per visit (unit: yen)
1	380
2 (*Kamitori* Arcade)	1084
3 (*Kamitori* Arcade)	1396
4	944
5 (*Tsuruya* Department)	2492
6 (Kumamoto PARCO)	1633
7	864
8 (*Shimotori* Arcade)	1456
9 (*Daiei* Supermarket)	500
10 (*Shimotori* Arcade)	2367
11 (*Shimotori* Arcade)	1211
12 (*Shinshigai* Arcade)	2939
13	3000
14	0
15	1600
16	2040
17	400
18	2660
19	600
21	225
25	0
27	0
28	2900
31	5000
32	2000
33	0
34	650
35	3100
37 (*Kumamoto-Hanshin* Department)	1356
38	1500
41	0
50	0
51	0
52 (Pedestrian paradise)	91

yen, respectively, at A and B. The total expenditure for this customer is 5000 yen. But the average expenditure per visit is 2500 yen per visit since this customer drops two visits to zone X. Here for the calculation of the average expenditure per visit, the visit with no expenditure is included.

In the next section, using the average expenditure per visit for each zone, we estimate the effect of the Kumamoto Castle Festival on expenditure.

4.2 The Effect of the Kumamoto Castle Festival on Expenditure

First, we measure the effect of the Kumamoto Castle Festival from the viewpoint of the amount of money spent by visitors who came for the festival purpose. Table 15.4 shows the result of the measurement. The effect of the Kumamoto Castle Festival on expenditure contributed by visitors with the festival purpose is the sum of the amount of money visitors with the festival purpose spent within the Kumamoto city center. The expenditures in each zone are calculated by multiplying the number of visitors with the festival purpose to that zone by the average expenditure per visit obtained in the previous section. As shown in the table, the total of the expenditure visitors with the festival purpose spent in the city center district of Kumamoto City turns out to be 51,320,000 yen.

On the other hand, the amount of money visitors with non-festival purposes spent at the pedestrian paradise, the venue of the Kumamoto Castle Festival, which should be thought of as a part of the monetary effect by the festival according to the above discussion, was calculated in the following way.

The amount of money spent by visitors with non-festival purposes at the festival venue = the number of visitors with non-festival purposes who visited to the pedestrian paradise × the average expenditure per visit at the pedestrian paradise.

Therefore, the effect contributed by visitors with non-festival purposes turns out as follows:

$$1,350,000 \text{ yen} = 14,845 \text{ visits} \times 91 \text{ yen per visit}$$

From the above, the total effect of the Kumamoto Castle Festival on expenditure turns out to be 52,670,000 yen (or 526,700 dollars).

$$52,670,000 \text{ yen} = 51,320,000 \text{ yen} + 1,350,000 \text{ yen}$$

5 Conclusion and Future Challenges

The most significant contribution of this study is to provide a methodology of how we theoretically estimate the net number of incoming visitors attracted by some event and how we measure the effect of the event which is possibly spread over other areas than that of the event venue, both of which are based on the micro-behavioral data obtained from the on-site survey of consumer shop-around behaviors. Thus we have presented a coherent methodology to evaluate the effects of attractions and events often adopted for revitalizing towns from the viewpoint of consumer *Kaiyu* behaviors with a typical example of the Kumamoto Castle Festival, in which the effects spread over other spots within the city center are shown to be tracked from the

Table 15.4 Expenditure by visitors with festival purpose spent at the city center district of Kumamoto City

Zone no.	The number of visitors who came for the festival purpose (unit: visits)	Average expenditure by zone per step (unit: yen)	Expenditure by visitors with festival purpose (unit: yen)
1	0	380	0
2 (*Kamitori* Arcade)	2432	1084	2,636,288
3 (*Kamitori* Arcade)	2888	1396	4,030,252
4	1694	944	1,596,499
5 (*Tsuruya* Department)	4185	2492	10,424,835
6 (Kumamoto PARCO)	1760	1633	2,872,447
7	0	864	0
8 (*Shimotori* Arcade)	2818	1456	4,103,008
9 (*Daiei* Supermarket)	0	500	0
10 (*Shimotori* Arcade)	2638	2367	6,241,508
11 (*Shimotori* Arcade)	1911	1211	2,311,100
12 (*Shinshigai* Arcade)	1429	2939	4,199,831
13	0	3000	0
14	0	0	0
15	0	1600	0
16	327	2040	667,080
17	0	400	0
18	0	2660	0
19	0	600	0
21	0	225	0
25	0	0	0
27	0	0	0
28	3057	2900	8,865,300
31	0	5000	0
32	99	2000	198,000
33	1870	0	0
34	66	650	42,900
35	0	3100	0
37 (*Kumamoto-Hanshin* Department)	1856	1356	2,513,525
38	0	1500	0
41	0	0	0
50	816	0	0
51	0	0	0
52 (Pedestrian paradise)	6900	91	621,000
Total			51,323,573

Kaiyu behaviors of the event participants both on the number of visitors basis and on monetary basis.

While this study might be said as a pilot study because of its small size of the samples, we see that visitors who came for the festival purpose mainly moved around the arcades near the festival venue and the effect of the Kumamoto Castle Festival on expenditure is roughly estimated as 52,670,000 yen.

Our future work would be for anything to find another event opportunity where we can expand the number of samples and perform a separate consistent estimation for the *Kaiyu* behaviors of event participants. Another future research topic would be to compare the cost and effect of holding the event such as the Kumamoto Castle Festival.

References

1. Saito S, Sato T, Yamashiro K (2010) Consistent estimation of shop-around behavior and measurement of the event effects spread over city center: an application to the measurement of the effect of Kumamoto Castle Festival, vol 26. Papers of the 26th annual meeting of The Japan Association for Real Estate Sciences. pp 175–182. (in Japanese)
2. Saito S (2010) Let's start urban development with accurately measuring people's flows: consumer-oriented planning and town equity. Mon Real Estate Distrib 8–9. 2010/04/. (in Japanese)
3. Saito S, Nakashima T, Kakoi M (2001) The consistent OD estimation for on-site person trip survey. Stud Reg Sci 31:191–208. (in Japanese)
4. Saito S, Nakashima T (2003) An application of the consistent OD estimation for on-site person trip survey: estimating the shop-around pattern of consumers at Daimyo district of Fukuoka City, Japan. Stud Reg Sci 33:173–203. (in Japanese)
5. Saito S (2008) Algorithm patent: method to accurately estimate shop-around pattern. Res Fukuoka Univ Cent Res Inst News Rep 13:1–2. (in Japanese)
6. Saito S, Kakoi M, Nakashima T (1999) On-site Poisson regression modeling for forecasting the number of visitors to city center retail environment and its evaluation. Stud Reg Sci 29:55–74. (in Japanese)
7. Saito S, Sato T, Yamashiro K (2010) Measuring event effects spread over inside city center based on consumer behavior: the effects of Kumamoto castle festival. Paper presented at the 47th annual meeting of Japan Section of Regional Science Association International (JSRSAI). (in Japanese)
8. Saito S, Kakoi M, Nakashima T, Igarashi Y, Kiguchi T (2008) A consumer behavior approach to estimating the economic effects of open cafes at city center retail district: how further do those open cafes accelerate the shop-around behavior of their customers? Fukuoka Univ Rev Econ 52:435–458. (in Japanese)
9. Saito S, Yamashiro K (2000) Economic impacts of the downtown one-dollar circuit bus estimated from consumer's shop-around behavior. Stud Reg Sci 31(1):57–75. (in Japanese)

Part V
Kaiyu Marketing and Value of Visit to City Center

Chapter 16
Did the Grand Renewal Opening of Department Store Enhance the Visit Value of Customers?

Saburo Saito, Kosuke Yamashiro, and Masakuni Iwami

Abstract In March 2004, Iwataya, which is a long-established local department store in Tenjin district at Fukuoka City, Japan, with lots of attachment from locals, made its grand renewal opening of its department store business with merged new two stores. The grand renewal opening this time is a drastic change not only in its physical renovation but also in its marketing strategies for reconfiguring its target customers. Thus it can be thought of as a social experiment to analyze what kinds of customers showed what types of responses and how they changed their purchasing behaviors and why. The purpose of this study is that, taking advantage of this opportunity, we investigate how the customers for Iwataya had evaluated its renewal opening of the New Iwataya and how they changed their purchasing behaviors at the New Iwataya. Specifically, we hypothesize that the drastic renewal change like Iwataya can bring to the customers the purchasing risk in the sense that while they surely could purchase the goods they wanted at the previous store, they might not be possible to find or purchase the goods they wanted at the new renewal shop. By focusing on consumers' perception on their purchasing risks at the old and the new

---

This chapter is based on the paper of Saburo Saito, Takaaki Nakashima, and Fumi Tsurumi [1], "Customers' Evaluation and Responses to the Relocated Renewal Opening of a Long-Established Local Department Store: A Panel Analysis of the Retrospective Micro-behavioral Data," presented at the 43rd Annual Meeting of Japan Section of Regional Science Association International (JSRSAI), 2006, which is modified for this chapter.

S. Saito (✉)
Faculty of Economics, Fukuoka University, Fukuoka, Japan

Fukuoka University Institute of Quantitative Behavioral Informatics for City and Space Economy (FQBIC), Fukuoka, Japan
e-mail: saito@fukuoka-u.ac.jp

K. Yamashiro
Department of Business and Economics, Nippon Bunri University, Oita City, Japan
e-mail: yamashiroks@nbu.ac.jp

M. Iwami
Fukuoka University Institute of Quantitative Behavioral Informatics for City and Space Economy (FQBIC), Fukuoka, Japan
e-mail: miwami@econ.fukuoka-u.ac.jp

Iwataya, we investigate how the change in the purpose realization rate for each customer has affected the changes in the visit frequency and expenditure for that customer at the new Iwataya. To do this, we need a panel data, so in the questionnaire survey conducted after the renewal opening, we devised the question items which ask the respondents about their behaviors before the opening as well as the behaviors at present and obtained the retrospective panel data for each sample. Using the retrospective panel data, we demonstrated the empirical facts that the changes in the purpose realization rate of the store perceived by consumers are strongly related to the changes in their visit frequency to the store and their expenditure at that store.

Keywords Renewal of department store · Gentrification strategy · Purpose realization rate · Purchasing risk · Iwataya · Fukuoka · Division ratio cross tabulation · Retrospective panel data · *Kaiyu*

1 Purpose

1.1 Background and Aim of This Study

Currently, as the Tenjin district at Fukuoka City, Japan, is said in the fourth distribution war, the opening of commercial facilities, such as the grand relocated renewal opening of Iwataya, a long-established local department store (Cf. [2]); Mina Tenjin opening, a renovated commercial complex located at the northern part of Tenjin district (Cf. [3]); etc., is continuing one after another. The Iwataya grand renewal opening was in March 2, 2004, and Mina Tenjin opened in October 29, 2005. In particular, the grand renewal opening of Iwataya has attracted much attention from local people since it is one of the two local department stores which has a lot of attachment from locals and its renewal opening had been symbolized as a landmark project for the management reconstruction from its bankruptcy in 2002.

Up to this opening, Iwataya had twists and turns. At the time when Tenjin district had a large-scale retail redevelopment from 1996 to 1998,[1] which coincided with the terminal building redevelopment of Fukuoka (Tenjin) Station of NNR (Nishi-Nippon Railroad Co.), NNR proposed Iwataya to enter into the north part of the new terminal building, which is situated in the middle between the former Iwataya main store[2] and the south part of the new terminal building and would serve as connecting the Iwataya main store and the south part of the new terminal building, in which a nationwide department store was expected to enter. Iwataya rejected NNR's proposal and opened a new department store, Z-side[3], in 1996. However, Iwataya fell into the first management difficulty in 1999 soon after establishing Z-side

---

[1] Refer to Chap. 2 of this book about the history and circumstances of this large retail redevelopment and its consequences on the consumer behaviors.

[2] The former Iwataya main store had the shop floor space of 32,546 m^2.

[3] The shop floor space of Z-side was 33,908 m^2.

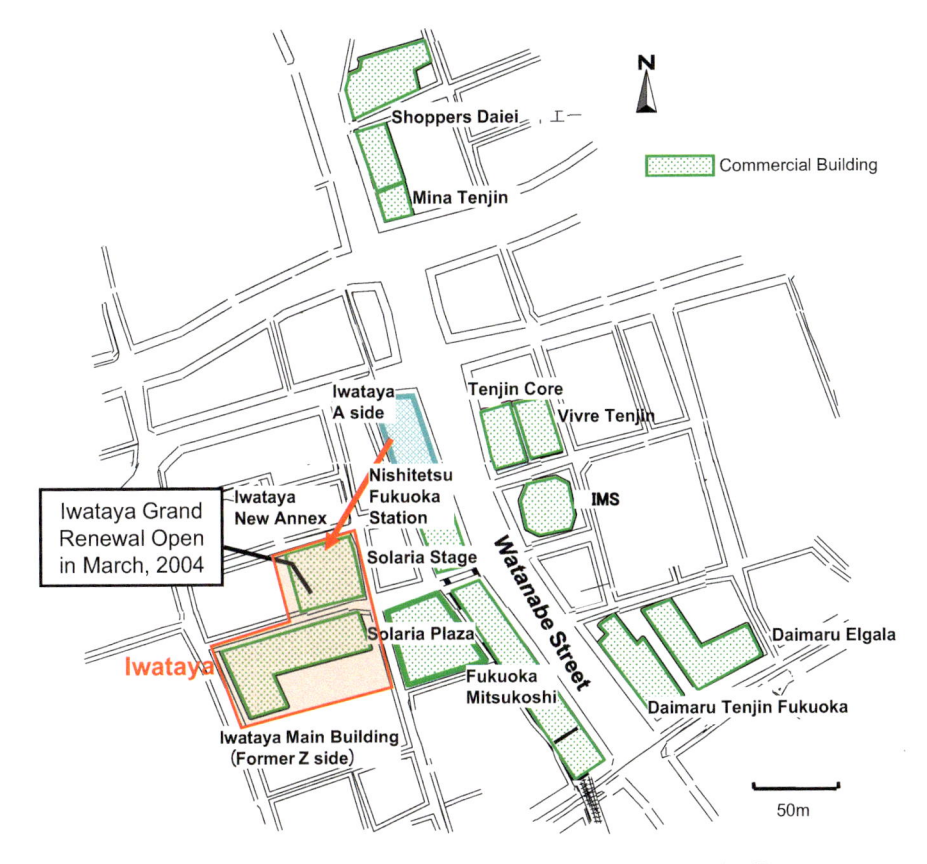

Fig. 16.1 Map of Iwataya grand renewal opening in the city center of Fukuoka City

allegedly because of its introduction of too innovative merchandizing system at that time and the change of their target of customers from the middle-aged to older customers into younger generations. Also raised as another reason is the severe competitions with the other two department stores, Hakata Daimaru Fukuoka Tenjin store and Mitsukoshi. The former expanded its shop floor space by about 20,000 m^2 in 1997, and the latter is the newly opened department store which entered in NNR's redeveloped terminal building with 38,000 m^2 of shop floor space in 1997.

While Iwataya were trying to reconstruct its management by itself, the second collapse occurred in 2002. Since the first collapse, the local business circles made a syndicate to embark on support by purchasing the NHK site neighboring to Z-side to enable Iwataya to implement its own reconstruction plan which is to construct a new building at this site, make Z-side into the main store, make the new building into a new store, and run its department store business with two stores merged. The above is the rough history and circumstances about the process up to the grand renewal opening of Iwataya. (See Fig. 16.1 for the map of Iwataya grand renewal opening in the Tenjin district.)

Along with the process up to its rebirth, we note that Iwataya had been trying to change its management strategy from the traditional marketing style for typical local department stores to a new style by implementing various strategies for attracting consumers such as restructuring its merchandise assortment and reconfiguring its target customers.

Speaking of how it resulted, the grand relocated renewal opening of Iwataya had been enthusiastically welcomed by locals with a fever of the opening of a new department store. And the turnover of the first month after the grand relocated renewal opening turned out to be a 12% increase from the previous year's record.

1.2 Purpose of This Study

From this case of Iwataya, we learn that not only the changes in the physical structure of commercial facilities but also the changes in the management strategies of commercial facilities affect purchasing behaviors of consumers who visit these commercial facilities. If there were no change in consumers' tastes, the effect of the changes in the physical and management strategies of commercial facilities must be reflected directly as the changes in consumers' purchasing behaviors. The cause of the changes in consumers' behaviors can be seen in the physical and management strategies. In a wider sense, the marketing strategy of one department store would change the behaviors of its customers, consequently would change *Kaiyu* behaviors of visitors, and thus would cause the change of the town.

In particular, the grand relocated renewal opening of Iwataya this time can be thought of as a social experiment to analyze what kinds of customers showed what types of responses and how they changed their purchasing behaviors and why, since this grand renewal was a very drastic renovation.

Thus the aim of this research is that, taking advantage of this opportunity of the relocated renewal opening of Iwataya and using consumer micro-behavioral data, we explore how the customers to Iwataya had evaluated its renewal opening of the New Iwataya and why they had changed their purchasing behaviors at the New Iwataya.

Two new features can be pointed out as the contribution of this study. The first is the data we devised to collect in our questionnaire survey, which we call the retrospective panel data. This point will be discussed in details later. The second is our hypothesis about why various customers responded differently to the same renewal opening of Iwataya. Consumers are assumed to choose their destination store to maximize their utility to shop at the destination store. We conceptualize this utility as their visit value to their destination. Furthermore, in this study, we hypothesize that consumers' visit value depends on whether or not their shopping purpose is realized, that is, whether or not they could purchase the goods they wanted.

From this viewpoint, the grand relocated renewal opening of Iwataya is interpreted as follows. The drastic renewal change of a long-established department store with lots of attachment from locals like Iwataya can be said to bring to the customers the purchasing risk in the sense that while they surely could purchase the goods they wanted at the previous store, they might not be possible to find or purchase the goods they wanted or sought at the new renewal shop. We generalize this conceptual framework. In general, if consumers visited a shop and failed to realize their shopping purpose and this continues to happen, their visit value to the shop decreases, and they will decrease their visit frequency and expenditure at the shop.

From the above, therefore, by focusing on consumers' perception on their purchasing risks at the old and the new Iwataya, the purpose of this study is to explore how the purpose realization rate changed from the old to the new Iwataya for each individual customer and to investigate how this change of the rate has affected the changes in the visit frequency and expenditure for that individual customer at the old and new Iwataya.[4]

2 Framework of the Analysis

2.1 *Data Used*

Outline of the 9th (2004) Survey of Consumer Shop-Around Behavior at the City Center of Fukuoka City

In this study, we use the data obtained from the 9th survey of consumer shop-around behaviors at city center of Fukuoka City conducted by the Fukuoka University Institute of Quantitative Behavioral Informatics for City and Space Economy (FQBIC) on July 10 (Saturday) and 11 (Sunday), 2004. Multiple survey points were set up at the city center district at Fukuoka City. The survey is an interview survey with questionnaire sheets taking about 15 min. The respondents are sampled at random from the visitors at survey points and asked about their shop-around history up to the survey point sampled and about their plan to shop around from there as well as their residence, transport means, and travel time and cost to the city center and so on. Their shop-around history is asked which place they visited, for what purpose, and how much they spent there if they spent any and is recorded as a sequence of the triples, each of which is composed of place visited, purpose done there, and money spent there, in the order of occurrence.

The main survey items include shop-around behavior history on the day of the survey, personal attributes of respondents, frequency of visits to the city center district of Fukuoka City, purchasing attitudes, purchasing behavior in the new and the old Iwataya, and brands purchased in Iwataya.

For the 9th survey of consumer shop-around behavior at the city center of Fukuoka City, we set up a total of ten survey points within the city center district:

---

[4]This study started from the following thesis research [4, 5].

six points at Tenjin district, two points at Hakata Riverain, one point at Canal City Hakata, and one point at Hakata Station. A total of 443 valid samples were collected.

Characteristics of Data Used: Retrospective Panel Data

We have been conducting the on-site survey of consumer shop-around behaviors at city center of Fukuoka City every year since 1996. However, each data obtained from different year constitutes a separate sample. Thus while these samples are obtained in different years, the interannual comparison between these samples just implies that it only compares different samples at different points of time. In other words, it is just the same as the comparison of different groups at different times or in different regions. A usual method of comparing these separate samples is the comparison of averages of separate samples at two points of time. Even if there is a difference in these averages at two time points, it never clarifies how each individual changes behaviors between two points of time. With respect to the renewal of Iwataya, we are concerned with how each customer responds to the renewal. We would like to know how individual customers have changed their behaviors when Iwataya changed its store from the old to the renewal one. For this purpose we need a panel data where the sample was sampled at different points of time, but it takes much cost.

In this study, while avoiding much cost, to obtain panel data, we devised question items in the survey after the renewal opening of Iwataya, which ask the respondents about their behaviors before the renewal opening. Hence, as of the survey day after the renewal opening, the respondents answer their present purchasing behaviors after the renewal opening and also answer, by recalling, their purchasing behaviors before the renewal opening. Thus we have collected respondents' panel data of purchasing behaviors at two points of time before and after the renewal opening of Iwataya.

In general, this panel data is characterized as follows. Standing at the present time when the policy, the renewal opening in this study, is implemented, the respondents give their present behaviors after the policy is implemented while giving their past behavior before the policy is implemented by recollecting. So we call this type of panel data as the retrospective panel data.

Utilizing the advantage of the retrospective panel data, we analyze what type of customers enhance their visit value to the Iwataya after the renewal opening.

2.2 Procedure for the Analysis

Framework for Analyzing Changes in Purchasing Behaviors Before and After Iwataya Renewal

This study analyzes how individual consumers change their purchasing behaviors before and after the relocated renewal opening of Iwataya.

16 Did the Grand Renewal Opening of Department Store Enhance the Visit... 323

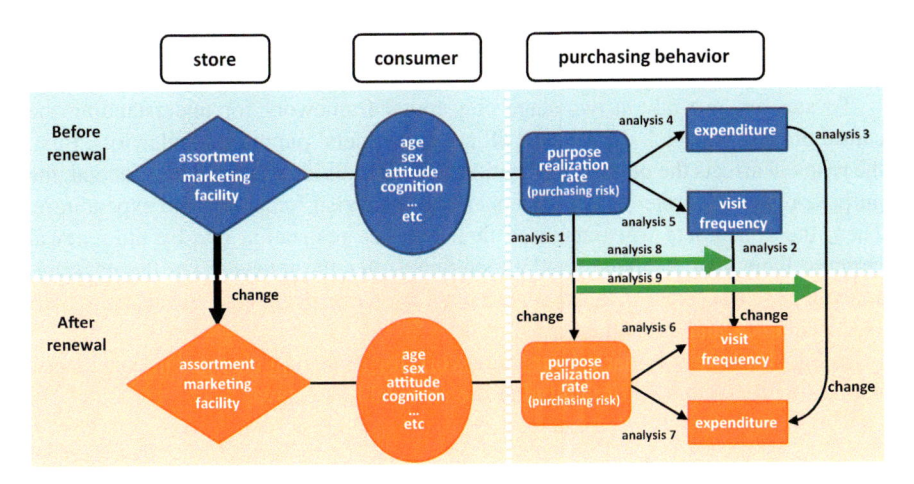

Fig. 16.2 Framework for analyzing changes in purchasing behaviors before and after the Iwataya renewal

We consider that purchasing behaviors of consumers at stores are outcomes from the interactions between stores and consumers. What kind of outcomes are resulted in depends on how consumers interacted with stores. The interactions between stores and consumers are determined by the attributes of stores and consumers. The attributes of stores include the product lineup, marketing strategy, physical facilities, and so on. On the other hand, consumers' attributes include gender, age, preferences, attitudes of purchasing, their cognition and familiarity with stores, and so on. Among the outcomes from the interactions between stores and consumers, we focus on the purpose realization rate of consumers as the most influential factor for purchasing behaviors of consumers. We hypothesize that it is their purpose realization rate that determines their purchasing behaviors such as the frequency of visits and their expenditure. This stream of causal flows is depicted in the upper and lower parts of Fig. 16.2.

When a store carries out its renewal, its product lineup, marketing strategy, and facility aspects are changed. The renewal of the store also brings about changes in the interaction between the store and consumers. For example, changes in tangible aspects of the facility might change the image consumers have about the store after its renewal. As a result, through the change of their perception of the store like their image of the store, we consider the renewal would finally lead to the changes in their purchasing behaviors.

To understand the process from consumers' perceptual changes leading to the changes in their visit frequency and expenditure, the leveraging concept, we regard, is the purchasing risk for consumers. For example, some consumers who can always purchase the goods they wanted at the store may now become unable to do so after its renewal. Such an increase in the purchasing risk for the consumers will appear as a change in their visit frequency and expenditure.

In this study, we measure the purchasing risks individual consumers have at commercial facilities by their perceived purpose realization rates at those facilities.

To sum up, we take a two-stage conceptual framework for understanding the causal process in which the renewal affects consumers' purchasing behaviors. First, the renewal affects the purpose realization rate for individual consumers. Second, the purpose realization rate for consumers affects their visit frequency and expenditure. Then, the consumers who increased their purpose realization rate by the renewal increase their visit frequency and expenditure after the renewal. On the contrary, consumers who decreased their purpose realization rate by the renewal decrease their visit frequency and expenditure.

In Fig. 16.2 we summarize our framework for analyzing changes in consumers' purchasing behaviors before and after the renewal of Iwataya.

Analyses to Be Done

In Fig. 16.2, we number the analyses that will be done in the following sections. The number of the analysis is attached to an arrow. Analyses 1–3 attaching to the arrows from before to after the renewal correspond to the analyses comparing the purpose realization rates, visit frequency, and expenditure between two points of time, before and after the renewal, respectively. They are panel data analyses. On the other hand, analyses 4 and 5 are cross-sectional analyses of the purpose realization rate with visit frequency and expenditure, respectively, before the renewal. Analyses 6 and 7 are those after the renewal. Our goal of the analyses in this study is to investigate whether there are significant relations between the change in the purpose realization rate and the changes in the visit frequency and expenditure. Note that arrows for analyses 8 and 9 are drawn between the arrows directed from before to after the renewal. From this, we have known that we must analyze the relation between the changes of two variables. More specifically, we must investigate how the change in the purpose realization rate affects the changes in visit frequency and in expenditure.

To say more concretely in this case, we investigate whether or not the individual customers who increased their purpose realization rate than before the renewal have increased their frequency of visits to Iwataya and their expenditure at Iwataya than before the renewal. To do this, the retrospective panel data stated above becomes indispensable.

Measurements of Purpose Realization Rate, Frequency of Visits, and Expenditure per Month

In the 9th survey of consumer shop-around behavior at city center of Fukuoka City, we set the following three question items to ask the respondents about their purchasing behaviors at Iwataya. Since Iwataya continues operating under two stores before and after the renewal, the three questions were asked of the respondents about each of the four stores, that is, for the old Iwataya before the renewal, A-side and

Z-side, and for the New Iwataya after the renewal, the main store (the former Z-side) and the new Annex store.

(i) Purpose realization rate: To measure the perceived purpose realization rate of consumers, we asked the respondents how many times out of ten times they can buy the goods they wanted at each of the four stores above. The question item is as follows. "Out of ten times you suppose to visit the store, how many times do you think you were able to buy the goods you wanted?"
(ii) Visit frequency: We asked the respondents about their frequency of visits to each of the four shops above for shopping, leisure, and dining purposes.
(iii) Average expenditure per visit: We asked the respondents about the average expenditure per visit at each of the four shops above.

In this study, we use 231 samples in the whole sample of the 9th survey, who answered at least the above three question items.

Integrated Measures for the Old and the New Iwataya

For each sample, we have obtained the purpose realization rates for four stores before and after the renewal. In order to simplify the comparison of purchasing behaviors between the old and the new Iwataya, we integrate two measures for two stores of the old Iwataya into the one measure for the old Iwataya. As for the new Iwataya, we do the same.

(i) *Purpose realization rate for the old and the new Iwataya*
   The purpose realization rate for each consumer at the old Iwataya is calculated as follows:

$$\text{Purpose realization rate} = \frac{f_1}{f_1 + f_2} r_1 + \frac{f_2}{f_1 + f_2} r_2$$

where

f_1: Visit frequency to the main store (A-side) of the old Iwataya
f_2: Visit frequency to Z-side of the old Iwataya
r_1: Purpose realization rate at the main store (A-side) of the old Iwataya
r_2: Purpose realization rate at Z-side of the old Iwataya

The purpose realization rate for the new Iwataya is calculated in the same way.

(ii) *Visit frequency to the old and the new Iwataya*
   The frequency of visits to the old Iwataya is defined as the sum of the visit frequency to the A-side and that to the Z-side of the old Iwataya. The visit frequency to the new Iwataya is calculated in the same way.

(iii) *Expenditure per month at the old and the new Iwataya*

The expenditure per visit at two stores of the old Iwataya are integrated and converted into a per month amount. The calculation is done as follows:

$$\text{Average expenditure per month} = f_1 M_1 + f_2 M_2$$

where

f_1: Visit frequency per month to the A-side of the old Iwataya
M_1: Average expenditure per visit at the A-side of the old Iwataya
f_2: Visit frequency per month to the Z-side of the old Iwataya
M_2: Average expenditure per visit at the Z-side of the old Iwataya

The average expenditure per month at the new Iwataya is calculated in the same way.

3 Changes in Purpose Realization Rate, Visit Frequency, and Expenditure per Month Before and After the Renewal

In this section we perform analyses 1–3 shown in the diagram in Fig. 16.2. We take up the age groups as consumer's attribute and see how different age groups differ in their responses to the renewal of Iwataya with respect to their purpose realization rate, visit frequency, and expenditure per month. The analyses are carried out by displaying the cumulative distribution functions (CDF) of before and after the renewal by age groups for each of the above three variables.

While it seems to be well known, for confirmation, we explain here how to read cumulative distribution functions.[5] Look at Fig. 16.3. For example, it displays two cumulative distribution functions of the purpose realization rate for shop A and B. The dark line corresponds to shop A and the bright to shop B. Cumulative distribution function is a function of X axis and the value of the function at x equals the percentage of samples whose value is less than x.

In Fig. 16.3 customers whose purpose realization rates are less than 0.4 are 80% for shop A and 65% for shop B. Hence, cases where the number of times that consumers can buy their desired products four out of ten times or less is 80% at shop A, while it is 65% at shop B. Thus, more customers can buy their desired products in shop B than in shop A. Note that the cumulative distribution function for shop A takes higher values than that for shop B on all values of x. In general, for such a case like this, it can be shown that the average of purpose realization rate for shop A is smaller than that for shop B.

---

[5]A cumulative distribution function F for a random variable X is defined as $F(x) = \Pr(X \leq x)$.

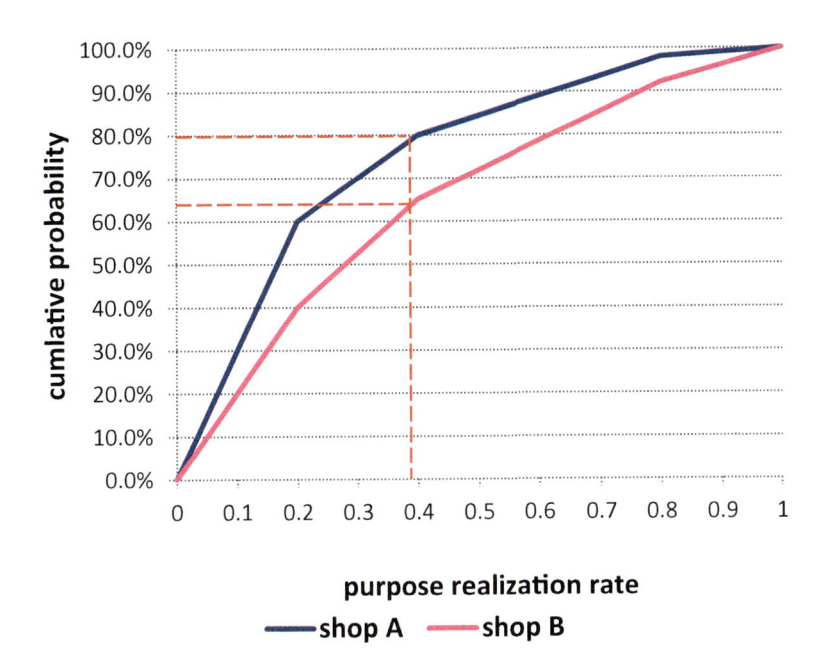

Fig. 16.3 Property of cumulative distribution function

3.1 Changes in Purpose Realization Rate Before and After the Renewal by Age Groups

For each of the seven age groups, we compare the two cumulative distribution functions of the purpose realization rate corresponding to before and after the renewal, that is, the old and the new Iwataya, to see how the purpose realization rate changed by the renewal of Iwataya for separate age groups. The results are shown in Fig. 16.4a–g. From these figures, almost all age groups except 25–29 age group clearly show decreases of the purpose realization rate after the renewal.

In particular, quite a large deviation between the purpose realization rates of the old and the new Iwataya can be seen for the age group of 50–59 and for that of older than 60. In both cases, the cumulative distribution function of purpose realization rate for the new Iwataya drastically shifted upward from the old one, which means drastic drops in purpose realization rate after the renewal.

In fact, if we note the intercepts of the old and the new Iwataya for the age group of older than 60, the intercept for the new turns out about 35% whereas that for the old Iwataya was 8%. Since the intercept can be interpreted as the percentage of the samples whose purpose realization rate is zero, this result symbolically implies that while before the renewal the probability that customers older than 60 could not purchase the goods they wanted was 8%, it rises to 35% after the renewal.

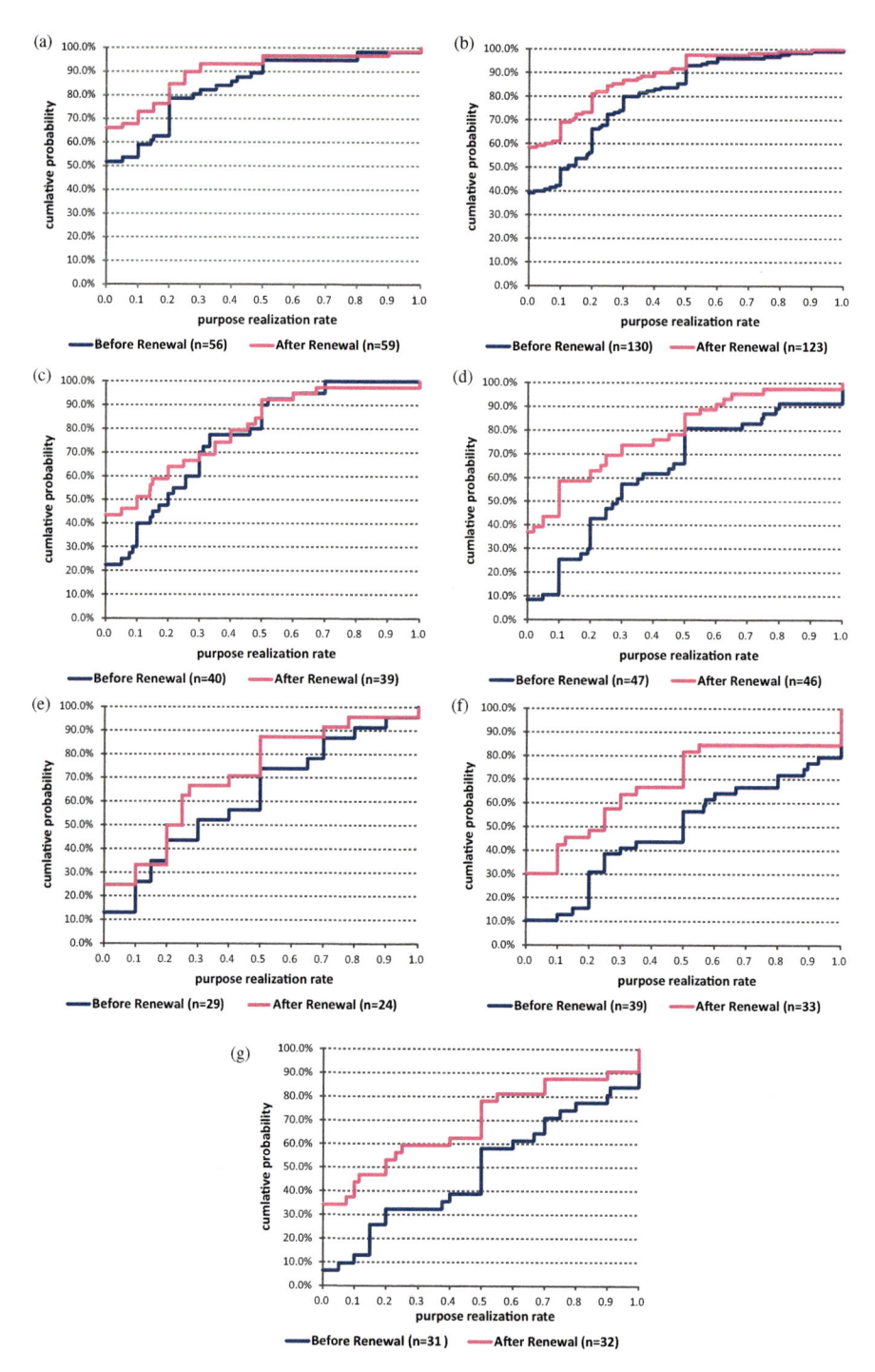

Fig. 16.4 Purpose realization rate before and after the renewal by age groups (**a**) Teenagers, (**b**) 20–24 years old, (**c**) 25–29 years old, (**d**) 30–39 years old, (**e**) 40–49 years old, (**f**) 50–59 years old, and (**g**) 60 years old and over

The above result can be said to be in accord with the marketing strategy the new Iwataya took at that time. The new Iwataya allegedly is said to focus their target customers on late 20s to early 30s young females with work.

3.2 Changes in Visit Frequency Before and After the Renewal by Age Groups

We carried out the same analyses as above for visit frequency to Iwataya. The results are shown in Fig. 16.5a–g. Overall, there was not a large change in visit frequency to Iwataya between before and after the renewal of Iwataya. The teenager group had a slightly higher visit frequency to the new Iwataya. However, for the age group of 50–59 years and for that of above 60, they slightly decreased their visit frequency to Iwataya after the renewal of Iwataya.

3.3 Changes in Expenditure per Month Before and After the Renewal by Age Groups

For each of the seven age groups, the same analyses by cumulative distribution functions of the old and the new Iwataya are performed for expenditure at Iwataya per month per customer. The results are shown in Fig. 16.6a–g. From these figures, we see that there was no significant change in expenditure due to the renewal of Iwataya, which is a similar finding for visit frequency. However, the expenditure at the old Iwataya was higher in the age groups such as 30s, 40s, 50s, and above 60, and especially, for the age group of above 60, the expenditure at the old Iwataya was much higher than at the new Iwataya. From the intercepts for the age group of 30–39, we also see that the proportion of those who did not make any purchase has risen from less than 20% to over 40% before and after the renewal.

4 Analysis of the Effect of Purpose Realization Rate on Visit Frequency and Expenditure

In this section, we try to empirically validate our hypothesis that the higher consumers' purpose realization rate of the store, the more frequently they visit that store and the more they spend at that store. To do this, we investigate whether or not there are some causal relationships from the purpose realization rate to visit frequency and expenditure by cross-sectional data analysis. These analyses correspond to analyses 4–7 in Fig. 16.2.

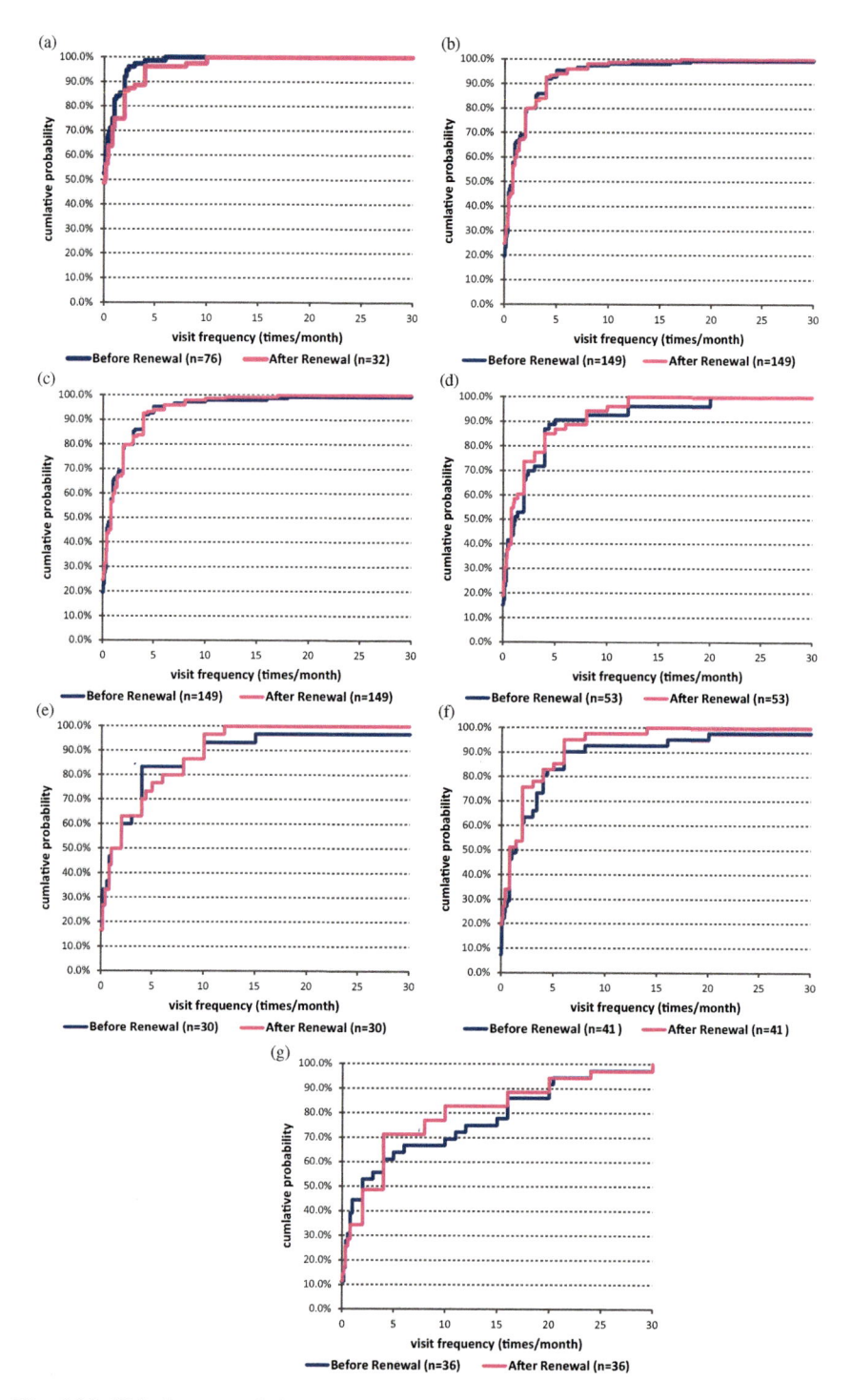

Fig. 16.5 Visit frequency before and after the renewal by age groups (**a**) Teenagers, (**b**) 20–24 years old, (**c**) 25–29 years old, (**d**) 30–39 years old, (**e**) 40–49 years old, (**f**) (50–59 years old, and (**g**) 60 years old and over

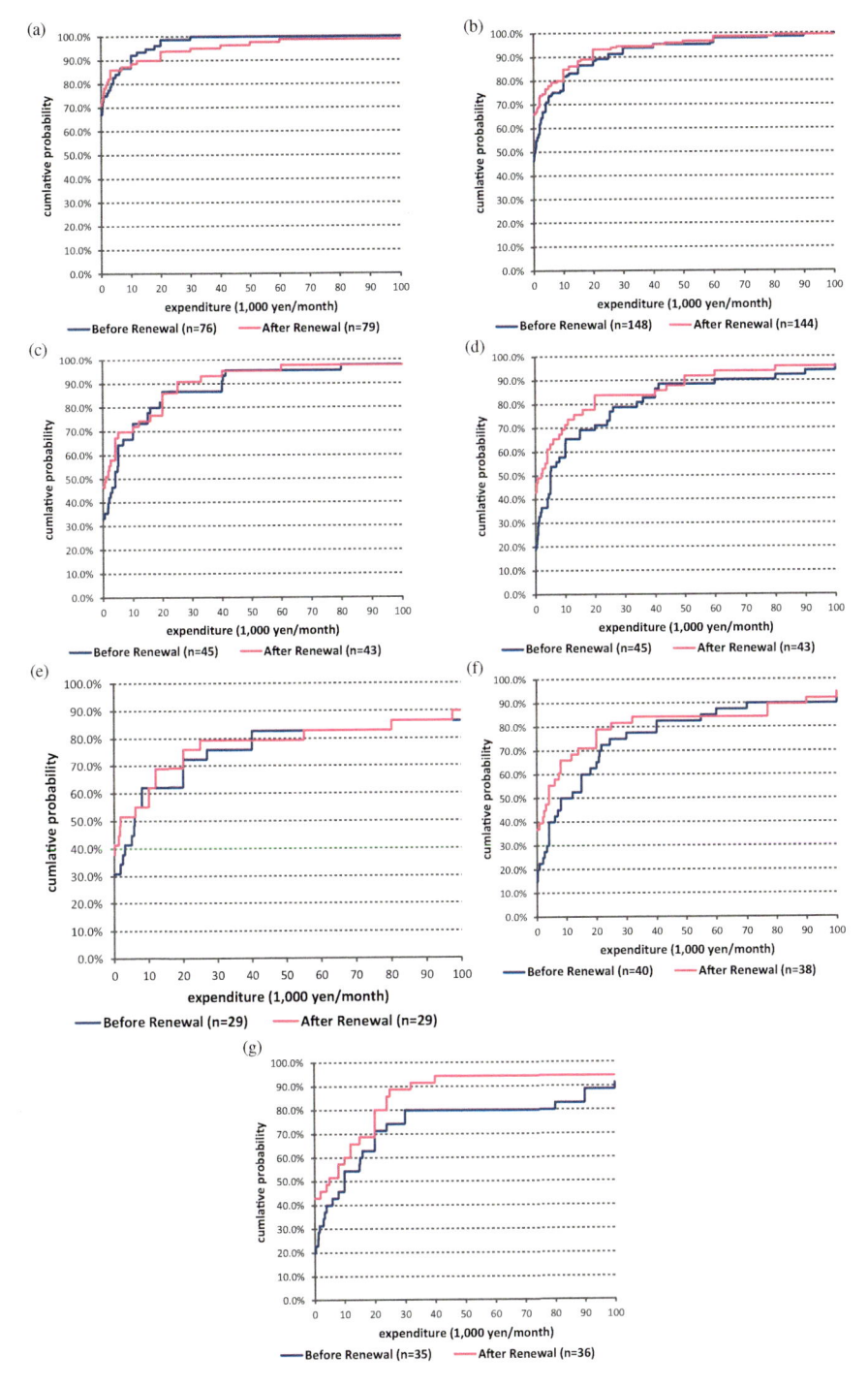

Fig. 16.6 Expenditure per month before and after the renewal by age groups (**a**) Teenagers, (**b**) 20–24 years old, (**c**) 25–29 years old, (**d**) 30–39 years old, (**e**) 40–49 years old, (**f**) 50–59 years old, and (**g**) 60 years old and over

The analyses proceed separately for two points of time, before and after the renewal. For each time point, we first calculate the average of the purpose realization rate for all samples. Then, according to this average, we divide the whole sample into two groups, samples whose purpose realization rate is above the average and those less than average. Next, we calculate the sample means of visit frequency and expenditure per month at Iwataya for these two groups. Then we perform t-tests to test whether or not there are statistically significant differences in these sample means of visit frequency and expenditure for these two groups with high and low purpose realization rate.

Finally, to confirm the obtained results, for each of two time points and for each of visit frequency and expenditure, we illustrate two cumulative distribution functions for the groups with high and low purpose realization rate and check whether the cumulative distribution function for the group with high purpose realization rate has shifted rightward. If this happens, we can have confirmed that the higher purpose realization rate increases the visit frequency and expenditure.

4.1 Effects of Purpose Realization Rate on Visit Frequency and Expenditure

Table 16.1a gives two averages of visit frequency to Iwataya per month for respective groups with high and low purpose realization rate before the renewal. Table 16.1b gives the same result after the renewal. From these tables, we see a clear relationship that the consumers with high purpose realization rate at Iwataya visit more frequently than those consumers with low purpose realization rate. This relationship holds for two points of time, before and after the renewal. Thus this relationship seems to hold for anytime independently of time.

On the other hand, Tables 16.2a and 16.2b are the results from the same analyses corresponding to expenditure per month per consumer. These results also show the same relationship that the customers with the higher purpose realization rate at Iwataya spend much more at Iwataya than those customers with the low purpose realization rate.

4.2 Cumulative Distribution Functions for Groups with High and Low Purpose Realization Rate with Respect to Visit Frequency and Expenditure

As we stated earlier, here we confirm the results obtained above by drawing the two cumulative distribution functions for the groups with high and low purpose realization rate to check whether the cumulative distribution function for the group with the high purpose realization rate has shifted toward the right.

Table 16.1a Averages of visit frequency per month by groups with high and low purpose realization rate (before the renewal)

	Average (times/month)	Number of obs.	Standard deviation
Consumers with low purpose realization rate	2.19	198	5.87
Consumers with high purpose realization rate	3.82	156	5.86
All	2.91	354	5.91

$t(352) = -6.287, p < 0.01$

Table 16.1b Averages of visit frequency per month by groups with high and low purpose realization rate (after the renewal)

	Average (times/month)	Number of obs.	Standard deviation
Consumers with low purpose realization rate	1.26	154	2.26
Consumers with high purpose realization rate	3.51	203	5.88
All	2.54	357	4.80

$t(355) = -10.130, p < 0.01$

Table 16.2a Averages of expenditure per month by groups with high and low purpose realization rate (before the renewal)

	Average (yen/month)	Number of obs.	Standard deviation
Consumers with low purpose realization rate	10,794	201	47,203
Consumers with high purpose realization rate	34,871	156	77,716
All	21,315	357	63,439

$t(355) = -917.189, p < 0.01$

Table 16.2b Averages of expenditure per month by groups with high and low purpose realization rate (after the renewal)

	Average (yen/month)	Number of obs.	Standard deviation
Consumers with low purpose realization rate	5245	154	20,931
Consumers with high purpose realization rate	24,178	203	56,444
Total	16,011	357	45,653

$t(355) = -873.514, p < 0.01$

The results are shown in Figs. 16.7 and 16.8. Figure 16.7a, b corresponds to the visit frequency before and after the renewal. Figure 16.8a, b corresponds to the expenditure before and after the renewal.

Note that the pink line indicates the cumulative distribution function for the group with the high purpose realization rate at Iwataya. From these figures, we have obtained a very clear result that all the cumulative distribution functions corresponding to this high group have shifted rightward from those corresponding to the low group.

While we obtained a very clear result by the cumulative distribution functions, the analyses have yet clarified whether each consumer who increased his/her purpose realization rate by the renewal actually has increased his/her visit frequency and expenditure at the new Iwataya. This analysis is involved with analyzing how the individual's change in purpose realization rate by the renewal relates to the other changes of that individual such as visit frequency and expenditure by the renewal. This is our next work in the next section.

5 The Increases in Purpose Realization Rate Cause the Increases in Visit Frequency and Expenditure Before and After the Renewal?

5.1 Division Ratio Cross Tabulation Method

This section investigates whether the increase in the purpose realization rate at the store perceived by an individual customer has increased his/her frequency of visits to that store and his/her expenditure at that store. The analyses in this section correspond to analyses 8 and 9 in Fig. 16.2.

For the analyses, we carry out the specific cross-tabulation analysis that utilizes the advantage of the retrospective panel micro-behavioral data. We call this method as the division ratio cross tabulation.

The key points we should note are the following. Since our data is panel data, we have two points of time, before and after the renewal, for each sample and for each variable. Thus for each sample we have two data corresponding to two points of time for each variable.

Let 1 and 2 denote two points of time, before and after the renewal, respectively. Now denote the data obtained from sample i for variable x at time 1 and 2 by x_{i1} and x_{i2}. For example, for the variable x of the purpose realization rate, we observe x_{i1} before the renewal and x_{i2} after the renewal for the same sample i. Hence we can divide the sample into two groups, those who increased the purpose realization rate and those who did not according to whether or not $x_{i2}/x_{i1} \geq 1$. Define a dichotomous variable v_i indicating which group sample i belongs to. For instance, v_i takes the value 1 if sample i decreases the purpose realization rate, i.e., $x_{i2}/x_{i1} < 1$ and the value 2 if sample i increases the rate, i.e., $x_{i2}/x_{i1} \geq 1$.

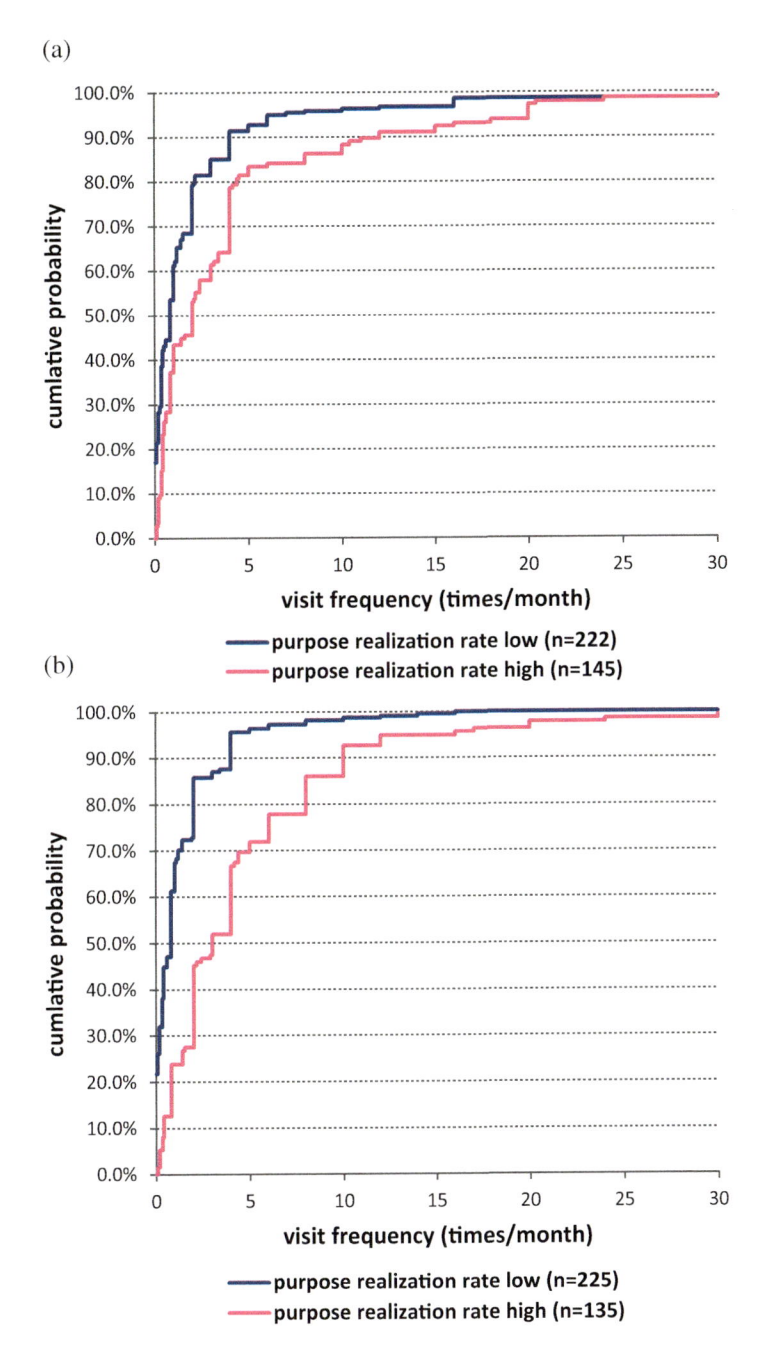

Fig. 16.7 Cumulative distribution functions of visit frequency for groups with high and low purpose realization rate (**a**) Before the renewal and (**b**) After the renewal

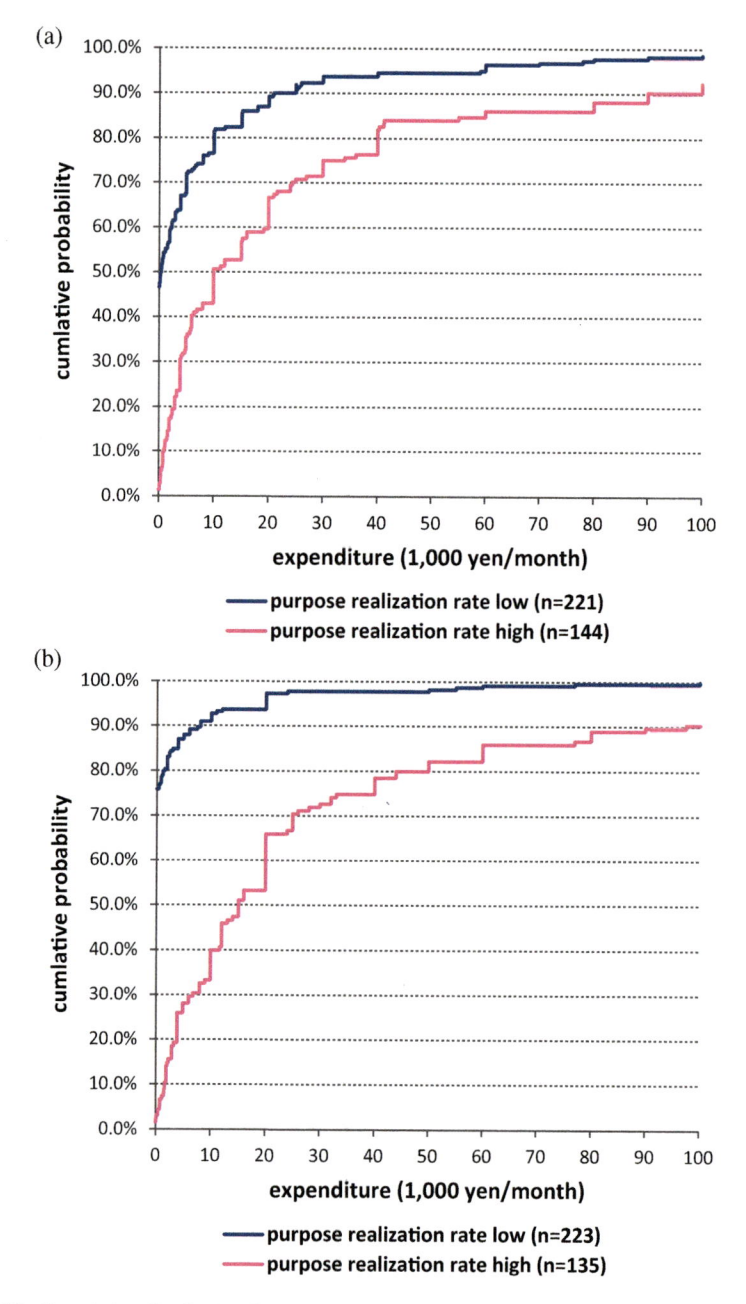

Fig. 16.8 Cumulative distribution functions of expenditure for groups with high and low purpose realization rate (**a**) Before the renewal and (**b**) After the renewal

In the same way, if we denote the visit frequency by y, we can divide the sample into two groups, those who increased the visit frequency and those who did not according to whether or not $y_{i2}/y_{i1} \geq 1$. We can also define a dichotomous variable w_i which takes the value 2 if sample i increases the visit frequency and the value 1 otherwise.

With this setup, we can perform the usual cross tabulation analysis between the two variables, v_i and w_i, which produce 2 by 2 table showing how the change in the purpose realization rate before and after the renewal affects the change in the visit frequency before and after the renewal. This is just the division ratio cross tabulation analysis between the purpose realization rate and visit frequency before and after the renewal.

5.2 The Increases in Purpose Realization Rate Cause the Increase in the Visit Frequency and the Expenditure?

Table 16.3 gives the result of the division ratio cross tabulation method applied to the analysis to explore how the purpose realization rate affects the visit frequency. From the table, we see the clear statistically significant relationship that the higher the purpose realization rate, the more the visit frequency, which is statistically significant at 10% level. In other words, the customers who increased their purpose realization rate of the new Iwataya have increased their frequency of visits to the new Iwataya.

Table 16.3 The higher the purpose realization rate, the higher the visit frequency? Changes in purpose realization rate by changes in visit frequency

			Changes in visit frequency before and after the renewal (ratio: after/before)		
			Visit frequency decreased group after/before <1	Visit frequency increased group after/before ≥ 1	Total
Changes in purpose realization rate before and after the renewal	Purpose realization rate decreased group after/before <1	%	40.5	59.5	100.0
		obs	64	94	158
	Purpose realization rate increased group after/before ≥ 1	%	30.0	70.0	100.0
		obs	27	63	90
Total		%	36.7	63.3	100.0
		obs	91	157	248

$\chi^2 = 2.7247, p = 0.0988$

Table 16.4 The higher the purpose realization rate, the more the expenditure per month? Changes in purpose realization rate by changes in expenditure per month

			Changes in expenditure per month per consumer before and after the renewal (ratio: after/before)		
			Expenditure per month decreased group after/before <1	Expenditure per month increased group after/before ≥1	Total
Changes in purpose realization rate before and after the renewal	Purpose realization rate decreased group after/before <1	%	68.6	31.4	100.0
		obs	105	48	153
	Purpose realization rate increased group after/before ≥1	%	34.9	65.1	100.0
		obs	30	56	86
Total		%	56.5	43.5	100.0
		obs	135	104	239

$\chi^2 = 25.504, p < 0.000$

Table 16.4 gives the result of the same division ratio cross tabulation method applied to the analysis to explore how the purpose realization rate affects the amount of expenditure per month per customer. Also this result shows the strong statistically significant relationship that the higher the purpose realization rate, the more the expenditure per month. This result is statistically significant at less than 0.1% level. The customers who have increased their purpose realization rate at the new Iwataya have increased their monthly expenditure at the new Iwataya.

Finally, Table 16.5 shows the result of the division ratio cross tabulation method applied to the analysis to see how the purpose realization rate affects the expenditure per visit. This result also shows an apparent statistically significant relationship that the higher the purpose realization rate, the larger the expenditure per visit, which is statistically significant at less than 0.1% level. In other words, the customers who have increased their purpose realization rate at the new Iwataya have increased their expenditure per visit at the new Iwataya.

The concept of the expenditure per visit is different from the concept of the expenditure per customer. The latter is concerned with how much a customer spends at one store or per 1 day or per month so that, for example, if a customer spends 2000 yen at the first floor and 3000 yen at the other floor, the expenditure for this customer is 5000 yen, whereas the expenditure per visit for this customer is 2500 yen. The expenditure per visit allows the double counting for the same visitors similar to the concept of the registered number of guests.

Thus the expenditure per visit can be said to be similar to average unit price per purchase. If we regard the expenditure per visit as a rough measure for the unit purchase price, the above result of the increase in the expenditure per visit for the

Table 16.5 The higher the purpose realization rate, the more the expenditure per visit? Changes in purpose realization rate by changes in expenditure per visit

			Changes in expenditure per visit before and after the renewal (ratio: after/before)		Total
			Expenditure per visit decreased group after/before <1	Expenditure per visit increased group after/before ≥1	
Changes in purpose realization rate before and after the renewal	Purpose realization rate decreased group after/before <1	%	59.5	40.5	100.0
		obs	94	64	158
	Purpose realization rate increased group after/before ≥1	%	17.6	82.4	100.0
		obs	15	70	85
Total		%	44.9	55.1	100.0
		obs	109	134	243

$\chi^2 = 39.127, p < 0.000$

customers with the high purpose realization rate might be the reflection of the gentrification strategy by the new Iwataya.

6 Conclusion and Future Challenges

The most critical contribution of this study is a devise to set the question items to collect the retrospective panel data from surveys of consumer *Kaiyu* behaviors and its related development of the division ratio cross tabulation method that utilizes the advantage of the retrospective panel data and makes it possible to investigate a causal relation by analyzing whether the changes in a causal variable in a particular individual affect the changes in the outcome variables for that individual.

With these methodological developments, we are able to find empirical facts that the purpose realization rate of the store which is perceived by consumers is strongly related to their visit frequency to the store and their expenditure at that store. This is another contribution of this study.

Since it might not be regarded as so explicit, we would like also to stress here that the conceptual construct of the purpose realization rate consumers perceive at stores played a critical and effective role to explain the changes of consumers' purchasing behaviors caused by the changes of the renewal of stores and their marketing strategies. Thus we would like to raise the conceptual construct of the purpose realization rate as another contribution of this study.

So far it has been our experience that it was very hard to confirm how the differences in policy performed by stores affect consumers' behaviors. To give one example, that is the atmosphere and images of shops perceived by consumers. It is very difficult to extract the effects of how these images affect consumers' choice behaviors. Unless the policy implemented by stores is quite a drastic one like this case of the renewal of Iwataya, and a large-scale retail redevelopment such as at Tenjin district of Fukuoka City, it has been almost impossible to detect the effect of policy by stores on choices of consumer behaviors.

When thinking back now, the variations caused by the store policy in choices of individual consumers may well be coupled with the variations caused by the heterogeneity in choices of heterogeneous individual consumers. Note that in the division ratio cross tabulation, we divide the value of a variable, say, expenditure at time 2 by the value of expenditure at time 1 for the same sample. Thus the dividing operation is equivalent to removing individual heterogeneity in the sense that the expenditure at time 1 differs among different samples by normalizing that to 1 and by converting the expenditure at time 2 into the ratio which is comparable among heterogeneous individuals.

Hence, it should be noticed that the division ratio cross tabulation method and the retrospective panel data first make it possible to perform the analysis which removes the heterogeneity of individuals only based on the sample survey of consumer *Kaiyu* behaviors conducted at one point of time.

As for the future challenges, there are many future works waiting to be explored. First of all, there might be many topics concerning the evaluation of various urban policy we can apply our division ratio cross tabulation method collecting the retrospective panel data. They are our direct future works. Further, we need to explore the theoretical foundation for the methodology of division ratio cross tabulation. Also what we need to investigate is to explore the possibility of the concept of our purpose realization rate to extend their effectiveness to explain consumer purchasing behaviors.

References

1. Saito S, Nakashima T, Tsurumi F (2006) Customers' evaluation and responses to the relocated renewal opening of a long-established local department store: a panel analysis of the retrospective micro-behavioral data. Paper presented at The 43rd annual meeting of Japan Section of Regional Science Association International (JSRSAI). (in Japanese)
2. Wikipedia. (2018, 2018/03/29). Iwataya. (in Japanese)
3. Wikipedia. (2018, 2018/03/29). mina tenjin. (in Japanese)
4. Ono S, Shibata K, Nishino A, Fujita M (2004) Changes in customers' purchasing behaviors caused by Renewal Opening of Iwataya (a long-established local department store): focusing on purpose realization rate, visit frequency, and expenditure. In: Proceedings for The 3rd Survey Research Conference on Marketing for Urban Development at City Center of Fukuoka, Japan. (in Japanese)
5. Ono S, Fujita M, Nishino A, Shibata K (2004) Customers' purchasing behaviors at a renewal commercial facility: focusing on the changes in purpose realization. In: Proceedings for The 3rd Survey Research Conference on Marketing for Urban Development at City Center of Fukuoka, Fukuoka. (in Japanese)

Chapter 17
A New Entry of Large Variety Shop Increases the Value of City Center?

Saburo Saito, Kosuke Yamashiro, and Masakuni Iwami

Abstract The large-scale variety store, Tenjin Loft, opened at the city center district of Fukuoka City, Tenjin district, in November 2007. This study explores how this entry of a new shop category to the city center has affected the consumer behaviors in this district, whether it increases the visit value of consumers to the city center, whether it enhances the attractiveness of the district, and consequently how it increases the value of town, more specifically, the value of Tenjin, from the viewpoint of consumer micro behaviors based on the actual on-site survey of *Kaiyu* behaviors at the city center of Fukuoka City.

Keywords Visit value · Value of town · Shop category · Variety shop · Store image · Store choice · *Kaiyu* · Shopping option · Retrospective panel data · Conditional logit model

---

This chapter is based on the paper, Saburo Saito, Kosuke Yamashiro, Erika Matsuda, Akiko Miyamoto, Chinatsu Torikai, and Masakuni Iwami [6], "How did the opening of Tenjin Loft (Large Variety Shop) increase the visit value of city center of Fukuoka City? Based on the analyses of relations among consumers' images, destination switching, purpose realization rate, visit frequency, and expenditure," Proceedings of The 25th Annual Meeting of The Japan Association for Real Estate Sciences, 2009, pp. 115–120. (in Japanese) The paper is modified for this chapter.

S. Saito (✉)
Faculty of Economics, Fukuoka University, Fukuoka, Japan

Fukuoka University Institute of Quantitative Behavioral Informatics for City and Space Economy (FQBIC), Fukuoka, Japan
e-mail: saito@fukuoka-u.ac.jp

K. Yamashiro
Department of Business and Economics, Nippon Bunri University, Oita City, Japan
e-mail: yamashiroks@nbu.ac.jp

M. Iwami
Fukuoka University Institute of Quantitative Behavioral Informatics for City and Space Economy (FQBIC), Fukuoka, Japan
e-mail: miwami@econ.fukuoka-u.ac.jp

1 Purpose of This Study

The large-scale variety store Tenjin Loft was opened in November 2007 at the Tenjin district, which is the city center commercial district of Fukuoka City, Japan. The opening of the store at Tenjin district is expected to increase the attractiveness of the city center of Fukuoka City as it can be regarded as widening consumers' purchasing options when they are shopping at this district.

The aim of this study is to identify what kinds of behavior changes individual consumers who came shopping at the city center of Fukuoka City have made before and after the opening of Tenjin Loft and to verify whether consumers who came shopping enhanced their visit value at the Tenjin district.

In particular, we would like to know what kinds of behavior changes were made by what types of consumers. In other words, we would like to identify these changes on an individual micro behavior basis. To do this, we use the "retrospective panel data," what we call, collected from our survey of consumer shop-around behaviors. We have already carried out some studies based on the retrospective panel data [1, 2, 4, 5]. The retrospective panel data were first devised to be obtained in our survey of consumer shop-around behaviors at city center of Fukuoka City in 2004. They were analyzed and discussed in Saito, Nakashima, and Tsurumi [4].[1]

Up to now, it has been very difficult to identify the effect of the changes in retail environment on individual consumers.[2] For example, take up the images consumers perceive on shops such as its physical space, style of exhibition, color, atmosphere, and so on. It seems apparent that these images perceived by consumers affect their choices of shops and brand switching. However, it has been very difficult to actually identify and verify their effect on their choices. This is because the variation of the cross-section data between different individuals, which we have mainly dealt with obtaining from our on-site survey of *Kaiyu* behaviors, is largely influenced by their income, time distance, and visit frequency to city center, etc., that is, their heterogeneity and this variation due to the heterogeneity is easily mixed up with the variation which should be caused consistently by the differences in the perceived images of consumers.

Thus it is known that it has been quite difficult to extract the effects with relatively small variation such as by perceived images from among the relatively large variations due to individual heterogeneity.

Therefore, in our survey conducted after the opening of Tenjin Loft, we included question items to ask the respondents retrospectively about their behaviors before the opening of Tenjin Loft. Thus we have collected panel data at two time points, before and after the opening of Tenjin Loft for each respondent. We call this type of data as the retrospective panel data. Its advantage is that the retrospective panel data makes it possible to measure the behavior changes in the same sample before and after an event, such as the opening of Tenjin Loft.

---

[1] See also Chap. 16 of this book, which is the revised version of the paper.

[2] Another effort to identify the effect of the renewal of a large retail shop on consumers' behaviors was to estimate the change of *Kaiyu* movements within the shop [3].

17 A New Entry of Large Variety Shop Increases the Value of City Center?

The purpose of this study is to ascertain what kinds of images perceive by consumers have the influences on their store choices between the existing similar variety store, INCUBE, and the new entrant, Tenjin Loft, by utilizing the advantage of the retrospective panel data. Furthermore, we attempt to investigate whether the visit value at the city center of Fukuoka City has been enhanced by the opening of new businesses which would give wider purchasing options for consumers in terms of their purpose realization rate, visit frequency, and amount of expenditure.

2 Data Used

In this study, we use the data obtained from the 13th survey of consumer shop-around behaviors at city center of Fukuoka City conducted by the Fukuoka University Institute of Quantitative Behavioral Informatics for City and Space Economy (FQBIC) on June 28th (Saturday) and 29th (Sunday) in 2008. Our on-site survey of consumer shop-around behaviors is an interview survey with questionnaire sheets taking about 15 min. Several survey points are set up in the city center of Fukuoka City; respondents are sampled at random from the visitors over 16 years of age visiting survey points. The respondents are asked about their shop-around behavior history on the day of the survey, that is, the places they visited, the purposes going there, and the expenditure spent there if any in the order of their occurrence.

Table 17.1 shows the outline of the survey including survey points, survey dates, survey time, question items and so on.

In Fig. 17.1 we give the map of the location of Tenjin Loft.

Table 17.1 Outline of the 13th survey of consumer shop-around behaviors at city center of Fukuoka City

Date of survey	6.29.2008 (Sat), 6.30.2008 (Sun)
Survey time	12:00–19:00
Survey points	Shoppers Daiei, Solaria Plaza, Iwataya, Daimaru Tenjin Fukuoka, Fukuoka Mitsukoshi, Canal City Hakata, Hakata Riverain, Hakata Station Concourse, Tenjin Loft
Survey method	Random sampling on-site interview survey with questionnaire sheets taking about 15 min. Respondents are randomly sampled from the visitors to the survey points over the age of 16
Main questionnaire items	1. Sample profiles (residence, age, gender, occupation, etc.)
	2. Shop-around history: the sequence of the triples; places visited, purposes done there, and expenditure there, if any, in the order of occurrence
	3. Travel time to the city center of Fukuoka city and travel mode
	4. Visit frequency to the city center of Fukuoka City, main commercial districts, and main commercial facilities in the city center of Fukuoka City
Number of samples	831 samples

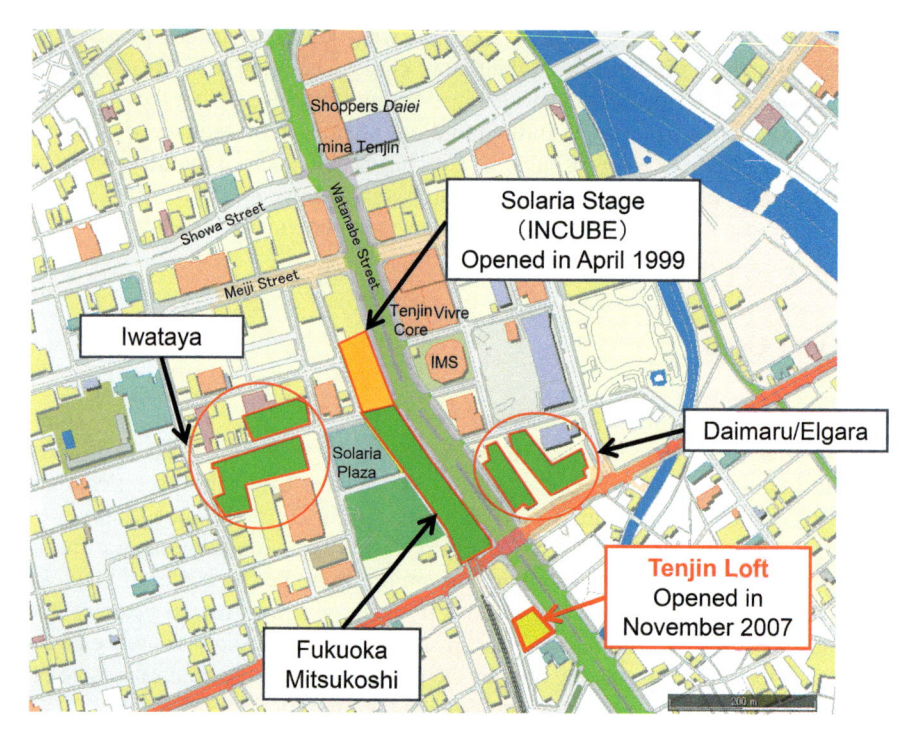

Fig. 17.1 Map of the location of Tenjin Loft (Cf. [7])

3 The Entry of Tenjin Loft Changes Consumer Purchasing Behaviors for Personal Items at Tenjin District?

In this section, we analyze how the opening of Tenjin Loft changed consumers' purchasing behaviors for personal items at the Tenjin district.

In the 13th survey of consumer shop-around behaviors, we asked the respondents where they bought six categories of products, namely, stationery, interior goods, variety products, household products, health-related products, and accessories, before and after the opening of Tenjin Loft at the city center commercial district of Fukuoka City. We call these six product categories as personal items. The alternatives of purchase destinations given to the respondents were INCUBE (the existing variety store mentioned above), Tenjin Loft (the new entrant only applicable after the opening), department stores, and other stores at the city center of Fukuoka City.

We have analyzed to what extent consumers have changed their purchase destinations before and after the opening of Tenjin Loft for each of six product categories. Considering the length of this paper, here we only report the results about the categories of stationery and health-related products, where consumers have shown relatively large changes in choices of their purchase destinations.

17 A New Entry of Large Variety Shop Increases the Value of City Center?

Table 17.2 Changes in purchase destinations before and after the opening of Tenjin Loft (stationery)

Before opening		After opening				
		INCUBE	Tenjin Loft	Department stores	Other stores in the city center of Fukuoka	All
INCUBE	obs	196	43	0	3	242
	%	81.0%	17.8%	0.0%	1.2%	100.0%
Department stores	obs	0	12	77	1	90
	%	0.0%	13.3%	85.6%	1.1%	100.0%
Other stores in the city center of Fukuoka	obs	2	20	1	197	220
	%	0.9%	9.1%	0.5%	89.5%	100.0%
All	obs	198	75	78	201	552
	%	35.9%	13.6%	14.1%	36.4%	100.0%

Table 17.3 Changes in purchase destinations before and after the opening of Tenjin Loft (health-related products)

Before opening		After opening				
		INCUBE	Tenjin Loft	Department stores	Other stores in the city center of Fukuoka	All
INCUBE	obs	40	20	3	3	66
	%	60.6%	30.3%	4.5%	4.5%	100.0%
Department stores	obs	0	11	160	3	174
	%	0.0%	6.3%	92.0%	1.7%	100.0%
Other stores in the city center of Fukuoka	obs	0	11	5	270	286
	%	0.0%	3.8%	1.7%	94.4%	100.0%
All	obs	40	42	168	276	526
	%	7.6%	8.0%	31.9%	52.5%	100.0%

We first look at stationery purchase. Table 17.2 shows its result. Of all the people who used "INCUBE" before the opening of Tenjin Loft, 17.8% changed to "Loft" after the opening. Also, 13.3% consumers changed from department stores to "Loft." On the other hand, it was found that more than 80% of shoppers did not change their destinations.

Next, we look at health-related products. Table 17.3 shows the result. Of all consumers who used "INCUBE" before the opening of Tenjin Loft, 30.3% of them changed their purchase destination to "Loft" after the opening, which was the highest among six product categories. On the other hand, there was almost no change away from the purchase destinations of "department stores" and "other stores in the city center of Fukuoka City."

4 Images of Tenjin Loft and INCUBE and Their Effects on Store Choices

In this section, we analyze the relationship between the images of Tenjin Loft and INCUBE consumers hold and store choices they make. We further verify to what extent the images consumers hold affect their store choices by using a conditional logit model.

4.1 Comparative Analysis of the Images Held by Consumers Who Changed Their Purchase Destinations to Loft and Those Who Did Not

Method

For each store of four purchase destinations, INCUBE, Tenjin Loft, Department stores, and other stores in the city center of Fukuoka City, the respondents were asked about the images of the store they hold with respect to 11 items. The 11 items are shown in Table 17.4. These items are asked as a multiple answering question, in which the respondents are requested to pick up all the image items that they think agree with the store.

In order to investigate the relationship between the store images consumers hold and their store choices, we restrict the samples to the ones who purchased at INCUBE before the opening of Tenjin Loft and purchased at INCUBE or Tenjin Loft after the opening.

Regarding each of the six product categories (stationery, interior goods, variety products, household products, health-related products, and accessories), the restricted samples are classified into two groups, those who changed their purchase destination from INCUBE to Tenjin Loft (INCUBE →Tenjin Loft) and those who

Table 17.4 The 11 items of store image

1	Price range is reasonable
2	Good products are available
3	A rich variety of products is assorted
4	Limited items are assorted
5	It is enriched with well-known brands
6	Customer services by clerks are good
7	It provides services specific to the store
8	The inside of the store is nice
9	Transportation is convenient
10	It can be shopped at on the way of commuting
11	It is an alternative when the desired product is not available in other stores

did not (INCUBE→INCUBE), and the perceived images of the two stores, INCUBE and Tenjin Loft, measured by the above 11 items are compared between these two groups.

Results of the Comparisons

Among the results corresponding to the six product categories, we present the results related to stationery and health-related products.

Figure 17.2 shows the result for stationary. As for stationery purchase, it is seen from the figure that consumers who changed from INCUBE to Tenjin Loft evaluate the store image of Tenjin Loft higher than that of INCUBE with respect to the image items such as "good products are available" and "the inside of the store is nice."

Next we consider health-related products. Figure 17.3 shows the result for health-related products. From the figure, it also is seen that those who changed from INCUBE to Tenjin Loft evaluate Tenjin Loft higher than INCUBE with respect to the store image items such as "good products are available," "a rich variety of products is assorted," and "the inside of the store is nice." Also Tenjin Loft is higher than INCUBE with respect to the store image items such as "it is enriched with well-known brands" and "it is an alternative when the desired product is not available in other stores."

	incube ⇒ incube	incube ⇒ Tenjin Loft
price range is reasonable	39.9% / 18.4%	34.1% / 27.5%
good products are available	22.3% / 24.7%	19.5% / 57.5%
a rich variety of products is assorted	67.4% / 57.0%	80.5% / 75.0%
limited items are assorted	3.6% / 9.5%	2.4% / 20.0%
it is enriched with well-known brands	1.0% / 7.6%	0.0% / 7.5%
customer services by clerks are good	5.7% / 5.1%	4.9% / 7.5%
it provides services specific to the store	1.0% / 3.2%	0.0% / 0.0%
the inside of the store is nice	9.8% / 30.4%	9.8% / 37.5%
transportation is convenient	37.3% / 11.4%	31.7% / 15.0%
it can be shopped at on the way of commuting	14.0% / 3.8%	24.4% / 10.0%
it is an alternative when the desired product is not available in other stores	9.3% / 10.1%	9.8% / 15.0%

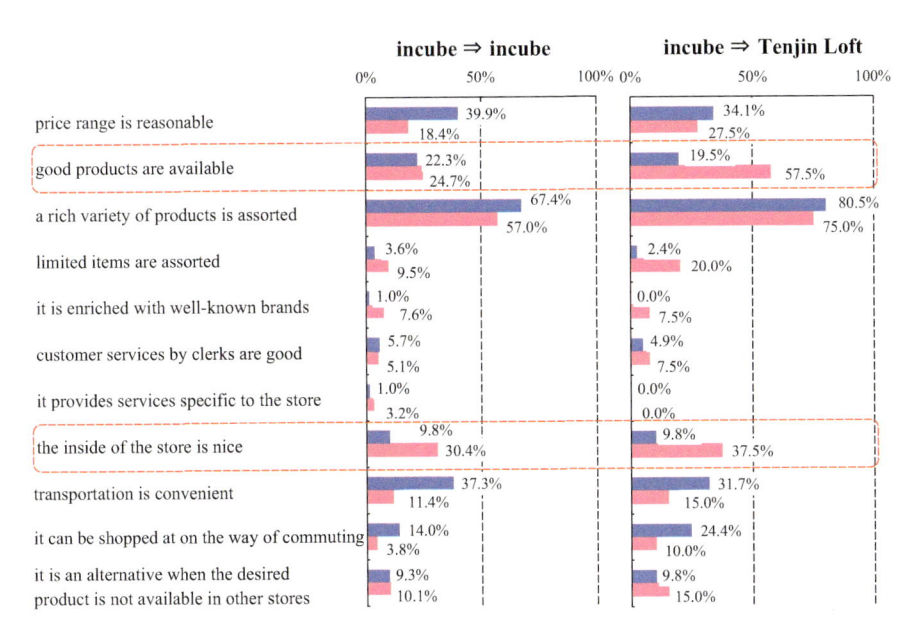

Fig. 17.2 Comparison of the images of INCUBE and Tenjin Loft held by consumers who changed from INCUBE to Tenjin Loft and those who did not (stationery). Upper bar: INCUBE, lower bar: Tenjin Loft

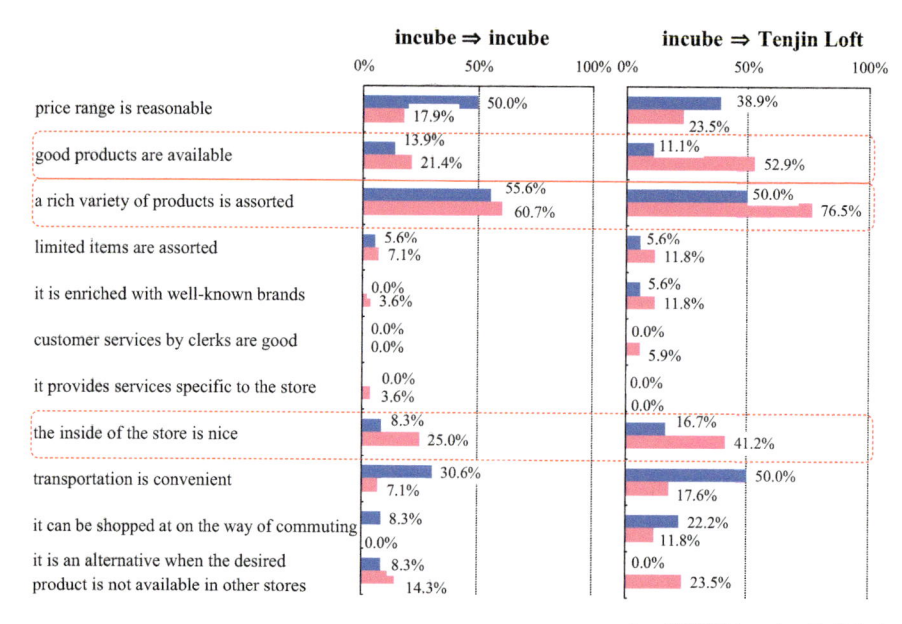

	incube ⇒ incube	incube ⇒ Tenjin Loft
price range is reasonable	50.0% / 17.9%	38.9% / 23.5%
good products are available	13.9% / 21.4%	11.1% / 52.9%
a rich variety of products is assorted	55.6% / 60.7%	50.0% / 76.5%
limited items are assorted	5.6% / 7.1%	5.6% / 11.8%
it is enriched with well-known brands	0.0% / 3.6%	5.6% / 11.8%
customer services by clerks are good	0.0% / 0.0%	0.0% / 5.9%
it provides services specific to the store	0.0% / 3.6%	0.0% / 0.0%
the inside of the store is nice	8.3% / 25.0%	16.7% / 41.2%
transportation is convenient	30.6% / 7.1%	50.0% / 17.6%
it can be shopped at on the way of commuting	8.3% / 0.0%	22.2% / 11.8%
it is an alternative when the desired product is not available in other stores	8.3% / 14.3%	0.0% / 23.5%

upper bar: INCUBE, lower bar: Tenjin Loft

Fig. 17.3 Comparison of the images of INCUBE and Tenjin Loft held by consumers who changed from INCUBE to Tenjin Loft and those who did not (health-related products). Upper bar: INCUBE; lower bar: Tenjin Loft

4.2 A Conditional Logit Model for Analyzing How Store Images Affect the Switch of Purchase Destinations

Method

The focus of the analysis is to explore what items of the images have influences on the store choices and to what extent. In order to look at the magnitude of the influence by the image, a conditional logit model is formulated, in which the explanatory variables are dummy variables corresponding to the above 11 image items, and the dependent variable is the dummy variable which represents the switch of purchase destinations.

More specifically, the dependent dummy variable or the decision variable takes the value 1 if the sample changed its purchase destination from INCUBE to Tenjin Loft and 0 otherwise. Recall that we restricted our samples to those who chose INCUBE before the opening of Tenjin Loft and chose either INCUBE or Tenjin Loft after the opening of Tenjin Loft. Thus after the opening of Tenjin Loft, the set of the alternatives for store choices consists of Tenjin Loft and INCUBE. We apply the conditional logit model here to the store choices after the opening of Tenjin Loft. In this setting of the conditional logit model, the choice of an alternative of Tenjin Loft can be regarded as the switch from INCUBE to Tenjin Loft and the other choice of the alternative of INCUBE as the no-switch from INCUBE. Hence the 11 items of store images for INCUBE and Tenjin Loft can be used as individual-alternative-specific explanatory variables corresponding to the respective alternatives of INCUBE and Tenjin Loft.

The magnitude of the influence by each image item is assessed by the statistical significance and by the estimated value of the parameter corresponding to each image item.

Results of the Analyses

Here, we report about stationery and health-related products out of the six product categories.

Table 17.5 gives the result for stationery purchase. Regarding stationery purchase, the explanatory variables corresponding to the store image items of "price

Table 17.5 Estimated results of the conditional logit model to explain how the store images affect the switch of purchase destinations (stationery)

Item	Estimate	t value	Standard deviation
1 Price range is reasonable	1.6223	4.5114***	0.3596
2 Good products are available	0.9533	3.0025***	0.3175
3 A rich variety of products is assorted	1.1705	3.8720***	0.3023
4 Limited items are assorted	0.3806	0.8117	0.4689
5 It is enriched with well-known brands	−0.2151	−0.3205	0.6711
6 Customer services by clerks are good	1.0554	1.4509	0.7274
7 It provides services specific to the store	0.5372	0.4686	1.1464
8 The inside of the store is nice	−0.3499	−1.0779	0.3246
9 Transportation is convenient	1.5718	4.1792***	0.3761
10 It can be shopped at on the way of commuting	0.0670	0.1292	0.5184
11 It is an alternative when the desired product is not available in other stores	0.0153	0.0331	0.4627
Log-likelihood with all parameters set to zeros	$L(0)$		−162.196
Log-likelihood with estimated values of parameters	$L(\hat{\beta})$		−119.916
−2 × Log-likelihood	$\rho = -2[L(0) - L(\hat{\beta})]$		84.560
McFadden's R-square	$\rho^2 = 1 - L(\hat{\beta})/L(0)$		0.261
Adjusted R-square	$\bar{\rho}^2 = 1 - (L(\hat{\beta}) - K)/L(0)$		0.193

*significant at 10%, **significant at 5%, ***significant at 1%, K:the number of parameters, $n = 234$

Table 17.6 Estimated results of the conditional logit model to explain how the store images affect the switch of purchase destinations (health-related products)

Item	Estimate	t value	Standard deviation
1 Price range is reasonable	2.0499	2.4856**	0.8247
2 Good products are available	1.4218	1.9805**	0.7179
3 A rich variety of products is assorted	1.2971	2.0088**	0.6457
4 Limited items are assorted	1.2131	0.9770	1.2416
5 It is enriched with well-known brands	−1.1051	−0.7258	1.5226
6 Customer services by clerks are good	14.2265	0.0306	465.3447
7 It provides services specific to the store	−10.5904	−0.0191	555.6601
8 The inside of the store is nice	0.1342	0.1944	0.6904
9 Transportation is convenient	0.8360	1.1790	0.7091
10 It can be shopped at on the way of commuting	1.4128	1.0859	1.3010
11 It is an alternative when the desired product is not available in other stores	1.8949	1.7266*	1.0975

$L(0) = -37.430$, $L(\hat{\beta}) = -26.589$, $\rho = 21.682$, $\rho^2 = 0.290$
*10% significant, **5% significant, ***1% significant, $n = 54$

range is reasonable," "good products are available," "a rich variety of products is assorted," and "transportation is convenient" turn out statistically significant. Thus these store images are known to have affected the switch of stationary purchase destinations from INCUBE to Tenjin Loft.

Table 17.6 presents the result for health-related products. As for health related products, the explanatory variables corresponding to the store image items of "price range is reasonable," "good products are available," "a rich variety of products is assorted," and "it is an alternative when the desired product is not available in other stores" turn out statistically significant. These store images can be said to have the influences on the switch of purchase destinations for health related products from INCUBE to Tenjin Loft.[3]

5 Changes in Consumer Behaviors Caused by the Opening of Tenjin Loft

5.1 A Framework for Analyzing Changes in Consumer Behaviors Before and After the Opening of Tenjin Loft

Figure 17.4 depicts a framework for analyzing the changes in consumer behaviors before and after the opening of Tenjin Loft.

---

[3] In the table, it is seen that the estimates for the store image items 6 and 7 have large absolute values. This is because the number of samples who selected each of two items was one. While we should delete these items, we leave them as they are here to include all 11 store image items.

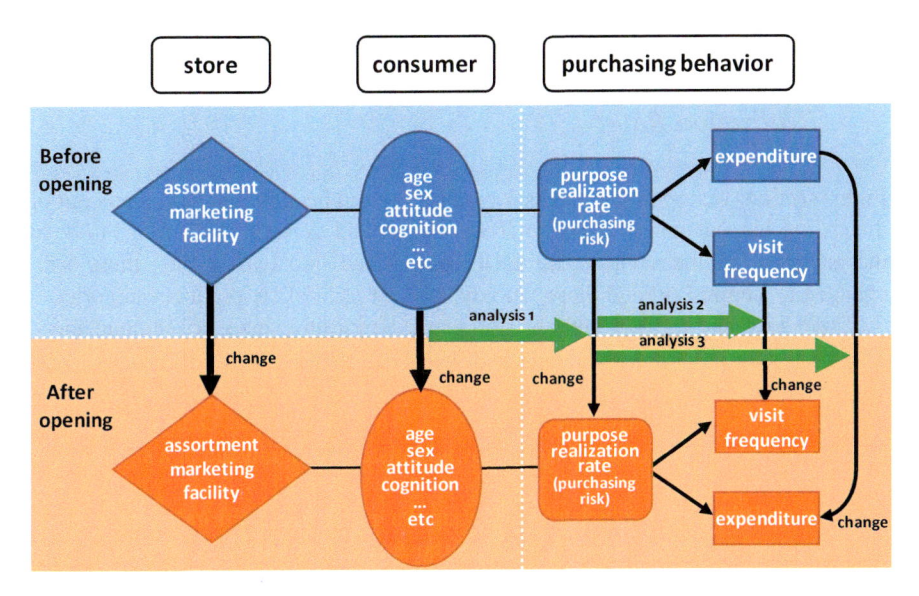

Fig. 17.4 Framework for analyzing changes in consumer behaviors before and after the opening of Tenjin Loft

By the opening of Tenjin Loft, a large-scale variety store, in the city center of Fukuoka City, an increase in a variety of products being sold at the city center such as miscellaneous goods is expected, consumers who visit the city center can have much more opportunities to purchase the products they desired to buy, and hence the purpose realization rate will increase. Here the purpose realization rate at a shopping site is defined as the probability that consumers can buy the product they wanted per one visit to that shopping site. Thus the opening of Tenjin Loft would enhance the purpose realization rate of consumers who visit at the city center.

So, the conceptual framework depicted in Fig. 17.3 hypothesizes that consumers who changed their purchase destinations from INCUBE to Tenjin Loft increased their purpose realization rate and consequently their visit value to Tenjin district, the city center of Fukuoka City. As a result, those who changed their purchase destinations from INCUBE to Tenjin Loft are hypothesized to increase their visit frequency to Tenjin district and their expenditure there.

The following three comparative analyses before and after the opening of Tenjin Loft are conducted in order to verify the above hypotheses.

Analysis 1. Those who changed their purchase destinations from INCUBE to Tenjin Loft increased their purpose realization rate.
Analysis 2. Those who increased their purpose realization rate increased their visit frequency to Tenjin district.
Analysis 3. Those who increased their purpose realization rate increase their expenditure at Tenjin district.

In Sect. 5, we conduct analysis 1, and in Sect. 6, we conduct analyses 2 and 3.

5.2 Consumers Who Changed Their Purchase Destinations from INCUBE to Tenjin Loft Enhanced Their Purpose Realization Rate?

In this section, we report the results for stationery and health-related products out of the six product categories. These analyses correspond to Analysis 1 in Fig. 17.4. In the analyses, all the samples are used and divided into two groups, those who changed to Tenjin Loft and those who did not, for each of six product categories.

Table 17.7 shows the result for stationery. Regarding stationery, among those who changed to Tenjin Loft, the percentage of people who increased their purpose realization rate was 33.3%, and in contrast, among those who did not change to Tenjin Loft, that percentage was only 0.7%. The difference in the change of purpose realization rate between these two groups is statistically significant at a significance level of 0.1%.

Look at Table 17.8, which gives the similar result for health-related products. Even for health-related products, among those who switched to Tenjin Loft, the percentage of people who increased their purpose realization rate rose to 23.1%. In contrast, among those who did not change to Tenjin Loft, that percentage was only 1.4%. In the case of health-related products too, the difference in the increases of

Table 17.7 Switch to Tenjin Loft and purpose realization rate (stationery)

Purchase destination switch		Purpose realization rate not increased (after/before ≤ 1)	Purpose realization rate increased (after/before >1)	All
Not changed to Tenjin Loft	obs	442	3	445
	%	99.3%	0.7%	100.0%
Changed to Tenjin Loft	obs	48	24	72
	%	66.7%	33.3%	100.0%
All	obs	490	27	517
	%	94.8%	5.2%	100.0%

$p < 0.001$

Table 17.8 Switch to Tenjin Loft and purpose realization rate (health-related products)

Purchase destination switch		Purpose realization rate not increased (after/before ≤ 1)	Purpose realization rate increased (after/before >1)	All
Not changed to Tenjin Loft	obs	438	6	444
	%	98.6%	1.4%	100.0%
Changed to Tenjin Loft	obs	30	9	39
	%	76.9%	23.1%	100.0%
All	obs	468	15	483
	%	96.9%	3.1%	100.0%

$p < 0.001$

purpose realization rate between these two groups was also statistically significant at a significance level of 0.1%.

6 Influence of Purpose Realization Rate on Visit Frequency and Expenditure

6.1 Influence of Purpose Realization Rate on Visit Frequency to Tenjin District

In this section, we investigate the relationship between the purpose realization rate and the visit frequency to the Tenjin district, which corresponds to the analysis 2 in Fig. 17.4.

Again, we report the results for stationery and health-related products out of the six product categories. In the analyses, all the samples are used and divided into two groups, those who increased their purpose realization rate and those who did not, for each of six product categories.

Table 17.9 gives the result for stationery. In order to see whether the increase in the purpose realization rate has the influence on the increase in the frequency of visit to Tenjin district, we divided the samples into two groups, those who increased their purpose realization rate and those who did not, and compared between these two groups whether they increased their visit frequency to Tenjin district before and after the opening of Tenjin Loft.

As for stationery, it is seen that among those who increased their purpose realization rate, the percentage of persons who increased the visit frequency to the Tenjin district was 22.2%. In contrast, among those who did not increase their purpose realization rate, that percentage was only 10.3%. The difference in the change of visit frequency between these two groups was statistically significant at a significance level of 10%.

Table 17.10 shows the result for health-related products. As for health-related products, among those who increased their purpose realization rate, the percentage of persons who increased their visit frequency to the Tenjin district was 26.7%, while

Table 17.9 Purpose realization rate and visit frequency to Tenjin district (stationery)

		Visit frequency not increased	Visit frequency increased	All
Purpose realization rate not increased	obs	442	51	493
	%	89.7%	10.3%	100.0%
Purpose realization rate increased	obs	21	6	27
	%	77.8%	22.2%	100.0%
All	obs	463	57	520
	%	89.0%	11.0%	100.0%

$p < 0.1$

Table 17.10 Purpose realization rate and visit frequency to Tenjin district (health related products)

		Visit frequency not increased	Visit frequency increased	All
Purpose realization rate not increased	obs	426	47	473
	%	90.1%	9.9%	100.0%
Purpose realization rate increased	obs	11	4	15
	%	73.3%	26.7%	100.0%
All	obs	437	51	488
	%	89.5%	10.5%	100.0%

$p < 0.1$

among those who did not increase their purpose realization rate, this percentage was 9.9%. The difference in the change of visit frequency between these two groups was also statistically significant at a significance level of 10%.

It should be noted that while the change of purpose realization rate, which corresponds to the causal variable here, is measured by each product category, the visit frequency, which corresponds to the outcome variable here, is the visit frequency to the Tenjin district, the city center of Fukuoka City. Considering it is usual that visitors visit the city center with various purposes, while a sharp effect to increase the realization rate for one specific purpose is shown to exist, it might be hard to detect its effect on the visit frequency to the city center which is related to the whole purposes since the city center is working to satisfy many visit purposes.

6.2 Influence of Purpose Realization Rate on Expenditure at Tenjin District

In this section, we analyze the relationship between purpose realization rate and expenditure at the Tenjin district for each of six product categories. This corresponds to the analysis 3 in Fig. 17.4.

Here we take up the results for interior goods and accessories. In the same way as in the previous section, all the samples are used and divided into two groups, those who increased their purpose realization rate and those who did not, for each of six product categories. For each product category, we compare between these two groups whether they increased their expenditure at the Tenjin district before and after the opening of Tenjin Loft.

Table 17.11 gives the result for interior goods. As for interior goods, among those who increased their purpose realization rate, the percentage of people who increased their average expenditure at the Tenjin district was 15.0%, while among those who did not increase their purpose realization rate, that percentage was 5.7%. The difference, however, in the change of expenditure at Tenjin district was not statistically significant because of the small number of samples who increased their purpose realization rate for interior goods.

Table 17.11 Purpose realization rate and average expenditure at Tenjin district (interior goods)

		Expenditure at Tenjin district not increased	Expenditure at Tenjin district increased	All
Purpose realization rate not increased	obs	364	22	386
	%	94.3%	5.7%	100.0%
Purpose realization rate increased	obs	17	3	20
	%	85.0%	15.0%	100.0%
All	obs	381	25	406
	%	93.8%	6.2%	100.0%

Table 17.12 Purpose realization rate and average expenditure at Tenjin district (accessories)

		Expenditure at Tenjin district not increased	Expenditure at Tenjin district increased	All
Purpose realization rate not increased	obs	419	29	448
	%	93.5%	6.5%	100.0%
Purpose realization rate increased	obs	7	2	9
	%	77.8%	22.2%	100.0%
All	obs	426	31	457
	%	93.2%	6.8%	100.0%

Table 17.12 shows the result for accessories. Regarding accessories, among those who increased their purpose realization rate for accessories, the percentage of persons who increase their average expenditure at Tenjin district was 22.2%. In contrast, that percentage was only 6.5% for those who did not increase their purpose realization rate. However, the difference in the change of expenditure at Tenjin district was not statistically significant.

Table 17.13 summarizes the results of the analyses done so far. In Analysis 1 where the influence of the switch to Tenjin Loft on the purpose realization rate is explored, those who changed their purchase destinations from INCUBE to Tenjin Loft had significantly increased their purpose realization rate for all six product categories. As shown in Analysis 2, the influence of purpose realization rate on the visit frequency to the Tenjin district was significant for two product categories, stationery and health-related products. As shown in Analysis 3, although the purpose realization rate had a positive influence on the average expenditure at the Tenjin district, the results were not significant.

Table 17.13 Summary of the results of analyses

Switch to Tenjin Loft by purpose realization rate by 6 item			Purpose realization rate by visit frequency to Tenjin district by 6 item			Purpose realization rate by average expenditure at Tenjin district by 6 item	
Stationary	***	0.000	Stationary	*	0.063	Stationary	0.439
Interior goods	***	0.000	Interior goods		0.153	Interior goods	0.117
Variety products	***	0.000	Variety products		0.689	Variety products	0.514
Household products	***	0.000	Household products		0.241	Household products	0.289
Health-related products	***	0.000	Health-related products	*	0.061	Health-related products	0.616
Accessories	***	0.000	Accessories		0.168	Accessories	0.119

***$p < 0.001$, **$p < 0.01$, *$p < 0.1$, unit: significant probability

7 Conclusion and Future Challenges

The significant contributions of this study are as follows. First, we have shown clearly what kinds of images of stores consumers hold affect their store choices by analyzing their purchase destination switches observed when the large variety store, Tenjin Loft, opened in the Tenjin district where an existing large variety store, INCUBE, was located.

This analysis was made possible by utilizing the retrospective panel data obtained from a sampling survey conducted at one point of time after the opening of Tenjin Loft where the respondents were asked retrospectively to answer their behaviors before the opening. Thus we can have micro behavioral panel data at two points of time for the same individual. This is another contribution of this study.

Also, we demonstrated that consumers who switched their purchase destinations from INCUBE to Tenjin Loft increased their purpose realization rate for all six product categories.

In addition, the study revealed that from the comparison between those consumers who increased their purpose realization rate and those who did not, the purpose realization rate has some positive influence on "visit frequency" and on "expenditure" for the entire town. The results were significant for some product categories, but we have not yet reached a strong relationship.

As for the future research topics, the relationship of the purpose realization rate to the visit frequency and expenditure should be explored further. Another future task is to measure the town equity of the Tenjin district that has been brought about by the increase in variety due to the opening of Tenjin Loft.

References

1. Saito S, Yamashiro K, Nakashima T (2006) An empirical verification of the increase of visit frequency to city center caused by the opening of a new subway line: a comparative analysis before and after opening based on retrospective panel data. Paper presented at the 43rd annual meeting of the Japan Section of Regional Science Association International (JSRSAI). (in Japanese)
2. Saito S, Nakashima T, Okuzono S, Mizokami T (2007) Changes in shop-around behaviors after the withdrawal of commercial facilities using retrospective pane data of visit frequency. In: Proceedings of the 23th annual meeting of the Japan Association for Real Estate Sciences 23: 107–112. (in Japanese)
3. Saito S, Yamashiro K, Nakashima K, Iwami M, Sato T (2007) Identifying the effect of shop floor renewals in a commercial facility: an application of *Kaiyu* Markov model. In: Proceedings of the 23th annual meeting of the Japan Association for Real Estate Sciences 23: 113–118. (in Japanese)
4. Saito S, Nakashima T, Tsurumi F (2006) Customers' evaluation and responses to the relocated renewal opening of a long-established local department store: a panel analysis of the retrospective micro-behavioral data. Paper presented at The 43rd Annual Meeting of Japan Section of Regional Science Association International (JSRSAI). (in Japanese)
5. Saito S, Yamashiro K, Sato H, Shichida T (2008) Analyzing changes in consumer behaviors caused by the opening of large retailer at city center using retrospective panel data: a case of Tenmonkan district at Kagoshima City, Japan. In: Proceedings of the 24th annual meeting of the Japan Association for Real Estate Sciences 24: 161–168. (in Japanese)
6. Saito S, Yamashiro K, Matsuda E, Miyamoto A, Torikai C, Iwami M (2009) How did the opening of Tenjin Loft (Large-scale Variety Store) increase the visit value of city center of Fukuoka City? Based on the analyses of relations among consumers' images, destination switching, purpose realization rate, visit frequency, and expenditure. In: Proceedings of the 25th annual meeting of the Japan Association for Real Estate Sciences, 115–120. (in Japanese)
7. Zenrin Co., Digital Map Z6, ZENRIN CO., LTD., 2003

Part VI
Emerging View of the Goal of Urban Development

Chapter 18
The Concept of Town Equity and the Goal of Urban Development

Saburo Saito

Abstract The concept of "town equity" is defined, and the meaning of "the goal of urban development" is clarified.

Keywords Town equity · Goal of urban development · Information evolution · Town walking · Hypertext city · Visit value · Brand equity · Asset of attractiveness · Consumer's mind

1 Information and Evaluation of Town

In this chapter, we define the concept of "town equity" and clarify the meaning of "the goal of urban development."

1.1 Information Evolution of Town Walking: Information and the Hypothesis on the Attractiveness of Town

The Hard Trick for Information Evolution: Canal City Hakata as an Example

We have a hypothesis about the attractiveness of towns. The commercial complex, Canal City Hakata, located in the city center district of Fukuoka City, is a very popular commercial establishment where people enjoy themselves by staying and spending their time. One of the unique characteristics of Canal City Hakata is that its

---

S. Saito (✉)
Faculty of Economics, Fukuoka University, Fukuoka, Japan

Fukuoka University Institute of Quantitative Behavioral Informatics for City and Space Economy (FQBIC), Fukuoka, Japan
e-mail: saito@fukuoka-u.ac.jp

Fig. 18.1 Information evolution of town walking

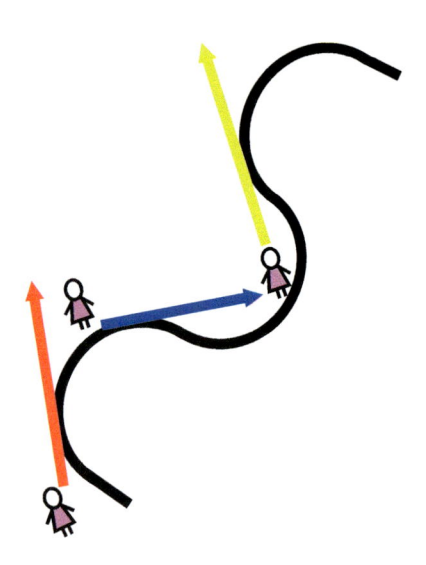

facility has a complex building design that uses many curves. The question then arises as to why there are so many curves.

I have interpreted the design as shown in Fig. 18.1. When people walk in the valley of a curve, their field of vision is limited. However, as they gradually walk toward the curve's mountain, visibility is regained, and the next valley appears for the first time. My interpretation is that this may be linked to "a feeling of excitement" when walking inside the facility, which then might become an element of attractiveness of the facility.

We call this mechanism that the new information jumps in one after another while walking around a town as the "information evolution of town walking."

Actually, our hypothesis is based on the idea that the attractiveness of town walking may consist in the information evolution of town walking. In other words, our hypothesis postulates that the attractiveness of a town lies in the possibility of discovering unexpected information and new information while walking around a town.

Therefore, our hypothesis infers that the facilities with many straight passages, lines, and good visibility would be less attractive than those with many curves because with straight passages, the things you can see ahead now from afar will just come closer to you as you walk through those passages without any unfolding of new or unexpected information.

According to this hypothesis, the fact that many curves were used throughout Canal City Hakata can be thought of as a trick that uses hard physical space as a means for effectively inducing the information evolution of town walking[1,2]. In

---

[1] As for the information evolution of town walking, refer to Saito [7].

[2] A new hotspot in Shanghai, Xintiandi, utilizes many cross intersections between division lots like traditional hutong, which is said to borrow the idea of the curves in Canal City Hakata.

short, many curves in Canal City Hakata can be regarded as a hard trick for information evolution.

The Soft Trick for Information Evolution: The Concept of Hypertext City

Thinking this way, the idea spreads. To enrich the information evolution of town walking, not only the hard tricks but also the various soft tricks can be utilized. Outdoor performances such as events, street performances, and musical and singing performances taking place inside a town can be regarded as examples of the soft tricks to stimulate the information evolution of town walking.

We remembered that in the early days of the Internet, the hypertext in which various WWW URL links were embedded in a plaintext had given us a great impact and appeal. Drawing the analogy between the hypertext and the information evolution of town walking, we once developed the idea of "Hypertext City." The Hypertext City is defined as a city in which real links or real agents are embedded in the city, and people who visit the city can make semantic interactions with the city through various information and intelligent mobile devices. We have advocated the implementation of the "Hypertext City Plan."[3]

While from the start real links in a town were supposed to be intelligent agents, the remarkable rapid progresses in technological innovations including smartphones, mobile ICTs, the IoT, and wearable technology are enhancing the feasibility of creating such a Hypertext City environment. However, as will be seen later in Chap. 20, recording and analyzing systematically what kinds of interactions are made between individual consumers and the information provided by the city is an attempt that has just begun.

Thus, it is still a future work to verify whether or not the individual consumers have increased their evaluation of the city through their experience of the information evolution of town walking inside the city while recording the detailed history with respect to what kinds of semantic interactions they have made with the information provided by various agents embedded in the city. In this sense, the hypothesis of "information evolution of town walking" is said to be a current working hypothesis of ours.[4]

---

[3]As for the idea of Hypertext City, refer to Saito [6, 7] and Saito, Nakashima, Kakoi, and Igarashi [10].

[4]As an indirect evidence for the hypothesis of information evolution of town walking, we have the result of the measurement of time value of shopping (Cf. Saito, Yamashiro, Kakoi, Nakashima [11] and Chap. 10 of this book). According to the result, when we compare consumers' time values by their purchase attitudes, among the three purchase attitudes, "coming to preview goods and prices," "thinking about buying if there are good products," and "a definite intention to buy the items," the highest time value is "thinking about buying if there are good products," the next is "coming to preview goods and prices," and the lowest is "a definite intention to buy the items." This means that consumers give a great value to the discovery of new good products on shopping, which indirectly supports our information evolution hypothesis.

2 The Concept of Town Equity

2.1 Dynamics of Changes of Hotspots Within a City Center District

Here, we take up an example of the city center of Fukuoka City. A town changes dynamically. Its changes are not limited to the construction and redevelopment of physical facilities located in the town. The value of the town has drastically changed according to the change of how people would evaluate the town.

In the city center of Fukuoka city, before 2000, the street of *Oyafuko-dori* located in the north of Tenjin district which is a core part of the city center of Fukuoka was quite a popular and vibrant hotspot. Young people had been gathering together there from all over Kyushu region every weekend night. *Oyafuko-dori* is a street not so wide with two sidewalks at both sides across two lanes for autos, which was once a path from the station to the preparation school for the high-school graduates who were preparing to take the entrance exam of the university. This is the origin of the street name, whose meaning is "not-filial-piety-street." At both sides of the street, there were and are many small restaurants, snack bars, live houses, mini theater, and night clubs. However, the popularity of the street had plummeted down because of factors such as the deterioration of security and the opening of adult entertainment businesses, and the hotspot had shifted from *Oyafuko-dori* to *Daimyo* district, another part of Tenjin district.

This instance demonstrates that city center hotspots have been changing dynamically depending on their popularity among people. If a town loses its popularity and people do not gather, the value of the town as real estate would also fall sharply. In other words, the changes in visitors' evaluations of *Oyafuko-dori* have greatly changed the value of the town. That is, people's evaluations of the town greatly affect its value.

Therefore, a town that is conscious of this fact has realized that people's evaluations should be regarded as a "town brand," and the town brand should be required to be maintained, nurtured, and managed.[5]

2.2 Brand Equity

From the concept of "brand," it is easy to extend to the concept of "brand equity." To clarify the goal of urban development, in the next section, we will introduce the concept of "town equity." Here, "town equity" is a term coined by the author borrowing from the analogical term, brand equity.

---

[5]*Marunouchi* district, a business district adjoining to Tokyo Station, is an instance. Also see Saito [5].

18 The Concept of Town Equity and the Goal of Urban Development

We explain the concept of brand equity using a hypothetical numerical example. Suppose that 10,000 people are willing to pay 10,000 yen more for cosmetics with a famous brand name than for cosmetics with no brand name. Suppose that these people buy the famous cosmetics 100 times a year. The annual sales of the company producing those cosmetics with the famous brand name will be increased by 10 billion yen.

In other words, just by being the cosmetics with a famous brand name, the increase of the revenue or the additional cash flow would be 10 billion yen per year. If the additional cash flow continues forever, it is the same as getting the annual interest, 10 billion yen every year forever. Dividing the amount of the annual cash flow by a discount rate, say 5%, gives the asset value of this cash flow, which equals 200 billion yen.

In short, if one has a deposit of 200 billion yen at a bank with interest of 5% each year forever, one will receive the interest income of 10 billion yen every year forever. This stream of the interest income is equivalent to the above cash flow of 10 billion yen per year generated by the famous brand name. Thus it is known that the name of the famous brand that generates the additional cash flow should be regarded as something that is the same as the monetary asset of the deposit of 200 billion yen. Therefore, the famous brand name, something like monetary asset, is called "intangible assets" in contrast to the visible assets like money, buildings, and lands.

Consequently, the value of the intangible asset corresponding to this additional cash flow accrued to the name of brand is referred to as "brand equity."[6]

As for the brand equity, the important point that should be noticed is that while the brand equity is the asset of the company which owns that brand, it is not the asset such as buildings or lands the company owns but the asset that is fostered in the minds of consumers who transact with this company.[7]

2.3 Definition of Town Equity[8]

Thinking in this way, we notice that the same is true for towns. Based on the concept of brand equity, "town equity" is defined as "the value of the asset of attractiveness of the town which is fostered in the minds of visitors who visit the town."[9]

---

[6]As for the brand equity, refer to Aaker [1].

[7]The recent uses of the concept of brand equity appear in terminology such as "place branding" or "tourism destination brand equity," which is similar to the concept of "town equity" (Cf. Gomez, Lopez, Molina [2]).

[8]The author used for the first time the word, "town equity," in the 20th anniversary symposium on "Urban Revitalization and Town Equity" held by Japan Association of Real Estate Sciences in 2004. (Cf. [3, 4])

[9]The term, "town equity" was first defined by Saito [4]. Also refer to Saito [7].

Several things to note about the concept of town equity are in order. The most important point to notice about the concept of town equity is that the value of the asset of attractiveness of the town is first defined at each individual micro-behavioral level in terms of the value of the intangible asset which is fostered in each mind of individual visitors. In other words, the town equity is first defined as a disaggregate concept. Thus the value of the asset of attractiveness of the town is first defined as the values which are fostered in the minds of visitors to the town, and then the town equity is defined as the aggregated value of them.

In addition, the concept of town equity may well be applied to a small district such as *Daimyo*, one district in the city center of Fukuoka City, or the whole area such as the entire Fukuoka City since the value of the intangible asset which is fostered in the minds of individuals does not restrict the area applicable. This is similar to the concept of brand equity which can be applied to one product or the line of products such as super brands.[10] Therefore, the concept is equally applicable to sightseeing spots and historical sites. It should be further noticed that since the concept of town equity is defined by the value of the asset, the town equity for a particular area can actually be measured in terms of money.

When constructing the concept of town equity as the values perceived by individual visitors, we notice that the goal of urban development, which had not been clearly conceived of so far, is in fact to maximize the values of the asset of attractiveness of the town in the minds of those visiting individuals.

If the goal of urban development is set to maximizing town equity, from the viewpoint of what kinds of functions, services, and facilities provided in the town enhance the values perceived by the visitors the most, we see that research on the evaluation of various urban development policies can be extended from the evaluation of the composition of stores in a commercial complex to that of the functional composition of the city.

3 The Goal of Urban Development

A town is not formed by a single entity. A town is a system which is formed by various entities. Considering this fact, the goal of urban development is redefined as "maximizing town equity regarding the town as one entity."

From the disaggregate definition of town equity, we see that the town equity at the micro behavioral level of individual visitors turns out the values of the asset of attractiveness of the town which are perceived, evaluated, and fostered in their minds through their experiences in the town while they visit the town.

---

[10]We have tried to measure the town equity of *Daimyo* district, a part of the city center of Fukuoka City (Cf. Saito, Iwami, Nakashima, Yamashiro, Sato [9]). In this measurement, the additional cash flow is defined as the willingness to pay additional expenditure to *Daimyo* in comparison with the average expenditure at the whole Tenjin district.

Thus from the perspective of individual visitors, the town equity can be seen the "values of the town" evaluated and perceived by the individual visitors. Assuming that individual visitors evaluate their "values of the town" according to how much the town is worth visiting for them, the "values of the town" become equivalent to their "visit values to the town."

Therefore, maximizing the value of the town is known to be equivalent to maximizing the visit values for individual visitors. Furthermore, the goal of urban development can be reduced to the question, "What kinds of visit values would be maximized for what types of visitors by providing what kinds of functions and facilities for constituting the town?"

The significance of the concept of town equity is that it clarifies the goal of urban development which remains ambiguous so far, formulates the value of the town as a measurable concept, and demonstrates that it actually can be measured based on *Kaiyu* behavior micro-data. In addition, town equity is measured in monetary terms as an asset value. Therefore, it makes it possible to compare in monetary terms the costs of implementing an urban development policy with the increase in the value of the town as town equity by this policy. Thus, the more cost-effective policies can be explicitly selected.[11]

In addition, noteworthy about the concept of town equity is that the attractiveness of the town is not homogeneous among visitors but differs heterogeneously individual by individual. Therefore, to identify how the value of the town would be increased by what policy, it is necessary to explore in detail what kinds of the value of the town would be increased for what types of visitors by taking what kinds of urban development policies.

The viewpoint of evaluating urban development policies focusing on how those policies would increase the values of the town perceived and experienced by visitors who visit the town with diverse preferences and motivation can be called as "consumer-oriented urban development."

While we defined the goal of urban development as maximizing town equity by considering the town as one entity, our research where parking lot policies in the city center have been examined from the viewpoint of maximizing town equity regarding the town as one entity will be presented in the next chapter.

References

1. Aaker DA (1991) Managing brand equity. Free Press, New York

---

[11]While gaining the additional cash flow is utilized to evaluate the town equity, the opposite case of waiving the additional cash flow is applied in Saito, Ishibashi, Kumata [8] to the valuation of the river flowing through the city center retail environment. In their study, the existence of the river decreases the turnover of retail sectors at both sides. This decrease of the turnover is regarded as the willingness to waive to preserve the natural asset of the river.

2. Gomez M, Lopez C, Molina A (2015) A model of tourism destination brand equity: the case of wine tourism destinations in Spain. Tour Manag 51:210–222
3. Ishibashi K, Morozumi M, Saito S (2005) Consumer behavior and town equity II: workshop held at the 20th annual meeting of the Japan Association for Real Estate Sciences (2004). Jpn J Real Estate Sci 19:152–161. (in Japanese)
4. Kakoi M, Deguchi A, Saito S (2005) Consumer behavior and town equity I: workshop held at the 20th annual meeting of the Japan Association of Real Estate Sciences. Jpn J Real Estate Sci 19:143–151. (in Japanese)
5. Saito S (2007) Valuation of town brand and consumer behavior: keynote speech at the open symposium on attractive urban development to inherit to descendants: fostering a distinctive town brand. Jpn J Real Estate Sci 21:12–20. (in Japanese)
6. Saito S (2001) Change town into hypertext: a commemorative lecture given at the opening ceremony of FQBIC (Fukuoka University Institute of quantitative behavioral informatics of city and space economy) on April 27th, 2001. Res: Cent Res Inst News Rep 17–21. (in Japanese)
7. Saito S (2005) Urban revitalization based on consumer shop-around: changing town into Hypertext city via mobile real link revolution – keynote speech at the 20th anniversary symposium on urban revitalization and town equity held in the 20th annual meeting of the Japan Association for Real Estate Sciences. Jpn J Real Estate Sci 19:8–17. (in Japanese)
8. Saito S, Ishibashi K, Kumata Y (2001) An opportunity cost approach to valuation of the river in a city center retail environment: an application of consumer's shop-around Markov model to the Murasaki river at Kitakyushu City. Stud Reg Sci 31:323–337. (in Japanese)
9. Saito S, Iwami M, Nakashima T, Yamashiro K, Sato T (2007) Measuring the town equity by Willingness To Pay: a case study at Daimyo district in the city center of Fukuoka City. In: Papers of the 23rd annual meeting of the Japan Association for Real Estate Sciences 23:97–102. (in Japanese)
10. Saito S, Nakashima T, Kakoi M, Igarashi Y (2008) Hypertext city plan: real GIS and town equity. In: Kumata Y, Yamamoto K (eds) Preservation of earth's environment resources by environment citizen. Kokon Shyoin, Tokyo, pp 131–146. (in Japanese)
11. Saito S, Yamashiro K, Kakoi M, Nakashima T (2003) Measuring time value of shoppers at city center retail environment and its application to forecast modal choice. Stud Reg Sci 33 (3):269–286. (in Japanese)

Chapter 19
City Center Parking Policy: A Business Model Approach

Saburo Saito, Kosuke Yamashiro, and Masakuni Iwami

Abstract This study takes up the Tenjin district, one core part of the city center commercial district of Fukuoka City, Japan, and clarifies what kind of parking policy should be adopted to increase the value of the Tenjin district as one business entity, through the analysis of micro data on visitors' parking-lot usage and *Kaiyu* behaviors in the Tenjin commercial district of Fukuoka City.

Keywords City center · Parking policy · Staying time · Economic loss · Cruising time · Fringe parking · Tenjin · Fukuoka · Retail turnover · Parking fee discount

---

This chapter is based on the paper, Saburo Saito, Takahiro Sato, Kosuke Yamashiro, Ken Takagi [1], "Estimating effects of establishment policy of parking lots at city center viewing from visitors' behaviors who used parking," *Papers of the 24th Annual Meeting of The Japan Association for Real Estate Sciences*, 2008, pp.169–176, which is modified for this chapter.

S. Saito (✉)
Faculty of Economics, Fukuoka University, Fukuoka, Japan

Fukuoka University Institute of Quantitative Behavioral Informatics for City and Space Economy (FQBIC), Fukuoka, Japan
e-mail: saito@fukuoka-u.ac.jp

K. Yamashiro
Department of Business and Economics, Nippon Bunri University, Oita City, Japan
e-mail: yamashiroks@nbu.ac.jp

M. Iwami
Fukuoka University Institute of Quantitative Behavioral Informatics for City and Space Economy (FQBIC), Fukuoka, Japan
e-mail: miwami@econ.fukuoka-u.ac.jp

1 Purpose of This Study

Today, chronic traffic congestions are a significant problem in central urban areas in Japan. Similar problems also occur in the city center of Fukuoka City. One of the factors to contribute to the congestions is considered to be shoppers who visit the city center by their private cars. To alleviate congestions and eliminate heavy traffics, policies such as taxation in the form of a traffic congestion tax for reducing the influx of vehicles into the city center area, the creation of transit malls, and so forth, have been drafted and introduced in various countries. However, are these policies of high value when the entire city center is regarded as one business entity? More specifically, is it really beneficial to exclude shoppers visiting by car when looking at the entire city center as one business entity?

The aim of this study is to address these questions. We first focus on the Tenjin district, which is the largest commercial agglomeration in western Japan and constitutes a core part of the city center of Fukuoka City, and clarify what kinds of parking policies should be adopted to increase the value of the Tenjin district as one business entity, through the analysis of micro behavioral data obtained from a survey of behaviors of shoppers using parking lots and another survey of consumer shop-around behaviors at the city center district of Fukuoka City.

Therefore, the purpose of this study is (1) to identify the economy size that car-use shoppers bring to the city center by estimating annual sales spent by car-use shoppers at the city center and annual sales of parking lots located in the city center, (2) to clarify how much time is spent for driving to find a vacant parking lot at the city center and to estimate the resulting economic loss due to cruising for parking lots, and (3) to explore what kinds of configurations and fee policies of parking lots would lead car-use shoppers to maximize their staying time and expenditure at the city center.

While this study focuses on city center parking policy from the angle of a business model to increase the value of the city center as one business entity, there exist several studies related to this research which have tried to evaluate urban policies from consumers' *Kaiyu* behaviors. Urban redevelopment and transport policies have been evaluated from *Kaiyu* movements [2–4] and from their economics effects [5–8]. As for the relationship between parking and *Kaiyu*, the relationships among parking lots choice, *Kaiyu*, and staying time were explored in [9]. In particular, Uchida and Kagaya [10] have investigated how the location of parking lots parked induces the unscheduled shopping visits or *Kaiyu* behaviors and shown that when the location of parking lots parked is 430 m away from the predetermined destination is most likely to induce *Kaiyu* behaviors. On the other hand, a mechanism in which the retailers and the chamber of commerce cooperate in subsidizing parking fees for car-use shoppers is shown as a social experiment in [11].

2 Data Used

In this study, we use data from the twelfth survey of consumer shop-around behavior at city center of Fukuoka City and the first survey of behaviors of shoppers using parking lots at city center of Fukuoka City. The outlines of both surveys are described below.

2.1 The Twelfth Survey of Consumer Shop-Around Behaviors at City Center of Fukuoka City

As for the 12th survey of consumer shop-around behaviors at city center of Fukuoka City (hereinafter abbreviated as the 2007 survey), several survey sampling points were set up at major commercial facilities in the city center district of Fukuoka City, which include the following eight sites: Solaria Plaza, Hakata Station Concourse, Canal City Hakata, Shoppers *Daiei*, *Iwataya*, *Daimaru* Elgala, Fukuoka *Mitsukoshi*, and Hakata Riverain. The survey is an on-site random sampling interview survey taking about 15 min, in which the respondents are randomly sampled from the visitors at the survey points and asked about their shop-around behavior on the survey day with questionnaire sheets.

The questionnaire items include the following: (1) items about micro behavioral history of shop-around on the survey day, that is, the sequence of the triples composed of places the respondents visited, purposes done there, and expenditure spent there if any in the order of their occurrence, (2) the frequency of visits to the city center of Fukuoka, and (3) the frequency of visits to major commercial facilities at the city center of Fukuoka.

The survey was conducted on June 30 (Saturday) and June 1 (Sunday), 2007, from 12:00 to 7:00 pm. The number of the obtained samples was 686 samples.

2.2 The First Survey of Behaviors of Shoppers Using Parking Lots at City Center of Fukuoka City

For the first survey of behaviors of shoppers using parking lots at city center of Fukuoka City (hereinafter abbreviated as the 2008 survey), the survey spots were set up at the following 11 parking lots in the Tenjin district. They are *Tikudo* Parking, *Ankoku* Parking Lot, F Parking North Tenjin, N Parking Tenjin, Tenjin Central Park Parking Lot, *Ayasugi* Multistory Parking Lot, Hakata Riverain Parking Lot, *Kamiyo* Parking Lot, Trust Park *Kego*, *Daiyoshi* Park Tenjin Big Tower, and Solaria Terminal Parking Lot. The survey was conducted as an on-site interview survey with questionnaire sheets, in which respondents were randomly sampled from shoppers who had parked their cars at the parking lots selected as the survey spots and visited there just before returning home from the city center.

The questionnaire items were similar to those of the survey of consumer shop-around behaviors at city center of Fukuoka City while adding such questions as the entrance route to the city center leading up to parking, and so on.

The survey dates were May 24 (Saturday) from 3:00 pm to 8:00 pm and May 25 (Sunday) from 2:00 pm to 7:00 pm, 2008. The obtained sample was 204 samples.

3 Size of Economy that Car-Use Shoppers Bring to the City Center of Fukuoka City

3.1 Expenditure Car-Use Shoppers Spend at the Tenjin District

We first estimate the annual sales that car-use shoppers bring at the Tenjin district as follows:

Annual sales by car−use shoppers = the number of visitors to Tenjin district
  × car utilization ratio
  × expenditure per car-use visitor per day
  × 365 days

Regarding the number of incoming visitors in the Tenjin district, we employ the estimate obtained in Saito [12]. It is estimated that the number of incoming visitors at the Tenjin district for the purposes of shopping, leisure, and having meals was 150,000 people per day in annual average. Regarding the car utilization ratio, we use the proportion of people who usually use cars as a transport means to travel to the Tenjin district obtained from the 2007 survey data. The ratio obtained turned out 14.1%. The 2008 survey data revealed that car-use shoppers spent in average 12,236 yen per day.

From the above, the annual sales by car-use shoppers become as follows:

94,458,860,000 yen = 150,000 (people/day) × 0.141 × 12,236 (yen/people)
  × 365 (days)

3.2 Annual Sales of Parking Lots

On the other hand, we calculate the annual sales of parking lots as follows:

Annual sales of parking lots = number of visitors to the Tenjin district
  × car-use ratio × average parking fee × 365days

Here, the number of visitors in the Tenjin district and the ratio of car-use shoppers are the same as those in the previous section, and the average parking fee is 898 yen per day, according to the 2008 survey data.

From the above, the annual sales of the parking lots become as follows:

$$6{,}932{,}335{,}500 \text{ yen} = 150{,}000 \text{ (person/day)} \times 0.141 \times 898 \text{ (yen/day)} \times 365 \text{ (days)}$$

3.3 Economy Size that Car-Use Shoppers Bring to Tenjin District

The size of the economy that car-use shoppers bring to the Tenjin district, which is thought of as the direct economic effect that car-use shoppers bring to the Tenjin district, is considered to be the total amount of expenditure that the car-use shoppers spend at the Tenjin district.

Economy size of car-use shoppers in Tenjin district
= annual sales contributed by car-use shoppers + annual sales of parking lots

From the above, the economy size of car-use shoppers in the Tenjin district becomes as follows:

$$101{,}391{,}190{,}000 \text{ yen} = 94{,}458{,}860{,}000 \text{ yen} + 6{,}932{,}330{,}000 \text{ yen}$$

From the result, we see that the contribution by car-use shoppers to the retail sector of the Tenjin district is not so small but a considerable amount of about 100 billion yen (or 1 billion dollars).

4 Size of Economic Loss due to Cruising Behaviors for Parking

4.1 Estimating Cruising Time for Parking by Car-Use Shoppers at the Tenjin District

In this section, we estimate the size of the economic loss caused by drivers' cruising behaviors to search for a vacant parking lot to park their cars in. The drivers' cruising behaviors are such ones as follows: When visitors come close to the city center, visitors must decide where to park their cars. Under the choice they decided, they drive to the target parking lot, but on the way there might be a vacant parking space.

Thus at this point, they must decide whether they choose that parking lot or continue to cruise. If they continue to drive and reach the target parking lot but it is full, they must choose whether they line up in a queue or continue to cruise again to search for another vacant parking lot. It is said that drivers' cruising behaviors are a main cause of the traffic congestions at city center [13].

Here we define the cruising time for parking as the sum of searching time for parking lots and waiting time for parking.

The economic loss due to cruising time for parking is defined as the loss of opportunity for car-use shoppers to do shopping. This implies that if car-use shoppers had allocated their cruising time for parking to their shopping, they would have been able to spend those times for shopping, and the retail sector at the city center would have had more retail sales.

From the viewpoint of the city center as one business entity, it loses the opportunity to gain the retail sales because of the cruising time for parking spent by car-use shoppers. Its economic loss is defined as the amount of the retail sales the city center has lost or the amount of the retail sales the city center would have gained if there were no cruising time for parking spent by car-use shoppers.

Average Cruising Time for Parking by Access Directions by Parking Blocks

In the 2008 survey, the respondents are asked how much time was taken from entering the Tenjin district until parking their cars at parking lots. This is the cruising time for parking defined above.

Table 19.1 shows the average cruising time for parking for car-use shoppers by access directions by parking blocks. As seen from the table, the average times for cruising for parking considerably vary by access directions and by parking blocks. This is the reason why we divide samples into groups classified by access direction and parking blocks when we calculate the average of cruising time for parking. For the Tenjin district, there are four directions to enter the Tenjin district which affect the easiness to find vacant parking lots since the imbalance of demand and supply of the number of parking lots differs among four directions to enter the Tenjin district. The same is applied to the five parking blocks at the Tenjin district: four parking blocks corresponding to four access directions and one block of the central part of the Tenjin district. Four access directions and five parking blocks were employed in our studies elsewhere[1] (Cf. [14–16]). The division of the five parking blocks is depicted in Fig. 19.2.

---

[1] Also see Chap. 8 of this book.

Table 19.1 Average cruising time for parking by access direction by parking blocks (unit: minute per auto/person)

Access direction	Parking block	obs	Cruising time average (unit: minute)	SD	Min	Max
North	North	35	6.2	3.7	1	15
	East	1	5.0	.	5	5
	West	9	9.7	3.2	2.5	15
	Middle	4	15.0	16.8	5	40
	Total	49	7.5	6.0	1	40
East	North	8	12.0	8.3	5	30
	East	16	10.4	8.0	2	30
	South	5	9.3	9.3	2.5	25
	West	6	10.6	6.9	1	20
	Middle	13	13.2	6.8	7	30
	Total	48	11.3	7.5	1	30
South	North	5	12.0	7.6	5	25
	East	4	27.5	5.0	20	30
	South	14	5.2	3.5	1.5	15
	West	7	11.4	6.3	5	20
	Middle	9	7.2	2.6	5	10
	Total	39	9.9	8.0	1.5	30
West	North	11	10.7	7.6	2	30
	East	6	22.5	18.9	10	60
	South	2	7.5	3.5	5	10
	West	10	4.2	2.7	1	10
	Middle	11	10.0	2.2	5	15
	Total	40	10.5	9.8	1	60
Total		176	9.8	7.9	1	60

Total Cruising Time for Parking Spent by Car-Use Shoppers at the Tenjin District

To calculate the total cruising time for parking, we need the actual numbers of car-use shoppers by groups classified by multiple classifications of access directions and parking blocks. Here we use the actual numbers of car-use shoppers by access direction by parking blocks per day in annual average estimated by Saito et al. [16].

From the average cruising time for parking and the estimated actual numbers of car-use shoppers by access directions by parking blocks, we obtain the total cruising time for parking at the Tenjin district spent by car-use shoppers. Table 19.2 shows the result.

Table 19.2 Total cruising time for parking by car-use shoppers at the Tenjin district (unit: minute)

Access direction	Parking block	Number of people per day	Cruising time average (unit: minute)	Cruising time total (unit: minute)
North	North	4182	6.2	25,925
	East	119	5.0	597
	West	1075	9.7	10,455
	Middle	478	15.0	7165
	Total	5854	7.5	44,143
East	North	956	12.0	11,474
	East	1912	10.4	19,835
	South	598	9.3	5559
	West	717	10.6	7588
	Middle	1553	13.2	20,551
	Total	5736	11.3	65,008
South	North	597	12.0	7167
	East	478	27.5	13,145
	South	1673	5.2	8662
	West	836	11.4	9558
	Middle	1075	7.2	7766
	Total	4659	9.9	46,298
West	North	1314	10.7	14,037
	East	836	22.5	18,812
	South	239	7.5	1793
	West	1195	4.2	4958
	Middle	1314	10.0	13,141
	Total	4898	10.8	52,741
Average			9.8	
Total		21,147		208,190

4.2 Calculating the Size of Economic Loss by Cruising Time for Parking

We calculate the size of economic loss caused by cruising time for parking spent by car-use shoppers at the Tenjin district.

The calculation formula is simply as follows:

The size of economic loss = total cruising time (minutes)
× average expenditure per minute of car-use shoppers

Here, as for the total cruising time for parking, we use the result obtained in the previous section. As for the average expenditure per minute of car-use shoppers, from the 2008 survey, it turns out to be 54.4 yen per minute, which was obtained by dividing the average observed expenditure of 12,235.5 yen per person by the average observed staying time of 224.6 min per person.

$$4,133,810,000 \text{ yen per year} = 208,189 \text{ (minutes)} \times 54.4 \text{ (yen/minute} \cdot \text{day)} \\ \times 365 \text{ (days)}$$

From the above, the total economic loss of cruising behaviors by car-use shoppers at the Tenjin district amounts to about 4.1 billion yen (or about 41 million dollars) per year.

4.3 Economic Effects by Shortening Cruising Time for Parking

Now let us turn our attention to the economic effects obtained from reducing cruising time for parking.

We set three cases of shortening the cruising time for parking: (1) 10% shortening, (2) 30% shortening, and (3) 50% shortening.

The results are as follows:

- Case of 10% shortening

$$413,381,000 \text{ yen/year} = 208,189 \text{ (minutes)} \times 54.4 \text{ (yen/minute} \cdot \text{day)} \\ \times 10 \text{ (\%)} \times 365 \text{ (days)}$$

- Case of 30% shortening

$$1,240,143,000 \text{ yen/year} = 208,189 \text{ (minutes)} \times 54.4 \text{ (yen/minute} \cdot \text{day)} \\ \times 30 \text{ (\%)} \times 365 \text{(days)}$$

- Case of 50% shortening

$$2,066,905,000 \text{ yen/year} = 208,189 \text{ (minutes)} \times 54.4 \text{ (yen/minute} \cdot \text{day)} \\ \times 50 \text{ (\%)} \times 365 \text{(day)}$$

Further, we look at these results by access directions and by parking blocks, respectively.

Table 19.3 gives the result by access directions and Table 19.4 by parking blocks. With respect to the access directions, the largest economic effects are obtained by the access from the east. As for the parking blocks, the north parking block attains the largest economic effects.

Table 19.3 Economic effects by shortening cruising time for parking by access directions

Access direction	Shortening rate		
	10%	30%	50%
North	87.65	262.95	438.25
East	129.07	387.23	645.39
South	91.92	275.78	459.64
West	104.72	314.16	523.61
Total	413.38	1240.14	2066.90

Unit: million yen

Table 19.4 Economic effect of shortening cruising time for parking by parking blocks

Parking blocks	Shortening rate		
	10%	30%	50%
North	116.36	349.09	581.82
East	104.02	312.07	520.12
South	31.79	95.38	158.97
West	64.64	193.94	323.24
Middle	96.54	289.64	483.74
Total	413.38	1240.14	2066.90

Unit: Million Yen

5 City Center Parking Policies: A Business Model

5.1 A Business Model for City Center Parking Policies Focusing on the Staying Time

By creating a new word, "town equity," we have arrived at an emerging view of the goal of urban development. The goal of urban development is conceived of as maximizing the town equity regarding the town as one entity.

Standing on this viewpoint, when considering and evaluating individual urban development policies, we should consider how these individual urban development policies contribute to the town equity or the value of the town.

Until now, fringe parking, park-and-ride, vehicle regulation in the city center area, and traffic congestion tax, among others, have been studied to solve traffic congestions in the city center area. However, few studies examined the city center transport policy from the viewpoint of increasing the value of the town while regarding the city center as one entity.

In this study, we would like to show how the city center parking policies can be formulated according to this viewpoint.

To invent creative city center parking policies, we need key ideas to build those policies consistently. Here we focus on the staying time car-use shoppers spend at the city center for shopping.

The idea is simple. First, according to the result of our 2008 survey, car-use shoppers spend at the Tenjin district an average of 54.4 yen per minute, including people who bought and those who did not, when they stay at the Tenjin district for 1 min. Therefore, by setting the locations of parking lots so as to make the staying time of car-use shoppers the longest and setting the parking fee at half the price, the staying time of car-use shoppers can be extended for increasing the retail sales at the entire Tenjin district.

While management entities of the parking lots and retailers such as commercial complexes and department stores are usually separate entities at the city center district, they both can cooperate to extend the staying time of car-use shoppers at the city center district to increase their business value at the city center district.

5.2 A Model to Determine the Staying Time of Car-Use Shoppers at the Tenjin District

We formulate the expression to represent how city center parking policies affect the staying time of car-use shoppers at the Tenjin district.

The staying time StT_i (minutes) in the Tenjin district is measured in the 2008 survey as the duration time from the time when the respondents parked their cars at the parking lots to the time when they left from those parking lots where they parked their cars. The dependent variable, the staying time StT_i of sample i, is expressed as a multiple regression model as follows:

$$StT_i = \alpha + \beta TDist_i + \gamma_1 Dest_i + \gamma_2 (Dest_i)^2 + \delta Fee_i \quad (19.1)$$

$TDist_i$ is the time distance (minutes) from the residential area to the Tenjin district for sample i, $Dest_i$ is the distance (meters) from the parking lot where sample i parked to the first destination in the Tenjin district sample i visited, and Fee_i is the parking fee per hour sample i paid. Further, α, β, γ_1, γ_2, δ are unknown parameters to be estimated.

The unique feature of this model is that we included as an explanatory variable the square of the distance from the parking lot parked to the first destination visited in the Tenjin district so as to enable one to find the distance from the parking lot parked to the first destination visited in the Tenjin district that maximizes the staying time. In this study, the distance from the parking lot to the first destination in the city center that maximizes the staying time of car-use shoppers is defined as the optimal location of the parking lots.

The parameters are estimated by the least squares method.

Table 19.5 Estimated results of parameters

		Parameter estimate	Standard error	t value	$Pr > \|t\|$
Constant		224.44761	34.97064	6.42	0.001
Time distance to Tenjin (minutes)	Tdist	0.10153	0.30265	0.34	0.738
Distance from parking lot to the destination (meters)	Dest	0.24832	0.09689	2.56	0.011
Square of the distance from the parking lot to the destination	$Dest^2$	−0.00017	0.00009	−1.84	0.067
Parking fee per hour (yen)	Fee	−0.26020	0.09306	−2.80	0.006
Root MSE		122.34416		R-Squared	0.0992
Dependent mean		219.75401		Adj R-Sq	0.0794

$n = 187$

5.3 Estimated Results of Parameters

Table 19.5 shows the results of the parameters estimated. From the table, we see that the sign of the parameter for the time distance from the residential area to the Tenjin district is positive and that of the parameter for the parking fee per hour is negative. These results satisfy the sign condition of our hypothesis.

Note that the sign of the parameter for the square of the distance from the parking lot parked to the first destination visited in the Tenjin district is negative. This estimated result means that we can calculate the optimal distance to maximize the staying time.

Additionally, except for the time distance from home to the Tenjin district, all parameter estimates are statistically different from 0 at significant levels less than 10%. However, when looking at the coefficient of determination with adjusted degrees of freedom, its value is 0.079, and the fit of the model is not so good.

5.4 How the Staying Time of Car-Use Shoppers Is Increased by Reducing Parking Fees and Relocating Parking Lots to Optimal Locations?

Cases to Forecast the Staying Time

We forecast the staying time for three cases. The first case is the present case without any policy. The second case is the case when the locations of parking lots are considered to be relocated in the optimal places that maximize the staying time. The third case is the case when the parking fee is made halved in addition to the second case.

As for the present case, we calculate the staying time at the Tenjin district for the hypothetical average sample who takes the sample means of the explanatory variables obtained from the 2008 survey as the values of the explanatory variables.

The sample means for the explanatory variables obtained from the 2008 survey are as follows.

- Time distance from the residential area to the Tenjin district: 46 min
- Distance from parking lot parked to the first destination visited in the Tenjin district: 391 m
- Parking fee per hour: 195 yen

The staying time for this hypothetical sample at the Tenjin district turns out to be 249.5 min.

Case When the Parking Lots Are Relocated to Optimal Locations

Let us see how much the staying time increases if the location of parking lot is moved to the optimal location.

From the estimated result of the staying time, we can get the optimal location of the parking lot that maximizes the staying time. As depicted in Figure 19.1, we obtained the result that the optimal location of the parking lot is the place where the distance from the parking lot parked to the first destination visited in the Tenjin district is 731 m.

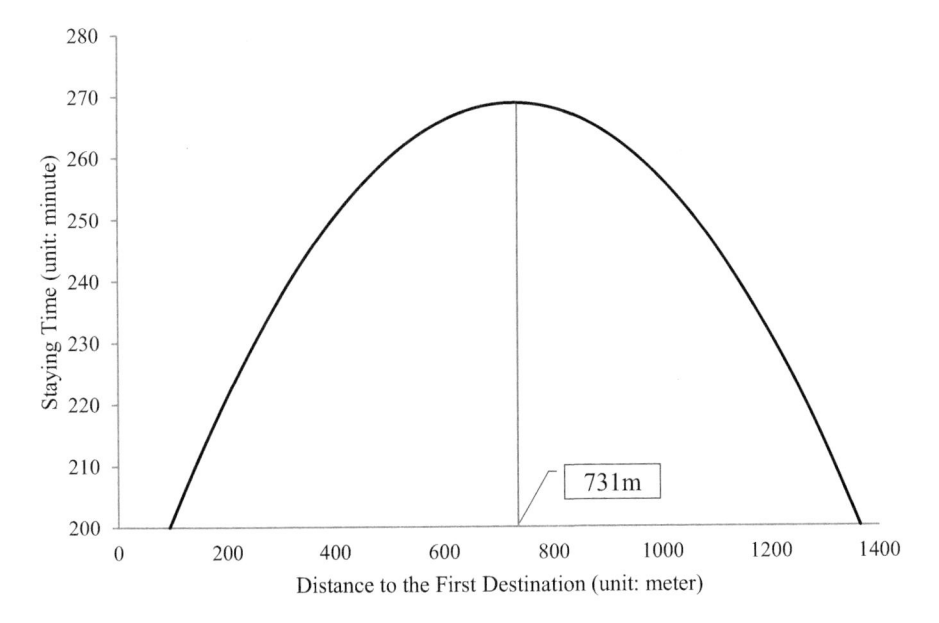

Fig. 19.1 Distance from the parking lot parked to the first destination visited that maximizes the staying time for car-use shoppers

This result implies that the fringe parking lot which is 731 m apart from the city center destination is the optimal parking location in the sense that it maximizes the staying time for the shoppers who use that parking lot.

In the present case, the distance from the parking lot parked to the first destination visited in the Tenjin district is assumed to be 391 m, which gives the staying time 249.4 min. On the other hand, when the parking lot was moved to the optimal location, 731 m away from the first destination in the Tenjin district, the staying time becomes 269 min.

Therefore, we found that there is an increase of 19.6 min in the staying time from the present case if the parking lot is moved to the optimum location. As for this point, it might be more precise to say that if car-use shoppers utilized the parking lots optimally located, their staying time at the city center would increase by 19.6 min.

Case When Parking Fees Are Made Halved

Furthermore, we estimate how long the staying time increases when reducing the parking fee from 195 yen per hour to 100 yen per hour (i.e., a 48.7% reduction).

In the case of reducing the parking fee to 100 yen per hour, the staying time is 274.2 min, which is 24.7 min longer than that in the present case.

Based on the results so far, it can be seen that when the parking lot is moved to the optimal location and the parking fee is reduced to 100 yen per hour, the total staying time increases by 44.3 min.

5.5 *A Business Model for Implementing the City Center Parking Policy to Halve Parking Fees Under Optimal Parking Lot Locations*

We saw that when we moved parking lots to the optimal locations and reduced the parking fees, the staying time of car-use shoppers at the city center would increase by 44.3 min. As seen earlier, the increase in the staying time would increase, in turn, the retail sales at the city center commercial district. Since there is no entity who plays a role for the city center as one entity to maximize the value of the city center, we need to devise some mechanism to attain the goal of urban development through the means of the city center parking policies. We call this mechanism as a business model.

To formulate a business model, we must first create a sustainable entity that autonomously implements the city center parking policy as its business. The sustainability of this entity here means that implementing the city center parking policy should be profitable as a business.

As the entity to implement the city center parking policy, we suppose some enterprise such as town management organizations.

In our city center parking policy, the enterprise first asks the companies of parking lots located at the city center to halve their parking fees by subsidizing or compensating their loss due to reducing their parking fees. Then the enterprise collects from the retailers at the city center the increased retail sales due to the increase in the staying time of car-use shoppers caused by halving parking fees.

While it is apparent that how much portion of the increased retail sales the enterprise can collect from the retailers depends on in what form the enterprise is organized and how the agreements among the enterprise and retailers are reached, for simplicity here we assume that all of the increased retail sales are collected by the enterprise to check the profitability for implementing the city center parking policy as a business.

This case can be considered to be probable when this enterprise is owned by the retailers.

The Amount of Compensation for Halving Parking Fees

In Sect. 3.2, the annual sales of companies of parking lots at the Tenjin district is estimated as 6,932,335,500 yen where the amount was obtained when the average parking fee per hour is assumed to be 195 yen. In our city center parking policy, the average parking fee is reduced to 100 yen per hour. It is almost half the price, 48.7% off the price. Thus the enterprise compensates the following amount to the parking lot companies:

$$3,376,047,389 \text{ (yen)} = 6,932,335,500 \text{ (yen)} \times 48.7\%$$

Increase of Retail Sales at the Tenjin District

The increase of retail sales at the Tenjin district is estimated as follows:

Increase in retail sales at Tenjin district = number of car-use shoppers at Tenjin district
× increase in staying time × expenditure per minute × 365 days

As used in Sect. 4.1, here we also employ the number of car-use shoppers at the Tenjin district estimated by Saito et al. [16]. The number of car-use shoppers at the Tenjin district is 21,150 people per day.

As seen from the above, if we assume that the parking fees are reduced to 100 yen per hour, and car-use shoppers utilize the optimally located parking lots, the increase of the staying time is 44.3 min. The expenditure per minute was 54.4 yen per minute from the 2008 survey.

Thus, the increase of retail sales at the Tenjin district becomes as follows:

$$18{,}603{,}970{,}000 \text{ yen} = 21{,}150 \text{ (people per day)} \times 44.3 \text{ (minutes)} \\ \times 54.4 \text{ (yen per minute)} \times 365 \text{ (days)}$$

Profits for the Enterprise Implementing the City Center Parking Policy

Clearly, the enterprise implementing the city center parking policy is autonomously sustainable because its profit surely becomes quite a large positive. The profit per year becomes as follows:

$$\text{Profit (yen)} = \text{revenue (yen)} - \text{cost (yen)}$$

$$15{,}227{,}922{,}611 \text{ yen} = 18{,}603{,}970{,}000 \text{ yen} - 3{,}376{,}047{,}389 \text{ yen}$$

The Net Increase of the Retail Sales at the Tenjin District: Economic Effect of City Center Parking Policy on the Tenjin District

Here we calculate the net increase of the retail sales at the Tenjin district caused by the city center parking policy. For the retailers at the Tenjin district, the net increase of the retail sales is shown to be 15.228 million yen per year. On the other hand, for the companies of the parking lots, their turnover before the reduction of parking fees are secured by the compensation from the enterprise for implementing the city center parking policy. However, the turnover corresponds to the parking fee that is charged for the parked time at the price before the discount so that the increase of the staying time due to the reduction of parking fee is not considered.

Thus the increase of turnover due to the increase of the staying time must be included in the net increase of the retail sales at the city center.

The formula to calculate the amount of the increased turnover is as follows:

$$\text{Increase of turnover due to the increase in staying time (yen)} \\ = \text{number of car-use shoppers at Tenjin district (people per day)} \\ \times \text{increase in staying time (minutes)}/60\text{(minutes)} \\ \times \text{parking fee halved (yen per hour)} \times 365 \text{ (days)}$$

The estimated increase of the turnover of parking lots due to the increase in the staying time amounts to the following:

$$569{,}974{,}875 \text{ yen per year} = 21{,}150 \text{ (people per day)} \times 44.3 \text{ (minutes)}/60 \\ \times 100 \text{ (yen per hour)} \times 365 \text{ (days)}$$

Therefore, the economic effect of our city center parking policy on the Tenjin district turns out as follows:

Economic effect on Tenjin district (yen)
= net increase of retail sales at Tenjin district (yen)
+ net increase of turnover of parking lots at Tenjin district (yen)

From the previous figures estimated, we obtain the following:

15,797,897,486 yen = 15,227,922,611 yen + 569,974,875 yen

This result shows that a great economic effect would be realized if we took the city center parking policy based on the perspective of the city center as a whole as one business entity.

6 Further Elaboration of a Business Model for Implementing the City Center Parking Policy

6.1 Further Elaboration of the Previous City Center Parking Policy

In the previous city center parking policy, we assumed that car-use shoppers utilize the parking lots that are optimal in the sense that the chosen parking lots are located at 731 m away from their first destination to visit in the city center. This is not a realistic assumption. In this section, we relax this assumption for the car-use shoppers who parked their cars at the central parking block at the Tenjin district. Other car-use shoppers who parked their cars at either of all other four parking blocks are assumed to utilize the optimally located parking lots.

As for the car-use shoppers who utilize the parking lots at the central parking block, we must give them an incentive that induces them to change their parked parking lots to the optimal ones. As the incentive, the enterprise for implementing the city center parking policy is assumed to employ the policy to reduce the parking fees of optimally located parking lots. Hereafter we refer the optimally located parking lots as the fringe parking lots.

Thus the policies we are concerned with are discounting the parking price of the fringe parking lots.

To do this, we must determine how much effective the discount rate induces the car-use shoppers to change the places of their parking lots to the optimal places. In order to determine the effectiveness of discount rates, we set the question items related to the fringe parking in the 2008 survey.

In the next section, we analyze the responses to these question items to investigate the effectiveness of discount rates.

6.2 A Model to Forecast the Shift to the Fringe Parking Lots

We formulate a conditional logit model to forecast the shift from the parking lots at the central parking block to fringe parking lots when the parking fees of fringe parking lots become cheaper.

In the conditional logit model, the probability to choose the fringe parking lots is expressed as follows:

$$P_1 = \frac{\exp(V_1)}{\exp(V_1) + \exp(V_2)}$$
$$P_2 = 1 - P_1$$

V_1 and V_2 are the deterministic utilities for car-use shoppers visiting the Tenjin district when they park their cars at the fringe parking lots and when they do so at the central parking block, respectively.

The deterministic utility is expressed as a linear function of the following explanatory variables: the parking fee, fee_i; the time distance, dist_i from home to the Tenjin district; and the staying time, stay_i, at the Tenjin district.

$$V_i = \alpha_1 \text{fee}_i + \alpha_2 \text{dist}_i + \alpha_3 \text{stay}_i, \quad i = 1, 2$$

$\alpha_1, \alpha_2, \alpha_3$ are unknown parameters to be estimated using the maximum likelihood estimation method.

In the estimation, only samples who parked their car at the parking lot located in the central block, from which the distance to the first destination they visited in the Tenjin district was within a 5-min walk, were used.

In the question items about the parking fees of the fringe parking lots, the respondents are asked how much they are willing to pay the parking fees for the parking lots when the parking lots are located at 1, 5, 10, 15, or 20 min walking distances to the first destinations the respondents visited in the city center. The data obtained as responses to these question items are used in the estimation of the conditional logit model.

To apply the conditional logit model, first we consider each response to each case of above five walking distances as one sample. That is, the five responses by one respondent are considered as five independent samples.

Furthermore, to create the dichotomous dependent variable of choice of whether to use the fringe parking lots or not, it is assumed that the responses to the cases when the walking distances are 1 min and 5 min correspond to the choice not to use the fringe parking lots, and the responses to the cases for 10, 15, and 20 min walking distances correspond to the choice to use the fringe parking lots. Note that we set the values of explanatory variables, fee, distance, and stay time corresponding to the choice of not using the fringe parking lots to 0.

Table 19.6 Estimated results of parameters

		Parameter estimate	Standard error	t value	Pr > |t|
Parking fee	fee	−0.00931	0.0020	−4.72***	0.0001
Time distance to Tenjin	dist	0.01270	0.0090	1.42	0.1570
Staying time	stay	0.00925	0.0026	3.61***	0.0003
Hit rate			75.79%		
Log-likelihood with all parameters set to zeros	$L(0)$		−65.85		
Log-likelihood with estimated values of parameters	$L(\hat{\beta})$		−45.67		
−2 × Log-likelihood	$\rho = -2\left[L(0) - L(\hat{\beta})\right]$		40.36		
McFadden's R-square	$\rho^2 = 1 - L(\hat{\beta})/L(0)$		0.3065		
Adjusted R-square	$\bar{\rho}^2 = 1 - \left(L(\hat{\beta}) - K\right)/L(0)$		0.2609		

***$p < 0.001$, $n = 95$

6.3 Results of Parameters Estimated

Table 19.6 shows the results of the parameters estimated by the conditional logit model. Looking at the estimation results, the sign of the parameter for parking fees is negative, and the signs of those for the time distance and the staying time are positive, which is in conformity with our hypothesis. Using the estimated results, we forecast the numbers of fringe parking users under several settings of parking fees.

6.4 Forecasted Results of the Numbers of Fringe Parking Users Under Several Settings of Parking Fees

We set four cases of fringe parking fees: 200 yen, 100 yen, 50 yen, and 0 yen per hour. We predict to what extent the car-use shoppers who parked their cars at the central parking block would shift to the fringe parking lots under these four schemes of the fringe parking fees.

As in the same way as before, the number of car-use shoppers who parked their cars at the central parking block is the one estimated by Saito et al. [16], which turns out 4420 people (cars) per day.[2]

Figure 19.2 shows the map of five parking blocks in the city center.

Table 19.7 shows the forecasted results of the numbers of car-use shoppers who would shift their parking lots to park their cars from the central parking block to the

---

[2] Also refer to Chap. 8 of this book.

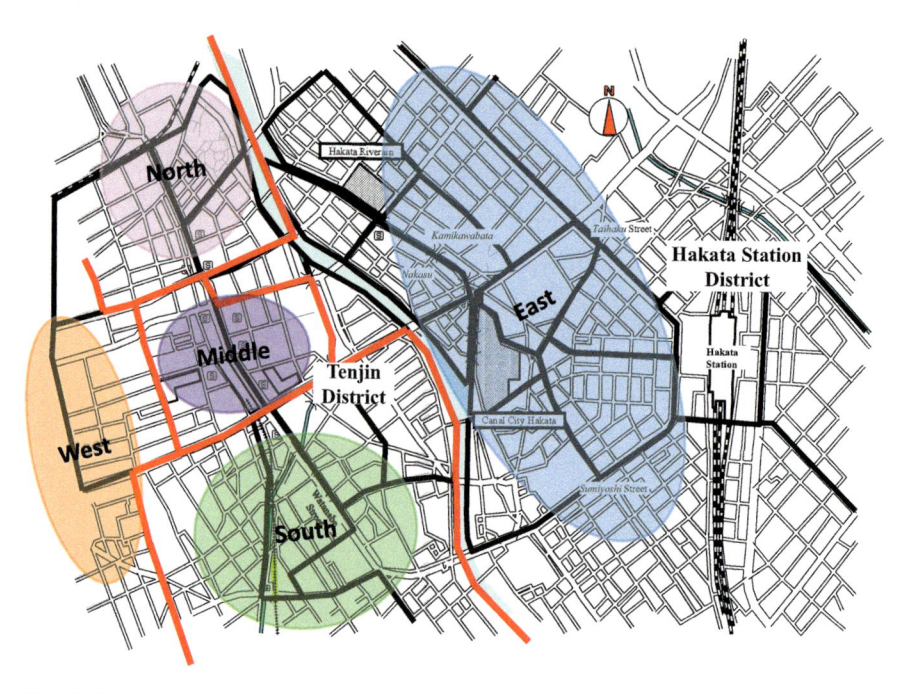

Fig. 19.2 Map of parking blocks in the city center district of Fukuoka City

Table 19.7 Forecasted results for numbers of fringe parking users

Parking fee (unit: yen/hour)	Number of car-use shoppers who used middle parking block (unit: people/day)	Rate to shift to fringe parking (%)	Forecasted numbers shifted to fringe parking (unit: people/day)
200	4420	56.2	2484
100	4420	74.0	3269
50	4420	81.1	3584
0	4420	86.8	3834

fringe parking lots under four cases of fringe parking fees. As the parking fee becomes cheaper, it turns out that the usage proportion of the fringe parking lots increases. In particular, when the fringe parking fee is 0 yen, 86.8% of car-use shoppers who had parked their cars at the central parking block shifted to the fringe parking lots, which indicates that making the fringe parking fee low might become an effective incentive for car-use shoppers to shift their parking lots to the fringe parking lots.

In our business model for implementing the city center parking policy, we will consider the case where car-use shoppers who parked their cars at the central parking block are given the incentives to move to the fringe parking lots by reducing the fringe parking fee to 50 yen, which halves the discounted parking fee, 100 yen, and refine the previous business model with respect to its profit and economic effects.

6.5 Refinement of the Business Model for Implementing the City Center Parking Policy

Cases Examined

We assume that car-use shoppers who parked their cars at four parking blocks corresponding to four access directions, i.e., east, west, north, and south parking blocks, are assumed to park their cars at the optimal parking lots. Also it is assumed that for car-use shoppers who parked their cars at the central parking block, the fringe parking lots are assumed to be the optimal locations that maximize their staying time.

Under these assumptions, we consider the two cases:

Case 1
Car-use shoppers who parked their cars at the central parking block do not use the fringe parking lots.
Case 2
By setting the fringe parking fee to 50 yen per hour, 81.1% of car-use shoppers who parked their cars at the central parking block move to the fringe parking lots.

Case 1

Profit for the Enterprise

In this case, the compensation by the enterprise, 3376 million yen per year, is the same as before. The revenue changes. It must be calculated separately for car-use shoppers parked at four parking blocks and for those parked at central parking block. For car-use shoppers parked at the four parking blocks, the amount of increase in the staying time, 44.3 min, is the same as before, but for those parked at the central block, it reduces to 24.7 min.

$$\begin{aligned}
&\text{Revenue from the increase in the staying time} \\
&= \text{Revenue from users parked at four blocks} \\
&\quad + \text{Revenue from users parked at the central block}
\end{aligned}$$

$$\begin{aligned}
16{,}883 \text{ million yen} &= 16{,}730 \text{ (people per day)} \times 44.3 \text{ (minutes)} \\
&\quad \times 54.4 \text{ (yen per minute)} \times 365 \text{ (days)} \\
&\quad + 4{,}420 \text{ (people per day)} \times 24.7 \text{ (minutes)} \\
&\quad \times 54.4 \text{ (yen per minute)} \times 365 \text{ (days)} \\
&= 14{,}716 \text{ million yen} + 2{,}167 \text{ million yen}
\end{aligned}$$

Therefore, the profit of the enterprise becomes as follows:

$$13{,}507 \text{ million yen} = 16{,}883 \text{ million yen} - 3{,}376 \text{ million yen}$$

Net Increase of Retail Sales at the Tenjin District

We must take into account the increase of turnover of parking lots.
The increase of turnover of parking lots at four parking blocks is estimated as follows:

$$\begin{aligned} 450 \text{ million yen per year} &= 16{,}730 \text{ (people per day)} \times 44.3 \text{ (minutes)}/60 \\ &\quad \times 100 \text{ (yen per hour)} \times 365 \text{ (days)} \end{aligned}$$

As for the central parking block, the increase of turnover becomes as follows:

$$\begin{aligned} 66 \text{ million yen per year} &= 4{,}420 \text{ (people per day)} \times 24.7 \text{(minutes)}/60 \\ &\quad \times 100 \text{ (yen per hour)} \times 365 \text{ (days)} \end{aligned}$$

Thus the net increase of turnover of parking lots turns out 516 million yen per year.

$$516 \text{ million yen} = 450 \text{ (million yen)} + 66 \text{ (million yen)}$$

Therefore, the net increase of retail sales at the Tenjin district, which equals the sum of the net increase of sales of retail sector at the Tenjin district and the net increase of turnover of parking lots, turns out as follows.

$$14{,}023 \text{ million yen} = 13{,}507 \text{ million yen} + 516 \text{ million yen}$$

Case 2

Profit for the Enterprise

Let us estimate the revenue of the enterprise, which is composed of contributions by three groups, car-use shoppers parked at four parking blocks (16,730 people), those moved from the central parking block and parked at fringe parking lots (3584 people), and those parked at the central parking block (836 people). The increase of staying time is 44.3 min for the first and the second group and that for the third group is 24.7 min.

Consequently, the revenue becomes 17,869 million yen per year.

$$17{,}869 \text{ million yen} = (16{,}730 + 3584) \text{ (people)} \times 44.3 \text{(minutes)}$$
$$\times\ 54.4 \text{ (yen per minute)} \times 365 \text{(days)} + 836 \text{ (people)}$$
$$\times\ 24.7 \text{(minutes)} \times 54.4 \text{ (yen per minute)} \times 365 \text{(days)}$$

The compensation paid by the enterprise is composed of two factors, the previous compensation and the additional compensation for the fringe parking lots. The previous compensation for reducing the parking fee from 195 yen per hour to 100 yen per hour is 3376 million yen, which was calculated by multiplying the total turnover of parking lots (6932 million yen) by discount rate ($0.487 = 95/195$).

The additional compensation for the fringe parking lots can be estimated as follows:

$$\text{Additional compensation for the fringe parking}$$
$$= \text{total turnover of parking lots } (6{,}932 \text{ million yen})$$
$$\times \text{ additional discount rate } (50/195)$$
$$\times \text{ portion applied to the fringe parking } (3{,}584/21{,}150)$$

The additional compensation for the fringe parking turns out 301 million yen.

$$301 \text{ million yen} = 6{,}932 \text{ million yen} \times (50/195) \times (3{,}584/21{,}150)$$

Thus the cost for the enterprise equals 3677 million yen. Hence, the profit for the enterprise turns out 14,192 million yen.

$$\text{Profit} = \text{revenue} - \text{cost}$$
$$14{,}192 \text{ million yen} = 17{,}869 \text{ million yen} - 3{,}677 \text{ million yen}$$

Net Increase of Retail Sales at the Tenjin District

As in the same way as before, we must take into account for the increase of turnover of parking lots. The turnover is composed of contributions by three groups, car-use shoppers parked at four parking blocks (16,730 people), those moved from the central parking block and parked at fringe parking lots (3584 people), and those parked at the central parking block (836 people). The increases of staying time for these three groups are, respectively, 44.3, 44.3, and 24.7 min, and the parking fees are, respectively, 100, 50, and 100 yen per hour. Thus the increase of turnover of parking lots is estimated as follows:

$$512 \text{ million yen} = 16{,}730(\text{people}) \times 44.3(\text{minutes})/60 \times 100(\text{yen}) \times 365 \text{ (days)}$$
$$+ 3{,}584 \text{ (people)} \times 44.3(\text{minutes})/60 \times 50(\text{yen}) \times 365(\text{days})$$
$$+ 836 \text{ (people)} \times 27.4(\text{minutes})/60 \times 100(\text{yen}) \times 365(\text{days})$$

Therefore, the net increase of retail sales turns out 14,704 million yen per year.

$$14{,}704 \text{ million yen} = 14{,}192 \text{ million yen} + 512 \text{ million yen}$$

From the above, we have seen that a considerable amount of economic effects would be brought to the city center economy by employing city center parking policies based on the viewpoint of regarding the city center as one business entity.

7 Conclusion and Future Challenges

In the city centers of many urban areas in Japan, traffic congestions are becoming a chronic problem. One of the factors to contribute the congestions is considered to be car-use shoppers who visit the city center with their cars. To alleviate congestions, policies to reduce the influx of cars into the city center have been drafted and introduced in several countries.

However, few studies, so far, have examined city center parking policies from the viewpoint of increasing the value of the town while regarding the city center as one business entity.

This study stands on the emerging view of the goal of urban development, which states that the goal of urban development is to maximize the town equity regarding the town as one entity, and addressed the problem.

For the purpose, taking up the Tenjin district, one core part of the city center of Fukuoka City, Japan, we first clarified the size of economy car-use shoppers bring to the city center and estimated the size of economic loss due to the cruising time for parking spent by car-use shoppers.

The most significant contribution of this study, we believe, is that we provided a conceptual scheme for a business model to implement the city center parking polices. In the business model, we created the enterprise implementing the city center parking policy as its business to enhance the value of the town.

Consequently, we have shown that the enterprise actually must be highly profitable and sustainable to continue to implement the city center parking policy while enhancing the value of the city center district regarding the city center as one entity based on the real micro behavioral data obtained from actual on-site surveys of consumer shop-around behaviors and behaviors of car-use shoppers conducted at the city center of Fukuoka City.

The biggest future challenge would be a further refinement of the business model presented in this study. Since the business model presented here is just thought of as a rough sketch of detailed plan, the way to realize the enterprise must be scrutinized while considering the possibility of applications of the advancing technologies of mobile ICT and IoT.

References

1. Saito S, Sato T, Yamashiro K, Takagi K (2008) Estimating effects of establishment policy of parking lots at city center viewing from visitors' behaviors who used parking. In: Papers of the 24th annual meeting of the Japan Association for Real Estate Sciences. pp 169–176. (in Japanese)
2. Saito S, Kakoi M, Nakashima T (1999) On-site Poisson regression modeling for forecasting the number of visitors to city center retail environment and its evaluation. Stud Reg Sci 29:55–74. (in Japanese)
3. Saito S, Kumata Y, Ishibashi K (1995) A choice-based Poisson regression model to forecast the number of shoppers: its application to evaluating changes of the number and shop-around pattern of shoppers after city center redevelopment at Kitakyushu City. Pap City Plan 30:523–528. (in Japanese)
4. Saito S, Nakashima T, Kakoi M (1999) Identifying the effect of city center retail redevelopment on consumer's shop-around behavior: an empirical study on structural changes of city center at Fukuoka City. Stud Reg Sci 29:107–130. (in Japanese)
5. Saito S, Yamashiro K (2001) Economic impacts of the downtown one-dollar circuit bus estimated from consumer's shop-around behavior: a case of the downtown one-dollar bus at Fukuoka City. Stud Reg Sci 31:57–75. (in Japanese)
6. Saito S, Yamashiro K, Kakoi M, Nakashima T (2003) Measuring time value of shoppers at city center retail environment and its application to forecast modal choice. Stud Reg Sci 33:269–286. (in Japanese)
7. Saito S, Yamashiro K, Nakashima T, Igarashi Y (2007) Predicting economic impacts of a new subway line on a city center retail sector: a case study of Fukuoka City based on a consumer behavior approach. Stud Reg Sci 37:841–854. (in Japanese)
8. Tamura M, Saito S, Nakashima T, Yamashiro K, Iwami M (2005) Forecasting the impacts of a new monorail line from the survey on consumers' stated prospective behavior changes: a case study in Naha City, JAPAN. Stud Reg Sci 35(1):125–142. (in Japanese)
9. Shimada T, Akamatsu H, Nakagawa Y (2003) Influence on parking-lot-selections with visiting-duration and shopping-behavior in Takasaki City-Center. Proc Infrastruct Plan (CD-ROM) 28. (in Japanese)
10. Uchida K, Kagaya S (2003) A study on the induced probability regarding customers' shop-around behavior by parking lots' situation in the center of Sapporo. Stud Reg Sci 33(1):99–114. (in Japanese)
11. Deguchi C, Kiyota K, Yoshitake T, Matsuyama S (2008) A case study on social experiments for cooperative operation system of parking lots in Miyazaki and problems. Infrastruct Plan Rev 25(2):373–384. (in Japanese)
12. Saito S (2000) Report on the survey of consumer shop-around behavior at city center of Fukuoka City, Japan 2000: with focusing on underground space and comparison of attractiveness of Japanese and Korean Cities, Fukuoka Asia Urban Research Center, Fukuoka City Government. (in Japanese)
13. Arnott R, Rave T, Schob R (2005) Alleviating urban traffic congestion. MIT Press, Cambridge, MA
14. Saito S, Sato T, Yamashiro K (2010) Little's formula and parking space policy viewed from consumers' parking behaviors at city center retail environment. Paper presented at the ninth International symposium on operations research and its applications held at Chengdu-Jiuzhaigou, China. Oper Res Appl 12:500–511

15. Saito S, Yamashiro K, Iwami M, Imanishi M (2014) Parking space policy for midtown commercial district and consumers' parking and shop-around behaviors: applying Little's formula to the analysis of demand-supply balances for parking capacity in Tenjin area, the midtown of Fukuoka, Japan. Fukuoka Univ Rev Econ 58(3–4):75–98. (in Japanese)
16. Saito S, Sato T, Yamashiro K (2008) An analysis of consumers' parking behaviors at city center commercial district. Paper presented at the 45th annual meeting of Japan Section of Regional Science Association International (JSRSAI). (in Japanese)

Part VII
Information and Consumer *Kaiyu* Behaviors

Chapter 20
Exploring Information Processing Behaviors of Consumers in the Middle of Their *Kaiyu* with Smartphone

Mamoru Imanishi, Kosuke Yamashiro, Masakuni Iwami, and Saburo Saito

Abstract At a year-end sale held in the Tenmonkan district, the city center commercial district of Kagoshima City, Japan, we carried out a social experiment that attempted to measure the effect of information provision on visitors by using a smartphone application developed by FQBIC that was able to simultaneously record users' positions and their interactions with information contents provided by the town such as flyers and the like. This study, as a first step, analyzes the logs obtained through this social experiment, which record the interactions between visitors and information provided by the town, and investigates what kinds of information contents and forms would most effectively induce visitors' *Kaiyu* within the city center district.

The Sections 1–3 of this chapter are based on the paper, Mamoru Imanishi, Kosuke Yamashiro, Masakuni Iwami, Saburo Saito [1], "Measurement and Analysis of Effects of Providing Information to Visitors during a year-end sale at a city center commercial district," *Papers of The 30th Annual Meeting of The Japan Association for Real Estate Sciences*, pp. 65–70, 2014, in Japanese, which is revised for this chapter. The Sections 4 and 5 of this chapter are based on the paper, Kosuke Yamashiro, Mamoru Imanishi, Masakuni Iwami, Saburo Saito [10], "What kind of information provision most effectively induces *Kaiyu*? A social experiment using smartphones during a year-end sale," the paper presented at the 52nd Annual Meeting of Japan Section of Regional Science Association International (JSRSAI), 2015, which is revised for this chapter.

M. Imanishi · K. Yamashiro
Department of Business and Economics, Nippon Bunri University, Oita City, Japan
e-mail: imanishimm@nbu.ac.jp; yamashiroks@nbu.ac.jp

M. Iwami
Fukuoka University Institute of Quantitative Behavioral Informatics for City and Space Economy (FQBIC), Fukuoka, Japan
e-mail: miwami@econ.fukuoka-u.ac.jp

S. Saito (✉)
Faculty of Economics, Fukuoka University, Fukuoka, Japan
Fukuoka University Institute of Quantitative Behavioral Informatics for City and Space Economy (FQBIC), Fukuoka, Japan
e-mail: saito@fukuoka-u.ac.jp

© Springer Nature Singapore Pte Ltd. 2018
S. Saito, K. Yamashiro (eds.), *Advances in Kaiyu Studies*, New Frontiers in Regional Science: Asian Perspectives 19, https://doi.org/10.1007/978-981-13-1739-2_20

Keywords *Kaiyu* · Shop-around behavior · Information provision · Location information · Smartphone app · Information transaction · Logs · GPS · Indoor Messaging System (IMES) · Forms of information contents · Shake · Tap

1 Purpose of This Study

In recent years, when using the GPS functionalities and accelerometers mounted on smartphones, obtaining "big data" has become much easier than before. Thus, also in the field of urban planning, it has begun to be gaining the attention on how to make use of big data (Cf. [2]). However, by and large, the attention has remained chiefly at the level of replacing conventional "hard" facility planning with planning related to information technologies that are likely to be implemented by ICT, such as traffic information and digital signage. The attention has rarely been talked in connection with the goal of urban development and the scientific evaluation of urban development policies.

On the other hand, the argument that big data should be connected with clarifying the goal of urban development and increasing the value of the town and should be used as a tool for the scientific evaluation of urban development policies has been highlighted by Saito [3–6]. In particular, it should be noticed that Saito has been paying much attention to the great possibility of big data generated from individual consumer's real-time micro-decision-making including interactions with information.

While this study stands on the same viewpoint as Saito's, in order to realize his perspective in a more concrete setting, we have decided to employ the following method. Taking up an actual city center commercial district at a regional core city as an experiment field, we have established there the temporal information environment equipped with a system that can generate a big data which can record the real-time micro-behavior history data of consumers who visited there concerning what kinds of interactions and what kinds of decisions they have made with the provided information, and then we analyze the particulars of the big data so obtained.

More specifically, we focused on *Tenmonkan*, which is the name of the city center commercial district at Kagoshima City,[1] Japan, and developed a smartphone application in conjunction with the year-end sale at Tenmonkan. By measuring the location of smartphones inside and outside of shops as well as providing flyer information for the year-end sale through the app, interactions between visiting consumers and provided information were prompted. With these devices, we established an integrated location/content information platform to enable one to store the logged records of consumers' transactions with provided information as big data.

---

[1]Kagoshima City is located at the southern part of Kyushu island, Japan, whose population is 599,814 (2015 census). Kagoshima City is a capital city of Kagoshima Prefecture and serves as a core city in this region. The 1.5% Kagoshima metropolitan area has 1,087,447 population (2010 census).

Furthermore, we recruited general participants among the shoppers visiting at the year-end sale who agreed to use the app on this platform in order to carry out the social experiment for collecting log data.

Until now, under the traditional information technology environment, it has been extremely difficult to measure and verify what kind of information provision has caused consumers to change their behaviors and how it has induced their *Kaiyu*.

The aim of this study is to take a major step toward overcoming this situation by making use of new information technologies.

Speaking further, our ultimate goal is to revitalize the town with stimulating visiting consumers' *Kaiyu* by providing them with real-time on-site information to support their on-site decision-makings. Toward that goal, we build an information platform that can scientifically verify the most effective ways of providing what information to which consumers for what purpose, and from there we intend to construct a model to explain the interactions between consumers and information provided using big data obtained there.

Here, as a first step toward this aim, the purpose of our study is set to carrying out a fundamental analysis of the information provision effect, which is related to the question as to what kind of information provision has what sort of effects on what type of visitors, based on data obtained from the social experiment of information provision through our smartphone app conducted in conjunction with the end-of-year sale of the Tenmonkan at Kagoshima City.

2 Overview of the Social Experiment

2.1 Outline of the Social Experiment

This study takes the Tenmonkan district, the city center commercial district of Kagoshima City, as the subject area for our social experiment. We obtained the cooperation of the local TMO (Town Management Organization), whose name is "We Love Tenmonkan Council." In conjunction with the year-end sale of Tenmonkan, our social experiment of information provision had been carried out for 3 days from Friday, December 13 to Sunday, December 15, 2013. The experiment utilized location information technologies such as GPS, the QZSS[2] (Quasi-Zenith Satellite System), and the IMES (Indoor MEssaging System). This social experiment was a part of our research supported by the competitive research fund from the Ministry of Internal Affairs and Communications (MIC) under the program, Strategic Information and Communications R&D Promotion Program (SCOPE). Our research title is "A development study on the system for measuring tourist movements around wide area using auto-GPS and IMES, and for an effective information provision to trigger *Kaiyu*."

---

[2]GNSS (global navigation satellite system) developed and operated by Japanese Government

Fig. 20.1 Smartphone app screen

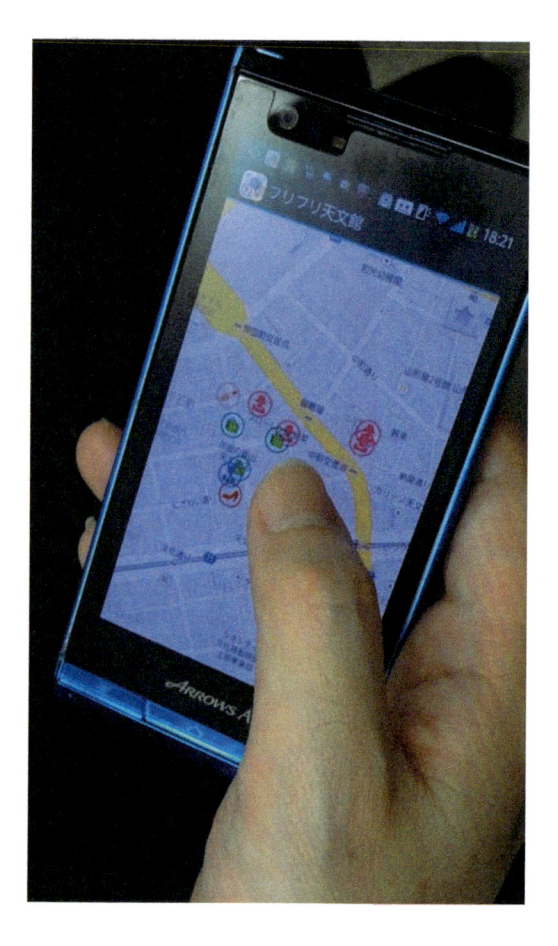

In the social experiment, we used a smartphone app we developed and named "*Furifuri* Tenmonkan" (Fig. 20.1), through which the sale information was randomly presented to the experiment participants, and investigated the information provision effects on the participants.

IMES[3] (Indoor MEssaging System)

IMES transmitters using IMES technology were installed in 67 individual shops in the Tenmonkan district, which allows location measurements to be taken seamlessly both inside and outside buildings. Until now, there were some cases in which

---

[3]Indoor location positioning system developed by JAXA (Japan Aerospace Exploration Agency). IMES installs an indoor GPS transmitter (module) using the same radio format as the GPS satellite and transmits "position information" of the transmitter instead of time information from the transmitter.

Fig. 20.2 Locations of IMES installed (Cf. [7])

multiple IMES transmitters had been installed within the same commercial facility, but this study represents the first attempt to install IMES on a community-wide scale (Fig. 20.2). Because positioning by conventional GPS can sometimes result in errors of up to 50 m, and smartphones are unable to perform GPS positioning inside buildings, we cannot easily check for visits to shops from the GPS position log. However, here by installing IMES in shops, we have had the benefit of obtaining a reliable history of shop visits, even to adjacent shops.

Participants in the Social Experiment

On the day of the social experiment, we set up a reception desk for the experiment on Tenmonkan-*Hondori* Street and asked visitors with android-based mobile phones to volunteer to participate in the social experiment. At the same time as participants were asked to install the social experiment app from the Google Play Services, we also provided them with the loan of IMES receivers. Afterward, as the participants loaded the app, we asked them to shop around the Tenmonkan district, and when they returned back to the reception point, we asked them to answer a post-questionnaire survey when they were returning the IMES receivers.

A total of 87 participants took part in our social experiment (the number of who responded to the post-questionnaire survey). In order to encourage participation in the social experiment, participants were provided with three raffle tickets for free. At the year-end sale, shoppers were given one raffle ticket for each of their 5000 yen

(50 dollars) purchases. While a sample size of 87 may seem small in relation to usual questionnaire surveys, when we include the various log data records – including search results, page browses, and favorites – as well as the interview post-questionnaire survey data collected by researchers for later verification, it can be said that the resulting data was quite rich. In fact, it contains 50,000 records.

Smartphone App

For this study, we developed a smartphone app that conducted simultaneous positioning using GPS, IMES, and Wi-Fi technology and provided information to users according to their location. The app, after obtaining participants' consent, transmitted the acquired position information to our database server at regular intervals (of 10 s) along with participants' anonymized and encrypted identification numbers and the date and time of acquisition.

On the other hand, when the participants shook their smartphones, they would be presented on the map with ten icons corresponding to randomly selected shops within a 100 m radius of their current location. This setting was set for this social experiment and can be changed.

The information provided was about 145 shops participating in the year-end sale whose discount information was listed on their application form for the year-end sale. Of these, we chose 54 shops which were also featured in a color flyer, and in order to verify this visual effect, the information of each of these 54 shops we trimmed from the color flyer was added to each shop information as a banner in the app.

The shops were divided into nine categories, including men's fashion, ladies' fashion, bags, jewelry, beauty, shoes, health, and others. The nine shop categories have different icons in the app, and each shop is displayed in the app by the icon corresponding to the category each shop belongs to.

Because there was a flood of information about restaurants and bars in a variety of different media formats, these shops were excluded from the scope of the experiment. As a result, we have eight shop categories.

2.2 Data Obtained by the Social Experiment

From our social experiment, we have obtained the following log data:

1. Participants' shop-around log (every 10 s)
2. Information search and display log (where did consumers search for information?)
3. Details and history of information search and display (which shops were displayed?); shake
4. Information browsing history (as to which shops did consumers view information?); tap

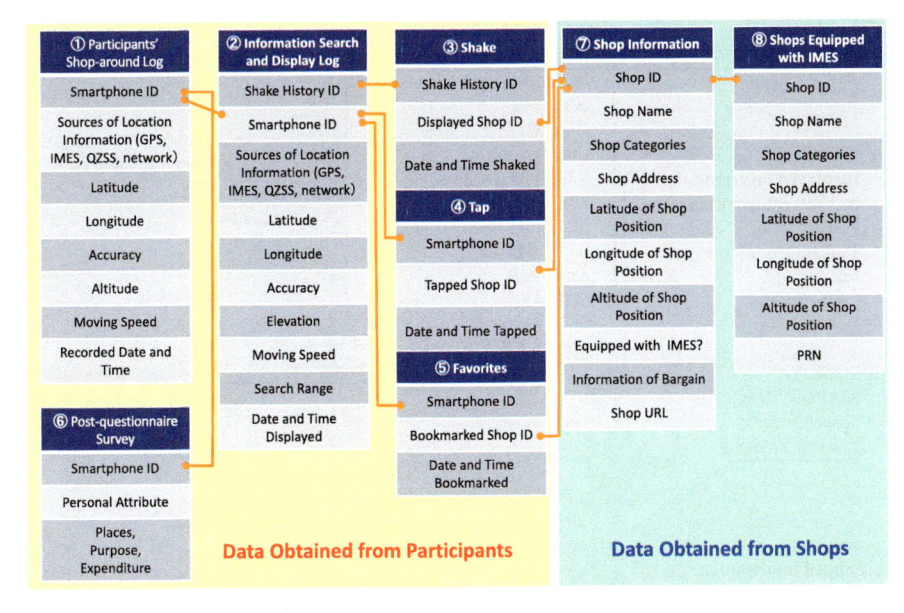

Fig. 20.3 Interrelationship among various log history data

5. Favorites (whether consumers saved a location in their favorites)
6. Post-questionnaire survey (interview survey)
7. Shop information
8. Shops equipped with IMES
9. User list

The above log data are interrelated with each other. A somewhat complicated interrelationship among these log data is displayed in Fig. 20.3. Key items that make these log data to be interrelated with each other are time, smartphone ID, and location (latitude, longitude, and altitude).

The behavioral and information transaction history of each consumer who visited and has interacted with information in the city center retail environment is recorded in movement, search, and display log data. By integrating these three logs, we can recover the history of information transactions with the retail environment for each consumer. In other words, we are able to know how consumers are transacting with the information provided by the retail environment, while they are moving around the city center.

Figure 20.4 displays the instance of history of one sample recovered by integrating above log data. From this figure, we see where the sample searched or wanted information by shaking and where the sample displayed or requested the detail information about shops by tapping icons.

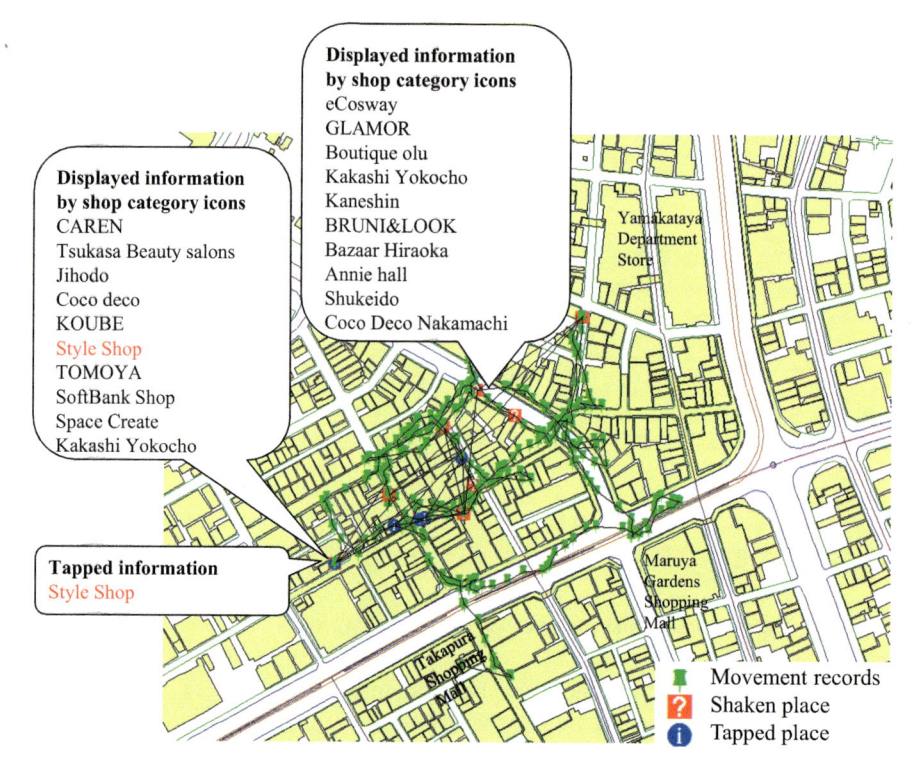

Fig. 20.4 History of information transactions by one sample (Cf. [7])

2.3 How to Measure Consumer Information Processing Behaviors?

In this section, for ease of understanding, we summarize the scheme of how we conceptualize the information processing behaviors of consumers which can be measured by our smartphone app, while consumers are using the app. Keywords are "shake" and "tap."

The scheme is succinctly depicted in Fig. 20.5. In our smartphone app, users shake their smartphones to request the app to provide the information of retail environment near them on the screen of their smartphones. Thus, the "shake' by consumers can be regarded as their "search for information" behavior in their information processing behaviors.

Under the setting in this social experiment, the result of "shake" is the randomly selected ten shops near the present location of the smartphone holder. The ten selected shops are displayed on the map at those locations with icons corresponding to the categories those shops belong to.

On the other hand, as for the "tap" in our app, users tap some icon displayed in their smartphone to request the app to show the detailed information about the shop

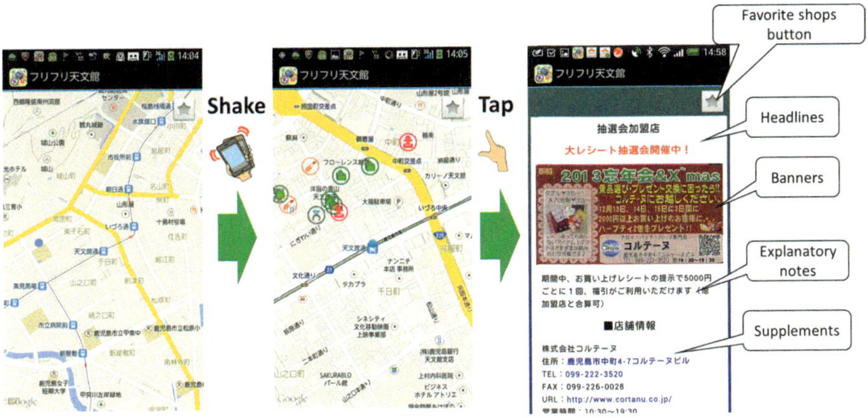

Fig. 20.5 Shake and tap for searching and focusing and the forms of information contents

of the tapped icon. Thus, the "tap" by consumers can be thought of as their "focus on the detail information" behavior in their information processing behaviors.

3 Analysis of Shake, Tap, and *Kaiyu* Visualization

3.1 Kaiyu *Visualization*

Figure 20.6 displays the log of spatial movements by shop-around behaviors of all participants within the Tenmonkan district. From the visualization in the figure, we see that their shop-around movements are centralized among major shopping streets and a department store such as between *Tenmonkan-Hondori* Street, *Haikara-Dori* Street, the *Senichi* Arcade, and *Yamakataya* department store.

Most previous studies traditionally have remained at the level of visualizing consumers' spatial movements recorded as location log by utilizing GIS and tried to apprehend consumers' shop-around behaviors according to this kind of visualization [8, 9]. However, visualization is just describing the results of consumers' decision to shop-around so that it does not explain why consumers decided to do such shop-around behaviors.

In order to carry out the urban development based on the scientific evidences using big data, it becomes important to analyze micro-behavioral data of individual consumers.

For this reason, the smartphone app we developed this time was designed to become a system that makes it possible to statistically verify what kind of information provision causes what kinds of changes to the shop-around behaviors of the app's users.

More specifically speaking, as shown in Fig. 20.3 about the interrelationship among logs, our app is designed to be a system that allows the statistical analysis of

Fig. 20.6 Behavior log of all participants within Tenmonkan district (Cf. [7])

Table 20.1 Average number of information searches by shakes

N	Mean	Std	Median	Mode	Max
91	18.13	29.982	9	7	243

why the selected result was chosen by users from among the presented alternatives by recording alternatives presented as a log in addition to the selected result.

Hence, it is necessary to focus not only on visualization, but also on the analysis of micro-behaviors of individual consumers. Therefore, in this study, we went beyond visualization to carry out the analysis of individual consumers' behaviors.

3.2 Feature of Shake for Searching

Table 20.1 shows how much information the participants retrieved by shaking their smartphones. Users did so 18 times on average. While the sample size of 91 was larger than the 87 participants in our social experiment, this is due to the presence of users who activated the app without enrolling in the social experiment. Although they did not take part in the post-questionnaire survey, they are included in the analysis since their log data is usable.

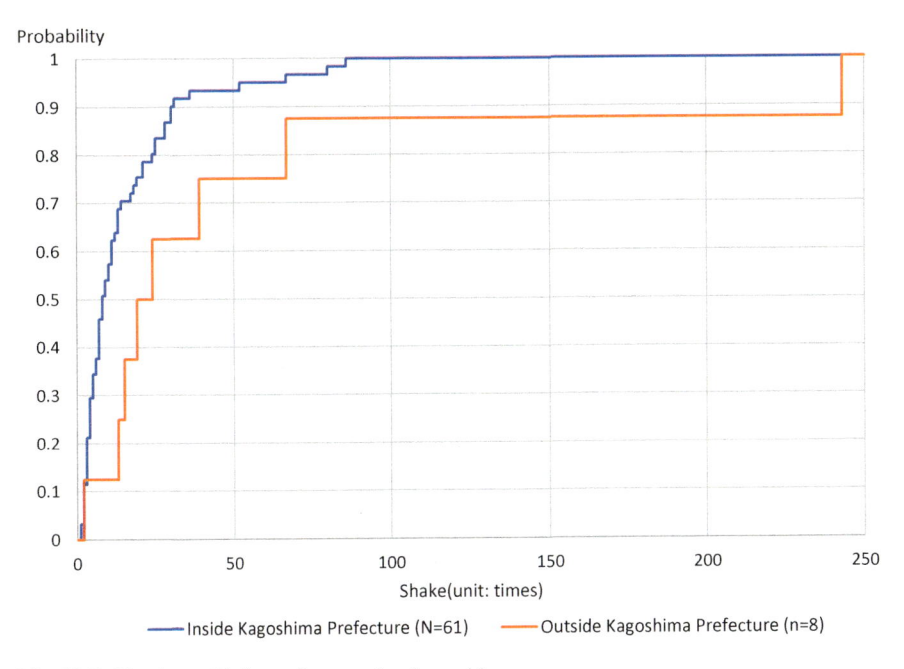

Fig. 20.7 Numbers of information searches by residence

In addition, from Fig. 20.7, although the sample size is small, we obtained a result which shows that participants from outside of the prefecture searched for information more often than did prefectural residents.

On the other hand, it is informative to see where participants shake more often. Figure 20.8 displays where participants shook their smartphones for searching for information.

3.3 Feature of Tap for Focusing on the Detail Information

Figure 20.9 illustrates the stacked bar chart showing the numbers of times icons were displayed and tapped with the number of shops registered in database belonging to each shop category. All of these numbers are expressed as percentages to the total numbers given in the lower part of the figure. The top bar in the stacked bars for each shop category expresses the percentage of the number of shops registered in database belonging to that category, the middle bar the number of times the icon of that shop category is displayed for the icons of the randomly selected shops by the app, and the bottom bar the number of times the icon of that shop category is tapped.

Since our app for this time randomly chooses shops and displays them on the map of the smartphone, the percentages of shop categories for displayed shops should be

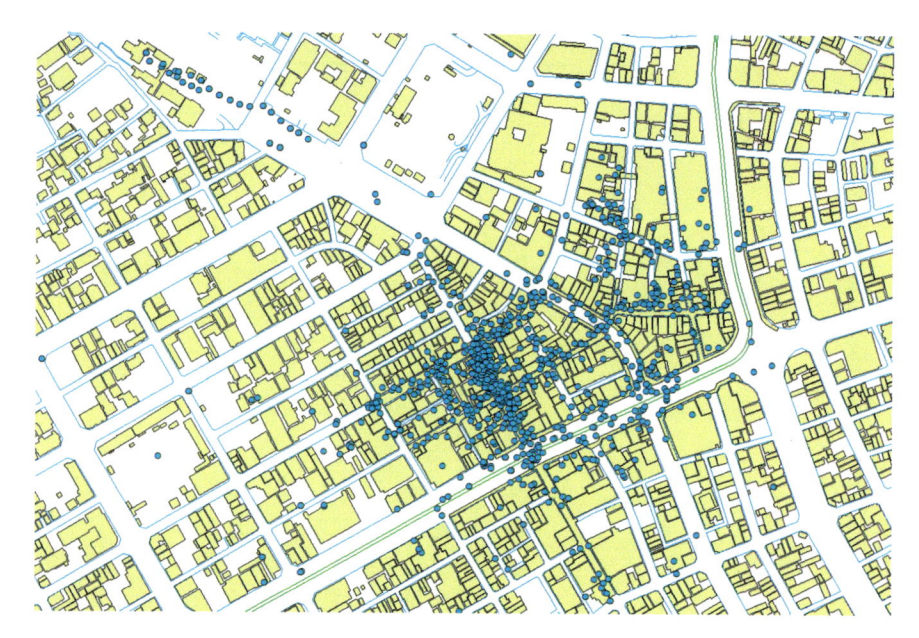

Fig. 20.8 Locations of shakes for searching (Cf. [7])

proportional to those registered in database. From the figure, we see this simple fact in the top and middle bars in the stacked bar chart.

Interesting is that the numbers of taps, while tapping is user's decision, are also proportional to those displayed by the app.

Table 20.2 displays the average number users tapped icons to view detail shop information. They tapped icons for browsing detail shop information for about seven shops on average. From Fig. 20.10, it may be seen that when we look at area of residence, visitors from outside of Kagoshima Prefecture tended to browse detail shop information more frequently.

3.4 Feature of Visit

Table 20.3 displays how often users visited a shop directly after browsing detail shop information on the app. Since IMES was installed in the interior of shops, and the system's radio waves do not reach outside the shops, as for the shops where IMES was installed, users can reliably be adjudged to have visited these shops when IMES location logs at these shops have been detected. As for shops that were not installed with IMES, these were determined from GPS logs. While the analysis of shop visits based on GPS logs has become a study in its own right, here, if users remained within 50 m of the central point of the shop for a 1-min period, it was simply judged

Shop Category Percentages

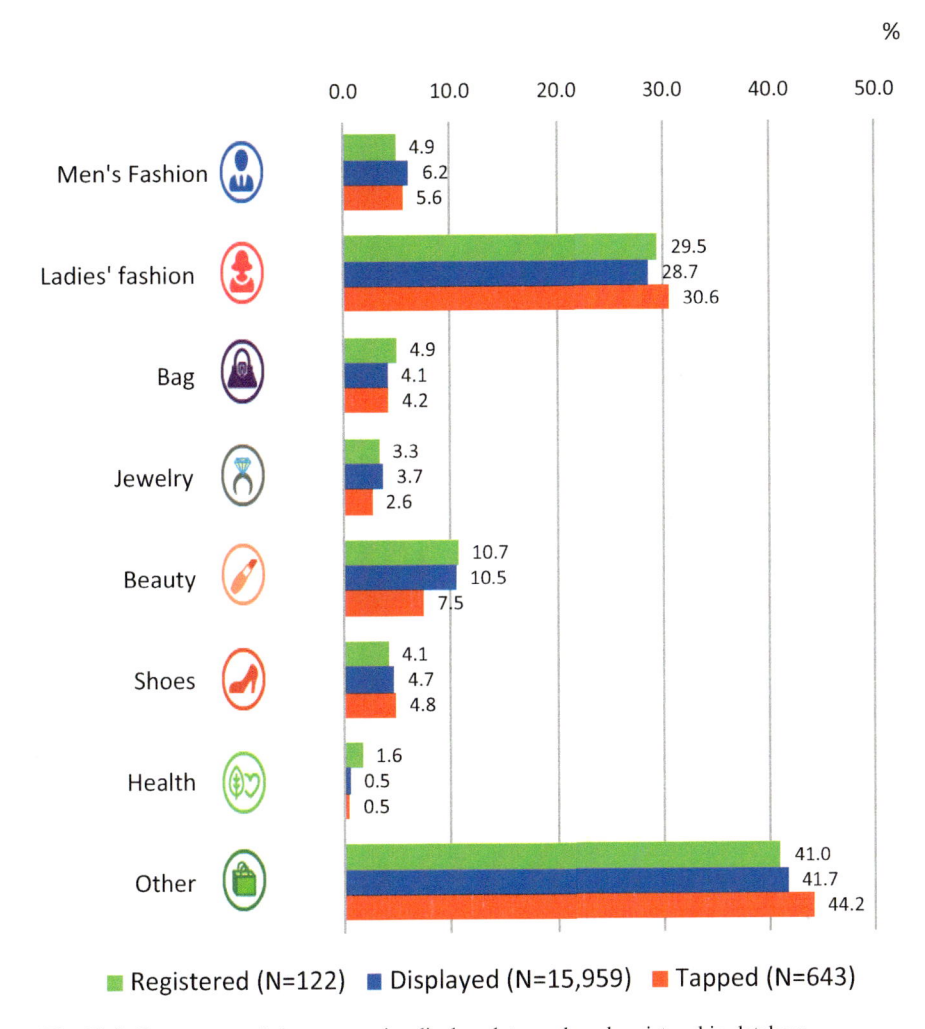

Category	Registered (N=122)	Displayed (N=15,959)	Tapped (N=643)
Men's Fashion	4.9	6.2	5.6
Ladies' fashion	29.5	28.7	30.6
Bag	4.9	4.1	4.2
Jewelry	3.3	3.7	2.6
Beauty	10.7	10.5	7.5
Shoes	4.1	4.7	4.8
Health	1.6	0.5	0.5
Other	41.0	41.7	44.2

Fig. 20.9 Percentages of shop categories displayed, tapped, and registered in database

Table 20.2 Average number of taps for detail shop information

N	Mean	Std	Median	Mode	Max
91	7.14	10.853	3	1	73

that they had visited the shop in question. Thirty-two people actually visited two shops on average.

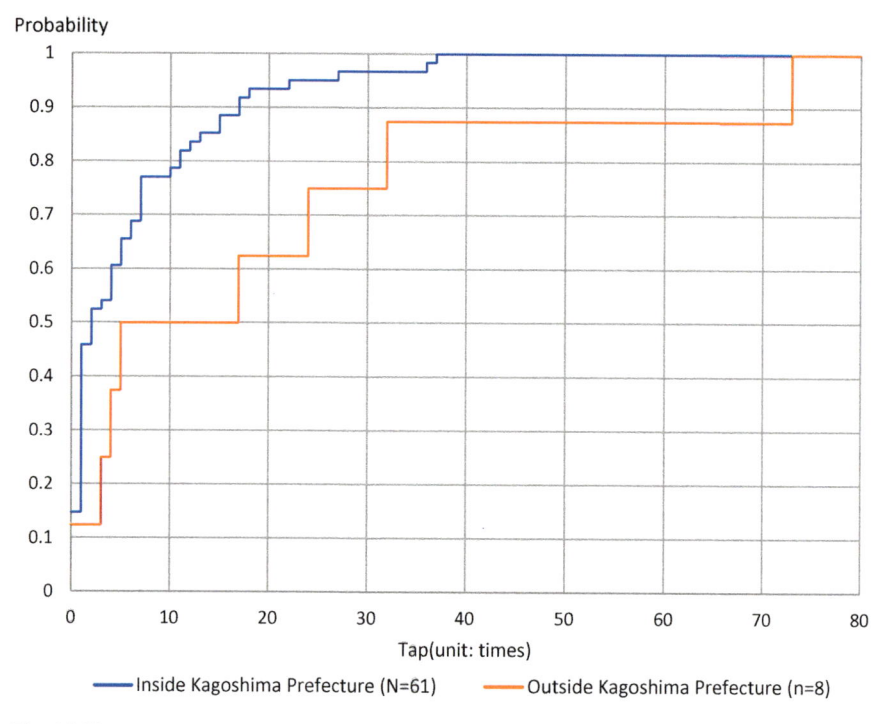

Fig. 20.10 Numbers of taps for detail shop information by residence

Table 20.3 Users actually visited the tapped shop?

N	Mean	Max	Min
32	2.25	15	0

Table 20.4 Percentage of the number of taps to that of shakes

N	Mean	Std	Median	Mode	Max
91	0.43	0.350	0.33	1.00	1.00

Table 20.5 Percentage of the number of taps to that of icons displayed by shakes

N	Mean	Std	Median	Mode	Max
91	0.057	0.07662	0.033	0.00	0.167

3.5 Feature of Transition Rates

Table 20.4 shows the percentage of how many times icons were tapped to the number of "shakes" for searches. The percentage of the number of taps to that of shakes was 43%.

Table 20.5 shows the percentage of the number of taps to the total number of icons displayed. In this social experiment, when the smartphone is shaken once, ten icons corresponding to the randomly selected shops come up by the app. Thus, the total number of displayed icons is about ten times as large as that of shakes. From the table, we see that the percentage was 5.7%, which is larger than 4.3%, tenth of the percentage in the previous Table 20.4.

This is because while one tap corresponds to one icon, users are possible to tap more than once by returning to the original shake result screen, or partly because when the size of the candidate shops from which the ten shops are randomly selected is less than 10, the number of icons to come up becomes smaller than 10.

4 What Kind of Information Provision Stimulates Tap by Consumers

4.1 A Logit Model to Investigate What Information Factors Affect "Tap" by Consumers

We are concerned with how to stimulate *Kaiyu* by information provision. For that purpose, we must explore the most effective way to provide information to induce consumers' *Kaiyu* behaviors. In this study, we divided consumers' information processing behaviors into three phases, shake, tap, and visit. In this framework, raising the transition rate becomes the key. In short, how to raise the transition rate from displaying to tapping and from tapping to visiting becomes the key point.

In this section, we investigate what kind of shop information is likely to be tapped by using a logit model.

What sorts of shop information should best be provided to consumers? As one hypothesis, it may conceivably be necessary for consumers to be provided with information about shops as near at hand as possible. Alternatively, it might be necessary for them to be provided with information about shops in their walking direction. So it might not be needed to provide them with shop information existing in the opposite direction to their walking direction.

Hence, in order to examine this question, we estimated the parameters relating to distance and direction using a logit model. We also add the explanatory dummy variables corresponding to the nine shop categories.

The choice probability for consumer i to tap icon m out of n displayed icons is expressed as the following logit model:

$$p_m^i = \frac{\exp(V_m^i)}{\sum_{j=1}^{n} \exp(V_j^i)}, \quad m = 1, \ldots, n \qquad (20.1)$$

where V_m^i represents the deterministic utility obtained by consumer i by tapping icon m.

In addition, this deterministic utility, as a linear function relating to direction dir, distance des from the current location to the shop, and dummy variables for nine shop categories, is expressed as follows:

$$V_m^i = \alpha \text{des}_m^i + \beta \text{dir}_m^i + \sum_{k=1}^{8} \gamma_k \delta \text{cat}_{mk} \qquad (20.2)$$

Here, des_m^i, dir_m^i represents the distance and direction from the location where consumer i shakes the smartphone to the shop represented by icon m.

For direction, we took inner product of the directional vector $\overrightarrow{AB}$ from a consumer's past position at coordinates A to the current position at coordinates B and the directional vector $\overrightarrow{BC}$ from the consumer's current position at coordinates B to the coordinates C of the icon that was displayed.

Dummy variables δcat_{mk} corresponding to the nine shop categories are defined as $\delta \text{cat}_{mk} = 1$ if the icon m belongs to shop category k, 0 otherwise.

Parameters α, β, γ_k, $k = 1, \ldots, 8$ are estimated by the maximum likelihood estimation method.

4.2 Estimated Results

Table 20.6 shows the estimated results of the parameters. From the table, we see that the distance to the destination and the direction to the icon strongly affect the tapping icon behaviors.[4]

These findings accord to our hypothesis. The two facts that the more the shops are closer, the more frequently those icons are tapped and that the shops existing in the same direction as the walking direction are more frequently tapped are simple but they are a starting point to explore further.

5 What Kind of Information Provision Most Effectively Induces *Kaiyu*?

5.1 Visit Ratios of Tapped Shops

In this section, we will perform the analysis of what kind of information provision about the shop effectively induces the visit to that shop.

---

[4]Three shop categories, beauty, health, and others, had not been tapped so that their dummy variables are deleted from the explanatory variables.

Table 20.6 Estimated results of parameters

Variable	Parameter estimate	SD	t value	Pr>|t|
Distance	−10.5684	0.404	−26.16	<0.0001
Direction	0.3678	0.089	4.15	<0.0001
Men's fashion	−0.5028	0.094	−5.38	<0.0001
Ladies' fashion	0.0345	0.045	0.77	0.4419
Bag	0.2593	0.090	2.88	0.004
Jewelry	−0.1659	0.098	−1.7	0.0889
Shoes	−0.1714	0.091	−1.88	0.0605
Loglikelihood with all parameters set to zeros	$L(0)$		−7069	
Loglikelihood with estimated values of parameters	$L(\widehat{\beta})$		−6674	
−2 × Likelihood ratio	$\rho = -2\left[L(0) - L(\widehat{\beta})\right]$		788.92	
McFadden's R-square	$\rho^2 = 1 - L(\widehat{\beta})/L(0)$		0.0558	
Adjusted R-square	$\overline{\rho}^2 = 1 - \left(L(\widehat{\beta}) - K\right)/L(0)$		0.0543	

$n = 3070$

From the analysis of users' tapping and visiting behaviors, we found that as a whole, among the 643 views about shops displayed with tapping by users, the number of shops users actually visited was 100 shops. The average percentage of transition from tapping to visiting was 15.6%.

In order to investigate more details about the transition, we analyzed the transition rate from tapping to visiting for each individual shop as the analysis of visit ratios for individual shops. The obtained results of visit ratios by shops are ordered from the highest in the decreasing order. Table 20.7 shows these results. The table shows only the shops which indicate their visit ratios are equal to or larger than 20%.

From the figure, we see that the highest visit ratio was 87.5% and a wide variation of visit ratios exists among shops.

5.2 How the Forms of Information Contents Affect Shop Visits?

Furthermore, we investigate whether or not the forms of information contents provided for the shop affect the visit ratio to that shop.

The forms of shop information contents we consider were depicted in the rightmost column of Fig. 20.5. They are banners, headlines, explanatory notes, and supplements.

Table 20.7 Visit ratios by shops with the numbers of taps and visits

ShopID	Shop Name	Category	Number of Taps	Number of Visited	Visit ratio
29	Baggage Higuchi	Bags	8	7	87.5%
32	coco deco	Ladies' fashion	8	6	75.0%
6	Daruma-ya Cosmetic Store	Beauty	8	5	62.5%
143	216 Junction STORE	Others	14	8	57.1%
130	Thank You Mart	Others	7	4	57.1%
94	Megane no Yonezawa (Tenmonkan)	Others	6	3	50.0%
135	Iki-ya	Others	4	2	50.0%
7	Futam-iya	Others	2	1	50.0%
59	BRUNI&LOOK	Ladies' fashion	2	1	50.0%
55	Minoru-en Green Tea Shop	Others	10	4	40.0%
96	Boushi-ya Hat Shop	Others	5	2	40.0%
116	chandelie	Ladies' fashion	6	2	33.3%
105	Futon no Kondo	Others	3	1	33.3%
108	Shobi-do Shoe Store	Shoes	3	1	33.3%
120	Paris Miki Glasses Store (Tenmonkan)	Others	13	4	30.8%
118	SWALLOW	Mens' fashion	7	2	28.6%
128	R's Stage	Ladies' fashion	18	5	27.8%
3	Jiho-do Clock Store	Jewelry	8	2	25.0%
113	Edo-ya (Shoe Store)	Shoes	8	2	25.0%
138	TOMOYA	Ladies' fashion	8	2	25.0%
9	Petit Bero	Ladies' fashion	4	1	25.0%
98	Megane Super (Naya Dori)	Others	4	1	25.0%
30	Daicyu	Others	13	3	23.1%
56	Makino	Others	13	3	23.1%
142	REGISTA Armadio	Mens' fashion	9	2	22.2%
58	Meishi-do	Others	10	2	20.0%
72	Day Light	Ladies' fashion	5	1	20.0%
123	Boutique OI	Ladies' fashion	5	1	20.0%
129	KOUBE	Others	5	1	20.0%

With or Without Banners and Visit Ratios

Table 20.8 shows that while for the shops without banners in their displayed information, the average of visit ratios to these shops is 8.6%, those shops with banners attain the average of visit ratio to their shops, 20.2%.

With or Without Headlines and Visit Ratios

Similarly, Table 20.9 gives the result of the case for with or without headlines. In this case the shops with headlines lowered their average visit ratio. This result might be contrary to the intuition. One possible reason for this result may be that the place where headlines are written is located above banners so that headlines had not so much appeal. At any rate, further investigations are needed.

With or Without Explanatory Notes and Visit Ratios

Table 20.10 shows the case for with or without explanatory notes. The shops with explanatory notes increased their average visit ratios. The average of visit ratio for the shops with explanatory notes is 14.5% in contrast to 9.0% for that without explanatory notes.

Table 20.8 With or without banners and visit ratios

Group	N	Mean	Std	Min	Max
With banners	28	0.202	0.3136	0	1
Without banners	117	0.086	0.1743	0	1

$t(143) = -2.65, p < 0.01$

Table 20.9 With or without headlines and visit ratios

Group	N	Mean	Std	Min	Max
With headlines	74	0.091	0.1879	0	1
Without headlines	71	0.126	0.2349	0	1

$t(143) = 0.99, p < 0.05$

Table 20.10 With or without explanatory notes and visit ratios

Group	N	Mean	Std	Min	Max
With explanatory notes	48	0.145	0.2574	0	1
Without explanatory notes	97	0.090	0.1845	0	1

$t(143) = -1.49, p < 0.05$

Table 20.11 With or without supplements and visit ratios

Group	N	Mean	Std	Min	Max
With supplements	46	0.151	0.2612	0	1
Without supplements	99	0.088	0.1831	0	1

$t(143) = -1.69, p < 0.1$

With or Without Supplements and Visit Ratios

In the same way, Table 20.11 gives the case for with or without supplements. Similar to the case of the explanatory notes, the shops with supplements indicate the higher average of visit ratio, 15.1% than the average of visit ratio for those without supplements, 8.8%.

6 Conclusion and Future Challenges

The most significant contribution of this study, we believe, is that though temporal, we developed an integrated location/content information platform that enables one to obtain a big data concerning consumers' real-time micro-behavioral choice history data of how they have interacted with on-site information provided by retail environment in a way that allows statistical analyses of these consumers' choices. Moreover, we have also demonstrated that by leveraging our original smartphone app, the interactions between consumers and information actually can be analyzed and the micro data such as obtained here has a great possibility to advance our understanding consumers' interactions with information.

In this study, as a first step, we found a simple fact that consumers are more likely to tap the information provision about shops existing closer to their present location and in the same direction as their walking direction.

As for our future challenges, since a big data such as obtained here is a rich data related to consumers' decision at the deep micro level concerning choice of information provided, shop information, and possibly purchase, the deep analysis of this kind of big data is indispensable for the development of effective information provision technologies to induce *Kaiyu*. Thus, one of our future works is to continue to carry out the detailed analyses of interactions between consumers and information while establishing a methodology to consistently analyze a rather complicated micro data obtained here.

Another future challenge is that we transform above various analyses into a toolkit to develop a cloud computing service that provides the consumers who visit the city center with real-time on-site decision-making support services to enhance the value of the city center.

References

1. Imanishi M, Iwami M, Yamashiro K, Saito S (2014) Measurement and analysis of effects of providing information to visitors during a year-end sale at a city center commercial district. In: Papers of the 30th annual meeting of The Japan Association for Real Estate Sciences. pp 65–70. (in Japanese)
2. Nikkei (2013) Urban development by big data. The Nikkei newspaper. Morning Ed. Date: 2013/10/06/. p 1. (in Japanese)
3. Saito S (2012) Incorporating big data sciences into strategic town management: town equity researches and the future of real estate sciences. Jpn J Real Estate Sci 26:38–46. (in Japanese)
4. Saito S (2012) Strategic town management and big data sciences: smart city and town equity. Statistics 63(9):10–19. (in Japanese)
5. Saito S (2013) Composing and distributing town equity indexes for promoting real estate Investment on urban areas: urban studies and real estate businesses in the era of big data. Real Estate Res 55:13–25. (in Japanese)
6. Saito S (2014) Analytics of shop-around behaviors enhances the value of town: big data and town equity. Urban Adv 62:20–29. (in Japanese)
7. The Geospatial Information Authority of Japan Fundamental Geospatial Data Site http://www.gsi.go.jp/kiban/
8. People Flow Project (2018) Evacuation visualization. Access date: 2018/04/30/. http://pflow.csis.u-tokyo.ac.jp/data-visualization/fukkou/
9. Jalan Research Center, Zenrin Datacom (2011) Sightseeing spot analysis using second generation location information. ToriMakashi (Terima kasih in Indonesian), No. 26. pp 4–9. (in Japanese)
10. Yamashiro K, Imanishi M, Iwami M, Saito S (2015) What kind of information provision most effectively induces Kaiyu? A social experiment using smartphones during a year-end sale, paper presented at the 52nd annual meeting of Japan Section of Regional Science Association International (JSRSAI). (in Japanese)

Part VIII
Urban Policy and Consumer Welfare

Chapter 21
Travel Demand Function of Korean Tourists to Kyushu Region, Japan

Saburo Saito, Hiroyuki Motomura, and Masakuni Iwami

Abstract The purpose of this paper is to show that we can accurately forecast the recent drastic increases of Korean tourists to Fukuoka and their sharp drops after 2008 based on the travel demand function estimated by the deliberate use of microdata obtained from the survey of inbound behaviors of Korean tourists conducted in 2000. Through showing this, we have demonstrated a way to solve the two inherent problems our questionnaire survey has; that is, one is the difficulty to get income data and the other the time limitation to perform detailed stated preference questions. To avoid detailed stated preference questions, we conceptualize respondents' responses to the price changes of each of two travel modes as the changes of demand for the composite travel service composed of sea line and air flight. Based on this conceptualization, we have estimated a travel demand function for the composite travel service from Busan and Fukuoka. To enable us to estimate the travel demand function without income data, we made ratios to cancel out the income term in the travel demand function. The estimated result clearly shows that we could have predicted the recent surges and drops of Korean tourists to Fukuoka almost accurately at the time of 2000.

This chapter is based on the paper, Saburo Saito, Hiroyuki Motomura [8], "Travel demand function of Korean tourists to Kyushu: Could we have accurately predicted the drastic increase of Korean visitors to Kyushu?" presented at the 46th Annual Meeting of JSRSAI (Japan Section of Reginal Science Association International), 2009, which is revised for this chapter.

S. Saito (✉)
Faculty of Economics, Fukuoka University, Fukuoka, Japan

Fukuoka University Institute of Quantitative Behavioral Informatics for City and Space Economy (FQBIC), Fukuoka, Japan
e-mail: saito@fukuoka-u.ac.jp

H. Motomura
Department of Business and Economics, Nippon Bunri University, Oita City, Japan
e-mail: motomura@nbu.ac.jp

M. Iwami
Fukuoka University Institute of Quantitative Behavioral Informatics for City and Space Economy (FQBIC), Fukuoka, Japan
e-mail: miwami@econ.fukuoka-u.ac.jp

Keywords Travel demand function · Stated preference · Tourism · Transport mode · Composite good · Nested utility function · Transport cost · Exchange rate · Korea tourists · Kyushu Island · Busan

1 Purpose of This Study

1.1 Background

The scenery of inbound tourism in Japan has completely changed in recent several years. In 2013, the number of foreign tourists exceeded 10 million people per year for the first time. Since then, the number of foreign tourists has been rapidly increasing by about 5 million people per year for every year and reached over 28 million people per year in 2017 (Cf. [10]).

In parallel to these drastic changes, the Cabinet of Japanese Government organized the Council of Ministers for Promoting Tourism Nation in 2013 and made an Action Program toward Realization of Tourism Nation, which continues to be revised every year to the present.

Japanese Government started in 2003 the new Tourism Policy: Visit Japan Campaign. They intended to attract 10 million foreign tourists to Japan annually and set it out as the goal that should be attained within 7 years. The policy was devised because there was a big imbalance such that while Japanese tourists visiting foreign countries were 16.5 million, incoming foreign tourists to Japan were just 5.2 million at that time. In 2007, foreign tourists who visited Japan have increased to 8.3 million.

If you look at not the whole Japan but a specific region, interesting phenomena had emerged that foreign visitors to the region have increased drastically around 2006. Specifically speaking, Kyushu Island is typical such a region. Kyushu Island, the southernmost among the four islands composing Japan, is located near Korea facing with Busan about 200 km apart from each other over Tsushima Strait.

The JR (Japan Railway) Kyushu introduced in 1991 an express sea line connecting Fukuoka (Hakata Port) and Busan in about 3 h by jetfoil ship called Beetle. Since its introduction, the number of passengers both from Japan and Korea has drastically increased. Most of the travelers between Fukuoka and Busan choose not the airline but the sea line. Thus, only two daily air flights between them remain in service in 2009, while four or five air flights once existed daily. It was quite interesting to explore the reasons why Korean tourists to Kyushu region have increased drastically in a way that dispels air flights between Fukuoka (Fukuoka Airport) and Busan.

However, this landscape of the inbound tourism in Kyushu at the year of 2009 had begun to drastically change since 2010. One cause of the change was frequent entry into Hakata Port of Large Cruise Ship with 3000 or 5000 foreign tourists. The actual result of year 2017 was 328 vessels of Large Cruise Ships berthed in Hakata

Port (Cf. [11]). The other cause of the change was the introduction of LCCs (low-cost carriers). During the years from 2010 to 2016, five LCCs set up the air routes between Fukuoka and Korea (Seoul or Busan) one after another[1] (Cf. [12]). The period of the introduction of LCCs accords with the period when the rapid increases in foreign tourists visiting Japan started.

1.2 Purpose

This study started around 2000 from Saito and Motomura [4], in which we were concerned with how accurately we can forecast the number of Korean tourists visiting Fukuoka. For the purpose, we employed consumer micro-behavior approach by carrying out the questionnaire survey of behaviors of Korean tourists in Kyushu, formulated and estimated Korean tourist travel demand function to Kyushu, and forecasted the number of Korean tourists visiting Fukuoka under various price conditions. While Saito and others elsewhere (Cf. [13]) utilized the on-site Poisson model to estimate and forecast the net number of incoming visitors to the city center at local core cities, we based our model formulation on the framework of spatial economics. (Cf. [1]) Our formulation is characterized as the direct estimation of utilities of Korean tourists to visit Kyushu. Under this framework, we have already carried out several related studies [2, 3, 5–7].

The purpose of this study is that by using our model of travel demand of Korean tourists to Kyushu, we show that we can accurately predict the increase of Korean tourists from 2005 to 2007 and also the sharp drop from 2008 to 2009 even based on the data and estimated result in 2000. Since our model was estimated based on micro-behavior data obtained from the survey of behaviors of Korean tourists in Kyushu conducted in 2000, this study also aims to show the potential value of utilization of inbound micro-behavior data of foreign tourists.

We also reconsider why our model has produced accurate predictions from the statistical viewpoint.

While we are concerned with whether we can forecast recent drastic growth of foreign tourists to Japan, more specifically, the recent rush of Korean tourists to Kyushu by our model, we leave it for our future work.

---

[1]As of 2018, five LCCs have air route between Fukuoka and Korea. Air Busan entered in 2010, T'way Airlines in 2011, Jeju Air in 2012, Jin Air in 2014, and Eastar Jet in 2016.

2 Travel Demand Function: A Composite Transport Goods Approach

2.1 How to Construct an Estimable Model of Travel Demand from Available Data

Below we estimate the travel demand function of Korean tourists to Kyushu using the micro-behavior data of Korean tourists obtained from the survey of behaviors of Korean tourists in Kyushu.

The survey was conducted in March of 2000 for Korean tourist passengers in the ship of Beetle on their way back to Busan. Questionnaire sheets were distributed, answered on board, and collected in the collection box until their disembarking. As usual in the questionnaire surveys, it is quite difficult to ask respondents about their income so that we did not ask them about their income in the questionnaire. Thus, we have no available data about the income of the samples.

While most of the question items were concerned with their actual behaviors during their stay in Kyushu, we also added the question items to ask about their stated preferences, that is, question items to ask about their intention, willingness, and choices under various hypothetical conditions.

More specifically, we asked the respondents about the following items. For each of the two travel modes, sea line and air flight, Korean respondents were asked how many times you think you will visit Fukuoka when the round travel charge would decrease to 100,000 and 50,000 won for Beetle and to 150,000 and 100,000 won for air flight. In another questionnaire item, we also asked the respondents how many times they visited Fukuoka in recent 2 years. Thus, we got from respondents the number of their visits to Fukuoka under five cases: two discounted prices for Beetle, two for air flight, and the present case.

Here it should be noted that we have not obtained from respondents the number of their visits to Fukuoka by two travel modes of sea line and air flight separately but obtained the whole number of visits to Fukuoka with either of the two travel modes. If we regard two travel modes as distinct services, we should have asked the respondents about each number of their visits to Fukuoka by each of the two travel modes for all cases of different travel charges. But following this procedure would have led to five more question items the respondents are forced to respond to so that we did not do so. Thus, we have no available data about the number of visits by each of the two travel modes, but we have the data about the total number of visits by either of the two travel modes.

To deal with this limitation of available data, we employed the following conceptual framework. Two travel modes jointly provide a composite travel service between Busan and Fukuoka. Thus, the numbers of visits to Fukuoka responded by the samples are thought of as their demand for this composite travel service. Based on this framework, the number of visits to Fukuoka is regarded as the demand for the composite service which changes as the price of the composite travel service changes.

In the next section, we will formalize this conceptual framework as a hierarchical nested utility function.

2.2 A Hierarchical Nested Utility Function

We consider Korean consumers who are facing with making a decision of how many times they travel to Kyushu as tourists under various conditions, in which only fares of travel modes are possible to change, while all other factors such as the attractiveness of Kyushu and the travel time distances of travel modes and so on are assumed to be fixed as they are at present.

Suppose that a Korean traveler i has the following utility function:

$$U_i = M_i^\beta A_i^\gamma \tag{21.1}$$

Here M_i denotes a quantity index for the composite service composed of several travel modes, A_i that for a composite good composed of all other goods, and β, γ parameters such that $\beta + \gamma = 1$.

Also suppose that the Korean traveler i has a sub-utility for the composite service of travel modes in such a form that

$$M_i = \prod_{j \in C} m_{ij}^{\alpha_j} \tag{21.2}$$

where C denotes a set of all travel modes, m_{ij} traveler i's demand for travel mode j, and α_j parameters. Let I_i denote the income for Korean traveler i. Let $p_j, j \in C$ denote the price for travel mode j. Set the price for A to 1 as a numeraire.

We assume that traveler i follows the following utility maximization process:

$$\max_{m_{ij}, j \in C, A_i} \left(\prod_{j \in C} m_{ij}^{\alpha_j} \right)^\beta A_i^\gamma \tag{21.3}$$

$$\text{Subject to } \sum_{j \in C} p_j m_{ij} + A_i = I_i$$

The above utility maximization problem can be interpreted as a two-stage maximization problem. At the first stage, traveler i decides how much budget is allocated to the composite good M_i under the hypothetical price q for the composite good M. In the second stage, given that allocated budget, traveler i decides how much are spent on m_{ij}, $j \in C$. We can solve the above two-stage utility maximization problem by the backward induction.

First, we solve the second-stage maximization problem. At the second stage, traveler i solves the following sub-utility maximization problem under the hypothetical budget $\bar{I}_i$:

$$\max_{m_{ij}} \prod_{j \in C} m_{ij}^{\alpha_j} \qquad (21.4)$$

$$\text{Subject to} \sum_{j \in C} p_j m_{ij} = \bar{I}_i$$

where p_j is the price of each travel mode j for the composite travel service M.

Then we obtain:

$$m_{ij} = \frac{\alpha_j}{\sum_{j \in C} \alpha_j p_j} \bar{I}_i \qquad (21.5)$$

Second, from the first stage, we must obtain $\bar{I}_i$. Speaking the result first, the hypothetical budget constraint $\bar{I}_i$ becomes as follows:

$$\bar{I}_i = \beta I_i \qquad (21.6)$$

To obtain this, we can follow a usual process solving Cobb-Douglas utility maximization. Setting q as the price for the composite travel service M, at the first stage, traveler i solves the following utility maximization problem:

$$\max_{M_i, A_i} M_i^\beta A_i^\gamma \qquad (21.7)$$

$$\text{Subject to } q M_i + A_i = I_i$$

As usual, the solution of this problem satisfies $qM_i = \beta I_i$, and qM_i also satisfies $qM_i = \bar{I}_i$ from the second stage. Thus, it follows $\bar{I}_i = \beta I_i$.

To follow the strict process of the backward induction, first we substitute the solution (21.5) into (21.2) to obtain the expression of M by $\bar{I}_i$, next solve for the hypothetical price q expressed by $\bar{I}_i$ using $qM_i = \bar{I}_i$, and express the maximization problem (21.7) with unknown $\bar{I}_i$. Then we obtain the same result.

After some calculations, we derive the following results:

$$\bar{I}_i = \beta I_i \qquad (21.8)$$

$$M_i = \prod_{j \in C} \left(\frac{\alpha_j}{\sum \alpha_j}\right)^{\alpha_j} p_j^{-\alpha_j} (\beta I_i)^{\alpha_j} \tag{21.9}$$

$$q = \prod_{j \in C} \left(\frac{\alpha_j}{\sum \alpha_j}\right)^{-\alpha_j} \prod_{j \in C} p_j^{\alpha_j} \beta^{1-\sum \alpha_j} I_i^{1-\sum \alpha_j} \tag{21.10}$$

$$m_{ij} = \frac{\alpha_j}{\sum_{j \in C} \alpha_j} \frac{\beta I_i}{p_j} \tag{21.11}$$

If we assume that $\sum_{j \in C} \alpha_j = 1$, these are simplified as

$$M_i = \prod_{j \in C} \alpha_j^{\alpha_j} \frac{\beta I_i}{\prod_{j \in C} p_j^{\alpha_j}} \tag{21.12}$$

$$q = \prod_{j \in C} \alpha_j^{-\alpha_j} \prod_{j \in C} p_j^{\alpha_j} \tag{21.13}$$

$$m_{ij} = \alpha_i \frac{\beta I_i}{p_j} \tag{21.14}$$

It is well known that if a utility function has a special form like the above, the usual utility maximization (21.3) becomes identical to the two-stage utility maximization. More specifically, suppose consumers have a utility function U of some group of commodities x and a composite good z for all other commodities. Assume that the utility function has a separable form such that $U = U(v(x), z)$ and the sub-utility function $v(x)$ is homothetic.[2] Then there exist a price index $e(p)$ and a quantity index $v(x)$ for the composite good x, and the usual utility maximization becomes equivalent to the two-stage utility maximization in which consumers maximize $U(v, z)$ under the price $e(p)$ at the first step to obtain the optimal expenditure $e(p)\tilde{v}$ and at the second step maximize sub-utility function $v(x)$ under the budget constraint $e(p)\tilde{v}$ to obtain the optimal consumption $\tilde{x}$.

In the above example, both $U(v, z)$ and $v(x)$ are Cobb-Douglas utility function with the sum of exponents equal to 1. Thus, they satisfy the separable and homothetic conditions (Cf. Varian [9] pp. 151–152).

---

[2] Let $h(\cdot)$ be a strictly increasing function. Let $g(x)$ be a function of homogeneous of degree 1. Then the function $h(g(x))$ is called a homothetic function.

3 Estimating Travel Demand Function Without Income Data

3.1 Data Used

As stated before, we conducted the survey of behaviors of Korean tourists in Kyushu. In Fukuoka City, there are two sea line routes, which are Camellia Line (Camellia) and JR Beetle II (Beetle) connecting Hakata Port in Fukuoka, Japan, and Busan Port in Busan, Korea. Camellia operates one overnight cruise per day. Passengers stay one night in the ship and it takes about 16 h each way. The round-trip fare is about 120,000 won. On the other hand, Beetle operates two cruises per day and takes about 3 h. This round-trip fare is about 170,000 won.

We carried out the survey to sample from the passengers on Beetle and Camellia on their way back to Busan. The outline of the survey is as follows:

1. Method: Questionnaire sheet survey.
2. Survey samples: Korean tourist passengers who visit Fukuoka using Beetle or Camelia.
3. Distribution and collection: Questionnaire sheets are distributed, answered on board, and collected in the collection box before disembarking.
4. Survey period: About 5 weeks from late March of 2000.
5. Number of collected samples: 365 samples (Camellia. 189 samples; Beetle, 176 samples)

We use the samples from Beetle for estimating travel demand function.

3.2 Estimation Method

In this section, we discuss the method to estimate foreign travel demand function without income data based on the previous results.

For simplicity, hereafter we only consider the two travel modes of sea line and air flight.

As stated above, we do not have data corresponding to m_{ij}, but the respondents' responses should be regarded as their demand M_i for the composite travel service between Busan and Fukuoka.

Let 1 represent Beetle and 2 Airline. From Eq. (21.9), demand M_i for the composite travel service by consumer i is expressed as follows:

$$M_i = \left(\frac{\alpha_1}{\alpha_1 + \alpha_2}\right)^{\alpha_1} \left(\frac{\alpha_2}{\alpha_1 + \alpha_2}\right)^{\alpha_2} p_1^{-\alpha_1} p_2^{-\alpha_2} \beta^{\alpha_1+\alpha_2} I_i^{\alpha_1+\alpha_2} \qquad (21.15)$$

Here p_1 and p_2 denote the prices of Beetle and Air Flight between Busan and Fukuoka, respectively.

Table 21.1 Analysis of variance

	DF	Sum of squares	Mean square	F value
Model	2	993.126	496.563	1008.981
Error	454	223.433	0.492	
Uncorrected total	456	1216.559		
Root MSE	0.70153	R-square	0.8163	
Dependent mean	1.43771	Adj R-square	0.8155	

Corresponding to the cases for stated preference questions, we consider five cases of prices of two travel modes including the present case and distinguish them by suffix k. Let $k = 0$ represent the present case and $k = 1,\ldots,4$ four other cases for different prices. Let M_{ik}, p_{1k}, and p_{2k} represent demand for the composite travel service, Beetle's price, and Air Flight's price corresponding to five cases, respectively.

Now to eliminate the income term from (21.15), we contrive a method to make the ratios of M_{ik}, $k = 1,\ldots,4$ to M_{i0} for the present case. This leads to the following:

$$\frac{M_{ik}}{M_{i0}} = \left(\frac{p_{1k}}{p_{10}}\right)^{-\alpha_1}\left(\frac{p_{2k}}{p_{20}}\right)^{-\alpha_2} \tag{21.16}$$

This expression shows that we can estimate the parameters α_1, α_2 by the following linear regression model:

$$\log\left(\frac{M_{ik}}{M_{i0}}\right) = -\alpha_1 \log\left(\frac{p_{1k}}{p_{10}}\right) - \alpha_2 \log\left(\frac{p_{2k}}{p_{20}}\right) + \varepsilon_{ik} \tag{21.17}$$

3.3 Estimated Results

Price data used for five cases are listed below.

$k = 0$: $p_{10} = 170{,}000$ won	$p_{20} = 245{,}600$ won
	The present case as of March 2000
$k = 1$: $p_{11} = 100{,}000$ won	$p_{21} = p_{20}$
$k = 2$: $p_{12} = 50{,}000$ won	$p_{22} = p_{20}$
$k = 3$: $p_{13} = p_{10}$	$p_{23} = 150{,}000$ won
$k = 4$: $p_{14} = p_{10}$	$p_{24} = 100{,}000$ won

The estimated results are shown in Tables 21.1 and 21.2. The estimated equation turns out as follows:

Table 21.2 Estimated parameters

Variable	DF	Parameter estimate	Standard error	t value	Pr > |t|
Beetle $-\alpha 1$	1	-1.6250	0.0493	-32.9890	0.0001
Air Flight $-\alpha 2$	1	-1.9546	0.0641	-30.4900	0.0001

$$\log \frac{M_{ik}}{M_{i0}} = \underset{(-32.989)}{-1.625 \log \frac{p_{1k}}{p_{10}}} - \underset{(-30.490)}{1.955 \log \frac{p_{2k}}{p_{20}}} \quad (21.18)$$

$$R^2 = 0.8164$$

Note that we did not employ the explanatory variables other than prices of travel modes. Hence, there are no explanatory variables that represent behaviors and characteristics of respondents. However, the R-squared value takes a considerably high value so that the model fitted very well to the actual data.

4 The Estimated Model Can Accurately Predict Drastic Increases and Drops of Korean Tourists?

4.1 Forecasting Method

Predicting the Annual Number of Visits to Fukuoka for Individuals

It should be noted that the estimated model explains the number of visits to Fukuoka by each individual Korean tourist. Thus, the estimated model is a micro-behavioral model for each individual of Korean tourists. To predict the number of visits to Fukuoka by each individual Korean tourist per year, we transform Eq. (21.17) into the following form:

$$\widehat{M}_{ik} = \bar{M}_{i0} \left(\frac{p_{1k}}{p_{10}}\right)^{-1.625} \left(\frac{p_{2k}}{p_{20}}\right)^{-1.955} \quad (21.19)$$

Here $\bar{M}_{i0}$ denotes the number of visits to Fukuoka per year for individual i as of 2000. The left hand side of $\widehat{M}_{ik}$ is the annual frequency of visits to Fukuoka for the individual i when the prices of Beetle and Air Flight are changed to p_{1k} and p_{2k} respectively.

Predicting the Aggregate Number of Korean Visitors to Fukuoka

To predict the total number of Korean tourists to Fukuoka per year, we have only to multiply the average of the frequency of visits to Fukuoka per individual per year by the net total number of Korean tourists visiting Fukuoka. From the survey data, the average of the frequency of visits to Fukuoka for all individuals was $M_{i0} = 1.1491$. In other words, the average number of visits to Fukuoka for all Korean respondents is 1.1491 times per year.

If the prices of Beetle and Air Flight were discounted to the above four cases, the average number of visits to Fukuoka, $\widehat{M}_{ik}, k = 1, \ldots, 4$ would increase to 2.722, 8.395, 3.021, and 6.654 times per year respectively. Hence, the aggregate annual total number of Korean tourists to Fukuoka also would increase 2.722, 8.395, 3.021, and 6.654 times as large as the present annual net total number of Korean tourists to Fukuoka respectively.

4.2 Can We Predict the Drastic Increases and Drops of Korean Tourists to Fukuoka from 2005 to 2009?

Now we predict the number of Korean visitors to Fukuoka in recent years from 2000 to 2009 based on the model estimated at the time of 2000.

To do this, first we need the annual net total number of Korean visitors to Fukuoka at the beginning of the year 2000. From the published data, we have 140,000 visitors from Korea to Fukuoka in the year 1999.

While it seems that this number will do for the net total number, the number includes the same person who visited Fukuoka more than once a year so that we need to correct the double counting. As stated above, according to the survey data, the average number of visits to Fukuoka per year is 1.1491 times for all the respondents. Hence, we must divide the number of total visitors to Fukuoka, 140,000, by this average number of visits, 1.1491. As a result, the annual net total number of Korean visitors to Fukuoka at the beginning of the year 2000 turns out to be 122 (=140/1.1491) thousands.

Next we need to know how prices of Beetle and Air Flight have changed through 2000–2009. Two main causes of changes should be considered. One is the discount by package tours. The other is the change of exchange rate.

The movement of exchange rate of Japanese yen to Korean won is shown in Table 21.3 and Fig. 21.1.

In Table 21.3, the figures express the value of Japanese yen in terms of won. We see that Korean won attained its highest value in 2007 taking 25% rise from 2000s price. We also see the recent sharp drop of value of Korean won due to the crisis of world economy in 2008 (Fig. 21.1).

Table 21.4 shows the transport cost included in the typical package tour from Busan to Fukuoka and its discount ratio to the regular fare 170,000 won for Beetle.

We assume that the air flight regular fare 245,600 won has not been discounted during the period from 2000 to 2009.

With these preparations, we have forecasted the annual total number of Korean tourists to Fukuoka from 2000 to 2009. The result is shown in Table 21.5 and Fig. 21.2.

Table 21.3 Exchange rate of won to yen

Year	KRW/JPY	Ratio to 1999
1999	10.485	1.000
2000	10.486	1.000
2001	10.614	1.012
2002	9.950	0.949
2003	10.292	0.981
2004	10.592	1.010
2005	9.254	0.882
2006	8.206	0.782
2007	7.898	0.753
2008	10.726	1.023
2009	13.636	1.301

Source: http://fx.sauder.ubc.ca/

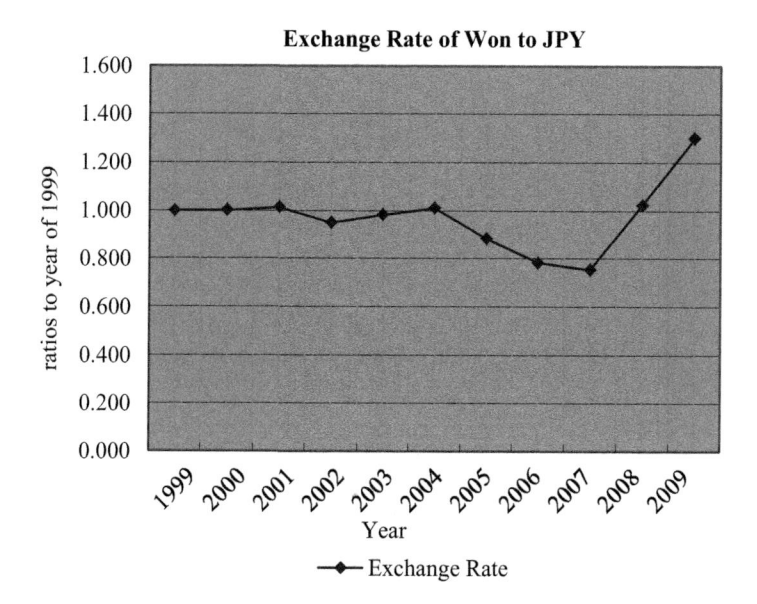

Fig. 21.1 Exchange rates of won to Japanese yen ratios in 1999

Figure 21.2 demonstrates that with just the information of changes of travel cost, we almost accurately have forecasted the movement of recent drastic increases and drops of Korean tourists to Fukuoka.

It should be noted that most of the changes of inflow of Korean tourists to Fukuoka can be explained by the changes of transport charges. While it is of quite importance to design the various tour products, the ratio of the transport charges to the total travel cost still is to be a key factor for designing the attractive tour products.

Table 21.4 Transport cost in package tour price from Busan to Fukuoka

Year	Transport cost in package tour (won)	Discount ratio*
2000	160,000	0.9412
2001	160,000	0.9412
2002	160,000	0.9412
2003	130,000	0.7647
2004	130,000	0.7647
2005	130,000	0.7647
2006	130,000	0.7647
2007	130,000	0.7647
2008	130,000	0.7647
2009	130,000	0.7647

Source: Interview to Net Japan Co.
*Ratio of the discounted transport cost to the regular fare 170,000 won by Beetle

Table 21.5 Numbers of Korean tourists to Fukuoka actual vs. predicted (unit: thousand)

Year	Numbers of visitors from Korea		Error %
	Actual	Predicted	
1999	140	140	
2000	166	154	7.48
2001	179	148	20.99
2002	200	186	7.37
2003	222	231	−3.97
2004	258	209	23.65
2005	298	338	−11.8
2006	413	519	−20.42
2007	509	595	−14.43
2008	448	199	124.70
2009	262	85	209.59

Error = (Actual−Predicted)/Predicted × 100

4.3 Some Property of Our Model

Here we reconsider why our model could accurately reproduce the increases and drops of Korean tourists from 2005 to 2009. First we note that the expression (21.9) of the demand of the composite travel service for the individual i is the indirect utility function corresponding to the sub-utility (21.2). Usually the indirect utility function is expressed by $V(p,I)$ as the function of price and income. Here, Eq. (21.9) also takes a multiplicative function form of prices and income.

Focusing on Eq. (21.15), we extend it to the one including other factors in the multiplicative form.

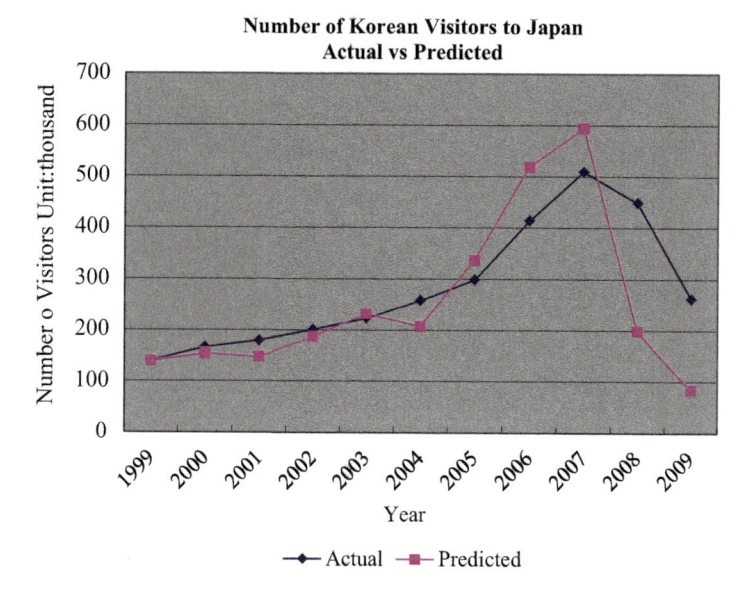

Number of Korean Visitors to Japan Actual vs Predicted

Fig. 21.2 Numbers of Korean tourists to Fukuoka actual vs. predicted

Let B denote the constant composed of parameters α_1, α_2, β. Let t represent the year in place of the index k. We extend the indirect sub-utility function (21.15) as the following form:

$$M_{it}(p(t), I_i(t), s_i, s_o, s_d) = Bp_{1t}^{\alpha_1} p_{2t}^{\alpha_2} I_{it}^{\alpha_1 + \alpha_2} V_i(s_i, s_o, s_d) \qquad (21.20)$$

Here the factors expressed by $V_i(s_i, s_o, s_d)$ are the function of the variable s_i; the personal traits of individual i; the variable s_o, the properties of the origin Busan; and the variable s_d, the properties of destination Fukuoka. If we assume that $V_i(s_i, s_o, s_d)$ does not depend on time t and takes a multiplicative form, the indirect sub-utility function can be rewritten as expression (21.20).

If we take the ratio of the value of the function at time point t to that at $t = 0$, we have:

$$\frac{M_{it}(p_{1t}, p_{2t}, I_{it}, s_i, s_o, s_d)}{M_{i0}(p_{10}, p_{20}, I_{i0}, s_i, s_o, s_d)} = \frac{Bp_{1t}^{\alpha_1} p_{2t}^{\alpha_2} I_{it}^{\alpha_1 + \alpha_2} V_i(s_i, s_o, s_d)}{Bp_{10}^{\alpha_1} p_{20}^{\alpha_2} I_{i0}^{\alpha_1 + \alpha_2} V_i(s_i, s_o, s_d)} \\ = \left(\frac{p_{1t}}{p_{10}}\right)^{\alpha_1} \left(\frac{p_{2t}}{p_{20}}\right)^{\alpha_2} \left(\frac{I_{it}}{I_{i0}}\right)^{\alpha_1 + \alpha_2} \qquad (21.21)$$

We see that all the factors which do not depend on t and all the factors which depend only on the time-independent traits of individual i are canceled out. In a short time period, it can be assumed that $I_{it} = I_{i0}$ so that Eq. (21.21) reduces to our previous estimation formula (21.16).

Our prediction formula can be rewritten as

$$M_{it}(p_{1t}, p_{2t}, I_{it}, s_i, s_o, s_d) \\ = M_{i0}(p_{10}, p_{20}, I_{i0}, s_i, s_o, s_d) \left(\frac{p_{1t}}{p_{10}}\right)^{\alpha_1} \left(\frac{p_{2t}}{p_{20}}\right)^{\alpha_2} \left(\frac{I_{it}}{I_{i0}}\right)^{\alpha_1 + \alpha_2} \quad (21.22)$$

From this, all the factors which are independent of t are said to be absorbed in the initial value of M_{i0}.

Therefore, our estimation method employed here has a similar property to that of fixed effect model for panel data, which removes the unobserved omitted variable biases. This might be one reason why our simple model can correctly predict the drastic increase and drops of Korean tourists to Kyushu.

5 Conclusion and Further Research

We have shown that based on the survey data of inbound behaviors of foreign tourists, we can construct a model to estimate a travel demand function of Korean tourists for visiting Fukuoka without using income data and detailed stated preference questions.

Through showing this, we believe that we have demonstrated a way to solve the two inherent problems usual questionnaire surveys have; that is, one is the difficulty to get income data and the other the time limitation to perform detailed stated preference questions.

To avoid detailed stated preference questions, we conceptualize respondents' responses to the price changes of each of the two travel modes as the changes of demand for the composite travel service composed of sea line and air flight. Based on this conceptualization, we have estimated a travel demand function for the composite travel service between Busan and Fukuoka. To enable us to estimate the travel demand function without income data, we made ratios to cancel out the income term in the travel demand function.

The estimated result clearly shows that we could have predicted the surges and drops of Korean tourists to Fukuoka from 2005 to 2009 almost accurately beforehand even at the time of 2000.

The fact that the estimated demand function can correctly reproduce the recent drastic changes though it uses only the information of prices of two travel modes implies that most of drastic increases and drops of Korean tourists to Fukuoka can be explained by the transport cost with considering the exchange rate. Thus, the model is simple as it is but a significant one for explaining the changes of tourism.

Furthermore, we give some explanation as to why our simple model produces a good forecast by showing some similarity to the fixed effect model which removes the heterogeneity among the samples by deleting the time-independent omitted variable biases related to each sample. We also note that all the time-independent factors related to origin and destination can be interpreted to be absorbed in the initial value. At any rate, the theoretical property of these characteristics should be explored further.

There remain many topics we should address in further research.

In this study we have considered only the factor of transportation expenses. We have assumed that there are no changes both in travel time distance and in the attractiveness in Kyushu. However, it is obvious that the distances and attractiveness of destination should be included in travelers' utility function. Also important is to account for travelers' heterogeneous preferences for designing promising tour products in Kyushu.

The value of utilizing micro-behavior data obtained from survey of inbound behaviors of foreign tourists is huge in the further research. In fact, we need to know how foreign tourists assess each component of tour products, that is, hotels, cultural heritages, natural scenery, city or rural tourism, shopping, and various activities.

Another potential value of the micro-behavior data is that we can estimate the economic impacts of inbound behaviors of foreign tourists to regional economies.

As shown in this paper, a deliberate use of micro-behavior data can be applied to policy study at macro level. Hence, the utilization of micro-behavioral data should be explored further.

References

1. Fujita M, Krugman PR, Venables AJ (2001) The spatial economy: cities, regions, and international trade. MIT Press, Cambridge, MA
2. Motomura H, Saito S (2007) Estimating the travel demand function in consideration of destination variables for inbound tourism. Presented at the 44th annual meeting of Japan Section of Regional Science Association International (JSRSAI) (in Japanese)
3. Saito S (ed) (2000) Report on the Survey of Consumer Shop-around Behavior at City Center of Fukuoka City, Japan 2000: with focusing on underground space and comparison of city attractiveness between Japan and Korea. Fukuoka Asia Urban Research Center, Fukuoka City Government. (in Japanese)
4. Saito S, Motomura H (2001a) A study on shopping and sightseeing behaviors of international tourists. Studies on Strait Area between Korea and Japan, No.1, pp 41–61 (in Japanese). pp 51–77. (in Korean)
5. Saito S, Motomura H (2001b) The number of Korean tourists to Fukuoka and transport cost: a simple method to estimate travel demand as a composite good of different travel modes and its applications. In: Proceedings of the 38th annual meeting of Japan Section of Regional Science Association International (JSRSAI), pp 333–340 (in Japanese)
6. Saito S, Motomura H, Kakoi M (2002) A study on shopping and sightseeing behaviors of international tourists II (part 1): analysis of Japanese tourists' behaviors in Korea and their evaluation of attractiveness of Japanese and Korean cities. Studies on Strait Area between Korea and Japan, No.2, pp 26–41 (in Japanese). pp 36–52 (in Korean)
7. Saito S, Motomura H (2002) A study on shopping and sightseeing behaviors of international tourists II (part 2): the number of Japanese tourists to Busan and transport cost. Studies on Strait Area between Korea and Japan, No.2, pp 42–50 (in Japanese). pp 52–63 (in Korean)

8. Saito S, Motomura H (2009) Travel demand function of Korean tourists to Kyushu: could we have accurately predicted the drastic increase of Korean visitors to Kyushu? paper presented at the 46th annual meeting of Japan Section of Regional Science Association International (JSRSAI)
9. Varian HR (1992) Microeconomic analysis, 3rd edn. W.W. Norton, New York
10. Japan Travel Bureau, Tourism Statistics. https://www.tourism.jp/tourism-database/stats/inbound/
11. Port of Hakata, Fukuoka City, Home Page. http://port-of-hakata.city.fukuoka.lg.jp/guide/cruise/cruise.htmls
12. Wikipedia, Fukuoka Airport. https://en.wikipedia.org/wiki/Fukuoka_Airport
13. Saito S, Kakoi M, Nakashima T (1999) On-site Poisson regression modeling for forecasting the number of visitors to city center retail environment and its evaluation. Stud Reg Sci 29:55–74. (in Japanese)

Chapter 22
Direct Approach to Estimating Welfare Changes Brought by a New Subway Line

Kosuke Yamashiro and Saburo Saito

Abstract In this study, taking up the opening of the new subway line as an example, we propose a new evaluation method for urban development policies based on the retrospective panel data concerning consumers' micro behaviors. It is a method to directly estimate the utility function of the residents along the subway line and to measure the economic effect of the introduction of the new subway line as welfare changes of those residents. The utility function is formulated as a hierarchical nested CES utility function, in which travel means of buses and subway are expressed as a composite service to provide a city center good and the shortening of travel time to the city center due to the opening of the subway is incorporated as the reduction of the generalized travel cost to the city center. Consequently, the residents along the subway line increase their frequency of visits to the city center and their welfare levels.

Keywords Retrospective panel data · Frequency of visits · City center · Hierarchical nested utility function · CES · Cobb-Douglas · Generalized travel cost · Welfare change · Evaluation of urban development policy

---

This chapter is based on the paper by Saburo Saito, Kosuke Yamashiro, Mamoru Imanishi, and Takaaki Nakashima [13], "Welfare Changes in Residents along New Subway Line By its Opening: Based on direct estimation of hierarchical nested CES travel demand function to city center" presented at the 44th Annual Meeting of Japan Section of Regional Science Association International (JSRSAI), 2007, which is revised for this chapter.

K. Yamashiro
Department of Business and Economics, Nippon Bunri University, Oita City, Japan
e-mail: yamashiroks@nbu.ac.jp

S. Saito (✉)
Faculty of Economics, Fukuoka University, Fukuoka, Japan

Fukuoka University Institute of Quantitative Behavioral Informatics for City and Space Economy (FQBIC), Fukuoka, Japan
e-mail: saito@fukuoka-u.ac.jp

1 Purpose of This Study

1.1 Background

The Fukuoka City Subway Nanakuma Line, which connects the Tenjin district, the city center of Fukuoka City to the southwest part of suburban areas of the city, was opened in February 2005. Since there was no rail transportation system until that time, the accessibility from the areas along the new subway line to the city center had been drastically improved.

Before the opening of this Nanakuma Subway Line, Saito, Yamashiro, Nakashima, and Igarashi [14] had predicted its economic effect on the city center retail sector of Fukuoka City. They considered that by shortening of the travel time to the city center, the residents along the subway line increased their frequency of visits to the city center, this increase in the visit frequency to the city center, in turn, caused the increase in their expenditure at the city center, and consequently, the retail sales at the city center increased. They regarded this increase in the retail sales at the city center as the economic effect on the city center retail sector caused by the opening of the new subway line.

According to this way of thinking, they actually conducted a forecast of changes in behaviors of residents along the subway line concerning choices of their travel means and their frequency of visits to the city center after the opening of the new subway line by carrying out the surveys of consumer shop-around behaviors at the city center of Fukuoka City. By their result, the economic effect on the city center commercial district of Fukuoka City was predicted to be 17,700,000,000 yen per year (or 177 million dollars per year). The above method is named as the "consumer behavior approach" by Yamashiro [16, 10, 11].

The behavioral hypothesis for the increase in the visit frequency to the city center by opening a new subway line behind the consumer behavior approach was conceptualized as follows.

There assume to be two kinds of goods. One is the city center goods. The other is the local goods. To purchase the city center goods, the residents along the subway line must travel to the city center, whereas the local goods are bought at their local areas. With the opening of the Nanakuma Line, the travel time to the city center is shortened so that the generalized transport costs for residents along the subway line to travel to the city center decrease. When the residents along the subway line purchase the city center good, they must pay the price which is the sum of their travel cost to the city center and the price charged by the retailer at the city center. Thus the decrease in their generalized travel cost to the city center reduces the prices of city center goods for them. Hence, the demand for city center goods increases due to the income effect. Under the assumption that one unit of the city center good is set to be the quantity those residents along the subway line purchase per visit at the city center, their frequency of visits to the city center is also derived to increase as well.

In addition to the prediction study, Saito, Yamashiro, and Nakashima [15] had also performed the verification study.[1] Actually, in order to verify whether the visit frequency of residents along the subway line increased after the opening of the new subway, they conducted a questionnaire survey in 2005 targeting Fukuoka University students who utilize the subway after its opening, collected retrospective panel data on the change in their frequency of visits to the city center, and performed the analysis to verify the increase in the visit frequency to the city center, which is derived from the above behavioral hypothesis.

From their analysis, it emerged that, after the opening of the new subway, the visit frequency to the city center increased in the whole area along the subway line. In particular, the visit frequency for the residents along the subway line who use the subway has markedly increased.

1.2 Purpose

To further our previous research, we embarked on directly formulating our behavioral hypothesis as a utility function which represents the behavioral mechanism to choose travel modes of buses and a new subway line and estimating the utility function to refine our behavioral hypothesis.[2]

For the purpose, based on the framework of spatial economics [4, 18], Saito, Yamashiro, and Imanishi [12] formulated a hierarchical nested CES utility function which represents the city center visit demand function by the residents along the new subway line and estimated its parameters using the retrospective panel data obtained from the above survey, which asked the respondents retrospectively about their behavior before the subway opening at the time after the subway opened.

However, their study was limited to the estimation of the city center visit demand function, and there remains an issue regarding to what extent the utility actually has increased for the residents along the subway line.

Therefore, the purpose of this study is not only to estimate the hierarchical nested CES utility function from the retrospective panel data but also to forecast the change of the city center visit demand in each of 278 district divisions of the residential areas along the new subway line.

Speaking further, another purpose of this study is to open a way to evaluate the urban development policies by the welfare change of consumers. In this study, we have done this by regarding the changes of the welfare for individual residents along the subway line as the changes of their city center visit demand. While our previous forecast study [14] on the economic effect of the new subway line was predicted as the increase in the sales at the city center retail sector, there might arise some

---

[1] Also see Chap. 13 of this book.
[2] Saito and Motomura [9] tried to forecast the number of foreign tourists by employing a similar approach to directly estimating consumer travel utility function, which is reproduced in Chapter 21 of this book.

objections to this definition. The final resolution to these arguments is to measure how much welfare such urban development policies would bring about to the consumers concerned on a monetary basis if possible.

While the previous efforts to do this traditionally have been employing a large-sized cost and organization, in contrast, our study intends to show that the same effort can be performed in a small scale for usual urban development policies if one utilizes the retrospective panel data though it is a posteriori evaluation.

2 A Hierarchical Nested CES (Constant Elasticity Substitution) Utility Function: Its Formulation and Estimation

2.1 Model

In this section, we formulate the hierarchical nested CES (constant elasticity substitution) utility function which derives the city center visit demand function for the residents along the new subway line (Cf. [12]).

A representative consumer i resides in the residential area i. Two kinds of goods exist. One is the local goods which is a composite good composed of all goods purchased at whole residential areas. The other is the city center goods which is a composite good composed of all goods purchased at the city center. We denote the consumption of local goods by x_0 and that of city center goods by x_1. For convenience of notation, we omit the suffix i which denotes that those consumptions are purchases by the consumer i.

We assume that one unit of the city center goods is the quantity the consumer purchases per visit at the city center. We interpret that the consumption quantity x_1 is the frequency of visits to the city center.

Before the subway opening, only the bus operator supplies the city center goods, while the city center goods are supplied by both bus and subway operators after the opening of the subway.

We set local goods as numeraire goods and set the price $p_0 = 1$. The generalized transport cost when using the bus is p_{11}, and the income of the consumer i is denoted by I_i. We define the utility function of consumers before the subway opening as the following Cobb-Douglas utility function:

$$U_i(x_0, x_1) = x_0^{1-a} x_1^a. \tag{22.1}$$

The parameter a represents the expenditure ratio for each good. Under the budget constraint, $x_0 + p_{11}x_{11} = I_i$, the utility maximization problem for consumer i becomes

$$\max_{x_0, x_1} U_i(x_0, x_1) \quad s.t. \quad x_0 + p_{11}x_{11} = I_i. \tag{22.2}$$

The solution is as follows:

$$x_0 = (1-a)I_i$$
$$x_{11} = a\left(\frac{I_i}{p_{11}}\right) \quad (22.3)$$

After the subway opening, the consumption of the city center goods when using the bus is represented by x_{11}, while that when using the subway is represented by x_{12}. The generalized transport cost when using the bus is expressed by p_{11}, while that when using the subway is expressed by p_{12}. The utility function of consumer i after the subway opening is defined similarly to the previous case as follows:

$$U_i(x_0, x_1) = x_0^{1-b} x_1^b. \quad (22.4)$$

However, since the city center goods are a composite good of buses and the subway, they are formulated as a CES function form:

$$x_1(x_{11}, x_{12}) = \left(x_{11}^{(\sigma-1)/\sigma} + x_{12}^{(\sigma-1)/\sigma}\right)^{\frac{\sigma}{\sigma-1}}. \quad (22.5)$$

Parameters a and b expressing the expenditure ratio are different before and after the subway opening because we assume that the expenditure ratio of local and downtown goods could change after the subway opening.

The utility maximization problem of consumer i after the subway opening becomes as follows:

$$\max_{x_0, x_1} U_i(x_0, x_1) \quad \text{s.t.} \quad x_0 + p_{11}x_{11} + p_{12}x_{12} = I_i, \quad (22.6)$$

where

$$x_1 = \left(x_{11}^{(\sigma-1)/\sigma} + x_{12}^{(\sigma-1)/\sigma}\right)^{\frac{\sigma}{\sigma-1}} \quad (22.7)$$

Since the utility function is nested, the utility maximization problem can be solved in two stages as shown by Kanemoto et al. [6]. The first stage determines the local goods x_0 and the city center goods x_1, while the second stage finds the allocation of the city center composite goods x_1 to buses x_{11} and the subway x_{12}. To solve this problem, we use backward induction.

First, we consider the second-stage problem, which is formulated as the maximization problem of the CES utility x_1 under the hypothetical budget constraint $\bar{I}_i$. The hypothetical budget $\bar{I}_i$ is determined after the first-stage problem is solved for the budget allocation to the city center composite goods from the budget I_i under the price index p_1 of the composite city center goods, which is obtained through solving the second-stage problem. Since the total expenditure on buses x_{11} and the subway x_{12} satisfies the budget $\bar{I}_i$, the following CES utility maximization problem must be solved:

$$\max_{x_{11},x_{12}} \left(x_{11}^{(\sigma-1)/\sigma} + x_{12}^{(\sigma-1)/\sigma} \right)^{\frac{\sigma}{1-\sigma}} \text{ s.t. } p_{11}x_{11} + p_{12}x_{12} = \bar{I}_i. \qquad (22.8)$$

The solution turns out as follows:

$$x_{11} = p_{11}^{-\sigma} p_1^{\sigma-1} \bar{I}_i, \quad x_{12} = p_{12}^{-\sigma} p_1^{\sigma-1} \bar{I}_i, \qquad (22.9)$$

$$p_1 = \left(p_{11}^{1-\sigma} + p_{12}^{1-\sigma} \right)^{\frac{1}{1-\sigma}}. \qquad (22.10)$$

Note that from Eq. (22.9) it holds that

$$\left(x_{11}^{(\sigma-1)/\sigma} + x_{12}^{(\sigma-1)/\sigma} \right)^{\frac{\sigma}{1-\sigma}} \cdot \left(p_{11}^{1-\sigma} + p_{12}^{1-\sigma} \right)^{\frac{1}{1-\sigma}} = x_1 p_1 = \bar{I}_i. \qquad (22.11)$$

Thus we see that p_1 is the price index for the city center composite goods.

Next, we solve the Cobb-Douglas utility maximization problem in the first stage under the above price index p_1.

$$\max_{x_0,x_1} x_0^{1-b} x_1^b \text{ s.t. } x_0 + p_1 x_1 = I_i. \qquad (22.12)$$

The solution becomes as follows:

$$x_1 = b\left(\frac{I_i}{p_1} \right). \qquad (22.13)$$

From the above, the hypothetical budget $\bar{I}_i$ is determined as

$$\bar{I}_i = p_1 x_1 = bI_i.$$

Hence, we have solved the original problem to maximize the hierarchical nested CES utility for consumer i after the opening of the new subway line. Its solution turns out to be as follows:

$$\begin{aligned} p_1 &= \left(p_{11}^{1-\sigma} + p_{12}^{1-\sigma} \right)^{\frac{1}{1-\sigma}} \\ x_1 &= \left(x_{11}^{(\sigma-1)/\sigma} + x_{12}^{(\sigma-1)/\sigma} \right)^{\frac{\sigma}{1-\sigma}} \\ x_1 &= b\left(\frac{I_i}{p_1} \right) \\ x_{11} &= p_{11}^{-\sigma} p_1^{\sigma-1} bI_i \\ x_{12} &= p_{12}^{-\sigma} p_1^{\sigma-1} bI_i. \end{aligned} \qquad (22.14)$$

2.2 Generalized Travel Cost

To make the model more realistic, we define generalized transport costs for buses and the subway. We express the generalized transport cost as the exponential function of travel time and travel cost.

Before the opening of the new subway line, the generalized transport cost for the buses is expressed as

$$p_{11} = e^{(\alpha t_1 + \beta c_1)}. \tag{22.15}$$

Here, α, β are unknown parameters, and t_1, c_1 represent the travel time and the travel cost, respectively, when consumers use the bus to travel to the city center.

After the subway opening, the subway is added to the available transportation modes. The generalized transport cost for the subway is

$$p_{12} = e^{(\alpha t_2 + \beta c_2)} \tag{22.16}$$

Here in the generalized transport costs, we omitted the subscript i for consumer i who resides in the residential area i. However, since the travel time to the city center and the bus fare are different among consumers depending on where they live, it should be noted that the generalized transport costs also vary across consumers.

2.3 Estimation Method

We describe the estimation method for the unknown parameters. Specifically, the estimation is performed according to the following procedure.

(i) Estimating parameter $\phi = (1 - \sigma)\alpha$, $\theta = (1 - \sigma)\beta$ from the logit model of transportation mode choice
(ii) Estimating $\delta = -\frac{\sigma}{1-\sigma}$ using the least squares method
(iii) Estimating σ using $\tilde{\delta}$ estimated in (ii)
(iv) Estimating α, β from the expressions in (i) by using $\tilde{\phi}, \tilde{\theta}$ estimated in (i), and $\tilde{\sigma}$ estimated in (iii)
(v) Estimating $\frac{b}{a}$ using the generalized transport costs p_{11}, p_{12} and the visit frequencies x_{11}, x_{12} before and after the opening of the new subway line, using $\tilde{\alpha}, \tilde{\beta}$ estimated in (iv)

Estimating Parameters ϕ, θ

The solution to the hierarchical nested CES utility maximization problem after the subway opening (22.14) is rewritten as follows:

$$x_{11} = \frac{p_{11}^{1-\sigma}}{p_{11}^{1-\sigma} + p_{12}^{1-\sigma}} \cdot \frac{bI_i}{p_{11}}$$
$$x_{12} = \frac{p_{12}^{1-\sigma}}{p_{11}^{1-\sigma} + p_{12}^{1-\sigma}} \cdot \frac{bI_i}{p_{12}}.$$
(22.17)

We can interpret this formula as consumer i chooses buses or the subway with the probabilities, $\frac{p_{1j}^{1-\sigma}}{p_{11}^{1-\sigma} + p_{12}^{1-\sigma}}, j = 1,2$, and then allocates the budget for the city center composite goods, bI_i, to each component of composite goods, buses and the subway, according to these probabilities. (Cf. [1–3] For another interpretation of CES, refer to [8]) Divisions of the budget, bI_i, by $p_{1j}, j = 1,2$ are transformations from money to quantities for buses and the subway components of the city center composite goods.

Substituting the expression of generalized transport costs into these probabilities leads us to a logit model for transportation mode choices after the opening of the new subway line as follows:

$$\frac{p_{1j}x_{1j}}{p_{11}x_{11} + p_{12}x_{12}} = \frac{p_{1j}^{1-\sigma}}{p_{11}^{1-\sigma} + p_{12}^{1-\sigma}}$$
$$= \frac{e^{(1-\sigma)at_j + (1-\sigma)\beta c_j}}{e^{(1-\sigma)at_1 + (1-\sigma)\beta c_1} + e^{(1-\sigma)at_2 + (1-\sigma)\beta c_2}}, j = 1,2.$$
(22.18)

Here, setting $\phi = (1-\sigma)\alpha$, $\theta = (1-\sigma)\beta$, the following is obtained.

$$p_{1j}^{1-\sigma} = e^{\phi t_j + \theta c_j} \quad j = 1,2.$$
(22.19)

Thus, it is seen that by using the estimates of the parameters ϕ, θ, we can obtain the estimates of $p_{1j}^{1-\sigma}, j = 1,2$ using Eq. (22.19). We denote the estimates of $p_{1j}^{1-\sigma}, j = 1,2$ by $\tilde{p}_{1j}, j = 1,2$.

Estimating Parameters δ, σ

From the definition of elasticity of substitution, we obtain the following equation.

$$\log \frac{x_{12}}{x_{11}} = -\sigma \log \frac{p_{12}}{p_{11}}.$$
(22.20)

We can estimate the elasticity of substitution, σ employing the following method. First, using the estimates, $\tilde{\phi}, \tilde{\theta}$ of the parameters, ϕ, θ, we get the estimated values of $p_{1j}^{1-\sigma}, j = 1,2$, which are denoted by $\tilde{p}_{1j}, j = 1,2$ as before. We use these values as the data for the right-hand side of Eq. (22.20) to estimate σ. Next, denote by x_{11}^B the frequency of visits to the city center by bus before the opening of the subway and by x_{12}^A the visit frequency to the city center by subway after the opening of the subway.

These observed values obtained from the retrospective panel data are used as the data for the left-hand side of Eq. (22.9) to estimate σ. Here, for the estimation, we restricted our samples to those residents who used buses before the subway opening and switched from buses to the subway after the opening of the subway.

The final stage is composed of two steps. As for the first step, we modify Eq. (22.20) to the following equation:

$$\log\frac{x_{12}^A}{x_{11}^B} = \delta\log\frac{\tilde{p}_{12}}{\tilde{p}_{11}} \qquad (22.21)$$

Here, while we omitted the subscript i for sample i as before, all the values of observed and estimated variables are meant to be those of sample i. The values of x_{11}^B, x_{12}^A are the observed data, and those of $\tilde{p}_{11}, \tilde{p}_{12}$ are the estimated values for $p_{11}^{1-\sigma}, p_{12}^{1-\sigma}$. More specifically, x_{11}^B is the frequency of visits to the city center using buses before the subway opening, while x_{12}^A is that to the city center using the subway after the opening of the subway.

From the above, we see that we can estimate δ by ordinary least square method. Denote its estimate by $\tilde{\delta}$.

As for the second step, note the relation (22.22) below.

$$\tilde{\delta}\log\frac{\tilde{p}_{12}}{\tilde{p}_{11}} = \tilde{\delta}(1-\sigma)\log\frac{p_{12}}{p_{11}} = -\sigma\log\frac{p_{12}}{p_{11}}. \qquad (22.22)$$

Since it holds $\tilde{\delta}(1-\sigma) = -\sigma$, we obtain the estimate of σ as $\tilde{\sigma} = \frac{\tilde{\delta}}{\tilde{\delta}-1}$.

Estimating Parameters α,β

We obtain the estimate of α,β from the relation $\phi = (1-\sigma)\alpha, \theta = (1-\sigma)\beta$, by using the estimated values $\tilde{\phi}, \tilde{\theta}, \tilde{\sigma}$.

Estimating Parameters a,b

We estimate the unknown parameters a,b in the Cobb-Douglas utility at the first stage in the hierarchical nested CES utility function. Specifically, we calculate the generalized transport costs of bus and subway in Eqs. (22.15) and (22.16), respectively, using the estimated values $\tilde{\alpha}, \tilde{\beta}$ and calculate the corresponding expenditures.

Before the opening, because of the nature of the Cobb-Douglas utility function, and since people purchase the city center goods by using buses at the rate of a out of income I_i, the following equation holds:

$$p_{11}x_{11}^B = aI_i \qquad (22.23)$$

On the other hand, after the opening, the city center goods are purchased at the rate of b out of income I_i. Thus, the following formula holds;

$$p_{11}x_{11}^A + p_{12}x_{12}^A = bI_i. \qquad (22.24)$$

Here, regarding the expenditure for the city center goods after the opening of the subway, only transportation modes which are chosen are used. By taking the ratio of the expenditure for generalized transport costs after the opening to that before the opening, we can obtain the estimate for $\frac{b}{a}$.

$$\frac{p_{11}x_{11}^A + p_{12}x_{12}^A}{p_{11}x_{11}^B} = \frac{b}{a} \qquad (22.25)$$

2.4 Data Used for Estimation

Ecole Card Usage Survey

The data used for the estimation are the data obtained from the second survey of the actual usage of the Ecole card by Fukuoka University students conducted by the Fukuoka University Institute of Quantitative Behavioral Informatics for City and Space Economy (FQBIC) for 5 days on November 17, 21, 30 and December 1 and 5, 2005.[3]

The survey was about 10-min questionnaire survey distributed and collected at lecture rooms for Fukuoka University students. The survey asked them about their behavior changes before and after the opening of Fukuoka City Subway Nanakuma Line, which passes through the university campus. Question items were items such as the last transportation means utilized to come to the university, changes in the frequency of school attendance, changes in the transportation modes to go to the city center and the corresponding frequency of visits to the city center, as well as the expenditure per visit at the city center, and so on.

The number of valid samples collected was 313 samples. The outline of the survey is shown in Table 22.1.

---

[3]The Ecole card is a commuter pass of unlimited bus rides, offered by Nishi-Nippon Railroad exclusively to students. If the area is within the Fukuoka metropolitan area, rides are unlimited at a price of 6000 yen per month.

Table 22.1 Outline of the second survey of the actual usage of the Ecole card by Fukuoka University students

Survey date	November 17, 2005 (Thursday)
	November 21, 2005 (Monday)
	November 30, 2005 (Monday)
	December 1, 2005 (Thursday)
	December 5, 2005 (Thursday)
Survey location	Carried out in the rooms of 6 lectures at Fukuoka University
Survey method	Distributed and collected the questionnaires at the lectures
Main question items	Personal attributes: place of residence, nearest bus stop/subway station
	Changes in transport means used for coming to university and the frequency of attendances at university before and after the opening of the subway
	Changes in the transport means and the frequency of visits to the city center
	Stated preference among Ecole card, commuter ticket, and *chika* passport
Number of samples collected	313 samples

Supplementary Data

The second survey of the actual usage of the Ecole card by Fukuoka University students asked about their frequency of visits to the city center of Fukuoka City, the transportation mode used to visit the city center before and after the subway opening, the nearest bus stop from home and the time distance up to there by walk, and the nearest subway station from home and the time distance up to there by walk. However, the survey did not ask about the travel time and transportation fare to the city center of Fukuoka City for each transportation mode. Therefore, supplementary data was collected using information about the travel time and transportation fare published on the website of each transportation operator. The collection method is described in the following.

(a) *Bus*

Travel time required: The time distance from home to the nearest bus stop (survey item) + the shortest travel time from the nearest bus stop to the Tenjin area bus stop (travel time described on the Nishi-Nippon Railroad's website [7]).

Transportation fare: Cheapest fare from the nearest bus stop to the Tenjin area bus stop (fare listed on the Nishi-Nippon Railroad's website [7]).

For Ecole card possessors who chose bus after the opening of the subway, the transportation fare was set at 200 yen per day. This amount was obtained by dividing the Ecole card monthly fee of 6000 yen by 30 days.

(b) *Subway*

Travel time required: The time distance from home to the nearest station (survey item) + travel time from the nearest station to the South Tenjin station (travel time described on the Fukuoka City Subway website [5]).

Transportation fare: Fare from nearest station to the South Tenjin station (fare listed on the Fukuoka City Subway website [5]).

Sample Analyzed

In this study, we conducted an analysis focusing on the residents living in the surroundings of Fukuoka University along the Nanakuma Subway Line among the samples in the second survey of the actual usage of the Ecole card by Fukuoka University students. Specifically, we restrict the samples to those who satisfy all the conditions below for the estimation.

1. From the question items, individuals that responded that the nearest station from home was *Kamo, Noke, Umebayashi, Fukudaimae, Nanakuma, Kanayama,* or *Chayama*.
2. Individuals who chose buses as the transportation means to the city center before the opening of subway and buses or the subway after the subway opening.
3. Individuals that responded to the question items about the frequency of visits to the city center both before and after the opening of the subway.

The respondents satisfying all the above conditions were 52.

2.5 Estimated Results of Parameters

Estimated Result of ϕ, θ

Table 22.2 shows the estimated results of parameters for the logit model. For both parameters, the t value is significant at 1%, the hit rate is also very high (92.3%), the freedom adjusted determination coefficient is 0.4899, and the fitness of the model is also good. From the estimated values $\tilde{\phi}, \tilde{\theta}$, we calculate the generalized transport costs p_{11}, p_{12} of buses and the subway, respectively.

Estimated Result of δ

The estimate $\tilde{\delta}$, obtained by the ordinary least squares method, is shown in Table 22.3.

Table 22.2 Estimated results of parameters $\tilde{\varphi}, \tilde{\theta}$

		Estimate	t value	SD
Time (minute) φ		−0.2123	−3.51***	0.0605
Cost (yen) θ		−0.0264	−3.00***	0.0088
Hit ratio		92.3%		
Log-likelihood with all parameters set to zeros	$L(0)$	−36.044		
Log-likelihood with .estimated values of parameters	$L(\hat{\beta})$	−17.680		
−2×Log-likelihood	$\rho = -[L(0) - L(\beta)]$	36.727		
McFadden's R-square	$\rho^2 = 1 - L(\hat{\beta})/L(0)$	0.5095		
Adjusted R-square	$\bar{\rho}^2 = 1 - \left(L(\hat{\beta}) - K\right)/L(0)$	0.4540		

***Significant at 1% K:the number of parameters, $n = 52$

Table 22.3 Estimated result of parameter $\tilde{\delta}$

	Estimate	t value	SD
$\tilde{\delta}$	2.9995	5.09***	0.5887
R-square	0.4058		
Adjusted R-square	0.3902		

***Significant at 1% $n = 52$

Estimated Result of Elasticity of Substitution σ

From the estimated value $\tilde{\delta}$, we calculate the estimate of the elasticity of substitution σ as follows:

$$\tilde{\sigma} = \frac{2.9995}{2.9995 - 1} = 1.5001.$$

Estimated Result of Parameters α, β

From the above estimated value of the elasticity of substitution σ, we obtain the estimates of α, β as follows:[4]

$$\tilde{\alpha} = \frac{\tilde{\varphi}}{1 - \tilde{\sigma}} = \frac{-0.2123}{-0.5001} = 0.4246, \tilde{\beta} = \frac{\tilde{\theta}}{1 - \tilde{\sigma}} = \frac{-0.0264}{-0.5001} = 0.0528$$

---

[4] As is well known, the quotient $\frac{\tilde{\alpha}}{\tilde{\beta}}$ indicates the value of time for the respondents. Since the unit of time and that of price are minute and yen, the time value becomes 8.04 yen/min or 482 yen/h, which seems to be reasonable for students.

Table 22.4 Average of $\frac{b}{a}$ for people who switched to the subway

Number of samples	Mean	SD	Minimum value	Maximum value
39	2.11	2.57	0.73	15.09

Estimated Result of $\frac{b}{a}$

Based on the results so far, the ratio $\frac{b}{a}$ can be calculated for each sample from the ratio of the expenditure for the city center composite goods after the opening of the subway to that before the opening. As noted above, here we focus on the samples who used buses before the opening but switched to the subway after the opening of the subway.

Thus the average of $\frac{b}{a}$ can be interpreted as the average of ratios of the expenditure for the subway city center composite goods after the opening to the expenditure for the buses city center composite goods before the opening among those samples who switched to the subway after the opening of the subway, where the expenditure is calculated based on the generalized transport costs. Table 22.4 gives the result.

In short, the average calculated here means that it is the average value obtained by averaging over people who switched to the subway the ratios of their expenditure for transport means after the opening to that before the opening with including their travel time values to their expenditure calculation.

2.6 The City Center Visit Demand Function After the Opening

Based on the estimated results so far, the city center visit demand functions of a typical consumer i in the residential area i before and after the opening of the subway can be formulated and identified as follows.

1. *Before the opening*

   From Eq. (22.3), we have

$$\widehat{x^i_{11}} = \frac{aI_i}{\widehat{p^i_{11}}} \tag{22.26}$$

2. *After the opening*

   From Eq. (22.14), we have

$$\widehat{x^{Ai}_{11}} = \widehat{p^i_{11}}^{-1.5001} \widehat{p^i_{1}}^{-0.5001} bI_i$$
$$\widehat{x^{Ai}_{12}} = \widehat{p^i_{12}}^{-1.5001} \widehat{p^i_{1}}^{0.5001} bI_i \tag{22.27}$$

By combining the above two expressions, we can obtain the city center visit demand function for each of two travel modes, buses and the subway, in a computable form for each consumer i.

$$\widehat{x_{11}^{Ai}} = \widehat{p_{11}^i}^{-1.5001} \widehat{p_1^i}^{0.5001} \widehat{\left(\frac{b}{a}\right)} \widehat{p_{11}^i x_{11}^i}$$

$$\widehat{x_{12}^{Ai}} = \widehat{p_{12}^i}^{-1.5001} \widehat{p_1^i}^{0.5001} \widehat{\left(\frac{b}{a}\right)} \widehat{p_{11}^i x_{11}^i}$$
(22.28)

3 Model Verification

In this section, we verify to what extent the city center visit demand function estimated reproduces actual data.

To do this, we compare the observed and the forecasted values of the frequency of visits for each of the 52 samples used for the estimation.

Figure 22.1 shows the result of the plot. From the figure we see that data points are very close to the 45-degree line. Thus the error is very small, and it can be said that the city center visit demand function estimated accurately explains the actual data.

Table 22.5 gives detailed data about the 52 samples used for the estimation. In the table, the prediction by the logit model of transportation mode choice is also given. The prediction of the model accurately predicted the choices of 48 samples out of 52 samples.

Fig. 22.1 The observed and predicted values of the frequency of visits to the city center for the samples used for estimation

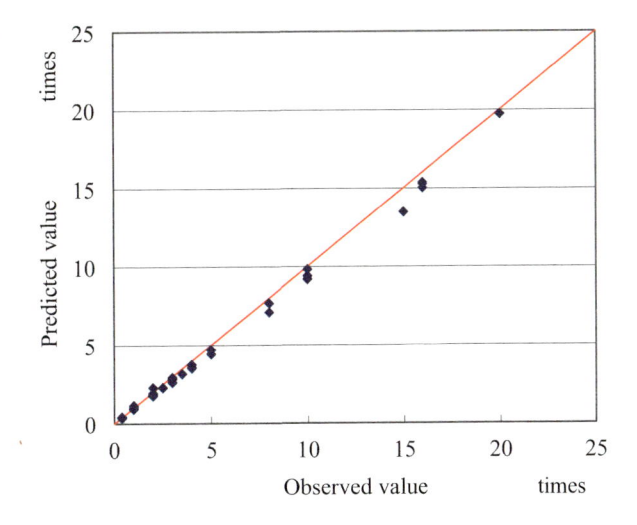

Table 22.5 Details of data and the observed and predicted modal choices and frequency of visits for the sample used for estimation

OBS	Gender (1: Male 2: Female)	Age	Ecol card (1: Have 0: Not)	Walking time to the nearest bus stop (minute)	Walking time to the nearest subway station (minute)	Bus Time to Tenjin (minute)	Bus Cost to Tenjin (yen)	Subway Time to Tenjin (minute)	Subway Cost to Tenjin (yen)	Generalized travel cost Bus	Generalized travel cost Subway	Mode choice after opening (observed) 1: Bus 2: Subway	Mode choice after opening (predicted) 1: Bus 2: Subway	Visit frequency before opening (times/month)	Visit frequency after opening (times/month)	Prediction b_i/a_i	Prediction x_{t1} Bus	Prediction x_{t2} Subway	Prediction $x_{t1}+x_{t2}$	Difference from the observed visit frequency	Absolute difference	Absolute percent error
1	1	20	0	3	3	27	340	17	290	557	427	2	2	1	2	1.532	0.715	1.067	1.782	0.219	0.219	10.925
2	1	24	0	3	20	35	340	34	290	621	563	1	2	0.4	0.4	1.000	0.195	0.226	0.421	−0.021	0.021	5.275
3	2	25	0	3	7	31	340	23	290	589	475	2	2	0.4	2	4.030	0.763	1.054	1.816	0.184	0.184	9.180
4	2	20	1	5	5	28	200	19	250	425	443	1	1	3	3	1.000	1.515	1.426	2.941	0.059	0.059	1.970
5	2	20	0	10	10	33	220	21	250	485	419	2	2	2	2	0.863	0.831	1.037	1.868	0.132	0.132	6.600
6	2	20	0	5	5	28	280	19	290	505	443	2	2	10	10	0.877	4.238	5.165	9.403	0.597	0.597	5.970
7	2	19	0	5	5	28	280	19	290	505	443	2	2	4	4	0.877	1.695	2.066	3.761	0.239	0.239	5.970
8	2	19	0	5	20	35	340	36	290	621	580	2	2	2	2	0.933	0.916	1.018	1.934	0.066	0.066	3.315
9	1	19	0	2	5	32	340	21	290	597	459	2	2	2	2	0.768	0.718	1.066	1.784	0.217	0.217	10.825
10	2	21	0	5	10	27	280	23	250	497	435	2	2	5	5	0.875	2.114	2.584	4.698	0.302	0.302	6.044
11	1	21	1	3	20	35	200	36	290	481	580	1	1	0.4	0.4	1.000	0.209	0.159	0.368	0.032	0.032	8.075
12	2	20	0	2	7	31	340	21	290	589	459	2	2	0.4	0.4	0.779	0.146	0.213	0.359	0.042	0.042	10.375
13	1	19	1	3	10	32	200	26	290	457	499	1	1	8	8	1.000	4.087	3.585	7.673	0.327	0.327	4.091
14	1	22	0	5	10	35	340	26	290	621	499	2	2	2	2	0.803	0.759	1.055	1.814	0.186	0.186	9.305
15	1	20	0	3	5	33	340	21	290	605	459	2	2	0.4	0.4	0.758	0.141	0.214	0.355	0.045	0.045	11.275
16	1	20	0	1	2	23	280	16	290	465	419	2	2	0.4	0.4	0.900	0.175	0.205	0.381	0.019	0.019	4.850
17	1	22	0	1	3	23	280	17	290	465	427	2	2	1	1	0.918	0.449	0.511	0.960	0.040	0.040	4.020
18	1	20	0	5	10	33	280	24	290	545	483	2	2	3	3	0.886	1.288	1.546	2.834	0.166	0.166	5.543
19	1	24	0	2	20	30	280	36	290	521	580	1	1	2	2	1.000	1.026	0.876	1.902	0.098	0.098	4.895
20	1	20	0	10	10	40	340	26	290	662	499	2	2	0.4	8	15.086	2.805	4.282	7.086	0.914	0.914	11.421
21	1	20	0	5	10	35	340	26	290	621	499	2	2	0.4	2	4.015	0.759	1.055	1.814	0.186	0.186	9.305
22	1	22	0	5	5	35	340	21	290	621	459	2	2	1	2	1.477	0.682	1.076	1.758	0.242	0.242	12.090
23	2	22	0	3	12	32	340	28	290	597	515	2	2	1	2	1.725	0.831	1.037	1.868	0.133	0.133	6.625
24	1	21	0	5	5	28	280	19	290	505	443	2	2	7	10	1.252	4.238	5.165	9.403	0.597	0.597	5.970

25	2	21	1	3	7	32	200	23	290	457	475	1	1	3	3	1.000	1.514	1.431	2.945	0.055	0.055	1.837
26	2	21	0	3	20	25	280	34	290	481	563	1	1	2	2	1.000	1.040	0.820	1.860	0.140	0.140	7.020
27	2	20	0	5	7	33	340	23	290	605	475	2	2	2	4	1.569	1.474	2.121	3.595	0.405	0.405	10.120
28	1	21	0	7	7	30	280	21	290	521	459	2	2	3	3	0.880	1.278	1.548	2.826	0.174	0.174	5.793
29	2	21	0	5	10	27	280	24	290	497	483	2	2	4	20	4.858	9.646	10.072	19.718	0.282	0.282	1.411
30	2	22	0	3	10	25	280	23	250	481	435	2	2	4	16	3.617	7.051	8.201	15.253	0.747	0.747	4.669
31	1	22	0	5	25	33	280	39	290	545	604	1	1	2	2	1.000	1.025	0.881	1.906	0.094	0.094	4.705
32	1	19	0	8	5	28	280	18	250	505	395	2	2	5	15	2.344	5.500	7.962	13.462	1.538	1.538	10.255
33	2	21	0	4	5	31	340	19	290	589	443	2	2	2	4	1.503	1.396	2.143	3.538	0.462	0.462	11.543
34	2	22	1	3	10	32	200	26	290	457	499	1	1	16	16	1.000	8.175	7.171	15.345	0.655	0.655	4.091
35	1	22	0	2	2	25	280	16	290	481	419	2	2	0.4	0.4	0.870	0.168	0.207	0.375	0.025	0.025	6.250
36	2	21	1	2	8	31	200	22	290	449	467	1	1	10	10	1.000	5.048	4.765	9.813	0.187	0.187	1.869
37	2	18	0	5	10	34	350	28	290	623	515	2	2	5	10	1.653	3.935	5.238	9.173	0.827	0.827	8.268
38	1	21	0	5	10	32	280	23	250	537	435	2	2	2	2	0.810	0.767	1.053	1.820	0.181	0.181	9.025
39	2	21	0	5	7	32	340	23	290	597	475	1	1	1	1	1.000	0.471	0.665	1.136	0.136	-0.136	13.620
40	2	22	0	10	7	39	340	23	290	654	475	2	2	3	3	0.727	1.003	1.620	2.623	0.377	0.377	12.580
41	2	21	0	3	3	20	240	14	250	401	363	2	2	1	1	0.905	0.441	0.513	0.954	0.047	0.047	4.650
42	2	20	0	7	6	34	350	25	290	623	491	2	2	0.4	2	3.938	0.741	1.060	1.800	0.200	0.200	9.985
43	2	21	0	3	8	25	280	21	250	481	419	2	2	16	16	0.871	6.725	8.277	15.002	0.998	0.998	6.238
44	2	21	0	1	8	30	340	24	290	581	483	2	2	5	10	1.662	3.963	5.231	9.194	0.806	0.806	8.061
45	1	22	0	5	5	27	280	18	250	497	395	1	2	0.4	3.5	6.948	1.310	1.851	3.160	0.340	0.340	9.703
46	1	19	0	5	10	34	340	24	290	613	483	2	2	2	2	1.000	0.940	1.346	2.286	0.286	-0.286	14.305
47	1	20	0	5	1	31	340	19	290	589	443	2	2	3	5	1.252	1.744	2.678	4.423	0.577	0.577	11.544
48	1	20	0	5	1	28	280	15	290	505	411	2	2	0.4	1	2.032	0.385	0.526	0.911	0.089	0.089	8.870
49	2	20	0	5	15	29	370	31	290	603	539	1	1	1	1	1.000	0.486	0.575	1.061	0.061	-0.061	6.090
50	2	19	0	5	10	29	340	26	290	573	499	2	2	0.4	2	4.353	0.841	1.035	1.875	0.125	0.125	6.240
51	1	20	0	5	10	34	340	26	290	613	499	2	2	2.5	2.5	0.814	0.965	1.314	2.279	0.221	0.221	8.840
52	1	22	0	3	3	26	280	17	290	489	427	2	2	7	10	1.246	4.214	5.171	9.384	0.616	0.616	6.159

4 Overview of Fukuoka City Subway Nanakuma Line

4.1 The City Subway Nanakuma Line

The Fukuoka City Subway Nanakuma Line opened in February 2005, and it goes from the South Tenjin station, passing through the wards of *Jonan*-ku and *Sawara*-ku, to the Hashimoto station in the *Nishi*-ku ward. It has a total length of 12 km. Figure 22.2 shows the route map of the Fukuoka City Subway Nanakuma Line.

While the southwest area of Fukuoka City, through which the Nanakuma Line passes, has about 0.5 million residents, which shares 40% of the total population of Fukuoka City, there had been no rail transportation system. Before the Nanakuma Line was opened, for the residents in this southwest area, there was no choice but to rely on buses and private cars to travel to the city center of Fukuoka City.

Therefore, by the opening of the new subway line, the travel time to the city center of Fukuoka City was drastically reduced. For example, the travel time from the Tenjin district, the city center of Fukuoka City, to the *Nanakuma* area where Fukuoka University campus is located was about 40 min by bus before the opening of the subway, but it takes just 16 min by the subway after the opening. Furthermore,

Fig. 22.2 Route map of Fukuoka City Subway Nanakuma Line (Cf. [17])

some areas along the subway line have cheaper fares than buses. For example, in the area around Fukuoka University, the fare is 340 yen by bus, but 290 yen by subway. Thus, the subway stands advantageous than buses with respect to both aspects, travel time and fare.

4.2 Definition of the Area Along the Subway Line

We defined the area along the subway line as those areas including 278 town-*chome* (corresponding to postal code) divisions located along the Nanakuma Line, which are the areas whose residents may use the new subway line, out of 672 town-*chome* divisions in 5 wards (*Chuo*-ku, *Minami*-ku, *Jonan*-ku, *Sawara*-ku, and *Nishi*-ku). The area along the subway line is shown in Fig. 22.3. The remaining town-*chome* divisions were excluded from the analysis because they are close to the preexisting subway line 1 or the *Nishitetsu Tenjin Omuta* line (railway line), and even after the opening of the Nanakuma Line, the travel time to the city center by the use of the new subway line was not expected to be shortened.

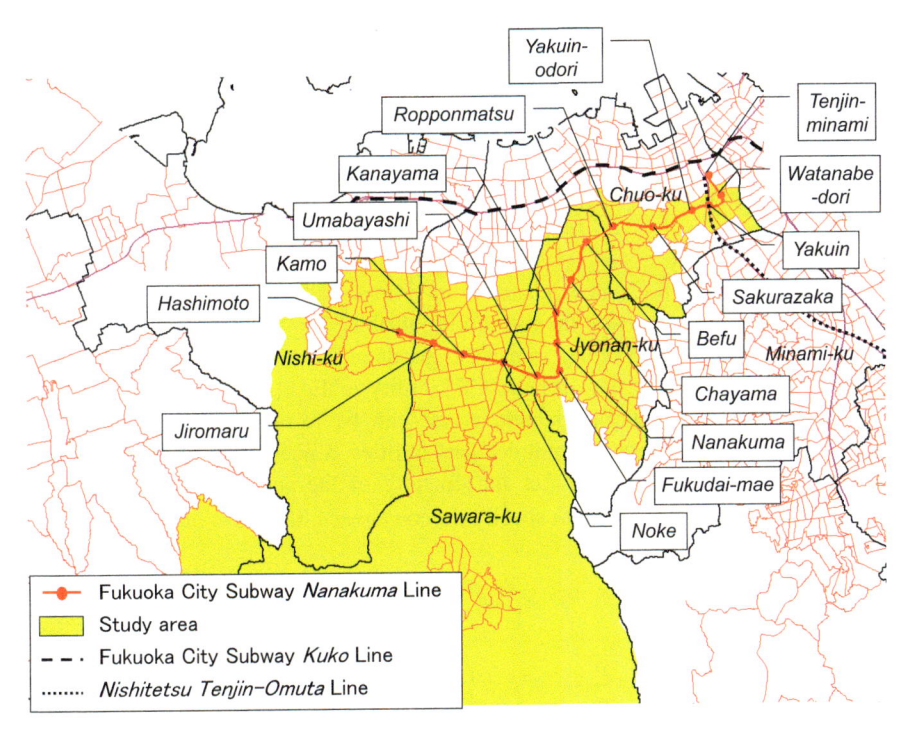

Fig. 22.3 Areas along the new subway line (Cf. [17])

5 Welfare Changes Brought to Residents Along Subway Line

5.1 Method to Forecast the Changes

In this section we use the estimated city center visit demand function to estimate to what extent residents along the subway line have changed their frequency of visits to the city center of Fukuoka City before and after the opening of the new subway line. In addition, we will consider how these changes in the residents' frequency of visits to the city center correspond to their welfare changes.

As for the estimation, we use Eq. (22.28). Taking the ratio of the city center visit demand using the subway to that using buses for each residential area along the subway line, we analyze how the ratio varies among the area along the subway line. The following equation is used for each area i:

$$\frac{x^i{}_{12}}{x^i{}_{11}} = \frac{p^i_{12} - 1.5001}{p^i_{11} - 1.5001} \tag{22.29}$$

To see how the welfare for the residents changed, we can directly compare the indirect utility for consumer i before and after the opening of the subway line. Denote the indirect utilities before and after the opening by $V^B(p_0, p_1^B)$ and $V^A(p_0, p_1^A)$, respectively. Here, the price of local goods is set to $p_0 = 1$. The prices p_1^B, p_1^A denote, respectively, the price indexes for the city center composite goods before and after the opening of the subway. The ratio of them becomes as follows:

$$\frac{V^A(p_0, p_1^A)}{V^B(p_0, p_1^B)} = \frac{b^b(1-b)^b}{a^a(1-a)^a} \frac{(p_1^A)^b}{(p_1^B)^a} \tag{22.30}$$

Since we know the value a and b only up to its ratio, to get the concrete evaluation, we will focus on the part utility attributed to the city center composite goods excluding the part utility from local goods in the whole utility.

From the usual procedure for the CES utility or Eq. (22.11), the indirect part utility for the city center composite goods can be formulated in the following way (Cf. [4, 6]). Denote the indirect part utilities contributed by the city center composite goods before and after the opening of the subway by $V^B(p_{11}^B)$ and $V^A(p_{11}^A, p_{12}^A)$, respectively. With specifying consumer i, they turn out as follows:

$$V_i^B(p_{11}^B i) = \frac{aI_i}{p_{11}^B i} \tag{22.31}$$

$$V_i^A(p_{11}^A i, p_{12}^A i) = \frac{bI_i}{p_1^A i}, \quad p_1^A i = \left((p_{11}^A i)^{1-\sigma} + (p_{12}^A i)^{1-\sigma} \right)^{\frac{1}{1-\sigma}} \tag{22.32}$$

To see how much the indirect part utility contributed by the city center composite goods after the opening of the subway increased from that before the opening, we take the ratio of them. By noting $p_{11}^B i = p_{11}^A i$ and transforming the ratio, we have the following expression of the welfare changes due to the introduction of a new subway line.

$$\frac{V_i^A(p_{11}^A i, p_{12}^A i)}{V_i^B(p_{11}^B i)} = \frac{b}{a} \frac{p_{11}^B i}{\left((p_{11}^A i)^{1-\sigma} + (p_{12}^A i)^{1-\sigma}\right)^{\frac{1}{1-\sigma}}} = \frac{b}{a}\left(1 + \left(\frac{p_{12}^A i}{p_{11}^A i}\right)^{1-\sigma}\right)^{\frac{1}{\sigma-1}} \quad (22.33)$$

The welfare change represented above is directly related to the changes in the ratio of the frequency of visits to the city center by using the subway to that by using buses. From Eqs. (22.14) and (22.28), the following holds:

$$\frac{x_{12}^A i}{x_{11}^A i} = \frac{(p_{12}^A i)^{-\sigma}}{(p_{11}^A i)^{-\sigma}} \quad (22.34)$$

Thus, we have

$$\frac{V_i^A(p_{11}^A i, p_{12}^A i)}{V_i^B(p_{11}^B i)} = \frac{b}{a}\left(1 + \left(\frac{x_{12}^A i}{x_{11}^A i}\right)^{\frac{\sigma-1}{\sigma}}\right)^{\frac{1}{\sigma-1}}. \quad (22.35)$$

From the above result, it is seen that the observed increase for the residents in their share of use of the subway in terms of their frequency of visits to the city center by the subway directly implies the increase in their welfare before and after the opening of the subway.

5.2 Data Used for Forecast

In the forecast, we consider both transportation modes, the subway and buses. In order to forecast the changes in all the 278 town-*chome* division, we need the travel time and the transportation fare for each of the two travel modes from all 278 town-*chome* divisions to the Tenjin district, the city center of Fukuoka City. However, we have no published data about these. Thus we had created these data by using GIS.

First of all, a reference point is set for each town-*chome* division. Then, as for buses, we identify the nearest bus stop closest to the reference point, measure the road distance, and obtain the time distance by walking from the reference point to the nearest bus stop. Regarding the travel time and fare from the nearest bus stop to the Tenjin district, we referred to the Nishi-Nippon Railroad's website [7]. As for the Nanakuma Subway Line, we also identify the nearest station from the reference point, measure the road distance, obtain the time distance by walking from the

Table 22.6 Average of increased ratios of demand for visit frequency to the city center due to the opening of the subway for all area divisions

	Number of town-*chome* divisions	Average
Average of increased ratios of visit frequency to the city center after the subway opening	278	3.46

reference point to the nearest station, and obtain the travel time and fare from the nearest station to the South Tenjin station by the Fukuoka City Subway's website [5].

5.3 Forecast Results

Table 22.6 shows the average of increased ratios of the demand for the frequency of visits to the city center by using the subway after the subway opening to that by using buses before the subway opening for all the areas along the subway line. It is found that the demand for visits to the city center by using the subway after the opening of the subway is 3.46 times as large as compared to that by using buses before the opening of the subway on average for all the 278 town-*chome* divisions.

Figure 22.4 displays these ratios for the 278 town-*chome* divisions. Depending on the value of the increased ratio, we classify five categories by different colors. There is almost no change in the city center area divisions near *Watanabe* street, *Yakuin*, and the surroundings of *Yakuin-odori*. However, from *Beppu* to *Hashimoto*, the ratios attain high values. Moreover, especially in the area divisions far from the subway station, the increased ratios also become larger. In the figure, we also indicate the increased ratios of welfare changes corresponding to the increased ratios of visit frequency to the city center calculated by Eq. (22.35).

6 Conclusion and Future Challenges

The significant contribution of this study is that we have shown a methodology in which the consumers' utility function is formulated as a hierarchical nested CES utility function, the hierarchical nested CES utility function is directly estimated based on the retrospective panel data concerning micro behavior history data of consumers, and the estimated results can be utilized for the evaluation of the urban development policies from the viewpoint of the welfare changes of individual consumers caused by the urban development policies.

As for our future challenges, to list our major future research, first, we must extend our functional form of the utility function from the present one with stratified by Cobb-Douglas and CES to the more flexible one such as stratified by CES and

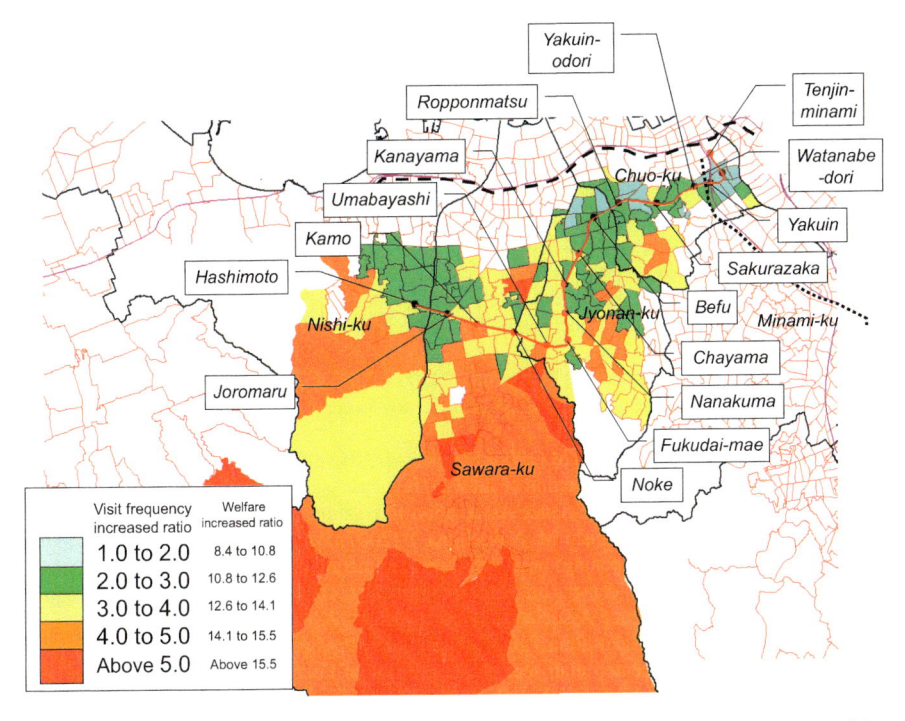

Fig. 22.4 Increased ratios of demand for visit frequency to the city center due to the opening of the subway for all area divisions (Cf. [17])

CES. Second, while this time we formulated the first-stage Cobb-Douglas functions as different forms for before and after the opening of the subway line to enable the model to account for the changes of the share of the expenditure on local and the city center composite goods, we must devise a method to evaluate the urban development policies on a monetary basis by making it possible to do EV or CV welfare analysis. Third, while we were concerned with transport policies here, we must show that our methodology can be extended to other kinds of urban development policies.

References

1. Anderson SP, De Palma A, Thisse J-F (1987) The CES is a discrete choice model? Econ Lett 24:139–140
2. Anderson SP, de Palma A, Thisse J-F (1988) The CES and the logit: two related models of heterogeneity. Reg Sci Urban Econ 18:155–164
3. Anderson SP, De Palma A, Thisse JF (1992) Discrete choice theory of product differentiation. MIT Press
4. Dixit AK, Stiglitz JE (1977) Monopolistic competition and optimum product diversity. Am Econ Rev 67:297–308

5. Fukuoka City Subway Home Page http://subway.city.fukuoka.jp/. (in Japanese)
6. Kanemoto Y, Hasuike K, Fujiwara T (2006) Microeconomic Modeling for Policy Analysis, Toyo Keizai. (in Japanese)
7. Nishi-Nippon Railroad Co., Ltd. Home Page http://www.nishitetsu.co.jp/. (in Japanese)
8. Okuda T (2002) A probabilistic approach to CES demand function, collected papers for presentation in the 39th annual meeting of Japan Section of Regional Science Association International (JSRSAI), pp. 391–398. (in Japanese)
9. Saito S, Motomura H (2001) A study on shopping and sightseeing behaviors of international tourists. Studies on Strait Area between Korea and Japan, Vol.1, pp. 41–61, (in Japanese), pp. 51–77, (in Korean)
10. Saito S, Yamashiro K (2001) Economic impacts of the downtown one-dollar circuit bus estimated from consumer's shop-around behavior: a case of the downtown one-dollar bus at Fukuoka City. Stud Reg Sci 31:57–75. (in Japanese)
11. Saito S, Yamashiro K, Kakoi M, Nakashima T (2003) Measuring time value of shoppers at city center retail environment and its application to forecast modal choice. Stud Reg Sci 33:269–286. (in Japanese)
12. Saito S, Yamashiro K, Imanishi M (2006) Estimation and application of travel demand function to city center: verification by opening of subway. In: paper presented at the 43rd annual meeting of Japan Section of Regional Science Association International (JSRSAI). (in Japanese)
13. Saito S, Yamashiro K, Imanishi M, Nakashima T (2007) Welfare changes in residents along new subway line by its opening: based on direct estimation of hirarchical nested CES travel demand function to city center. In: paper presented at The 44th annual meeting of Japan Section of Regional Science Association International (JSRSAI). (in Japanese)
14. Saito S, Yamashiro K, Nakashima T, Igarashi Y (2007) Predicting economic impacts of a new subway line on a city center retail sector: a case study of Fukuoka City based on a consumer behavior approach. Stud Reg Sci 37:841–854. (in Japanese)
15. Saito S, Yamashiro K, Nakashima T (2006) An empirical verification of the increase of visit frequency to city center caused by the opening of a new subway line: a comparative analysis before and after opening based on retrospective data. In: paper presented at 43rd annual meeting of Japan Section of Regional Science Association International (JSRSAI). (in Japanese)
16. Yamashiro K (2012) A study on the evaluation of transport policy based on consumer behavior approach. Doctoral dissertation, Graduate School of Economics, PhD in Economics (Economics-1402), Fukuoka University, Fukuoka. (in Japanese)
17. Zenrin Co. (2003) Digital Map Z6, ZENRIN CO., LTD.
18. Fujita M, Krugman PR, Venables AJ (2001) The spatial economy: cities, regions, and international trade. MIT Press, Cambridge, MA

Index

A
Absorbing stationary Markov chain, 19, 52, 70
Accessories, 134, 139, 344, 346, 354–356
Action Program toward Realization of Tourism Nation, 420
Activity effect type evaluation scheme, 6–8
Activity system, 2–5, 7–9
Additional cash flow, 365–367
Air Busan, 421
Artificial panel data, 290
Asset of attractiveness, 365, 366
Association matrix of shop-around purpose transition, 95, 105
Attraction effect, 275–277, 284–290, 294
Attraction nodes, 176–178, 198, 220
Average expenditure, 169, 180–181, 183, 225, 227, 231, 233–234, 236, 248, 257, 282, 306–312, 325, 326, 354–356, 366, 376
Average numbers of *Kaiyu* steps, 232, 233

B
Backward induction, 423, 424, 441
Backward variable selection method, 135, 138, 141
Basic form of activity effect evaluation scheme, 6–8
Becker, 190, 192–194
Beetle, 420, 422, 426–431
Behavioral hypothesis, 245, 438, 439
Behavior purpose, 19, 114
Bicycle post, 190
Bootstrap, 34
Brand equity, 364–366

Bus, 8, 99, 115, 165–184, 191, 219, 245, 262, 276, 299, 392, 443
Busan, 420–422, 426, 429, 431–433
Business model, 369–393

C
Canal City Hakata, 16–18, 25–27, 29–31, 33, 35–37, 39–41, 73, 77–80, 149, 168, 170, 171, 173, 174, 197, 343, 360–363, 371
Castle, 49, 299–311, 313
Center of gravity, 14, 16, 23–24, 43, 44, 168
CES, 440–451, 456, 458, 459
Chuo-ku, 175, 246, 247, 263, 266, 455
Circuit bus, 8, 165–184
City center, 8, 13–45, 69–87, 113, 131–142, 146, 165–184, 190, 217–237, 241–259, 261–271, 276, 297–313, 319, 341–356, 361, 369–393, 398, 421, 438
City center cafe, 232, 234
City center cafes in the middle of *Kaiyu*, 232
City center commercial district, 18, 33, 41, 44, 132, 141, 166, 167, 173, 241–259, 262, 263, 269, 270, 342, 344, 382, 398, 399
City center district, 13–45, 113, 114, 120, 126, 127, 166–171, 173, 176–178, 181, 184, 190, 191, 196, 197, 204, 206, 208–212, 214, 218–220, 223, 225–228, 231, 235, 237, 242, 244, 248, 249, 257, 258, 262, 263, 266–267, 270, 271, 284, 289, 290, 294, 299–308, 311, 312, 321, 361, 370, 371, 379, 388, 392
City center parking policy, 369–393

City center retail district, 147, 150, 151, 153, 161
City center revitalization, 8, 299
City center structure, 14, 22
City Center 100-yen Circuit Bus, 8, 165–184
City formation system, 2–6, 8, 9
City formation system type evaluation scheme, 8–9
City management, 1
City planning, 1, 3–5
City system, 4
Cobb-Douglas, 424, 425, 440, 442, 445, 458, 459
Composite good, 258, 423, 425, 440–442, 444, 450, 456, 459
Conditional logit model, 348–350, 386
Consistent estimation method, 24, 52, 298, 299, 302–305
Constant elasticity of substitution (CES), 440–451, 456, 458, 459
Consumer behavior, 8, 15, 16, 18, 91, 112, 191, 214, 217–237, 242, 244, 245, 258, 262, 318, 320, 339, 340, 342, 350–353, 406, 438
Consumer behavior approach, 8, 217–237, 242, 244, 438
Consumer-oriented urban development, 367
Consumer shop-around behavior, 13–45, 70–75, 80, 87, 92, 93, 112, 132, 133, 141, 148–150, 218, 299–303, 308, 311, 321–322, 324, 342–344, 371, 392, 405, 438
Consumer's mind, 147, 365
Consumers multistage choices, 92
Consumer's shop-around behavior, 13–45, 70–73, 75, 80, 87, 91–93, 112, 132, 133, 141, 148–150, 218, 299–303, 308, 311, 321–322, 324, 342–344, 371, 392, 405, 438
Cruising time, 155, 160, 161, 373–378, 392
Cumulative distributions, 116, 156, 157, 280, 282, 326, 327, 329, 332–336

D
Daimaru Elgala, 14, 17, 18, 25–40, 71, 174, 249, 371
Daimaru Tenjin Fukuoka, 26–27, 29–31, 35–37, 39, 40, 71, 74, 77–79, 81–85, 220, 319, 343
Daimyo, 77–79, 81–83, 85, 364, 366
Dazaifu city, 47–67, 176

Dazaifu Tenman-gu Shrine, 49, 51, 52, 54–57, 59–61, 64, 68
Decision-making process of behavioral entity, 92
Decomposition, 101–107
Definition of town equity, 365–366
Department store, 14, 16, 43, 71, 72, 84, 87, 176, 190, 302, 304–306, 317–340, 405
DeSerpa, 192, 193, 196
Development effect on consumer behaviors, 23
Difference in difference, 270
Direct approach, 437–459
Disaggregate Huff model, 92, 167
Distance distribution function, 111–129
Distance from the parking lot parked to the first destination, 379, 381, 382
Distance resistance, 112
Division ratio cross tabulation, 334–340
Doutor, 218, 227, 229–231
Downtown
 commercial space structure, 92
 space, 92, 93, 111–129

E
Eastar Jet, 421
Ecole Card, 264, 265, 446–448
Economic effect, 8, 132, 165–184, 217–237, 241–259, 262, 276, 373, 377–378, 385, 388, 392, 438, 439
Economic effect of the city center cafés, 235–237
Economic effect of the city center 100-yen bus, 169, 182
Economic effects of a new subway line, 241–259
Economic loss, 370, 373–378, 392
Effects spread over other districts within city center, 311
Entrance area, 95
Evaluation
 scheme, 1, 2, 4–9
 of urban development policy, 9
Event effects, 276
Exchange rate, 429, 430, 433
Exit area, 95, 97
Ex post forecast, 191, 192, 201, 212–214

F
Feeder transport, 257, 259
Fixed effect model, 433

Index

Forms of information contents, 405, 413–415
Four step models, 262
FQBIC, *see* Fukuoka University Institute of Quantitative Behavioral Informatics for City and Space Economy (FQBIC)
Frequency of visits, 51, 53, 57, 133, 135, 149, 173, 197, 225, 227, 242, 244, 245, 248, 252, 253, 258, 261–271, 287, 288, 290–292, 300, 301, 321, 323–325, 334, 337, 371, 428, 429, 438–440, 444–448, 451, 452, 456, 457
Fringe parking, 378, 382, 385–391
Frobenius root, 104
Fukuoka City, 13–45, 132, 146, 166, 190, 218, 242, 262, 274, 318, 342, 361, 370, 426, 438
Fukuoka City Subway, 171, 199, 250
Fukuoka City Subway Nanakuma Line, 242, 245–247, 253, 255, 263, 438, 454–455
Fukuoka Dental College, 263
Fukuoka Mitsukoshi, 14, 17, 27, 31, 33, 34, 37, 38, 40, 43, 73, 76–86, 174, 249, 343, 371
Fukuoka Shop-Around Survey, 133
Fukuoka Tenjin Daimaru, 25–27, 29–31, 35–40, 71–73, 77–79, 82–85, 220, 319, 343
Fukuoka University, 72, 171, 196, 219, 247, 263–265, 271, 300, 321, 446–449, 454, 455
Fukuoka University Institute of Quantitative Behavioral Informatics for City and Space Economy (FQBIC), 72, 219, 247–249, 264, 300, 321, 343, 368
Furniture industry, 274, 275

G

Gaining the additional cash flow, 367
Generalized travel cost, 190, 191, 242, 438, 443, 451
Gentrification strategy, 339
Global positioning system (GPS), 398–402, 408
Goal of urban development, 361–367, 378, 382, 392, 398
GPS, *see* Global positioning system (GPS)

H

Hakata Riverain, 14, 73, 76–80, 149, 173, 174, 197–201, 204, 213, 214, 322, 343, 371

Hakata Station, 16, 18, 25–27, 29–31, 33, 35–41, 73, 80, 149, 168, 170, 171, 173, 174, 197–200, 204, 213, 214, 322, 323, 343, 371
Hakataza, 142
Hard trick for information evolution, 361–363
Hashimoto Station, 246, 263, 454
Having meals, 94, 97, 98, 101, 105, 107, 115, 286, 372
Hazard, 97, 99, 112, 113, 116–120, 123–128
Health related products, 344–350, 352–356
Heterogeneity of individuals, 340
Hierarchical nested utility function, 423–425
Historical heritage, 47–67
History of information transactions, 403, 404
Home-based surveys, 167
Household products, 142, 344, 346, 356
Huff model, 15, 92, 93, 112, 166, 167
Hutong, 362
Hypertext city, 363

I

ICT, *see* Information anc Communication Technologies (ICT)
Ideal city type evaluation scheme, 5–7
Imaizumi district, 77
Implementing the city center parking policy as a business, 382–392
IMS, 25–27, 29–31, 35–40, 78–80, 82–87
Increase, 14, 55, 71, 96, 112, 132, 166, 201, 220, 242, 262, 275, 298, 320, 341–356, 365, 370, 414, 420, 438
Increase in expenditure in the city center commercial district, 169
Increase in the number of consumer *Kaiyu* steps, 169, 183, 184
Increase in visit frequency, 337
INCUBE, 343–356
Indoor Messaging System (IMES), 400–401
Information anc Communication Technologies (ICT), 4, 66, 393, 398
Information evolution, 361–363
Information processing behavior, 397–416
Information provision, 8, 191, 399, 400, 411, 412, 416
Information transaction, 403
Institutions, 3
Interior goods, 344, 346, 354–356
Internet Of Things (IoT), 4
Interrelationship among various log history data, 403

Introduction of subway, 261–271
IoT, *see* Internet Of Things (IoT)
11 Items of store image, 346
Iwataya
   A-side, 25–27, 29–31, 35–39, 45, 72, 78, 82, 84
   Z-Side, 14, 16, 18, 26–28, 30–34, 36–40, 72–74, 78, 82, 84, 174, 220, 249

J
Japan Railways (JR), 16
Japan Railways (JR) Hakata Station, 16, 168
Jeju Air, 421
Jin Air, 421
Jonan-ku, 175, 246, 247, 263, 265, 266, 454, 455

K
Kabuki, 142
Kagoshima City, 398, 399
Kaidan-in Temple, 49
Kaiyu
   distance distribution function, 111–129
   flow, 70, 86
   index, 87
   Markov model, 15, 16, 18–22, 41, 44, 52–53, 65, 70, 73–75, 80, 87, 167, 168
Kamitori, 300, 301, 308, 310, 312
Kanzeon-ji Temple, 51, 57, 60, 62, 64, 66
Kaplan-Meier non-parametric estimations, 116, 118, 120
Korean tourists, 419–434
Kumamoto Castle Festival, 299–311
Kumamoto city, 299–308, 311, 312
Kyushu Island, 48, 299, 398, 420
Kyushu University, 263

L
Large cruise ship, 420
Large-scale variety store, 342, 351
LCC, *see* Low Cost Carrier (LCC)
Least squares, 379, 443, 445, 448
Length of time staying at city center cafes, 227, 228, 230
Linked Trip, 20, 75, 94
Little's formula, 145–162
Location information, 399
Logit model, 167, 193, 195, 243, 251, 346, 348–350, 386, 411, 443, 444, 447
Logs, 403, 405, 408
Low Cost Carrier (LCC), 421

M
Markov chain, 19, 21, 52, 70, 73, 92
Markov model, 15, 16, 18–22, 33, 41, 44, 52, 65, 70, 73–75, 80, 87, 167, 168
Markov-type analysis, 119, 126
Marunouchi, 364
Maximum eigenvalue, 104
Maximum likelihood, 117, 195, 204, 251, 287, 386, 412
Meta-theoretic evaluation framework, 1–9
Metropolitan areas, 92
MIC, *see* Ministry of Internal Affairs and Communications (MIC)
Minami-ku, 175, 246, 247, 266, 455
Ministry of Internal Affairs and Communications (MIC), 399
Mizuki Ruins, 49
Mobile ICT devices, 66
Modal choice, 95, 114, 190, 191, 193, 194, 196–204, 214, 245, 247–252, 254–256, 262, 451
Modal share, 191, 192, 198, 199, 201, 212–214, 255
Modified consistent estimation method without the order of sites visited, 305
Multiple regression analysis for exploring factors, 138–141
Municipal tourism policy, 47–67

N
Nakamura Gakuen University, 263
Nested utility function, 423–425
Net incoming visitors, 51, 52, 59, 60, 66, 67
Net number of incoming visitors, 52, 59, 60, 66, 67, 298, 299, 302–305, 311, 421
New subway line, 80, 241–259, 261–271, 437–459
Nishi-ku, 175, 246, 247, 263, 266, 454, 455
Nishi-Nippon Railroad Co. Ltd. (NNR), 166
Nishitetsu Fukuoka Station, 16, 43
Nishitetsu Tenjin Omuta line, 247, 250, 455
Nishiyama, 7
NNR, *see* Nishi-Nippon Railroad Co. Ltd. (NNR)
Non-derived behavior purposes, 96, 115

O
Occurrence order, 91–109, 112, 118
OD, *see* Origin Destination (OD)
Okawa City, 274, 276, 277, 280–281, 286–292, 294, 295

Okawa Woodworks Festival, 274–287, 289–295
One-dollar, 276, 299
One-dollar circuit bus, 199
On-site decision-making, 399, 416
On-site *Kaiyu* surveys, 167, 168
On-site Poisson model, 421
On-site survey of *Kaiyu* behaviors at the city center of Fukuoka City, 173, 196, 208, 219
Open café, 218
Origin Destination (OD), 52, 55–57, 59, 60, 168, 190, 197, 302

P
Pairwise comparison, 104
Panel data, 265, 290, 322, 324, 334, 339, 342, 356, 433
Park and ride, 253, 255–259
Parking capacity analysis, 147, 150, 151, 153, 156–162
Parking fee discount, 383
Parking policy, 369–393
Parking space, 147, 149, 151, 162
Partitioned transition probability matrix, 75
Part utility, 456, 457
Passing by, 94–98, 114, 115
Physical=activity interdependence extension form, 7–8
Physical system, 2–9
Poisson regression model, 33, 80, 276, 277
Policy extension form, 8
Positive reciprocal matrix, 104, 105
Probability matrix, 20, 52, 53, 60, 73, 75–77, 97
Probability of returning home, 96–100
Probability of visiting each commercial facility node, 75
Probability to continue the shop-around, 96
Profitable as a business, 382
Purchasing risk, 321, 323, 324
Purpose realization rate, 321, 323–339, 343, 351–356
Purpose transition, 91, 95–98, 100–101, 105, 109, 115
Purpose transition step, 95

Q
Quasi-Zenith Satellite System (QZSS), 399
Quitting rate, 93
QZSS, *see* Quasi-Zenith Satellite System (QZSS)

R
Radius of attraction spatial area, 294
Ratio scale, 96, 104–108
Ratio scale decomposition theorem, 101–109
Renewal of department store, 317–340
Reproducibility theorem, 20, 21
Retail district, 147, 150, 151, 153, 161, 190
Retail redevelopment, 13–45, 168, 318, 340
Retail sector, 93, 167, 169, 219, 237, 242, 244, 367, 374, 390, 438, 439
Retrospective panel data, 263, 265, 320, 322, 324, 339, 340, 342, 343, 356, 439, 440, 445, 458
Revealed-preference (RP), 190, 191, 193
Revitalization, 8, 70, 168, 298, 299, 365
RP, *see* Revealed-preference (RP)
Ruins, 49, 50, 56, 58, 62, 65, 68
Ruins of Chikuzen Kokubun-ji Temple, 49
Ruins of the old Dazaifu Government Office, 50, 56, 58, 62, 64

S
Saga city, 93, 113, 114
Saga City Citizen Survey, 93, 113
Saga city government, 93, 113
Sawara-ku, 175, 246, 263, 266, 455
Shake, 402, 404–412
Shanghai, 362
Shimokawabata, 14
Shimotori, 300, 301, 308, 310, 312
Shinshigai, 300, 301, 306, 308, 310, 312
Shop-around
   behavior, 13–45, 70–73, 75, 80, 87, 91–93, 112–114, 132, 133, 141, 148–150, 166, 218, 299–303, 308, 311, 321–322, 324, 342–344, 371, 392, 405, 438
   choice probability, 21–28, 34, 38, 44, 52, 53
   distance, 112, 113, 116–119, 121–129
   effect, 19–22, 28–34, 45, 70, 75–80, 87, 92, 112
   Markov model, 15, 16, 18–24, 33, 41, 44, 52, 70, 73, 75, 80, 87, 167
   pattern, 15, 21, 24, 41–43, 299, 302–304
   purpose (transition) step, 91–109, 112, 113, 115
   route, 95, 115, 302–304
   step, 21, 95–99, 101–107, 109, 112, 115, 118–121
Shop category, 407, 412
Shoppers Daiei, 25–27, 29–31, 35–37, 39, 40, 73, 77–84, 149, 174, 249, 343, 371
Shopping district, 18, 33

Shopping for clothing, 94, 97, 98, 101, 105, 109, 115
Shopping mall, 25–27, 29–31, 35–37, 39, 69–87, 132, 142, 173, 174, 284, 298
Shopping option, 342, 343
Showa Street, 16, 169, 197
Six categories of products, 344
Smartphone app, 398–402, 404, 405
Social decision making system, 3–5, 8, 9
Social experiment, 166, 320, 370, 398–406, 410, 412
Soft trick for information evolution, 363
Sojourn, 28, 32, 35–37, 41, 81–87, 132
Solaria Plaza, 18, 25–27, 29–31, 35–37, 39, 40, 43, 73, 76–79, 81–83, 85, 149, 174, 249, 343, 371
Spread over, 87, 297–313
Square of the distance from the parking lot parked to the first destination, 379, 380
Starbucks, 218, 227, 229–231
Stated preference (SP), 192, 193, 265, 422, 427, 433, 447
Stationery, 344–347, 349, 352, 353, 355
Staying time, 131–142, 147, 152, 156, 157, 159, 162, 370, 377–384, 386, 387, 389–391
Store choice, 343, 346–350, 356
Store image, 346, 347, 349, 350
Subway, 16, 70, 80, 168, 194, 241–259, 261–271, 437–459
Sumiyoshi Street, 16, 169, 197
Supplementary data, 247, 447–448
Survey of consumer parking behavior, 146, 148, 149, 152, 153, 161
Survey of consumer shop-around behavior at city center of Fukuoka City, 73, 74, 133, 148–150, 321–322, 324, 343, 371
Survival analysis, 97, 113, 116, 118, 127
Survival time, 116–118
Sustainable entity, 382
Switch of purchase destinations, 348–350

T
Taihaku Street, 16, 169, 197
Tap, 402, 404–412, 414, 416
Tenjin area, 24, 43, 71–72, 80, 150, 251, 447
Tenjin core, 25–27, 29–31, 35–37, 39, 40, 77–79, 82–85
Tenjin district, 14, 16, 23, 24, 28, 33, 34, 41, 43, 44, 70–73, 76, 80, 85, 168–170, 173, 174, 197, 198, 200, 204, 213, 249, 250, 254, 255, 318, 319, 322, 340, 342, 344–345, 351, 353–356, 364, 366, 370–377, 379–386, 390–392, 438, 454, 457
Tenjin Loft, 342–356
Tenjin Minami Station, 70, 246, 263
Tenmonkan, 398–401, 405, 406
Tenmonkan-hondori, 401, 405
Three evaluation schemes, 4
Time allocation model, 192, 193
Time value of shopping, 189–214, 363
Time values by day of week, 203
Time values by main purpose, 208–210
Time values by purchasing attitude, 210–211
Time values by travel fare, 190, 205–206
Tokyo Station, 364
Total visit frequency (TVST), 20, 22–24, 33–41, 43, 44, 53, 59, 76
Tourism, 47–67, 276, 365, 420, 433, 434
Town brand, 364
Town-chome division, 245–248, 253–256, 455, 457, 458
Town equity, 356, 361–367, 378, 392
Town management organizations, 382, 399
Town walking, 361–363
Transition probability among all purposes becomes non-stationary, 96
Transportation nodes, 176–178, 198, 220
Transport cost, 245, 250, 262, 429, 431, 433, 438, 440, 441, 443–446, 450
Transport mode, 196, 212, 248, 254
Travel demand function, 419–434
Trip, 15, 19, 20, 28–32, 52, 53, 75, 92–94, 96, 99, 109, 112–114, 117–119, 121, 125, 126, 128, 146, 148, 149, 167–169, 244, 247, 252, 262, 426
Trip chain, 15, 92, 112, 166, 197, 302, 303
Tsuruya, 304–306, 308, 310, 312
Turnover, 93, 132, 235, 237, 242, 244, 248, 320, 367, 384, 385, 390, 391
TV commercial effect, 292–294
TVST, *see* Total visit frequency
T'way, 421

U
Underground shopping mall, 25–27, 29–31, 35–37, 39, 40, 69–87, 174
Unobserved omitted variable biase, 433
Urban development, 1–9, 132, 166, 218, 361–367, 378, 382, 392, 398, 405, 439, 440, 458, 459

Urban development policy evaluation, 9
Urban Renaissance Headquarters of Japanese Government, 70

V

Value of the intangible asset, 365, 366
Value of town, 341
Variety products, 344, 356
Variety shop, 341–356
Visited places, 51, 73, 94, 95, 133, 149, 198, 218, 249, 279, 284, 301, 343
Visit frequency, 20, 53, 73, 92, 242, 262, 277, 304, 321, 342, 438
Visit Japan Campaign, 420
Visitor attraction area, 290–294
Visitors with festival purpose, 279–286, 288, 290, 302, 309, 312
Visit value, 317–340, 342, 343, 351, 367
Visit values to town, 367

W

Waiving the additional cash flow, 367
Walk-around, 14, 18, 47–67, 73, 94, 95, 114–116, 120, 122, 178
Walking Path of History, 47–67
Watanabe Street, 169, 197, 458
Welfare change, 437–459
Woodworks Festival, 273–295

X

XinTianDi, 362

Y

Yakuin-station, 247
Yakuin-Watanabe-Dori-Tenjin-Minami station, 247
Yanagawa City, 286, 287, 289
100 Yen bus, 165–184, 191, 192, 212–214, 219